Lecture Notes in Computer S

T0238801

Commenced Publication in 1973
Founding and Former Series Editors:
Gerhard Goos, Juris Hartmanis, and Jan van Leeuwen

Editorial Board

Beniamino Murgante   Sanjay Misra
Maurizio Carlini   Carmelo M. Torre
Hong-Quang Nguyen   David Taniar
Bernady O. Apduhan   Osvaldo Gervasi (Eds.)

Computational Science and Its Applications – ICCSA 2013

13th International Conference
Ho Chi Minh City, Vietnam, June 24-27, 2013
Proceedings, Part III

 Springer

Volume Editors

Beniamino Murgante, Università degli Studi della Basilicata, Potenza, Italy
E-mail: beniamino.murgante@unibas.it

Sanjay Misra, Covenant University, Canaanland OTA, Nigeria
E-mail: sanjay.misra@covenantuniversity.edu.ng

Maurizio Carlini, Università degli Studi della Tuscia, Viterbo, Italy
E-mail: maurizio.carlini@unitus.it

Carmelo M. Torre, Politecnico di Bari, Italy
E-mail: torre@poliba.it

Hong-Quang Nguyen, Int. University VNU-HCM, Ho Chi Minh City, Vietnam
E-mail: htphong@hcmiu.edu.vn

David Taniar, Monash University, Clayton, VIC, Australia
E-mail: david.taniar@infotech.monash.edu.au

Bernady O. Apduhan, Kyushu Sangyo University, Fukuoka, Japan
E-mail: bob@is.kyusan-u.ac.jp

Osvaldo Gervasi, University of Perugia, Italy
E-mail: osvaldo@unipg.it

ISSN 0302-9743   e-ISSN 1611-3349
ISBN 978-3-642-39645-8   e-ISBN 978-3-642-39646-5
DOI 10.1007/978-3-642-39646-5
Springer Heidelberg Dordrecht London New York

Library of Congress Control Number: 2013942720

CR Subject Classification (1998): C.2.4, C.2, H.4, F.2, H.3, D.2, F.1, H.5, H.2.8, K.6.5, I.3

LNCS Sublibrary: SL 1 – Theoretical Computer Science and General Issues

Typesetting: Camera-ready by author, data conversion by Scientific Publishing Services, Chennai, India

Printed on acid-free paper

Springer is part of Springer Science+Business Media (www.springer.com)

Preface

These multiple volumes (LNCS volumes 7971, 7972, 7973, 7974, and 7975) consist of the peer-reviewed papers from the 2013 International Conference on Computational Science and Its Applications (ICCSA2013) held in Ho Chi Minh City, Vietnam, during June 24–27, 2013.

ICCSA 2013 was a successful event in the International Conferences on Computational Science and Its Applications (ICCSA) conference series, previously held in Salvador, Brazil (2012), Santander, Spain (2011), Fukuoka, Japan (2010), Suwon, South Korea (2009), Perugia, Italy (2008), Kuala Lumpur, Malaysia (2007), Glasgow, UK (2006), Singapore (2005), Assisi, Italy (2004), Montreal, Canada (2003), (as ICCS) Amsterdam, The Netherlands (2002), and San Francisco, USA (2001).

Computational science is a main pillar of most of the present research, industrial, and commercial activities and plays a unique role in exploiting ICT innovative technologies; the ICCSA conference series have been providing a venue to researchers and industry practitioners to discuss new ideas, to share complex problems and their solutions, and to shape new trends in computational science.

Apart from the general track, ICCSA 2013 also included 33 special sessions and workshops, in various areas of computational sciences, ranging from computational science technologies, to specific areas of computational sciences, such as computer graphics and virtual reality. We accepted 46 papers for the general track, and 202 in special sessions and workshops, with an acceptance rate of 29.8%. We would like to express our appreciation to the Workshops and Special Sessions Chairs and Co-chairs.

The success of the ICCSA conference series, in general, and ICCSA 2013, in particular, is due to the support of many people: authors, presenters, participants, keynote speakers, Workshop Chairs, Organizing Committee members, student volunteers, Program Committee members, International Liaison Chairs, and people in other various roles. We would like to thank them all. We would also like to thank Springer for their continuous support in publishing ICCSA conference proceedings.

May 2013

David Taniar
Beniamino Murgante
Hong-Quang Nguyen

Message from the General Chairs

On behalf of the ICCSA Organizing Committee it is our great pleasure to welcome you to the proceedings of the 13th International Conference on Computational Science and Its Applications (ICCSA 2013), held June 24–27, 2013, in Ho Chi Minh City, Vietnam.

ICCSA is one of the most successful international conferences in the field of computational sciences, and ICCSA 2013 was the 13th conference of this series previously held in Salvador da Bahia, Brazil (2012), in Santander, Spain (2011), Fukuoka, Japan (2010), Suwon, Korea (2009), Perugia, Italy (2008), Kuala Lumpur, Malaysia (2007), Glasgow, UK (2006), Singapore (2005), Assisi, Italy (2004), Montreal, Canada (2003), (as ICCS) Amsterdam, The Netherlands (2002), and San Francisco, USA (2001).

The computational science community has enthusiastically embraced the successive editions of ICCSA, thus contributing to making ICCSA a focal meeting point for those interested in innovative, cutting-edge research about the latest and most exciting developments in the field. It provides a major forum for researchers and scientists from academia, industry and government to share their views on many challenging research problems, and to present and discuss their novel ideas, research results, new applications and experience on all aspects of computational science and its applications. We are grateful to all those who have contributed to the ICCSA conference series.

For the successful organization of ICCSA 2013, an international conference of this size and diversity, we counted on the great support of many people and organizations.

We would like to thank all the workshop organizers for their diligent work, which further enhanced the conference level and all reviewers for their expertise and generous effort, which led to a very high quality event with excellent papers and presentations.

We especially recognize the contribution of the Program Committee and local Organizing Committee members for their tremendous support, the faculty members of the School of Computer Science and Engineering and authorities of the International University (HCM-VNU), Vietnam, for allowing us to use the venue and facilities to realize this highly successful event. Further, we would like to express our gratitude to the Office of the Naval Research, US Navy, and other institutions/organizations that supported our efforts to bring the conference to fruition.

We would like to sincerely thank our keynote speakers who willingly accepted our invitation and shared their expertise.

We also thank our publisher, Springer-Verlag, for accepting to publish the proceedings and for their kind assistance and cooperation during the editing process.

Finally, we thank all authors for their submissions and all conference attendees for making ICCSA 2013 truly an excellent forum on computational science, facilitating an exchange of ideas, fostering new collaborations and shaping the future of this exciting field.

We thank you all for participating in ICCSA 2013, and hope that you find the proceedings stimulating and interesting for your research and professional activities.

Osvaldo Gervasi
Bernady O. Apduhan
Duc Cuong Nguyen

Organization

ICCSA 2013 was organized by The Ho Chi Minh City International University (Vietnam), University of Perugia (Italy), University of Basilicata (Italy), Monash University (Australia), and Kyushu Sangyo University (Japan).

Honorary General Chairs

Phong Thanh Ho	International University (VNU-HCM), Vietnam
Antonio Laganà	University of Perugia, Italy
Norio Shiratori	Tohoku University, Japan
Kenneth C.J. Tan	Qontix, UK

General Chairs

Osvaldo Gervasi	University of Perugia, Italy
Bernady O. Apduhan	Kyushu Sangyo University, Japan
Duc Cuong Nguyen	International University (VNU-HCM), Vietnam

Program Committee Chairs

David Taniar	Monash University, Australia
Beniamino Murgante	University of Basilicata, Italy
Hong-Quang Nguyen	International University (VNU-HCM), Vietnam

Workshop and Session Organizing Chair

Beniamino Murgante	University of Basilicata, Italy

Local Organizing Committee

Hong Quang Nguyen	International University (VNU-HCM), Vietnam (Chair)
Bao Ngoc Phan	International University (VNU-HCM), Vietnam

Van Hoang   International University (VNU-HCM),
                                     Vietnam
Ly Le   International University (VNU-HCM),
                                     Vietnam

International Liaison Chairs

Jemal Abawajy   Deakin University, Australia
Ana Carla P. Bitencourt   Universidade Federal do Reconcavo da Bahia,
                                     Brazil
Claudia Bauzer Medeiros   University of Campinas, Brazil
Alfredo Cuzzocrea   ICAR-CNR and University of Calabria, Italy
Marina L. Gavrilova   University of Calgary, Canada
Robert C.H. Hsu   Chung Hua University, Taiwan
Andrés Iglesias   University of Cantabria, Spain
Tai-Hoon Kim   Hannam University, Korea
Sanjay Misra   University of Minna, Nigeria
Takashi Naka   Kyushu Sangyo University, Japan
Ana Maria A.C. Rocha   University of Minho, Portugal
Rafael D.C. Santos   National Institute for Space Research, Brazil

Workshop Organizers

Advances in Web-Based Learning (AWBL 2013)

Mustafa Murat Inceoglu   Ege University, Turkey

Big Data: Management, Analysis, and Applications (Big-Data 2013)

Wenny Rahayu   La Trobe University, Australia

Bio-inspired Computing and Applications (BIOCA 2013)

Nadia Nedjah   State University of Rio de Janeiro, Brazil
Luiza de Macedo Mourell   State University of Rio de Janeiro, Brazil

Computational and Applied Mathematics (CAM 2013)

Ana Maria Rocha   University of Minho, Portugal
Maria Irene Falcao   University of Minho, Portugal

Computer-Aided Modeling, Simulation, and Analysis (CAMSA 2013)

Jie Shen   University of Michigan, USA
Yanhui Wang   Beijing Jiaotong University, China
Hao Chen   Shanghai University of Engineering Science,
                                     China

Computer Algebra Systems and Their Applications (CASA 2013)

Andres Iglesias   University of Cantabria, Spain
Akemi Galvez   University of Cantabria, Spain

Computational Geometry and Applications (CGA 2013)

Marina L. Gavrilova   University of Calgary, Canada
Han Ming Huang   Guangxi Normal University, China

Chemistry and Materials Sciences and Technologies (CMST 2013)

Antonio Laganà   University of Perugia, Italy

Cities, Technologies and Planning (CTP 2013)

Giuseppe Borruso   University of Trieste, Italy
Beniamino Murgante   University of Basilicata, Italy

Computational Tools and Techniques for Citizen Science and Scientific Outreach (CTTCS 2013)

Rafael Santos   National Institute for Space Research, Brazil
Jordan Raddickand   Johns Hopkins University, USA
Ani Thakar   Johns Hopkins University, USA

Econometrics and Multidimensional Evaluation in the Urban Environment (EMEUE 2013)

Carmelo M. Torre   Polytechnic of Bari, Italy
Maria Cerreta   Università Federico II of Naples, Italy
Paola Perchinunno   University of Bari, Italy

Energy and Environment - Scientific, Engineering and Computational Aspects of Renewable Energy Sources, Energy Saving and Recycling of Waste Materials (ENEENV 2013)

Maurizio Carlini   University of Viterbo, Italy
Carlo Cattani   University of Salerno, Italy

Future Computing Systems, Technologies, and Applications (FISTA 2013)

Bernady O. Apduhan   Kyushu Sangyo University, Japan
Rafael Santos   National Institute for Space Research, Brazil
Jianhua Ma   Hosei University, Japan
Qun Jin   Waseda University, Japan

Geographical Analysis, Urban Modeling, Spatial Statistics (GEOG-AN-MOD 2013)

Giuseppe Borruso   University of Trieste, Italy
Beniamino Murgante   University of Basilicata, Italy
Hartmut Asche   University of Potsdam, Germany

International Workshop on Biomathematics, Bioinformatics and Biostatistics (IBBB 2013)

Unal Ufuktepe   Izmir University of Economics, Turkey
Andres Iglesias   University of Cantabria, Spain

International Workshop on Agricultural and Environmental Information and Decision Support Systems (IAEIDSS 2013)

Sandro Bimonte   IRSTEA, France
Andr Miralles   IRSTEA, France
Franois Pinet   IRSTEA, France
Frederic Flouvat   University of New Caledonia, New Caledonia

International Workshop on Collective Evolutionary Systems (IWCES 2013)

Alfredo Milani   University of Perugia, Italy
Clement Leung   Hong Kong Baptist University, Hong Kong

Mobile Communications (MC 2013)

Hyunseung Choo   Sungkyunkwan University, Korea

Mobile Computing, Sensing, and Actuation for Cyber Physical Systems (MSA4CPS 2013)

Moonseong Kim   Korean Intellectual Property Office, Korea
Saad Qaisar   NUST School of Electrical Engineering and
                           Computer Science, Pakistan

Mining Social Media (MSM 2013)

Robert M. Patton   Oak Ridge National Laboratory, USA
Chad A. Steed   Oak Ridge National Laboratory, USA
David R. Resseguie   Oak Ridge National Laboratory, USA
Robert M. Patton   Oak Ridge National Laboratory, USA

Parallel and Mobile Computing in Future Networks (PMCFUN 2013)

Al-Sakib Khan Pathan   International Islamic University Malaysia,
                             Malaysia

Quantum Mechanics: Computational Strategies and Applications (QMCSA 2013)

Mirco Ragni   Universidad Federal de Bahia, Brazil
Vincenzo Aquilanti   University of Perugia, Italy
Ana Carla Peixoto Bitencourt Universidade Federal do Reconcavo da Bahia,
                             Brazil
Roger Anderson   University of California, USA
Frederico Vasconcellos
    Prudente   Universidad Federal de Bahia, Brazil

Remote Sensing Data Analysis, Modeling, Interpretation and Applications: From a Global View to a Local Analysis (RS 2013)

Rosa Lasaponara   Institute of Methodologies for Environmental
                             Analysis - National Research Council, Italy
Nicola Masini   Archaeological and Monumental Heritage
                             Institute - National Research Council, Italy

Soft Computing for Knowledge Discovery in Databases (SCKDD 2013)

Tutut Herawan   Universitas Ahmad Dahlan, Indonesia

Software Engineering Processes and Applications (SEPA 2013)

Sanjay Misra   Covenant University, Nigeria

Spatial Data Structures and Algorithms for Geoinformatics (SDSAG 2013)

Farid Karimipour   University of Tehran, Iran and
                             Vienna University of Technology, Austria

Software Quality (SQ 2013)

Sanjay Misra   Covenant University, Nigeria

Security and Privacy in Computational Sciences (SPCS 2013)

Arijit Ukil   Tata Consultancy Services, India

Technical Session on Computer Graphics and Geometric Modeling (TSCG 2013)

Andres Iglesias   University of Cantabria, Spain

Tools and Techniques in Software Development Processes (TTSDP 2013)

Sanjay Misra   Covenant University, Nigeria

Virtual Reality and Its Applications (VRA 2013)

Osvaldo Gervasi   University of Perugia, Italy
Lucio Depaolis   University of Salento, Italy

Wireless and Ad-Hoc Networking (WADNet 2013)

Jongchan Lee   Kunsan National University, Korea
Sangjoon Park   Kunsan National University, Korea

Warehousing and OLAPing Complex, Spatial and Spatio-Temporal Data (WOCD 2013)

Alfredo Cuzzocrea   Istituto di Calcolo e Reti ad Alte Prestazioni -
                                   National Research Council, Italy and
                                   University of Calabria, Italy

Program Committee

Jemal Abawajy   Deakin University, Australia
Kenny Adamson   University of Ulster, UK
Filipe Alvelos   University of Minho, Portugal
Hartmut Asche   University of Potsdam, Germany
Md. Abul Kalam Azad   University of Minho, Portugal
Assis Azevedo   University of Minho, Portugal
Michela Bertolotto   University College Dublin, Ireland
Sandro Bimonte   CEMAGREF, TSCF, France
Rod Blais   University of Calgary, Canada
Ivan Blecic   University of Sassari, Italy
Giuseppe Borruso   University of Trieste, Italy
Yves Caniou   Lyon University, France
José A. Cardoso e Cunha   Universidade Nova de Lisboa, Portugal
Carlo Cattani   University of Salerno, Italy
Mete Celik   Erciyes University, Turkey
Alexander Chemeris   National Technical University of Ukraine
                                   "KPI", Ukraine
Min Young Chung   Sungkyunkwan University, Korea
Gilberto Corso Pereira   Federal University of Bahia, Brazil
M. Fernanda Costa   University of Minho, Portugal

Wenny Rahayu	La Trobe University, Australia
Jerzy Respondek	Silesian University of Technology, Poland
Ana Maria A.C. Rocha	University of Minho, Portugal
Humberto Rocha	INESC-Coimbra, Portugal
Alexey Rodionov	Institute of Computational Mathematics and Mathematical Geophysics, Russia
Cristina S. Rodrigues	University of Minho, Portugal
Haiduke Sarafian	The Pennsylvania State University, USA
Ricardo Severino	University of Minho, Portugal
Jie Shen	University of Michigan, USA
Qi Shi	Liverpool John Moores University, UK
Dale Shires	U.S. Army Research Laboratory, USA
Ana Paula Teixeira	University of Tras-os-Montes and Alto Douro, Portugal
Senhorinha Teixeira	University of Minho, Portugal
Graça Tomaz	University of Aveiro, Portugal
Carmelo Torre	Polytechnic of Bari, Italy
Javier Martinez Torres	Centro Universitario de la Defensa Zaragoza, Spain
Giuseppe A. Trunfio	University of Sassari, Italy
Unal Ufuktepe	Izmir University of Economics, Turkey
Mario Valle	Swiss National Supercomputing Centre, Switzerland
Pablo Vanegas	University of Cuenca, Equador
Paulo Vasconcelos	University of Porto, Portugal
Piero Giorgio Verdini	INFN Pisa and CERN, Italy
Marco Vizzari	University of Perugia, Italy
Krzysztof Walkowiak	Wroclaw University of Technology, Poland
Robert Weibel	University of Zurich, Switzerland
Roland Wismüller	Universität Siegen, Germany
Xin-She Yang	National Physical Laboratory, UK
Haifeng Zhao	University of California, Davis, USA
Kewen Zhao	University of Qiongzhou, China

Additional Reviewers

Antonio Aguilar	Universitat de Barcelona, Spain
José Alfonso Aguilar Caldern	Universidad Autnoma de Sinaloa, Mexico
Vladimir Alarcon	Geosystems Research Institute, Mississippi State University, USA
Margarita Alberti	Universitat de Barcelona, Spain
Vincenzo Aquilanti	University of Perugia, Italy
Takefusa Atsuko	National Institute of Advanced Industrial Science and Technology, Japan
Raffaele Attardi	University of Napoli Federico II, Italy

Yong-Wan Roh	Korean Intellectual Property Office, South Korea
Luiz Roncaratti	Universidade de Brasilia, Brazil
Marzio Rosi	University of Perugia, Italy
Francesco Rotondo	Polytechnic of Bari, Italy
Catherine Roussey	National Research Institute of Science and Technology for Environment and Agriculture, France
Rafael Oliva Santos	Universidad de La Habana, Cuba
Valentino Santucci	University of Perugia, Italy
Dario Schirone	University of Bari, Italy
Michel Schneider	Institut Supérieur d'Informatique de Modélisation et de leurs Applications, France
Gabriella Schoier	University of Trieste, Italy
Francesco Scorza	University of Basilicata, Italy
Nazha Selmaoui	Université de la Nouvelle-Calédonie, New Caledonia
Ricardo Severino	University of Minho, Portugal
Vladimir V. Shakhov	Institute of Computational Mathematics and Mathematical Geophysics SB RAS, Russia
Sungyun Shin	National University Kunsan, South Korea
Minhan Shon	Sungkyunkwan University, South Korea
Ruchi Shukla	University of Johannesburg, South Africa
Luneque Silva Jr.	State University of Rio de Janeiro, Brazil
V.B. Singh	University of Delhi, India
Michel Soares	Federal University of Uberlândia, Brazil
Changhwan Son	Sungkyunkwan University, South Korea
Henning Sten Hansen	Aalborg University, Denmark
Emanuele Strano	University of the West of England, UK
Madeena Sultana	Jahangirnagar University, Bangladesh
Setsuo Takato	Toho University, Japan
Kazuaki Tanaka	Kyushu Institute of Technology, Japan
Xueyan Tang	Nanyang Technological University, Singapore
Sergio Tasso	University of Perugia, Italy
Luciano Telesca	IMAA National Research Council, Italy
Lucia Tilio	University of Basilicata, Italy
Graça Tomaz	Instituto Politécnico da Guarda, Portugal
Melanie Tomintz	Carinthia University of Applied Sciences, Austria
Javier Torres	Universidad de Zaragoza, Spain
Csaba Toth	University of Calgari, Canada
Hai Tran	U.S. Government Accountability Office, USA
Jim Treadwell	Oak Ridge National Laboratory, USA

Chih-Hsiao Tsai	Takming University of Science and Technology, Taiwan
Devis Tuia	Laboratory of Geographic Information Systems, Switzerland
Arijit Ukil	Tata Consultancy Services, India
Paulo Vasconcelos	University of Porto, Portugal
Flavio Vella	University of Perugia, Italy
Mauro Villarini	University of Tuscia, Italy
Christine Voiron-Canicio	Université Nice Sophia Antipolis, France
Kira Vyatkina	Saint Petersburg State University, Russia
Jian-Da Wu	National Changhua University of Education, Taiwan
Toshihiro Yamauchi	Okayama University, Japan
Iwan Tri Riyadi Yanto	Universitas Ahmad Dahlan, Indonesia
Syed Shan-e-Hyder Zaidi	Sungkyunkwan University, South Korea
Vyacheslav Zalyubouskiy	Sungkyunkwan University, South Korea
Alejandro Zunino	National University of the Center of the Buenos Aires Province, Argentina

Sponsoring Organizations

ICCSA 2013 would not have been possible without tremendous support of many organizations and institutions, for which all organizers and participants of ICCSA 2013 express their sincere gratitude:

Ho CHi Minh City International University, Vietnam
(http://www.hcmiu.edu.vn/HomePage.aspx)

University of Perugia, Italy
(http://www.unipg.it)

 MONASH University

Monash University, Australia
(http://monash.edu)

Kyushu Sangyo University, Japan
(www.kyusan-u.ac.jp)

University of Basilicata, Italy (http://www.unibas.it)

The Office of Naval Research, USA
(http://www.onr.navy.mil/Science-technology/onr-global.aspx)

ICCSA 2013 Invited Speakers

Dharma Agrawal
University of Cincinnati, USA

Manfred M. Fisher
Vienna University of Economics and Business, Austria

Wenny Rahayu
La Trobe University, Australia

Selecting LTE and Wireless Mesh Networks for Indoor/Outdoor Applications

Dharma Agrawal*

School of Computing Sciences and Informatics, University of Cincinnati, USA
dharmaagrawal@gmail.com

Abstract. The smart phone usage and multimedia devices have been increasing yearly and predictions indicate drastic increase in the upcoming years. Recently, various wireless technologies have been introduced to add flexibility to these gadgets. As data plans offered by the network service providers are expensive, users are inclined to utilize freely accessible and commonly available Wi-Fi networks indoors.

LTE (Long Term Evolution) has been a topic of discussion in providing high data rates outdoors and various service providers are planning to roll out LTE networks all over the world. The objective of this presentation is to compare usefulness of these two leading wireless schemes based on LTE and Wireless Mesh Networks (WMN) and bring forward their advantages for indoor and outdoor environments. We also investigate to see if a hybrid LTE-WMN network may be feasible. Both these networks are heterogeneous in nature, employ cognitive approach and support multi hop communication. The main motivation behind this work is to utilize similarities in these networks, explore their capability of offering high data rates and generally have large coverage areas.

In this work, we compare both these networks in terms of their data rates, range, cost, throughput, and power consumption. We also compare 802.11n based WMN with Femto cell in an indoor coverage scenario, while for outdoors; 802.16 based WMN is compared with LTE. The main objective is to help users select a network that could provide enhanced performance in a cost effective manner.

---

* More information can be found at http://www.iccsa.org/invited-speakers

Neoclassical Growth Theory, Regions and Spatial Externalities

Manfred M. Fisher*

Vienna University of Economics and Business, Austria
manfred.fischer@wu.ac.at

Abstract. The presentation considers the standard neoclassical growth model in a Mankiw-Romer-Weil world with externalities across regions.

The reduced form of this theoretical model and its associated empirical model lead to a spatial Durbin model, and this model provides very rich own- and cross-partial derivatives that quantify the magnitude of direct and indirect (spillover or externalities) effects that arise from changes in regions characteristics (human and physical capital investment or population growth rates) at the outset in the theoretical model.

A logical consequence of the simple dependence on a small number of nearby regions in the initial theoretical specification leads to a final-form model outcome where changes in a single region can potentially impact all other regions. This is perhaps surprising, but of course we must temper this result by noting that there is a decay of influence as we move to more distant or less connected regions.

Using the scalar summary impact measures introduced by LeSage and Pace (2009) we can quantify and summarize the complicated set of non-linear impacts that fall on all regions as a result of changes in the physical and human capital in any region. We can decompose these impacts into direct and indirect (or externality) effects. Data for a system of 198 regions across 22 European countries over the period 1995 to 2004 are used to test the predictions of the model and to draw inferences regarding the magnitude of regional output responses to changes in physical and human capital endowments.

The results reveal that technological interdependence among regions works through physical capital externalities crossing regional borders.

---

* More information can be found at http://www.iccsa.org/invited-speakers

Global Spatial-Temporal Data Integration to Support Collaborative Decision Making

Wenny Rahayu*

La Trobe University, Australia
W.Rahayu@latrobe.edu.au

Abstract. There has been a huge effort in the recent years to establish a standard vocabulary and data representation for the areas where a collaborative decision support is required. The development of global standards for data interchange in time critical domains such as air traffic control, transportation systems, and medical informatics, have enabled the general industry in these areas to move into a more data-centric operations and services. The main aim of the standards is to support integration and collaborative decision support systems that are operationally driven by the underlying data.

The problem that impedes rapid and correct decision-making is that information is often segregated in many different formats and domains, and integrating them has been recognised as one of the major problems. For example, in the aviation industry, weather data given to flight en-route has different formats and standards from those of the airport notification messages. The fact that messages are exchanged using different standards has been an inherent problem in data integration in many spatial-temporal domains. The solution is to provide seamless data integration so that a sequence of information can be analysed on the fly.

Our aim is to develop an integration method for data that comes from different domains that operationally need to interact together. We especially focus on those domains that have temporal and spatial characteristics as their main properties. For example, in a flight plan from Melbourne to Ho Chi Minh City which comprises of multiple international airspace segments, a pilot can get an integrated view of the flight route with the weather forecast and airport notifications at each segment. This is only achievable if flight route, airport notifications, and weather forecast at each segment are integrated in a spatial temporal system.

In this talk, our recent efforts in large data integration, filtering, and visualisation will be presented. These integration efforts are often required to support real-time decision making processes in emergency situations, flight delays, and severe weather conditions.

---

* More information can be found at http://www.iccsa.org/invited-speakers

Table of Contents – Part III

Workshop on Software Engineering Processes and Applications (SEPA 2013)

Role "Intellectual Processor" in Conceptual Designing of Software Intensive Systems 1
Petr Sosnin

Modeling and Verification of Change Processes in Collaborative Software Engineering 17
Phan Thi Thanh Huyen, Kunihiko Hiraishi, and Koichiro Ochimizu

Relating Goal Modeling with BPCM Models in a Combined Framework 33
Shang Gao

Increasing the Rigorousness of Measures Definition through a UML/OCL Model Based on the Briand et al.'s Framework 43
Luis Reynoso, Marcelo Amaolo, Daniel Dolz, Claudio Vaucheret, and Mabel Álvarez

Improving Requirements Specification in WebREd-Tool by Using a NFR's Classification 59
José Alfonso Aguilar, Sanjay Misra, Anibal Zaldívar, and Roberto Bernal

Application of an Extended SysML Requirements Diagram to Model Real-Time Control Systems 70
Fabíola Goncalves C. Ribeiro, Sanjay Misra, and Michel S. Soares

Frequent Statement and De-reference Elimination for Distributed Programs 82
Mohamed A. El-Zawawy

Agile Software Development: It Is about Knowledge Management and Creativity 98
Claudio León de la Barra, Broderick Crawford, Ricardo Soto, Sanjay Misra, and Eric Monfroy

Formalization and Model Checking of SysML State Machine Diagrams
by CSP# 114
Takahiro Ando, Hirokazu Yatsu, Weiqiang Kong,
Kenji Hisazumi, and Akira Fukuda

From Arrows to Netlists Describing Hardware 128
Matthias Brettschneider and Tobias Häberlein

Evaluation of Process Architecture Design Methods 144
Mery Pesantes, Hugo A. Mitre, and Cuauhtémoc Lemus

Multi Back-Ends for a Model Library Abstraction Layer 160
Ngoc Viet Tran, Andreas Ganser, and Horst Lichter

Explicit Untainting to Reduce Shadow Memory Usage and Access
Frequency in Taint Analysis 175
Jae-Won Min, Young-Hyun Choi, Jung-Ho Eom, and
Tai-Myoung Chung

A Framework for Security Testing 187
Daya Gupta, Kakali Chatterjee, and Shruti Jaiswal

Comparing Software Architecture Descriptions and Raw Source-Code:
A Statistical Analysis of Maintainability Metrics 199
Eudisley Anjos, Fernando Castor, and Mário Zenha-Rela

Systematic Mapping of Architectures for Telemedicine Systems 214
Glauco de Sousa e Silva, Ana Paula Nunes Guimarães,
Hugo Neves de Oliveira, Tatiana Aires Tavares, and
Eudisley Gomes dos Anjos

Architectural Model for Generating User Interfaces Based on Class
Metadata 230
Luiz Azevedo, Clovis Torres Fernandes, and Eduardo Martins Guerra

Workshop on Computer-Aided Modeling, Simulation, and Analysis (CAMSA 2013)

A Novel Fuzzy Co-occurrence Matrix for Texture Feature Extraction ... 246
Yutthana Munklang, Sansanee Auephanwiriyakul, and
Nipon Theera-Umpon

Improved Flow Shop Schedules with Total Completion Time
Criterion... 258
Dipak Laha and Dhiren Kumar Behera

Workshop on Wireless and Ad Hoc Networking (WADNet 2013)

Towards a Real Architecture of Wireless Ad-Hoc Router on
Open-Source Linux Platform   271
    Quan Le-Trung and Minh-Son Nguyen

A Study of Robot-Based Context-Aware Fire Escape Service Model   287
    *NamJin Bae, KyungHun Kwak, Sivamani Saraswathi,
    JangWoo Park, ChangSun Shin, and YongYun Cho*

Workshop on Cities, Technologies and Planning (CTP 2013)

Co-creating Urban Development: A Living Lab for Community
Regeneration in the Second District of Palermo   294
    Jesse Marsh, Francesco Molinari, and Ferdinando Trapani

Semantic Interoperability of German and European Land-Use
Information   309
    Hartmut Müller and Falk Würriehausen

The Representation for All Model: An Agent-Based Collaborative
Method for More Meaningful Citizen Participation in Urban
Planning   324
    Maria-Lluïsa Marsal-Llacuna and Josep-Lluís de la Rosa-Esteva

Smart Cities as "EnvironMental" Cities   340
    Luciano De Bonis

Impact of Urban Development and Vegetation on Land Surface
Temperature of Dhaka City   351
    Debasish Roy Raja and Meher Nigar Neema

Design of a Team-Based Relocation Scheme in Electric Vehicle Sharing
Systems   368
    Junghoon Lee and Gyung-Leen Park

Qualitative Analysis of Volunteered Geographic Information in a
Spatially Enabled Society Project   378
    *Jarbas Nunes Vidal-Filho, Jugurta Lisboa-Filho,
    Wagner Dias de Souza, and Gerson Rodrigues dos Santos*

An Innovative Approach to Assess the Quality of Major Parks in
Environmentally Degraded Mega-City Dhaka   394
    *Antora Mohsena Haque, Md. Rifat Hossain,
    Md. Hasan Murshed Farhan, and Meher Nigar Neema*

Analysis of Potential Factors Bringing Disparity in House Rent of
Dhaka City 408
Taslima Akter, Md. Mehedi Hasan, Akter Uz Zaman,
Md. Rifat Hossain, and Meher Nigar Neema

Integrated GIS and Remote Sensing Techniques to Support PV
Potential Assessment of Roofs in Urban Areas 422
Flavio Borfecchia, Maurizio Pollino, Luigi De Cecco,
Sandro Martini, Luigi La Porta, Alessandro Marucci, and
Emanuela Caiaffa

Deriving Mobility Practices and Patterns from Mobile Phone Data 438
Fabio Manfredini, Paola Pucci, and Paolo Tagliolato

GIS Based Urban Design for Sustainable Transport and Sustainable
Growth for Two-Wheeler Related Mega Cities like HANOI 452
Martin Ruhé, Hans-Peter Thamm, Leif Fornauf, and
Matias Ruiz Lorbacher

New Concepts for Structuring 3D City Models – An Extended Level
of Detail Concept for CityGML Buildings 466
Marc-O. Löwner, Joachim Benner, Gerhard Gröger, and
Karl-Heinz Häfele

Walking into the Past: Design Mobile App for the Geo-referred and the
Multimodal User Experience in the Context of Cultural Heritage 481
Letizia Bollini, Rinaldo De Palma, and Rossella Nota

Building Investments for the Revitalization of the Territory:
A Multisectoral Model of Economic Analysis 493
Gianluigi De Mare, Antonio Nesticò, and Francesco Tajani

Dynamic Analysis of the Property Market in the City of Avellino
(Italy): The Wheaton-Di Pasquale Model Applied to the Residential
Segment 509
Gianluigi De Mare, Benedetto Manganelli, and Antonio Nesticò

Spatial Representation: City and Digital Spaces 524
Gilberto Corso Pereira, Maria Célia Furtado Rocha, and
Pablo Vieira Florentino

Web 3D Service Implementation 538
Nuno Oliveira and Jorge Gustavo Rocha

The e-Participation in Tranquillity Areas Identification as a Key Factor
for Sustainable Landscape Planning 550
Giuseppe Modica, Paolo Zoccali, and Salvatore Di Fazio

Free Web Mapping Tools to Characterise Landscape Dynamics and to
Favour e-Participation 566
    Maurizio Pollino and Giuseppe Modica

Improving EU Cohesion Policy: The Spatial Distribution Analysis of
Regional Development Investments Funded by EU Structural Funds
2007/2013 in Italy 582
    Francesco Scorza

A Web-Based Participatory Management and Evaluation Support
System for Urban Maintenance 594
    Ivan Blecic, Dario Canu, Arnaldo Cecchini, and
    Giuseppe Andrea Trunfio

Web 3.0 and Knowledge Management: Opportunities for Spatial
Planning and Decision Making.................................... 606
    Beniamino Murgante and Vito Garramone

Enhancing the Spatial Dimensions of Open Data: Geocoding Open PA
Information Using Geo Platform Fusion to Support Planning Process ... 622
    Francesco Izzi, Giuseppe La Scaleia, Dimitri Dello Buono,
    Francesco Scorza, and Giuseppe Las Casas

Cities and Smartness: A Critical Analysis of Opportunities and Risks ... 630
    Beniamino Murgante and Giuseppe Borruso

Author Index 643

Role "Intellectual Processor" in Conceptual Designing of Software Intensive Systems

Petr Sosnin

Ulyanovsk State Technical University, Severny Venetc str. 32,
432027 Ulyanovsk, Russia
sosnin@ulstu.ru

Abstract. The paper presents a question-answer approach to programming of designers' activity in collaborative designing of software intensive systems. Efficiency of a conceptual work can be substantially increased if a human part of collaborative works will be fulfilled by designers in a form of an execution of programs written in a specialized pseudo-code language. Such programs simulate an experimental activity of designers investigating the own behavior in solution processes of project tasks. Acting similarly a scientist any designer will play a role named in the approach as "intellectual processor". Such approach was investigated and evolved till an instrumental system providing the pseudo-code programming of intellectual processors combined with computer processors.

Keywords: Conceptual designing, precedent, pseudo-code programming, question-answering, software intensive system.

1 Introduction

Very often designers of a software intensive system (SIS) should take into account a triple intensity – an intensive own activity, intensity embedded to SIS and intense activity of its potential users. The first and the third types of indicated intensities have the type of a human intensive activity. The essential feature of this type of the intensity is a necessity of using the experiential behavior in an important part of a work being executed by a human in a computerized system.

The intensity problem should be analyzed first of all from the side of the activity of designers working with software units. It is necessary to note that software engineering has an empirical nature [1]. Creating the software the designers should experiment and use the accessible experience in the real time. Furthermore, solving the definite project tasks the designers should simulate the future user behavior which also can be the intensive human activity. Thus designers have relations to any indicated intensity. By other words there are many situations in designing of SIS when designers should use experiential behavior in the real time. Such behavior should be supported by effective tools but such tools are practically absent in technologies used for designing of SIS.

B. Murgante et al. (Eds.): ICCSA 2013, Part III, LNCS 7973, pp. 1–16, 2013.

This paper presents question-answer approach to modeling of the experiential behavior of designers working at the conceptual stage of designing. The conceptual stage is chosen because the price of misunderstanding and errors on the given stage is very high. Negative influences of these human factors can be decreased if designers will be working with a necessary experience as scientists. Processes of question-answering are reliable mechanisms for the access to the accessible experience for its real time using and creating in designing. In the approach the special attention is being given to the complexity of project situations with which the designers should interact.

The approach is constructively specified and materialized for the activity of designers playing a role named "intellectual processor" (I-processor). This role is oriented on the designer behavior based on using the experience and its models. Playing the role is being supported by the specialized toolkit WIQA (Working In Questions and Answers) [2].

2   Preliminary Statements

The basic cause of the intensive designer activity in conceptual designing is a complexity as the process of designing so and SIS being created. In general sense the complexity (or simplicity) of SIS reflects the degree of difficulty for designers in their interactions with definite models of SIS (or its components) in solving the definite tasks. The system or its any component is complex, if the designer (interacting with the system) does not have sufficient resources for the achievement of the necessary level of understanding or achieving the other planned purposes.

Often enough various interpretations of Kolmogorov measure [3] are applied for estimations of a degree of the system complexity. This measure is connected with the minimal length of program P providing the construction of system S from its initial description D. Distinctions in interpretations are caused usually by features of system S   and formal descriptions to be used for objects P and D.

In creating of SIS the objects of the P-type are being built in step by step into the process of designing with using the certain means of metaprogramming M. Actuality of such work demonstrates that the complexity of "P-object" no less than the complexity of SIS in its any used state. Moreover, M-program providing the construction of P-object is being built on the base of the same initial description D of the system S. It can be presented by the following chain $D \rightarrow M \rightarrow P \rightarrow S$.

Named relations between D, M, P can be used by designers for disuniting the process of designing on stages $[D(t_0) \rightarrow M_0 \rightarrow P_0 \rightarrow S(t_1)]$, $[D(t_1) \rightarrow M_1 \rightarrow P_1 \rightarrow S(t_2)]$, ..., $[D(t_i) \rightarrow M_i \rightarrow P_i \rightarrow S(t_{i+1})]$, ..., $[D(t_{n-1}) \rightarrow M_{n-1} \rightarrow P_{n-1} \rightarrow S(t_n)]$ where a set $\{S(t_i)\}$ is being created   with using "the programs" of M- and P-types.

We suggest a constructive way for using means of programming for units of M- and P-types. But first of all we shall explain our understanding of these units which have direct relations to the experiential activity of designers solving the tasks of conceptual designing.

Designers who will be testing the solutions should have the possibilities for experimenting with them. In this sense P-programs correspond to plans of experiments being fulfilled by designers. M-programs are intended for creating the P-programs in forms which are effective for their use by designers. Programs of this type are responsible for collaborative solving the tasks in coordination.

Below we shall come back to the diversity and similarity of M- and P-programs but at first we shall indicate means used for the named purposes in the modern technologies providing the development of SIS. For this we shall analyze means embedded to the Rational Unified Process (RUP) [4].

The conformity to requirements and understandability are being reached in the RUP with the help of "block and line" diagrams expressed in the Unified Modeling Language (UML). The content of diagrams built by designers is being clarified by necessary textual descriptions. But UML is not the language of the executable type and therefore diagrams are not suitable for experimenting with them as with programs of P-type.

For collaborative solving the tasks in coordination the RUP suggests the means of normative workflows the relations between which are being regulated by a set of rules. For any task of the definite normative workflow the RUP suggests its interactive diagrammatic model with a set of components the use of which can help in solving the task. Forms of programming are not used also in all of these means. The similar state of affairs with conceptual designing exists in other known technologies supporting the development of SIS.

Our approach to using the programming forms of designers' activity in conceptual designing is based on the experiential behavior of the designer playing the role named "intellectual processor". Comparing the actions of designers in interactions with experience with human actions in applications of Model Human Processor [5] has led to the decision to specify the human processor based on question-answering [6]. Such decision took into account additionally the related works describing: experience of human-computer interactions [5], experience factory [7] and experience base [8], experiential learning [9] and practice of workflows [10].

3   Experiential Behavior of Designers

The responsibility for solutions of tasks is being distributed between of designers in accordance with the competence of each of them. The team competence being needed for conceptual designing should include:

1. Certain competencies which can help to designers in their attempts to solve the necessary tasks $Z^S = \{Z^S_i\}$ of the SIS subject area.
2. A system of competencies which is necessary for solving the normative tasks $Z^N = \{Z^N_j\}$ of technology used by designers.

Told above emphasizes that designers should be very qualified specialists in the technology domain but that is not sufficient for the successful designing. Normative tasks are invariant to the SIS domain and therefore designers should gain certain experience needed for solving the definite tasks of the SIS subject area. The most part of the additional experience is being acquired by designers in the experiential learning when tasks of Z^S-type are being solved. Solving of any task Z^S_i is similar to its decomposing into a series on the base of normative tasks. Such decomposition is no more than point of view, helping in the detachment of a very important designer activity connected with the adaptation of the chosen normative task Z^N_j to the content of the task $Z^S_i(t)$. It is rational to consider that any similar decomposition requires solving the corresponding task of adaptation Z^A_j.

Any task of Z^A-type should be determined, analyzed and solved by the corresponding designer on the base of the accessible experience in the real time. Functions of the accessible experience can be being fulfilled by the personal and collective experience and useful experience models also. Moreover, solving any task of Z^A-type any designer gains new units of experience by conducting the certain experiment. It is useful to conduct similar experiments, explicitly emulating the experimental work of scientists.

Forms of the experiential behavior are being used by designers not only in their real work with tasks of Z^A-type but also in solutions of tasks:

- from the set $\{Z^W_m\}$ providing the works with tasks of Z^S-type in corresponding workflows $\{W_m\}$ in SIS;
- from the set $\{Z^W_n\}$ providing the works with tasks of Z^N-type in corresponding workflows $\{W_n\}$ in the used technology;
- from sets $\{Z^G_p\}$ and $\{Z^G_r\}$ any of which corresponds to the definite group of workflows in SIS or technology.

Analysis of all indicated tasks from the side of their automation used in conceptual designing leads to following conclusions:

1. The current level of the automation is being achieved by the use means of automation for normative practices (corresponding to Z^N-tasks) embedded to the used technology, for example, to the RUP.
2. So the automation of Z^S-tasks is being expressed only by their decomposition on the basis of Z^N-tasks.
3. Automating the tasks of Z^W- and Z^G-types is provided usually by specialized means implemented on the base of executable business process languages for example on the base of languages BPEL (Business Process Executable Language) or YAWL (Yet Another Workflow Language) [10] which do not take into account the experiential behavior of designers.
4. The automation of the designer activity which corresponds to solving of Z^A-tasks is not being supported technologically.

Thus there is a potential for the automation of the designer activity. The automation of the experiential behavior of designers can simplify the complexity of conceptual designing. Such way of the automation will be presented below in the context of our understanding the nature of the experiential behavior.

On our deep persuasion the automation should be coordinated with the nature of human interactions with the personal experience. This nature reflects the own existence in the dialog forms in consciousness. Natural question-answering is a reliable mechanism of human interactions with the own experience.

A question is a natural phenomenon which appears when a human should use or wants to use the experience. If the phenomenon of such type is revealed by the human the revealed question can be described in the natural language. Any of such description is no more than a linguistic (symbolic) model of the corresponding question.

Interactions with the symbolic model of the question can renew the phenomenon of this question for its reuse.

The adequacy of any model of any question should be carefully tested because such model is used for creating the necessary answer. The definite question and corresponding answer are bound as "cause and effect". They supplement each other. The answer form depends on the use of the answer.

Questions have different types. A very important class of questions is tasks. Any statement of any task is its symbolic model also. There are several useful forms of the answer for the tasks, for example, "method of task solving", "solution idea", "conceptual solution" and "programmed solution".

Thus questions and answer are used not only for interactions with experience. They can be used, for example, for the access to the appropriate unit of experience or for modeling the unit of experience or for reasoning in different cases. Any human has the rich experience of using the questions and answers and their models.

The definite task and behavioral answer on it can be fulfilled the role of the experience unit if they are combined and prepared for the future reuse. Very often the reused typical behavior is named "precedent". Models of precedents as models of corresponding tasks are models of experience units which can be united in the special library "Experience Base". Such library is a model of the definite part or parts of the natural experience.

4   Means for Interactions with Experience

The described understanding has led us to the constructive scheme of interactions with experience the generalized view of which is presented in Fig. 1.

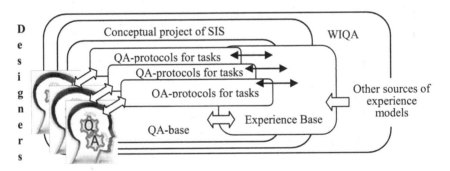

Fig. 1. Interactions with the accessible experience

Solving any appointed task, the designer registers the used question-answer reasoning (QA-reasoning) in the specialized protocol (QA-protocol) so that this QA-protocol can be used as the task model (QA-model). Such models can be used by designers for experimenting in the real time with tasks being solved.

Units of experiential behavior extracted from solution processes are being modeled on the base of QA-models of tasks. Created behavioral models are being loaded in the Experience Base of the specialized processor WIQA. After that they can be used by designers as units of the accessible experience. Experience models from the other sources can be uploaded in the Experience Base also.

The toolkit interpretation as the processor has used analogies with text processors giving the opportunities to create texts with the help of text elements and a system of operations. In WIQA the functions of text elements fulfilled interactive objects of "question" (Q) and "answer" (A) types therefore the toolkit has been specified as "question-answer processor" (QA-processor). "Questions" and "answers", being processed by this processor, were models of natural questions and answers extracted from reasoning of designers. Such functions of the toolkit are provided in the current version of QA-processor but its opportunities are evolved till the work with programs of M- and P-types. QA-processor is a client-server system with database (QA-base) oriented on interactions with the accessible experience including Experience Base.

QA-processor helps to create QA-model for a task of any type used in conceptual designing the SIS. This opportunity is being supported by the framework taking into account the RUP tasks being solved in conceptual designing. The structure presented in Fig. 2 reflects this framework in the form of a system of architectural views on the task. Views of this framework are presented in detail in our publication [11].

Applying of the framework to the definite task supposes its adjusting to this task. Adjusting is being implemented in forms of filling the framework by the useful content till the adequate task model which should help in solving the task (for example for the task Z). The adjusted model opens the opportunity for experimenting with the task solution from sufficient viewpoints.

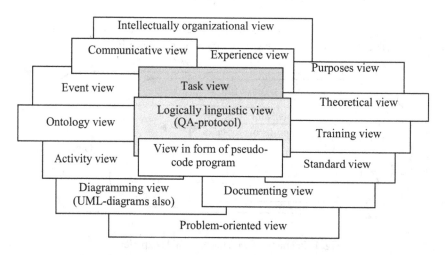

Fig. 2. Framework for QA-model of task

The framework helps to define more exactly the notion of "QA-model". The simplest version of QA-model includes the statement of the task Z and corresponding QA-protocol. More complicated versions include additionally necessary views. Additional views are evolved the content of the simplest QA-model.

Thus the adjustment of the framework it is the choice of necessary views and filling them by the adequate content correlated with names of views. Reasoning of this paper concerns only the task view and QA-protocols therefore additional views will remain below without explanations.

So if designers of SIS use the toolkit WIQA they have the opportunity for conceptual modeling the tasks of different indicated types. In this case the current state of tasks being solved collaboratively is being registered in QA-base of the toolkit and this state is visually accessible in forms of the tree of tasks and QA-models for corresponding tasks. The named opportunity is presented figuratively in Fig. 3 where QA-base is interpreted as a specialized QA-memory.

In the toolkit its QA-base is materialized with using Date Base Management System MS SQL 2008 Standard Edition. But for its users the QA-base is opened as a memory storing the interactive objects presenting "tasks" Z, "questions" Q and "answers" A. Moreover, these objects are combined in hierarchical structures.

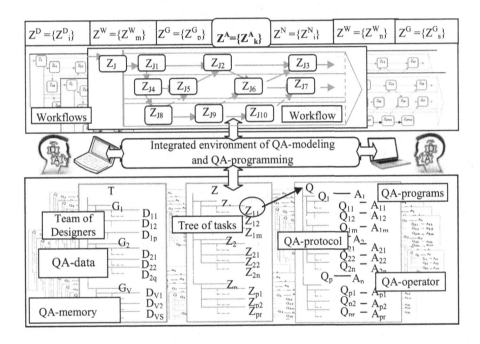

Fig. 3. Sinking of tasks in WIQA-environment

In their real time work the designers interact with such objects. They process them with the help of appropriate operations helping to find and test the solution of tasks. All of that has led to the idea of programming the designer activity with storing of created programs in QA-memory. So all indicated tasks are being uploaded to QA-memory with the rich system of operations with interactive objects of Z-, Q- and A- types. Designers have the possibility to program the interactions with necessary objects. Such programs are similar to plans of the experimental activity in conceptual designing of SIS.

Creating the programs for experimenting with conceptual models of tasks should be fulfilled by designers the competence of which can be far from the experience of professional programmers. It was a main cause to embed the means of pseudo-code programming in the toolkit WIQA. The similarity of the language of (pseudo-code) programming and the natural language in its algorithmic usage was the second and no less important cause. The created language is oriented on QA-memory and named "WIQA". Below this language will be designated L^{WIQA}.

Let's generalize intermediate results of reasoning about interactions of designers with interactive objects included in QA-model. This generalization is reflected on the scheme presented in Fig 4.

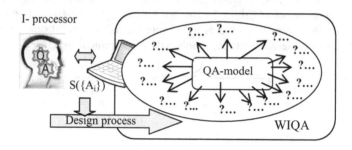

Fig. 4. Interactions with QA-model of task

The description of the definite behavioral unit composed of designer interactions with QA-model can fulfill the role of a model of such designer activity. In order to distinguish this type of models from other types of models they will be named below "QA-models of the designer activity". Such model is a text which consists of instructions indicating the designer actions which should be executed in the reuse of behavioral unit in the WIQA-medium.

Similar models can be created for acting the human not only in the WIQA-medium. Their content, form and appointment are demonstrated by the following technique:

```
//Reset of Outlook Express
01. Quit all programs.
02. Start On the menu Run, click.
03. Open In the box regedit, type, and then OK the click.
04. Move to and select the following key:
HKEY_CURRENT_USER/Software/Microsoft/Office/9.0/Outlook
05. In the Name list, FirstRunDialog select.
06. If you want to enable only the Welcome to Microsoft Outlook
greeting, on the Edit menu Modify, click the type True in the
Value Data box, and then OK the click.
07. If you also want to re-create all sample welcome items, move
to and select the following key:
```

```
HKEY_CURRENT_USER/Software/Microsoft/Office/9.0/Outlook/Setup
O8. In the Name list, select and delete the following keys:
CreateWelcome First-Run
O9. In the Confirm Value Delete dialog box click Yes, for each
entry.
O.10. On the Registry menu, click Exit.
O11.  End.
```

This technique is chosen to emphasize the following:

1. There are many behavior units describing the human activity in different computerized mediums.
2. Descriptions of similar typical activities help in the reuse of these precedents.
3. Descriptions of techniques have forms of programs (N-programs) written in the natural language L^N in its algorithmic usage.
4. Such N-programs consist of operators being fulfilled by the human interacting with the definite computerized system. In the example of N-program its operators are marked by the symbol "O" with the corresponding digital index.

There are no obstacles for uploading the N-programs in QA-memory. This way is used for uploading the techniques supporting the designer activity in the WIQA-medium. Operators of techniques are placed in Q-objects. Corresponding A-objects are used for registering the facts or features of executed operations. Thus any QA-operator is registered in pair of corresponding Q- and A-objects or shortly in QA-unit.

The operator interpretation as the question has a cause. The designer activates the operator to receive the necessary answer from QA-model being investigated. Let us mark that any name (for example, the name of variable) can be interpreted as "question" while the corresponding answer is a meaning (value) of this name.

5 Pseudo-Code Language L^{WIQA}

The specialized pseudo-code language L^{WIQA} has been developed for the use of QA-reasoning in programming the plans for solutions of tasks. This language is oriented on its use in experiential interactions of designers with the accessible experience when they create programs of the own activity and investigate them. Step by step L^{WIQA} has been evolved till the state included following components:

- **traditional types of data** such as scalars, lists, records, sets, stacks, queues and the other data types;
- **data model** of the relational type describing the structure of database;
- **basic operators** including traditional pseudo-code operators, for example, Appoint, Input, Output, If-Then-Else, While-Do, GOTO, Call, Interrupt, Finish and the others operators;
- **SQL-operators** in their simplified subset including Create Database, Create Table, Drop Table, Select, Delete From, Insert Into, Update;
- **operators for managing the workflows** oriented on collaborative designing (Seize, Interrupt, Wait, Cancel, Queue);

- **operators of visualization** developed now for the creation of dynamic view of cards presenting QA-units in the direct access of the designer to objects of QA-memory.

The important type of basic operators is an explicit or implicit command aimed at the execution by the designer the definite action. The explicit commands are being written in the natural language. It is possible to consider, that the example with the reset of Outlook Express is written in the language L^{WIQA}.

But the traditional meaning of named data and operators is only one side of their content. The other side is bound with attributes of QA-units in which data and operators are uploaded. Definite data or operator inherits the attributes of the corresponding QA-unit as a cell of QA-memory. They inherit not only attributes of QA-units but their understanding as "questions" and "answers" also.

Thus data and operators of L^{WIQA} inherit a very useful set of attributes of QA-units in which they are registered. For example, such set includes "type or sub-type of QA-unit", "name of creator", "time of last saving" and any set of any additional attributes appointed by designers with using the specialized plug-ins [2]. One of basic attributes is a "unique unit name" which can use as a unit address in QA-memory. The unique name includes the name of unit type (for example "Q") and the compound unique index appointed automatically. Thus addresses of cells reflect the hierarchy relations between related QA-units.

Attributes of QA-units essentially influence on data (QA-data) and operators (QA-operators) of pseudo-code programs. Such influence can be used effectively in executions of pseudo-code programs. Named specificity of QA-data and QA-operators emphasizes the essential difference between L^{WIQA} and other known pseudo-code languages. Therefore, pseudo-code programs written in L^{WIQA} have received the name "QA-programs".

Interpreting the current operator, I-processor can fulfill any actions till its activation (for example, to test circumstances) and after its execution (for example, to estimate the results), using any means of the toolkit WIQA.

6 Basic Features of Intellectual Processor

Told above is sufficient for holistic specifying the features of I-processor. Indicated interactions with the accessible experience are a kind of the specialized designer activity which can be implemented as the role played by any designer in parallel with other appointed roles.

Any modern technology used for designing the SIS includes modeling the work of designers with the help of roles. For example the current version of Rational Unified Process (RUP) supports activities of designers playing approximately 40 roles. RUP is a "heavy" technology therefore for small teams the quantity and specifications of used roles are decreased and simplified [4].

In existing technologies the role provided interactions with experience is being fulfilled by designers without its explicit specifying and special instrumental support. Therefore in this paper the content of the role is defined for its use by designers in WIQA-medium but the role can be materialized by the other way.

In our works such role is being named "intellectual processor" because the responsibility of Human Intelligence is bound with discovering and mastering of experience units for their reuse in the human life. The essence of I-processor is being defined by features enumerated below:

1. First of all I-processor is a role being played by any designer solving the appointed tasks in conceptual designing of SIS.
   In any case the role is a special version of a designer behavior which satisfies to the definite set of rules. Role specifications depend on appointed tasks and tools supporting their preliminary solving and reuses of solutions.
2. I-processors are intended for experimenting with tasks the solving of which are problematically without explicit real time access to the personal experience and/or collective experience and/or models of useful experience.
3. The role of I-processor is oriented on the behavior of a scientist who prepares and conducts an experiment, registering its results in the understandable and checkable form. Experimenting, the designer investigates units of the behavior which have led to solutions of corresponding project tasks in conceptual designing.
4. In its activity any I-processor interprets investigated tasks as precedents and interacts with experience units as with models of precedents.
5. The real time work of any I-processor is being accompanied by QA-reasoning and its models being registered in QA-memory of the toolkit WIQA.
   The choice of QA-reasoning as the basic form for I-processor is determined by the intention to model the dialog nature of consciousness. For that the implicit QA-reasoning accompanying the cognitive processes inside I-processor should "be translated" and transferred to QA-memory as the model QA-reasoning.
6. QA-reasoning is used by I-processor for creating QA-models of tasks in different forms including their versions as QA-programs.
7. The use of QA-reasoning in interactions of I-processor with QA-programs is being implemented in the pseudo-code language L^{WIQA}. Knowing and effective using of this language by I-processor is a very important its feature.

Thus, in general case the activity of I-processor is similar to the experimental activity of the designer who creates QA-model of the task for experimenting with the prototype of its solution. In such experimenting with tasks I-processor can use as means of QA-modeling so and means of QA-programming.

As told above the use of QA-programs by I-processor for experimenting with tasks is its very essential feature. That was a cause to create the library of the specialized QA-programs providing some versions of such experimenting. This library includes a number of QA-techniques for cognitive tasks analysis, decision-making and typical procedures of estimations. For example, the library section of QA-techniques for cognitive task analysis includes widely known methods and the other methods.

7   Specimens of QA-Programs

To clarify the features and syntax of QA-programs a number of their examples will be presented below. The basic applicability of QA-programs is the description of precedents in program forms. The toolkit WIQA supports the creation of precedent models in accordance with its logical model presented in Fig. 5.

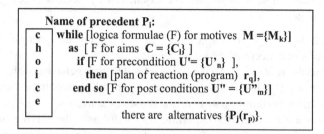

Name of precedent P_i:
c | **while** [logica formulae (F) for motives $M = \{M_k\}$]
h | **as** [F for aims $C = \{C_l\}$]
o | **if** [F for precondition $U' = \{U'_n\}$],
i | **then** [plan of reaction (program) r_q],
c | **end so** [F for post conditions $U'' = \{U''_m\}$]
e | --------------------------------------
                        there are alternatives $\{P_j(r_p)\}$.

Fig. 5. Logical model of precedent

The logical scheme of the model describes the precedent from the viewpoint of a conditioned reflex which is intellectually processed in accordance with the solved task and its QA-model.

Hence, it is necessary to program the access to the precedent and also the reuse of its reaction. The logical model should help to solve the task of the precedent choice from the Experience Base. The typical scheme of QA-programs for the conditioned parts has the following view:

```
QA-PROGRAM_1 (condition for the access to the precedent):
N1. Variable &V_1& / symbol "N" indicates type "Name of
variable"
V1. Value of V_1 / symbol "V" indicates type "Value of
variable"
N2. Variable &V_2& / symbol "&" is used as tag
V2. Value of V_2.
.......................................
NM. Variable &V_M& / Comment_M?
VM. Value of V_M.
ON. F = Logical expression (&V_1&, &V_2&, …, &V_M&)
EN. Value of Expression.
OP. Finish.
```

It is necessary to notice that the designer can build or to modify or to fulfill (step by step) the definite example of similar programs in the real time work with the corresponding precedent which designer creates or reuses. The reaction part of the precedent has very often the technique type similar technique for Reset of Outlook Express described above. In other cases the designer can program the reaction part in the prototype form.

The next QA-program is used in plug-ins of WIQA named "System of Interruptions" for the calculation of a priority of the interrupted QA-programs being executed by the designer in parallel:

```
O 1.11 Procedure &DiscardPriority&
O 1.11.1 &P& := &Pmax&/
O 1.11.2 Label &DP1&/ "O 1.11.2"-unique index name (Address)
O 1.11.3 &Priority& := &P&
O 1.11.4 CALL &GetTaskByPr&
O 1.11.5 &base& -> &TaskPriority& := &base& -> &TaskPriority&+1
O 1.11.6 CALL &ChangeTask&
O 1.11.7 &P& := &P& - 1
O 1.11.8 IF &P& < &base& -> &Pmin& THEN &base& -> &NewPriority&
:= &Pmin& ELSE GOTO &DP1&
O 1.11.9 ENDPROC &DiscardPriority&
```

This QA-procedure is being translated by the compiler (not by the interpreter) because they are being processed by the computer processor (not by I-processor). Therefore, A-lines of operators are excluded from the source code.

One more example from the subsystem "Controlling of assignments" provides the access of designers to the model of team uploaded in QA-memory. The access for editing of the assignment of tasks is described by the following QA-procedure:

```
Q 2.3.3 PROCEDURE &Edit_Assignment&// index names are inherited
from the full QA-program
   Q 2.3.3.1 &DbQAId& := QA_GetQAId (&current_project&,
"DB_assignment")
   Q 2.3.3.2 &DbId& := OpenDB(&current_project&. &DbQaId&)
   Q 2.3.3.3 &groups& := ExecuteSQL(&DbId&, "SELECT ID,
Name_group FROM Group")
   Q 2.3.3.4 &people& := ExecuteSQL(&DbId&, "SELECT ID, Design-
er-name FROM Designer")
   Q 2.3.3.5 &ToFindQuery& := AC_FindTask(&groups&, &people&)
   Q 2.3.3.6 &FoundTasks& := ExecuteSQL (&DBId&, &ToFindQuery&)
   Q 2.3.3.7 &TaskId& := AC_SELECTTASK(&FoundTasks&)
   Q 2.3.3.8 &TaskQuery& := "SELECT * FROM Assignment WHERE ID="
   Q 2.3.3.9 &TaskQuery& := STRCAT(&TaskQuery&, &TaskId&)
   Q 2.3.3.10 &FoundTask& := ExecuteSQL (&DbId&, &TaskQuery&)
   Q 2.3.3.11 AC_VIEWTASK (&FoundTask&, &Groups&, &People&)
Q 2.3.3.12 ENDPROC & Edit_Assignment &
```

This example demonstrates the use of operators which support QA-programming the tasks interacting with prototypes of databases.

It is necessary to mark that a creation and execution of any QA-program is being implemented in an instrumental environment including two translators (interpreter and compiler), editor, debugger and a number of specialized utilities for working with data declarations [2].

8   Collaborative Activity of I-Processors

Conceptual designing of SIS is the collaborative activity of designers working in coordination. A number of plug-ins is embedded to the toolkit WIQA for supporting the collaborative activity of I-processors.

The toolkit supports the relation between I-processors through solving of a number of communication tasks in their collaborative experimenting with project tasks. And as told above the toolkit provides the real time management of workflows the tasks of which are being executed by groups of I-processors. The general scheme of the management is presented in Fig. 6.

In this scheme the subsystem "Organizational structure" supports the real time assigning of tasks to members of the designer team. The copy of the team model $(T^*, \{G_v\}, \{D_{vs}\})$ specifying groups $\{G_v\}$ of designers D_{vs} can be uploaded in QA-memory in the form presented in Fig.3. Subsystem "Controlling of assignments" is used for binding any assignment with planned time of its fulfillment by the designer who is responsible for the assigned task. Subsystem "Kanban" [13] automatically reflects steps of workflows' execution with the help of visualizing the current state of tasks' queues. It helps to control the process of designing.

Subsystem of interruptions gives the possibility to interrupt any task being solved or corresponding QA-program (if it is necessary) for working with other tasks or QA-programs. The interruption subsystem supports the return to any interrupted task or QA-program in its point of interruption.

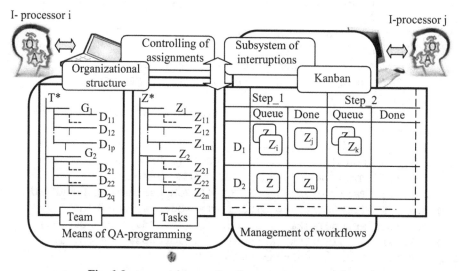

Fig. 6. Instrumental supporting the management workflows

Pseudo-code programming of workflows is based on using the library of workflow patterns. Coding the units of this library illustrates the example of the workflow pattern "Simple Merge". This pattern is described by the statement: *The convergence of two or more branches into a single subsequent branch such that each enablement of an incoming branch results in the thread of control being passed to the subsequent branch.* For three tasks Z1, Z2 and Z3 the QA-program of this pattern has the following view:

```
Q.3.5 PROCEDURE &Simple_Merge&// index name are inherited also
but from library of pattern
Q 3.5.1SET &out&, 4; &ins[0]&, 1; &ins[1]&, 2; &outgroup[0]&, 1;
&outgroup[1]&, 2; &cnt&, 0
Q 3.5.2 LABEL &L1&
Q 3.5.3 SEIZE &outs[&cnt&]&, &outgroup[&cnt&]&
Q 3.5.4 INC &cnt&
Q 3.5.5 TEST L, &cnt&, &ins&.length &L1&
Q 3.5.6 LABEL &L2_1&
Q 3.5.7 SET &cnt&, 0;
Q 3.5.8 LABEL &L2&
Q 3.5.9 TEST E, &ins[&cnt&]&, &ins[&cnt&]&.state, DONE &L3&
Q 3.5.10 INC &cnt&
Q 3.5.11 TEST L, &cnt&, &ins&.length &L2_1&
Q 3.5.12 TRANSFER &L2&
Q 3.5.13 LABEL &L3&
Q 3.5.14 QUEUE &out&, TRUE
Q 3.5.15 SET &ins[&cnt&]&.state, WAITING
Q 3.5.16 TRANSFER &L2_1&
Q 3.5.17 ENDPROC &Edit_Assignment&
```

The last example also as previous QA-programs is presented only for demonstrating the syntax of L^{WIQA} and therefore without explanations. But all presented specimens of QA-programs are not so difficult for their understanding.

9 Conclusion

The approach described in this paper suggests the system of means simplifying the complexity of designers' interactions with project tasks in conceptual designing of SIS. Simplifying is caused by the use in designers' activity useful analogies with the work of scientists conducting the experiments. Emulating to scientists the designers investigate own behavior in processes of tasks' solutions. Moreover they simulate such behavior with the help of pseudo-code programs describing the plans of experimenting. Thus designers investigate programmed plans of experiments which they prepare, conduct and describe in understandable and checkable forms for reuse.

In experimenting the investigated behavioral units are modeled as precedents. Such a form of a human activity is natural because intellectual processing of precedents lays in the base of the human experience. Experimenting the designers evolve the accessible experience by using real time interactions with its current state. This feature has found its normative specifications in the role "intellectual processor" playing of which by designers is being supported by the toolkit "WIQA". In collaborative way-of-working this role can be used additionally to any other role of the technology applied in conceptual designing. The toolkit opens the possibility for the separate execution of any operator by the designer playing the role of the intellectual processor. Before and after the execution of any operator of any QA-program the designer can check or investigate its preconditions and post-conditions. Moreover the investigated operator can be changed and evolved as syntactically so semantically, for example with the help of additional attributes.

Debugged QA-programs are the source of resources of the M- and P-types promoting the simplification of the complexity in their reuse. The possibility of experimenting is supported by the special library of QA-programs destined for cognitive task analysis, problem-solving and decision-making included to the named toolkit. Suggested means are used in one project organization creating the family of SISs.

References

1. Basili, V.R., Rombach, H.D., Schneider, K., Kitchenham, B., Pfahl, D., Selby, R.W. (eds.): Empirical Software Engineering Issues: Critical Assessment and Future Directions. LNCS, vol. 4336. Springer, Heidelberg (2007)
2. Sosnin, P.: Pseudo-code Programming of Designer Activities in Development of Software Intensive Systems. In: Jiang, H., Ding, W., Ali, M., Wu, X. (eds.) IEA/AIE 2012. LNCS, vol. 7345, pp. 457–466. Springer, Heidelberg (2012)
3. Li, M., Vitanui, P.M.B.: An Introduction to Kolmogorov Complexity and Its Applications, 3rd edn. Text in Computer Science. Springer (2008)
4. Borges, P., Machado, R.J., Ribeiro, P.: Mapping RUP Roles to Small Software Development Teams. In: International Conference on Software and System Process (ICSSP 2012), pp. 190–199 (2012)
5. Karray, F., Alemzadeh, M., Saleh, J.A., Arab, M.N.: Human-Computer Interaction: Overview on State of the Art. Smart Sensing and Intelligent Systems 1(1), 138–159 (2008)
6. Webber, B., Webb, N.: Question Answering. In: Clark, Fox, Lappin (eds.) Handbook of Computational Linguistics and Natural Language Processing. Blackwells (2010)
7. Ras, E., Rech, J., Weber, S.: Knowledge services for experience factories. In: Proc. of the 5th Conference on Professional Knowledge Management, pp. 232–241 (2009)
8. Henninger, S.: Tool Support for Experience-based Software Development Methodologies. Advances in Computers 59, 29–82 (2003)
9. Passarelli, A.M., Kolb, D.A.: The learning way – Learning from experience as the path to lifelong learning and development. In: London, M. (ed.) Oxford Handbook of Lifelong Learning, pp. 70–90. Oxford University Press (2011)
10. Van der Aalst, W.M.P., Hofstede, A.H.M.: Workflow Patterns Put into Context. Software and Systems Modeling 11(3), 319–323 (2012)
11. Sosnin, P.: Conceptual Solution of the Tasks in Designing the Software Intensive Systems. In: Proc. of the 14th IEEE Mediterranean Electrotechnical Conference (MELECON 2008), Ajaccio, France, pp. 293–298 (2008)
12. Wang, J.X.: Lean Manufacturing Business Bottom-Line Based. In: Kanban: Align Manufacturing Flow with Demand Pull, pp. 185–204. CRC Press (2010)

Modeling and Verification of Change Processes in Collaborative Software Engineering

Phan Thi Thanh Huyen, Kunihiko Hiraishi, and Koichiro Ochimizu

School of Information Science, Japan Advanced Institute of Science and Technology
Nomi-shi, Ishikawa, 923-1292 Japan
{huyenttp,hira,ochimizu}@jaist.ac.jp

Abstract. In collaborative software engineering, many change processes implementing change requests are executed concurrently by different workers. However, the fact that the workers do not have sufficient information about the others' work and complicated dependencies among artifacts can lead to unexpected inconsistencies among the change-impacted artifacts. By focusing on the contexts of the changes, i.e. the change processes containing the changes, rather than the concurrent changes only like the previous works, we have proposed an approach that helps the workers detect and resolve the inconsistencies more effectively [1]. Our approach is to build a Change Support Environment (CSE) that represents the change processes explicitly as the Change Support Worflows (CSWs) and manages their execution based on our patterns of inconsistency, including many patterns besides the conflict patterns mentioned in the previous works. To evaluate the feasibility of our proposed approach, this paper presents a formal model of CSE using Colored Petri Nets (CPN) to model the artifacts, and both data flow and control flow of CSWs. CPN Tools is used to edit, simulate, and verify the CPN model of CSE to detect data-related abnormalities, in particular the patterns of inconsistency. Differently from the previous works in workflow modeling, our method for modeling CSWs using CPN can represent many aspects of a workflow, including data flow, control structure, and execution time, in one single model. Data and changes on the value of data are also represented explicitly. In addition, our modeling and verification method can be applied to other types of workflow.

Keywords: Inconsistency awareness, Change process, Collaborative software engineering, Colored Petri Nets, Modeling and verification.

1 Introduction

Collaboration is indispensable in software engineering along with the increased scale and complexity of projects. In a collaborative work, software artifacts are created based on the collaboration of many workers. However, because of the fact that the workers do not always have sufficient information about the others' work, and complicated *dependencies* among the artifacts, changes to some artifacts of a worker may affect the changes of other workers. Therefore inconsistencies may happen among the change-impacted artifacts.

B. Murgante et al. (Eds.): ICCSA 2013, Part III, LNCS 7973, pp. 17–32, 2013.
© Springer-Verlag Berlin Heidelberg 2013

There were several attempts to detect conflicts, a type of inconsistency caused by concurrent changes to the same artifact (direct-conflict) or to dependency-related artifacts (indirect-conflict) [2]. Version control systems (VCSs) [3], widely used in collaborative software development, could detect only direct-conflicts at check-in time after the changes have been finished. To detect conflicts earlier, when the changes are being implemented, some recent studies [2, 4–7] have concentrated on workspace awareness, namely continuously sharing information of ongoing parallel activities across the workspaces.

The above mentioned studies [2–7] focused on conflicts and concurrent changes. However, because of the dependencies among artifacts, when a requirement changes, it is often necessary to implement a *change process* that is a sequence of tasks applying changes to a set of artifacts to implement a change request. Ignoring the *context* of the change, i.e. the change process containing the change, may neglect some inconsistencies that may only be detected much later in the development process (Sec. 2.1). Also, in these studies, when a (potential) conflict is detected, workers are aware of concurrent changes and the related artifacts only. Nevertheless, to resolve an inconsistency, a worker will need to consider the contexts of the changes causing the inconsistency rather than the changes only. Based on these considerations, in [1], we have presented an approach to dealing with the inconsistency problem more effectively. We have proposed a Change Support Environment (CSE) where the change processes are represented explicitly as the Change Support Workflows (CSWs) and their execution is managed based on the patterns of inconsistency. We have also identified and categorized these patterns, including the conflict patterns, considering both concurrent and non-concurrent changes and the contexts of the changes. If the change processes in the system can be managed, a worker could have a clearer view of the contexts of the changes of his and the related workers, and hence inconsistencies, including conflicts, can be detected earlier and resolved more easily.

To prove the feasibility of our proposed approach in [1], this paper presents a formal model of CSE and verifies this model to detect the patterns of inconsistency. As mentioned earlier, inconsistencies among CSWs are directly related to data, and hence, data flow verification is required in CSE. However, previous works in workflow verification mostly concentrated on structure verification, temporal verification, and resource verification [8]. Although data are an important aspect of workflows, only little research in data-flow verification can be observed and they focused on data-flow errors in a single workflow instance. Unlike previous works, we model the essential behaviors of CSE using Colored Petri Nets (CP-nets or CPN) [9]. Then, we use CPN Tools [10] to edit, simulate, and verify the CPN model of CSE to detect data-flow abnormalities including the patterns of inconsistency. We make the following contributions:

- Using Colored Petri Nets to model workflows instead of extending existing languages to represent the data factor like the works in [12–15].
- Being able to represent data and changes on the properties of data explicitly.
- Being able to represent both control flow and data flow in one single model.

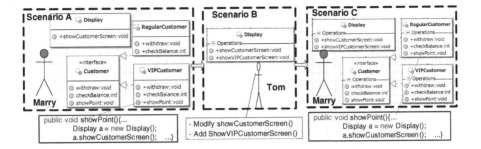

Fig. 1. Motivating example [1]

- Considering the data-flow errors caused by mutual influences among the data-related workflows, in addition to the interactions among concurrent activities in a single workflow.

Furthermore, our method can be applied to model and verify data-related abnormalities of other types of workflows.

The remainder of this paper is organized as follows. Section 2 introduces the important contents of [1] and some basic concepts of CPN. Section 3 describes the CPN model of CSE. Section 4 analyzes the CPN model of CSE to detect data abnormalities, specially the patterns of inconsistency. Section 5 presents related work in data-flow verification. Finally, Section 6 discusses future work and concludes the paper.

2 Background

2.1 Change Support Environment (CSE)

For the sake of completeness, we re-introduce the important contents of [1] that are the motivating example, an overview of the patterns of inconsistency, and our approach to dealing with inconsistency in collaborative software engineering.

Motivating Example. Marry and Tom are the developers of a hypothetical airline-ticket sale software system. To implement a change request that adds a signature *showPoint()* to the *Customer* interface, Mary implemented the *show-Point()* method for both *VIPCustomer* and *RegularCustomer* classes that implement the *Customer* interface. Two empty *showPoint()* methods were added to the *VIPCustomer* and *RegularCustomer* classes at the beginning to avoid compilation errors. In Fig. 1, these methods are visible when their implementations are already finished. In the morning, she finished the *showPoint()* method of the *VIPCustomer* class using the *showCustomerScreen()* method of the *Display* class (Fig. 1 - Scenario A). Because she had an urgent meeting with a customer, she checked-in her changes, left the office, and continued the work at home at night. She checked-out the project to her workspace and also used

Table 1. Inconsistency category

		Tasks	
		Concurrent	*Non-concurrent*
Artifacts	*Same*	Direct-Conflict	Direct-Revision-Inconsistency
	Different	Indirect-Conflict	Indirect-Revision-Inconsistency Interleaving-Inconsistency (RWR, WWR)

the *showCustomerScreen()* method for implementing the *showPoint()* method of the *RegularCustomer* class (Fig. 1 - Scenario C). Meanwhile, in the afternoon of that day, Tom, the author of the *Display* class, decided to distinguish the display screen of VIP customers from that of regular customers by modifying the *showCustomerScreen()* method, and adding a new method, *showVIPCustomerScreen()*, to display the screen of VIP customers (Fig. 1 - Scenario B). He also checked-in his update successfully. VCSs or conflict awareness techniques do not report any errors in this situation because of no syntax conflict error and sequential property of the change activities. Unfortunately, there is an inconsistency in the implementations of the *showPoint()* method in the *VIPCustomer* class and the *RegularCustomer* class.

This inconsistency could have been detected before it propagated further, if Mary had recognized this inconsistency situation and revised the implementation of the *showPoint()* method in the *VIPCustomer* class with regard to the new method in the *Display* class, the *showVIPCustomerScreen()* method. Nonetheless, it is difficult for a worker to remember all the changes she has made before. It will be more difficult to recognize this situation if the task of updating the *VIPCustomer* class is assigned to a different worker, for example Peter. Also, Mary cannot lock all the involved artifacts because of her long-term change process. If Tom had notified Mary of his change, the inconsistency could have been avoided. However, performing all the required tasks, for instance, finding the works impacted by his change, specifying the change, and sending to the impacted workers, by himself is tricky and time-consuming. Therefore, it is necessary to have a system that supports the workers to detect and resolve the inconsistencies effectively based on the contexts of the changes.

Inconsistency Category. We define **inconsistency** as a situation in which a worker is unaware of the changes or the impact of the changes, made by tasks in different change processes, to some artifacts to which his change process refers.

We have identified the patterns of inconsistency and classified them based on the time orders of the tasks, concurrent or not, the relationships between the changed artifacts, same artifact or different artifacts, and the contexts of the changes that are the change processes containing the tasks (Table 1).

- **Direct-Conflict** happens when concurrent tasks in different change processes change the same artifact (Fig. 2.a).
- **Indirect-Conflict** happens when a change to one artifact affects a concurrent change applied to other artifact in different change process (Fig. 2.b).

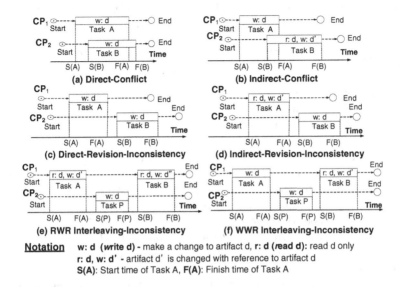

Fig. 2. Illustrations of the patterns of inconsistency

- **Direct-Revision-Inconsistency** happens when there are contradictory intentions in revising the same artifact at different times by different workers in different change processes (Fig. 2.c).
- **Indirect-Revision-Inconsistency** happens when a later change to an artifact affects the earlier changes to other artifacts in different change processes (Fig. 2.d).
- **Interleaving-Inconsistency** happens when there are inconsistent views of a shared artifact by two tasks in the same change process, because the shared artifact is modified by a task in another change process at sometime in the interval between these two tasks (Fig. 2.e, Fig. 2.f).

The method for detecting these patterns will be presented in Sec. 4.

Approach. Our approach to the inconsistency problem is a combination of the process support approach and awareness support approach, which are the main collaboration techniques in software engineering [11]. We have proposed to build a Change Support Environment (CSE) where each change process is represented explicitly as a Change Support Workflow (CSW) and each worker follows a CSW when he implements a change request (process support approach). By managing the execution of CSWs based on the patterns of inconsistency and sharing information about the work of each other (awareness support approach), the workers can detect and resolve inconsistencies more effectively.

A CSW is a sequence of change activities defined to carry out a change request. Activities in a CSW take care of creating new artifacts or changing the existing ones. A CSW can be generated using the elements impacted by a change request, and the dependency relationships among them. A worker can also define a CSW by himself. Some CSW primitives and how to model them in CPN are presented in Sec. 3. Following is the formal definition of a CSW.

Definition 1 (Change Support Workflow). A **CSW** is a tuple $< id, A, E, D, C, W, GD, GW, GT >$, where

- id is the workflow identifier.
- A is a set of change activities.
- $E \subseteq (A \times A)$ is a set of directed edges that represent the orders of the change activities.
- D is a set of artifacts accessed by the activities of the CSW.
- $C = \{r, w^a, w^m, w^d\}$ is a set of types of change to the artifacts (r: *read*, w: *write*, a: *add*, m: *modify*, d: *delete*). *read* means that this artifact is for reference only. *write* means that this artifact will be changed. *add*, *delete*, *modify* are subtypes of *write*.
- W is a set of workers who execute the activities of the CSW.
- $GD : A \times C \to 2^D$ is a function that returns a set of artifacts associated with an activity, and a change type.
- $GW : A \to 2^W$ is a function that returns a set of workers associated with a change activity of the CSW.
- $GT : A \to R^+ \times R^+$ is a time interval function that returns the Start Time (S) and the Finish Time (F) of an activity. R^+ is the set of positive real numbers. 0 denotes an undecided start time or finish time. The interval between S and F is called Execution Time, $E = F - S$.

2.2   Colored Petri Net Preliminaries

This section gives a brief introduction of Colored Petri Nets [9, 16] that are used for modeling the essential behaviors of CSE in Sec. 3.

Definition 2 (Colored Petri Nets). A **CP-net** is a tuple $(P, T, A, \Sigma, N, C, G, E, I)$, where

1. P is a finite set of **places**.
2. T is a finite set of **transitions** such that $P \cap T = \emptyset$.
3. $A \subseteq (P \times T) \cup (T \times P)$ is a set of directed **arcs**.
4. Σ is a finite set of non-empty types, called **color sets**.
5. V is a finite set of **typed variables** such that $Type[v] \in \Sigma$ for all variables $v \in V$.
6. $C : P \to \Sigma$ is a **color set function** that assigns a color set to each place.
7. $G : T \to EXPR_V$ is a **guard function** that assigns a guard to each transition t such that $Type[G(t)] = Bool$.
8. $E : A \to EXPR_V$ is an **arc expression function** that assigns an arc expression to each arc a such that $Type[E(a)] = C(p)_{MS}$, where p is the place connected to the arc a.
9. $I : P \to EXPR_\emptyset$ is an **initialization function** that assigns an initialization expression to each place p such that $Type[I(p)] = C(p)_{MS}$.

Definition 3. For a Colored Petri Net CPN = $(P, T, A, \Sigma, N, C, G, E, I)$:

1. A **marking** is a function M that maps each place $p \in P$ into a multiset of tokens $M(p) \in C(p)_{MS}$.
2. The **initial marking** M_0 is defined by $M_0(p) = I(p)()$ for all $p \in P$.

A CP-net can be extended with a *time concept*. A token in a CP-net has a concrete value, *color*, and in a timed CP-net, it may, in addition, carry a *time stamp* telling when it can be used. Also, a transition may produce tokens with a *delay*. To model tokens carrying a time stamp, the corresponding color set must be made a *timed color set* by adding the term *timed*.

In addition to the *time concept*, a CP-net is also extended with a *hierarchy concept*. The idea is to decompose a CP-net into modules, *CP-net modules*. A module has an interface, *port place*, and is replaced by a *substitution transition*. A module is also referred to as a *page*. A substitution transition refers to a *subpage* that corresponds to a module contained in another module, named *superpage*. To connect a subpage with its superpage, each port place in the subpage is linked with a place in the superpage, named *socket*. By relating a socket to a port, the two places are semantically merged into one place.

3   Modeling CSE by CP-Nets for Abnormality Detection

In this section, we describe how to model the essential behaviors of CSE. Then, this formal model will be analyzed in the next section to detect abnormalities, specially the patterns of inconsistency.

CP-nets are used to model the necessary behaviors of CSE because:

1. CP-nets combine the capabilities of Petri nets, a basic model widely used for modeling and analyzing control flow of workflows, with the high-level language.
2. We can use CPN Tools to edit, simulate, and analyze the CPN model of CSE.
3. We can avoid defining a new language by extending existing languages to represent unsupported factors like [12–15], and avoid the need of proving the correctness of the new language [17] as well.
4. Aspects of a CSW including control (activity, edge), data (artifact), resource (worker), and time (Execution time) can be represented by CP-nets. Because modeling resources by CP-nets was mentioned in some previous studies, this paper ignores modeling resources to reduce the complexity of the CPN model.

- Allowing tokens to be associated with colors in CP-nets helps us represent the changes on the properties of data, for example, version and content, during the execution of CSWs easily.
- CP-nets include the *time concept* which helps us specify the execution time of activities in a CSW.
- CP-nets support the *hierarchy concept* which allows us to model large systems easily.

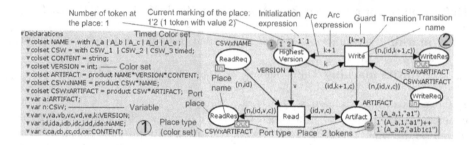

Fig. 3. 1. Color sets and variables used in CPN Model of CSE. **2.** Modeling Artifact a as a simple VCS.

3.1   Color Set

Fig. 3.1 summarizes the color sets and variables used for modeling CSE by CP-nets. We represent the sets of artifacts and CSWs as color sets *ARTIFACT*, and *CSW* respectively. Other color sets are composed from the basic ones.

The color set *ARTIFACT* is a product type representing the artifacts in CSE. The first element of the color set *ARTIFACT*, *NAME*, is an enumeration type that enumerates the artifact names. The second element, *VERSION*, is an integer type that denotes the version of an artifact. The third element, *CONTENT*, is a string type that denotes the content of an artifact. Value $(A\_b, 1, "b1")$ is an example of a color belonging to the color set *ARTIFACT* in which A_b, 1, and "b1" are colors of the color sets *NAME*, *VERSION*, and *CONTENT*, respectively.

The color set *CSW* is an enumeration type that enumerates the identifications of CSWs. To represent the execution times E of the change activities, *CSW* is declared as a timed color set by adding the term *timed* at the end of its declaration.

3.2   Modeling Artifacts

In software development environments, VCSs are fundamental tools for enabling workers to work together in parallel on a single data repository. Based on this observation, we model each artifact as a simplified VCS to allow CSWs to access the artifacts concurrently (see Fig. 3.2). Each artifact is represented by a CP-net module with four port places: *ReadReq*, *ReadRes*, *WriteReq*, and *WriteRes* that represent a read (check-out) request, read response, write (check-in) request, and write response, respectively. CSWs will connect to these port places to access the corresponding artifacts. The state of an artifact is modeled by a place *Artifact*. This place contains tokens that represent all versions of the artifact created during the execution of CSWs involved. Tokens in the place *Artifact* will have the same name, but different version number. We assume that all CSWs will access the latest version of the artifact. Therefore, we use another place, *Highest Version*, containing only one token, to denote the highest version number of the artifact. When there is a read request, the transition *Read* will return the version that has the highest version number. When there is a write request, the transition *Write* will update the highest version number, and store and return the new version with the updated version number.

Fig. 3.2 shows a *marking* of the CP-net module modeling the artifact *a*. The place *Artifact* contains two tokens: one token with color *(A_a,1,"a1")*, *1'(A_a,1,"a1")*, and one token with color $(A\_a, 2, "a1b1\ c1")$, $1'(A\_a, 2, "a1b1c1")$. The place *Highest Version* contains one token with value 2, $1'2$, which indicates that the latest version of the artifact *a* is 2. The markings of the places *Artifact* and *Highest Version* are different from their *initial markings* that are represented by the corresponding initial expressions. In the initial marking, the place *Artifact* contains only one token with value *(A_a,1,"a1")*, *1'(A_a,1,"a1")*, hence the token in the place *Highest Version* has value 1, $1'1$.

To model other artifacts, for example the artifact *b* having only one version in which the version number is 1 and the content is "b1", we just need to change the initial expressions of the places *Artifact* and *Highest Version* to $1'(A\_b, 1, "b1")$ and $1'1$ respectively (See **Subpage:Accessb** in Fig. 5.2). If we want to represent an artifact that has not been created yet, we set the initial marking of the place *Highest Version* to *0* and do not set the initial marking for the place *Artifact*.

3.3   Modeling CSWs

Ignoring the data factor, each activity A_i can be modeled by a transition T_i. The execution time of A_i, $E_i = F(A_i) - S(A_i)$, is modeled by the delay time of transition T_i: $@+E_i$ (Fig. 4.a).

In the case of data-related activities (Fig. 4.h), each activity A_i in a CSW is modeled by four transitions: $preT_i$ (sending read requests), $rcvdT_i$ (receiving read responses), T_i (changing data and sending write requests), and $postT_i$ (receiving write responses). First, the transition $preT_i$ sends read requests by specifying the names of the data needed to be read, including the data in the read data set and the data in the write data set (except for the data not yet created), as the inscriptions of the arcs connecting $preT_i$ to the port places *ReadReq* of the read artifacts. Next, the transition $rcvdT_i$ receives the tokens returned from the port places *ReadRes* and stores them as local artifacts in some specific places, for example the place P_i_RA. P_i_RA means a place used for modeling an activity A_i, P_i, and containing the *read* value of an artifact *a*, *RA*. Then, the transition T_i sends tokens representing the updated versions of the changed artifacts to the port places *WriteReq* of the corresponding artifacts. In this paper, we do not focus on the contents of artifacts, and we assume that the content of a written artifact is a concatenation of the contents of its read artifacts. Finally, the transition $postT_i$ receives tokens from the port places *WriteRes* and stores them as local written artifacts in some specific places, for example the place P_i_WA. P_i_WA means a place used for modeling an activity A_i, P_i, and containing the *write* value of an artifact *a*, *WA*. The identification of the CSW containing the activity is also included in read/write requests and read/write responses to ensure responses are delivered correctly. The execution time E_i of A_i is modeled by the delay time of the transition $rcvdT_i$: $@+E_i$.

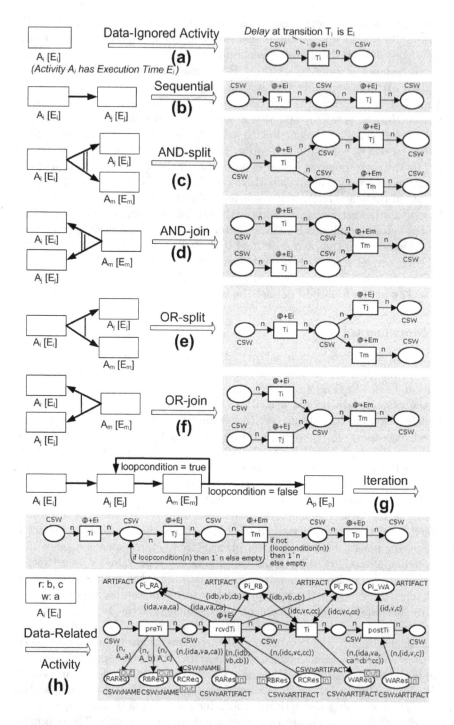

Fig. 4. Modeling CSW primitives by CP-nets

Fig. 4.b-g shows how the basic constructions of a CSW are modeled by CPN: sequential structure (Fig. 4.b), AND-split (Fig. 4.c), AND-join (Fig. 4.d), OR-split (Fig. 4.e), OR-join (Fig. 4.f), and iteration (Fig. 4.g). Based on these structures, we can develop more complicated structures using inscriptions and expressions supplied by CPN to specify the conditions of route choices, type and quantity of tokens (resources, data) transmitted, and execution condition of transitions, etc. For simplicity, we only represent the control flow and time factor in these structures. Therefore, when representing a CSW in CPN, one should model the control flow first using the above structures. Then, one can model the data factor of each data-related activity by replacing its corresponding transition by the four transitions as described in Fig. 4.h.

Fig. 5 gives an example of modeling a CSE with three CSWs, $CSW1$, $CSW2$, $CSW3$, connected by five artifacts a, b, c, d, e in terms of CP-net. $CSW2$ and $CSW3$ represent the CSWs of Tom and Mary presented in Motivating Example of Sec. 2.1. The artifacts d, e, b represent the $VIPCustomer$, $RegularCustomer$, and $Display$ classes respectively. The hierarchical structure is used in the CPN model of this CSE. The superpage contains eight subpages: $csw1$, $csw2$, $csw3$, $Accessa$, $Accessb$, $Accessc$, $Accessd$, and $Accesse$ that represent $CSW1$, $CSW2$, $CSW3$, a, b, c, d, and e, respectively. Due to the limited space of the paper, we do not show the subpages representing a, c, d, and e here. However, their structures are the same as the structure of the subpage $Accessb$ except for the initial markings of the places $Artifact$. As described in Sec. 2.2, the port places of the subpages will be linked to the sockets of their superpages. In this example, the sockets $RBReq$, $RBRes$, $WBReq$, and $WBRes$ of the superpage will link to the port places $ReadReq$, $ReadRes$, $WriteReq$, and $WriteRes$ of the subpage $AccessB$, respectively. And so are the remaining sockets.

4   Detecting Abnormalities in the CPN Model of CSE

In the CPN model of CSE, data-related CSWs are connected and create a *synthesized CSW-net* (See the **Superpage** in Fig. 5.2). We can simulate and analyze the synthesized CSW-net using the simulation tools, and state space tools with state space querying functions, CTL (Computation Tree Logic) state formula operators, and model checking functions, supplied by CPN Tools [10], to detect abnormalities.

4.1   Missing Data and Direct Conflict

Missing data is the situation in which an artifact needs to be read, but either it has never been created or it has been deleted. In CSE, missing data happens on an artifact a if the place $Artifact$ corresponding with it contains no token. Therefore, there is a deadlock at a transition $rcvdT_i$ that is waiting for a read response from a.

A direct conflict happens on an artifact a if there is a deadlock at a transition $postT_i$ that is waiting for a write response from a. This deadlock happens because

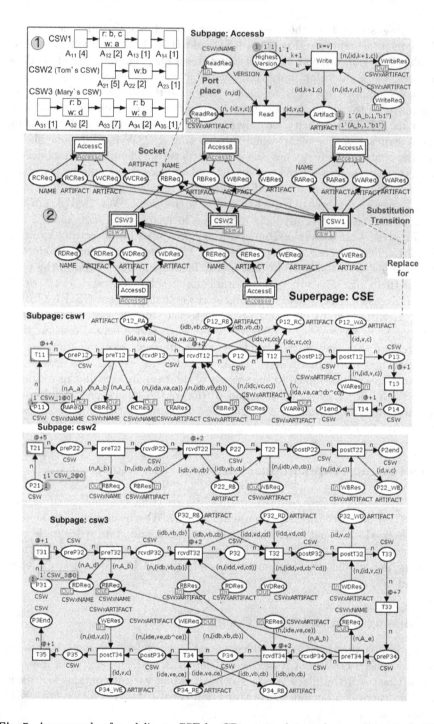

Fig. 5. An example of modeling a CSE by CP-nets to detect abnormalities **1.** A CSE with 3 CSWs and 5 artifacts **2.** CPN model of the CSE

Statistics			
State Space	Scc Graph		Home Properties
Nodes: 49	Nodes: 49		
Arcs: 63	Arcs: 63		Home Markings: Initial Marking is not a home marking
Secs: 0	Secs: 0		
Status: Full			Liveness Properties
Boundedness Properties			
			Dead Markings: [49]
Best Integer Bounds			Dead Transition Instances: Accessc'Write 1
	Upper	Lower	Live Transition Instances: None
Accessa'Artifact 1	2	1	
Accessa'Highest_Version 1	1	1	Fairness Properties
Accessb'Artifact 1	2	1	
			No infinite occurrence sequences.

Fig. 6. The state space analysis report of the CPN model described in Fig. 5

a write response is returned only if the version number in the write request is equal to the number stored in the place *Highest Version* corresponding to the artifact a (See the guard of the transition *Write* in Fig. 3.2 or Fig. 5.2). This condition simulates the check-in condition of VCSs. Our modeling method helps detect conflicts caused by not only concurrent activities in different CSWs (Direct-Conflict in Sec. 2.1) but also concurrent activities in the same CSW.

Using the simulation tools or state space analysis report of CPN Tools, we can detect **missing data** and **direct conflict** easily. For example, Fig. 6 shows the standard report generated for the state space analysis of the CPN model described in Fig. 5.2. We observe the absence of the home markings. There are no infinite occurrence sequences and the CSW-net terminates at the dead marking 49. The *Accessc'Write* transition is dead because there is no write request for the artifact c. Standard reports can help us detect abnormalities easily, especially by observing dead markings, or boundedness properties. In this case, no abnormalities are detected through the standard report. This means that there are no errors relating to **missing data** or **direct conflicts**.

4.2   Direct-Revision-Inconsistency

Direct-Revision-Inconsistency can be recognized by observing the superpage of the CPN model of CSE. If there is more than one connection to a socket *WAReq* linking with a port place *WriteReq* of a subpage *Accessa* that models an artifact a, there may be a direct conflict or a Direct-Revision-Inconsistency related to the artifact a. If no deadlock happens, we can eliminate the direct conflict case. For example, in the Superpage of the CPN model described in Fig. 5.2, there is at most one connection to the sockets *WAReq, WBReq, WCReq, WDReq, WEReq*. Therefore, a **Direct-Revision-Inconsistency** does not happen.

4.3   Other Patterns of Inconsistency

State space querying and model checking are necessary to detect the remaining patterns of inconsistency. We will show a concrete example of applying state space queries and CTL model checking functions on the CPN model in Fig. 5.2 to detect these patterns.

```
fun IsConcurrentRW n = (
(Mark.csw2'P22_RB 1 n) == (Mark.csw1'P12_RB 1 n)) andalso
          ((Mark.csw2'P22_RB 1 n) <><> empty) andalso (
((Mark.csw2'P22_WB 1 n) = empty) orelse
((Mark.csw1'P12_WA 1 n) = empty) );

val IndirectConflict = POS(NF("IndirectConflict",IsConcurrentRW));
eval_node IndirectConflict InitNode;

val IndirectConflictStates = PredAllNodes(     val IsConcurrentRW = fn : Node -> bool
fn n =>                                        val IndirectConflict = EXIST_UNTIL (TT,NF ("IndirectConflict",fn)) : A
          IsConcurrentRW n                     val it = true : bool
);                                             val IndirectConflictStates = [38,37,36,35,34,33,32,31,30,29] : Node list

fun IsRWR n =
((Mark.csw3'P32_RB 1 n) <><> (Mark.csw3'P34_RB 1 n)) andalso
          ((Mark.csw3'P32_RB 1 n) <><> empty) andalso (
          ((Mark.csw3'P34_RB 1 n) <><> empty);

val RWRInconsistency = POS(NF("RWR Inconsistency",IsRWR));
eval_node RWRInconsistency InitNode;              val IsRWR = fn : Node -> bool
                                                  val RWRInconsistency = EXIST_UNTIL (TT,NF ("RWR Inconsistency",fn)) : A
val RWRInconsistencyStates = PredAllNodes(        val it = true : bool
fn n =>                                           val RWRInconsistencyStates = [49,48,47,46,45] : Node list
          IsRWR n
);
```

Fig. 7. Example of model checking and query on the CPN model described in Fig. 5

The Superpage in Fig. 5.2 reveals that the artifact b is read by CSW1 and CSW2, and read/written by CSW3. Hence, we will do some verifications with b.

As shown in Fig. 7, we first check the possibility of an **Indirect-Conflict**. The function *IsConcurrentRW* checks whether the artifact b is concurrently read and written by the activities A_{12} and A_{22} respectively. Because *IndirectConflict* is evaluated as *true*, we conclude that an Indirect-Conflict happens between A_{12} and A_{22} at some states. We can find these states by a query function, *IndirectConflictStates*, that traverses the entire state space to find the states satisfying the *IsConcurrentRW* function. Also, an Indirect-Conflict between A_{12} and A_{22} means that an **Indirect-Revision-Inconsistency** does not happen between these two activities.

Next, we check the possibility of a **RWR Interleaving-Inconsistency**. The function *IsRWR* checks whether the activities A_{32} and A_{34} read the same value of b. Similar to the previous case, RWR Interleaving-Inconsistency happens because of the *true* value of *RWRInconsistency*. *RWRIncosistencyStates* returns a list of states that make the function *IsRWR* true.

Although a **WWR Interleaving-Inconsistency** does not appear in this example, we can imitate the method checking RWR Interleaving-Inconsistency to detect this inconsistency. We will define a function *IsWWR* for checking whether an activity A_j in a CSW *csw* reads the value created before by an activity A_i in the same CSW. The remaining functions are defined similarly to those of RWR Interleaving-Inconsistency except that *RWR* is replaced with *WWR*.

fun isWWR n = ((Mark.csw'Pi_WB 1 n) <><> (Mark.csw'Pj_RB 1 n))

and also (Mark.csw'Pi_WB 1 n) <><> empty)

and also ((Mark.csw'Pj_RB 1 n) <><> empty);

5   Related Work in Data-Flow Verification

Workflow verification has received a lot of attention, specially structure verification, resource verification, and temporal verification. However, there is little work for data-flow verification [8]. In addition, although there are many ways to model a workflow, such as directed graph, BPMN, WF-Net, UML activity diagram, they do not specify data flow formally or their specifications do not help with data-flow verification. Therefore, suitable data modeling must be conducted before data verification. In [12], the authors used a data-flow matrix and an extension of the UML activity diagram that incorporates data input and output, to specify data. Then, a dependency-based algorithm, was proposed to detect three basic types of data-flow anomalies: missing data, redundant data, and conflicting data. Reference [13] proposed another workflow modeling technique, named Dual Workflow Nets, that enabled describing both data flow and control flow by introducing new types of nodes and distinguishing data token from control token. A formal verification method for detecting the control/data-flow inconsistencies was also presented. Another approach for modeling the data flow of BPMN was given in [14]. The authors formalized the basic data object processing in BPMN using Petri nets, and presented a technique for repairing some data anomalies corresponding to deadlocks in the composed Petri nets. WorkFlow net with Data (WFD-net) was introduced in [15] as an extension of WF-Net in which a transition can have a guard, and can read from, write to, or delete data elements. Based on this net, data-flow anti-patterns comprising missing data, redundant data, lost data, conflicting data, never destroyed, twice destroyed, and not deleted on time, were defined in terms of temporal logic CTL.

Differently from the previous works, we exploit the modeling power of CP-nets to provide a method for representing many aspects of a workflow, including data flow, control structure, and execution time, in one single model. Using CP-nets, we can avoid extending existing languages to represent the data factor, unlike the previous works. Data and changes on the properties of data can be represented explicitly too. We also show how to analyze the generated model to detect data abnormalities concerning the mutual influences among the data-related workflows instead of the interactions among the activities in a single workflow only. The method we use to model CSWs can be applied to model business workflows effectively.

6   Conclusion

This paper has presented a formal model of our proposed Change Support Environment (CSE) [1], which represents the change processes as Change Support Workflows (CSWs) and manages their execution for dealing with inconsistencies in collaborative environments more effectively. CP-nets are used to model the necessary behaviors of CSE, in particular the data flow of CSWs. CPN Tools is used for verifying the generated model to detect the patterns of inconsistency.

Due to the modeling cost and the state explosion problem in model checking, our proposed method is suitable for the small size CSE with the CSWs that need to be designed carefully and to verify data-flow related errors in advance, before executing them. Despite the limitations, successful modeling and verification of CSE are the initial achievements in proving the feasibility and correctness of the approach of CSE. In addition, this modeling and verification method with some advantages compared to the previous works in data-flow verification can be applied to other types of workflows.

In future work, we will continue improving the formal model of CSE with regard to modeling other common operations on artifacts such as branching and merging, and representing the indirect dependencies among artifacts. Automating the inconsistency analysis is also worth considering. Moreover, how to derive the correct CSWs from the inconsistency-involved CSWs is under considertation.

References

1. Huyen, P.T.T., Ochimizu, K.: Toward Inconsistency Awareness in Collaborative Software Development. In: Proc. of APSEC 2011, pp. 154–162 (2011)
2. Sarma, A., et al.: Towards Supporting Awareness of Indirect Conflicts across Software Configuration Management Workspaces. In: Proc. of ASE 2007, pp. 94–103 (2007)
3. Altmanninger, K., et al.: A Survey on Model Versioning Approach. IJWIS 5(3), 271–304 (2009)
4. Servant, F., et al.: CASI: Preventing Indirect Conflicts through a Live Visualization. In: Proc. of CHASE 2010, pp. 39–46 (2010)
5. Dewan, P., Hegde, R.: Semi-synchronous Conflict Detection and Resolution in Asynchronous Software Development. In: Proc. of ECSCW 2007, pp. 159–178 (2007)
6. Hattori, L., Lanza, M.: Syde: A Tool for Collaborative Software Development. In: Proc. of ICSE 2010, pp. 235–238 (2010)
7. Brun, Y., et al.: Proactive Detection of Collaboration Conflicts. In: Proc. of ESEC/FSE, pp. 168–178 (2011)
8. Sadiq, S., et al.: Data flow and validation in workflow modeling. In: Proc. of ADC 2004, pp. 207–214 (2004)
9. Jensen, K., Kristensen, L.M.: Colored Petri Nets - Modeling and Validation of Concurrent Systems. Springer (2009)
10. CPN Tools, http://cpntools.org/
11. Whitehead, J.: Collaboration in Software Engineering: A Roadmap. In: Proc. of FOSE 2007, pp. 214–225 (2007)
12. Sun, S.X., et al.: Formulating the Data Flow Perspective for Business Process Management. Information Systems Research 17(4), 374–391 (2006)
13. Fan, S., Dou, W.-C., Chen, J.: Dual Workflow Nets: Mixed Control/Data-Flow Representation for Workflow Modeling and Verification. In: Chang, K.C.-C., Wang, W., Chen, L., Ellis, C.A., Hsu, C.-H., Tsoi, A.C., Wang, H. (eds.) APWeb/WAIM 2007. LNCS, vol. 4537, pp. 433–444. Springer, Heidelberg (2007)
14. Awad, A., et al.: Diagnosing and Repairing Data Anomalies in Process Models. In: Proc. of BPD 2009. LNBIP, pp. 1–24 (2009)
15. Trčka, N., van der Aalst, W.M.P., Sidorova, N.: Data-Flow Anti-Patterns: Discovering Data-Flow Errors in Workflows. In: van Eck, P., Gordijn, J., Wieringa, R. (eds.) CAiSE 2009. LNCS, vol. 5565, pp. 425–439. Springer, Heidelberg (2009)
16. van der Aalst, W., Stahl, C.: Modeling Business Processes: A Petri net-Oriented Approach. The MIT Press (2011)
17. Sidorova, N., et al.: Soundness verification for conceptual workflow nets with data: Early detection of errors with the most precision possible. Information Systems 37(7), 1026–1043 (2009)

Relating Goal Modeling
with BPCM Models in a Combined Framework

Shang Gao

Department of Business Administration,
Zhongnan University of Economics and Law, Wuhan, China
shangkth@gmail.com

Abstract. In this paper, the issue of relating goal models with BPCM models in a combined framework is addressed. A business process characterizing model (BPCM) can be seen as a business-oriented model for the use in the early stages of a project, both for traditional development, but also for the development of multi-channel solutions working across a set of contexts. The combined modeling framework consists of goal modeling, process modeling and business process characterizing modeling. The framework is meant to guide both business stakeholders and model developers during modeling-based development. A development methodology to guide the development of goal models in terms of i* from business process characterizing model is proposed. Furthermore, the development methodology is illustrated by an exemplar of call for sponsors case in the field of scientific conference organization.

Keywords: Goal Modeling, BPCM, Process Modeling.

1   Introduction

Different enterprise models can be used to derive business requirements and system requirements to build business process support systems. The goal models, business models and process models are some typical models used for this purposes. A business model helps identifying value exchange information in the business. A goal model is directed at describing intention and strategies of different actors involved in the business. A process model mainly focuses on activities in the business. In this paper, we look at how to derive goal models from a business oriented modeling languages in terms of BPCM models.

As illustrated in [6], one of the main ways of utilizing models is to describe some essential information of a business as informal support in order to facilitate communication among stakeholders. As information systems have moved from office-based systems to mobile and multi-channel systems (e.g.,[9-11, 15]), the potentially essential information to take into account has been extended to support the work process in a number of new contexts.

In [8], by taking inspiration from this idea, we proposed a business process characterizing model (BPCM), which can be seen as a business-oriented model for the use in the early stages of a project, both for traditional development, but also for the

B. Murgante et al. (Eds.): ICCSA 2013, Part III, LNCS 7973, pp. 33–42, 2013.
© Springer-Verlag Berlin Heidelberg 2013

development of multi-channel solutions working across a set of contexts.. Business stakeholders might not be familiar with traditional modeling, but should be able to produce a BPCM model that can capture the knowledge about major business processes. Furthermore, BPCM can help to bridge the gaps between business stakeholders and technical model developers ensuring better business process models.

The combined framework for constructing business process support system is present in this paper. This framework mainly consists of goal modeling, process modeling, and business process characterizing modeling. It provides a structured way to develop a business process support system from a BPCM model.

The objective of this paper is to study how to relate goal modeling with BPCM models in the combined framework. To address this, we propose some guides to derive goal models in terms of i* from BPCM models, and illustrate these guides by an exemplar of call for sponsors case in the field of scientific conference organization.

The remainder of this paper is organized as follows. Section 2 briefly describes BPCM. Section 3 describes the combined framework used for developing business process support systems. Guidelines for mapping from a BPCM to a goal model in terms of i* are provided in Section 4. In Section 5, these preliminary guidelines are illustrated by an exemplar in the field conference arrangement process. The use of exemplars is widely recognized as a technique for early evaluation of modeling approaches [5]. Finally, section 6 concludes the paper.

2   Business Process Characterizing Model (BPCM)

In the course of business process support systems development, model developers often focus on operational and procedural aspects of business process systems, while various business stakeholders are more likely to express different concerns with regard to process models in terms of business oriented concepts. The business process characterizing model (BPCM) is a model used to represent a high level knowledge of business processes.

In an effort to facilitate readers' understanding of the BPCM modeling, we summarize the general definition of the elements for business process characterizing modeling (see Table 1). Since some elements in the BPCM model refer to other ontologies or concepts as presented in the Table 1, a brief description of those concepts and the motivation to incorporte those concepts is provided as follows.

Concerning the element context, there is no universal or absolute definition for context. [4] describes context as "typically the location, identity, and state of people, groups and computational and physical objects". Context is the reification of the environment, that is, whatever provides a surrounding in which the system operates. People can base their own perceptions or understanding to define context in different ways. In order to better design business process support systems, it is crucial to understand the working contxet and collect and deliver contextual information in a better way. By including a context element in a BPCM model, the correspondent business process support system can be made to serve people better in both mobile and more stationary computing settings.

Table 1. Definition of the elements in BPCM

Element	General Definition
Process	The business process people want to characterize. This element can be related to a common business process ontology such as SCOR [3].
Resource	This element is inspired by the resource concept in the REA framework [17]. This element can clearly address what are consumed and what are gained in a business process.
Actor	This element describes the people and organizations with different roles involve in a business process. This element can illustrate who are important to which business process.
Context	It includes contextual characteristics in terms of devices, software on the devices and networks providing connections between the devices and others.
Business Domain	This element classifies the related business domain(s). We link this element to the North American Industry Classification System (NAICS).
Goal	This element can address what goals need to be fulfilled in the business process. The goals may be related to operational goals and strategical goals. Operational goals are related to hard-goals, usually covering functional requirements; while strategic goals are related to soft-goals, which set the basis for non-functional requirements.
Process Type	According to REA [17], REA does not model only exchanges but also conversions. Exchange and conversion can be seen as two typical process types.
Version	This element is designed for keeping track revisions of BPCM models.

W3C has released a draft version of the Delivery Context Ontology (DCO). This ontology constructed in OWL provides a model of characteristics of the environment in which a device interacts with the web or other services. In this research work, we incorporate some key entities of DCO into the context element of BPCM. Some other research work has also started addressing the relationship between context and system development at the requirement level. For instance, [2] investigates the relation between context and requirements at the beginning of goal oriented analysis, and [19] extends the application of the problem frames approach with context monitoring and switching problems.

REA [17] [12] was originally conceived as a framework for accounting information systems, but it has subsequently broadened its scope and developed into an enterprise domain ontology [13] and e-commerce framework [1]. The core concepts in the REA ontology are *resource*, *event* and *agent*. The intuition behind this approach is that every business transaction can be described as an event where two agents exchange resources. In order to acquire a resource from other agents, an agent has to give up some of its own resource. The duality of resource transfer is essential in commerce. It never happens that one agent simply gives away a resource to another without expecting another resource back as compensation. Basically, there are two types of events: *exchange* and *conversion* [13, 17]. An exchange occurs when an agent receives economic resources from another agent and gives resource back to that agent. A conversion occurs when an agent consumes resources to produce other resources. Annotating process with process type enable users to identify the attribute of business processes around the resource lifecycle. As illustrated in [7], the element resource of BPCM is important in identifying relevant tasks or activities for the construction of process models. For each resource in a BPCM, it should include a message flow which links two associated tasks in a BPMN process model, whereby the source of the message flow connected to the dependee's task and the destination of the message flow connected to the depender's task.

Last but not least, the North American Industry Classification System (NAICS) is a standard for the collection, tabulation, presentation, and analysis of statistical data describing the U.S. economy. NAICS is based on a production-oriented concept, meaning that it groups establishments into industries according to similarity in the processes used to produce goods or services. Each business process is labeled with a business domain. This is of help for model users to search or retrieve business processes within specific business domain.

3   The Combined Framework

The overall development framework is presented in Fig.1. The goal of this framework is to build a method to design a business process support system that conforms to business stakeholders' views on the business process. In this framework, we consider a set of BPCMs as a starting point for developing an IT system. It is possible to have other approaches, e.g., starting with a goal model and then deriving use cases from a goal model [20]. In [16], the authors also argued that it is often beneficial to start a modeling-based project with an informal model and then develop the visual models. Incidentally, it could also be beneficial to also link to more elaborate business modeling frameworks, such as Osterwalder's Business Model Canvas [18]. We have not yet investigated this possibility.

In our combined framework, as a first step, data about all components of BPCM are gathered and recorded in a textual table according to the elements presented in the last section. Some early stage requirement engineering techniques [14] can be used to gather this information. As a result, a BPCM model is created. Then, we try to derive goal models and process models from the BPCM model. Lastly, those models can be used as inputs for deriving a candidate IT system.

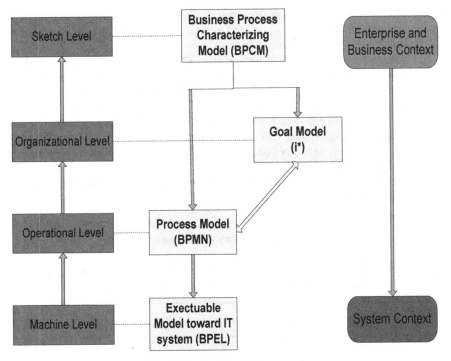

Fig. 1. The Combined Framework

4   Ideas to Guide Goal Modeling Based on a BPCM

Since i* is chosen as the modeling language for goal modeling in our combined framework, transiting from a BPCM model to an i* goal model is essentially about specifying the dependencies between actors involved in a business process. We define the following preliminary guidelines for constructing an i* model from a BPCM model.

1)   Identify the actors in a BPCM model. Each actor within a BPCM model can be represented as a candidate to be an agent in an i* model.
2)   Map actors. Translate the identified actors to the agents in an i* model.
3)   Identify the resources in a BPCM model. Since we do not have an element in a BPCM model to address the relevant tasks in the business process domain, some additional effort need to be put into tasks discovery in process modeling. Therefore, we proposed the complementary requirement table in Table 2. For each resource in a BPCM, it should associate with two actors and two tasks respectively as it shows in the Complementary Requirement Table (Table 2).

4) Map resources. Represent the resources in an i* model. The complementary requirement table can also be used as a primary heuristic for making possible linkages between resources and tasks in an i* model.

5) Assign tasks to equivalent agents in an i* model. After getting the defined task names from Table 2 in the step 3, those tasks can be assigned to equivalent agents in an i* model. The intention of the tasks in the same agent needs to be checked to see any possible dependency between them.

6) Map the hard goal(s) in a BPCM model to the equivalent agent(s) in an i* model.

7) Make linkages between tasks and goals in an i* model if any dependency between them is found.

8) Identify soft-goal dependencies in an i* model. This is based on the soft goal(s) in a BPCM model.

Table 2. Complementary Requirement Table

Resource	Source Pool or Lane	Destination Pool or Lane	Related Task in Source Pool	Related Task in Destination Pool

5   Exemplar

In this section, we apply our guidelines to a small exemplar. The exemplar is an extension of a part of the original conference case used e.g., in the IFIP CRIS conferences in the eighties. This case primarily focuses on the call for conference sponsorship process. This case involves the interaction between the conference organization team and potential sponsors. The overall case description is presented here: Prior to holding a conference, a conference organization team is established. All services involved in a conference cost money, which have to be balanced by the income from registration fee from participants and sponsors. Further, the conference organization would like to gather sufficient sponsoring from potential sponsors to avoid a high conference registration cost. Thus, in order to get funds from sponsors, a call for sponsor is sent out to potential sponsors by the conference organization team. Some potential sponsors decide to sponsor the conference and file the sponsorship form to the organization team. Then, the organization team acknowledges their sponsorships by sending out the acknowledgment letters.

After characterizing this call for sponsors case using the characterizing model proposed in Section 2, we summarize the derived BPCM model in Table 3. We assume that no change would occur to the case presented above. This means that we do not need to keep track of any changes involved in the case description. Therefore, the version element of BPCM is not included in Table 3.

Table 3. Call for Sponsors Case in a BPCM Model

BPCM Name / BPCM Elements	Call for Sponsors Case
Process	Attract Sponsors (P2 Plan Source)
Resource	Call for Sponsor, Sponsorship Form, Sponsor Acknowledgement
Actors	Conference Organization Team, Potential Sponsors
Context	Over LAN Internet Communication
Business Domain	561920 Convention and Trade Show Organizers
Goal	Hard Goals: get sufficient sponsorship to avoid a high conference fee
Process Type	Conversion

Table 4. Complementary Requirement Table: Call For Sponsors Case

Resource	Source Pool or Lane	Destination Pool or Lane	Related Task in Source Pool	Related Task in Destination Pool
Call for Sponsor	Conference Organization Team	Potential Sponsors	Publish Call for Sponsor	Receive Call for Sponsor
Sponsorship Form	Sponsor	Conference Organization Team	Apply for the conference sponsorship	Receive the form
Sponsorship Acknowledgement	Organizational Committee	Potential Sponsors	Acknowledge the sponsorship	Receive the acknowledgement

Fig.2 is the i* model thus extracted according the BPCM model of the call for sponsors case as presented in Table 3.

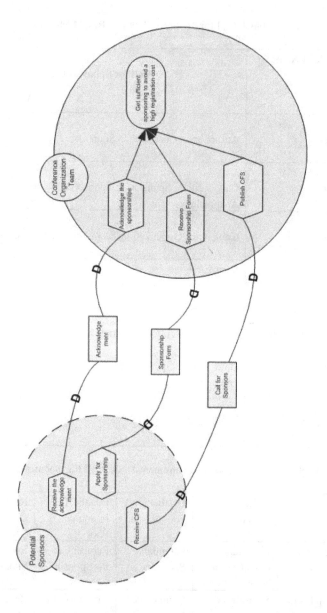

Fig. 2. The i* Model

The two actors (conference organization team and potential sponsors) in the BPCM model can directly map to two agents in the i* model. Three identified resources are placed between two agents in the i* model. A complementary requirement table (see Table 4) is used to identify related tasks in this business case. Then, the identified tasks are mapped to the correspondent agents in the i* model. The dependencies between resources and related tasks can be made according to the complementary requirement table. Since only one hard goal is identified in the BPCM model, this goal can be easily mapped to the agent conference organization team. The identified three tasks in the agent (conference organization team) can be regarded as three means to achieve the hard goal. Therefore, the dependencies are made between them. Lastly, as soft goal is not indicated in the BPCM model, soft goal dependencies are not applicable to this i* model.

6   Conclusion

In this paper, we firstly introduced a business-oriented modeling language, named BPCM, and a combined framework for constructing a business process support system. The framework was created to guide both business stakeholders and model developers during the life cycle of a modeling-based project development. We considered a BPCM model created by business stakeholders as a starting point for developing an IT system. Then, the start of goal models and process models can be derived from a BPCM model. Process models were then used as inputs for deriving a candidate IT system.

Further, some guidelines on transiting a BPCM model to a goal model in terms of i* were discussed. Furthermore, these guidelines were applied to an exemplar in a small business case: Call for Sponsors. It is evident from the mappings presented in the exemplar study that a BPCM model was able to provide useful input to build goals models which conform to business stakeholders' perspective on a business process scenario.

Acknowledgement. This research is supported by the Fundamental Research Funds for the Central Universities, China (Project No. ZUEL. 2012065).

References

1. UN/CEFACT Modeling Methodology (UMM) User Guide (2007)
2. Ali, R., Dalpiaz, F., Giorgini, P.: A Goal Modeling Framework for Self-contextualizable Software. In: Halpin, T., Krogstie, J., Nurcan, S., Proper, E., Schmidt, R., Soffer, P., Ukor, R. (eds.) Enterprise, Business-Process and Information Systems Modeling. LNBIP, vol. 29, pp. 326–338. Springer, Heidelberg (2009)
3. Council, S.-C.: SCOR Model 8.0 Quick Reference Guide (2006)
4. Dey, A.K., Abowd, G.D., Salber, D.: A conceptual framework and a toolkit for supporting the rapid prototyping of context-aware applications. Human-Computer Interaction 16(2), 97–166 (2001)
5. Feather, M.S., Fickas, S., Finkelstein, A., et al.: Requirements and Specification Exemplars. Automated Software Engineering 4(4), 419–438 (1997)
6. Fowler, M.: UML Distilled: A Brief Guide to the Standard Object Modeling Language. Addison-Wesley, Reading (2003)
7. Gao, S., Krogstie, J.: A Combined Framework for Development of Business Process Support Systems. In: Persson, A., Stirna, J. (eds.) PoEM 2009. LNBIP, vol. 39, pp. 115–129. Springer, Heidelberg (2009)
8. Gao, S., Krogstie, J.: Facilitating Business Process Development via a Process Characterizing Model. In: International Symposium on Knowledge Acquisition and Modeling 2008. IEEE CS (2008)

9. Gao, S., Krogstie, J., Gransæther, P.A.: Mobile Services Acceptance Model. In: Proceedings of ICHIT. IEEE Computer Society (2008)
10. Gao, S., Krogstie, J., Siau, K.: Developing an instrument to measure the adoption of mobile services. Mobile Information Systems 7(1), 45–67 (2011)
11. Gao, S., Moe, S.P., Krogstie, J.: An Empirical Test of the Mobile Services Acceptance Model. In: 2010 Ninth International Conference on Mobile Business and 2010 Ninth Global Mobility Roundtable (ICMB-GMR), pp. 168–175 (2010)
12. Geerts, G.L., McCarthy, W.E.: An Accounting Object Infrastructure for Knowledge-Based Enterprise Models. IEEE Intelligent Systems 14(4), 89–94 (1999)
13. Hruby, P.: Model-Driven Design Using Business Patterns. Springer, New York (2006)
14. Hull, E., Jackson, K., Dick, J.: Requirements Engineering. Springer, New York (2004)
15. Krogstie, J., Lyytinen, K., Opdahl, A.L., et al.: Research areas and challenges for mobile information systems. Int. J. Mob. Commun. 2(3), 220–234 (2004)
16. Maiden, N.A.M., Jones, S.V., Manning, S., Greenwood, J., Renou, L.: Model-Driven Requirements Engineering: Synchronising Models in an Air Traffic Management Case Study. In: Persson, A., Stirna, J. (eds.) CAiSE 2004. LNCS, vol. 3084, pp. 368–383. Springer, Heidelberg (2004)
17. McCarthy, W.E.: The REA accounting model: A generalized framework for accounting systems in a shared data environment 57, 554-578 (1982)
18. Osterwalder, A., Pigneur, Y.: Business Model Generation. John Wiley & Sons, New Jersey (2010)
19. Salifu, M., Yu, Y., Nuseibeh, B.: Specifying Monitoring and Switching Problems in Context. In: RE 2007. IEEE CS Press (2007)
20. Santander, V.F.A., Castro, J.: Deriving Use Cases from Organizational Modeling. In: RE 2002, pp. 32–42. IEEE CS (2002)

Increasing the Rigorousness
of Measures Definition through a UML/OCL Model
Based on the Briand et al.'s Framework

Luis Reynoso[1,*], Marcelo Amaolo[1], Daniel Dolz[1],
Claudio Vaucheret[1], and Mabel Álvarez[2]

[1] University of Comahue
Buenos Aires 1400, Neuquén
{Luis.Reynoso,Marcelo.Amaolo,Daniel.Dolz,
Claudio.Vaucheret}@fai.uncoma.edu.ar
[2] Patagonia San Juan Bosco University, Argentina
Belgrano y Rawson (9100) Trelew, Chubut
mablop@speedy.com.ar

Abstract. The use of a formal definition of measures upon a metamodel assures that measures capture the software artifacts they intend for, improve repeatability and could facilitate the implementation of measures extraction tools. However, it does not assure that the measure captures the measurement concept it claims (like size, coupling, etc). For that purpose many formal frameworks had been defined. The well-known property-based framework proposed by Briand et al. defines the most important measurement concepts regardless the specific software artifacts to which these concepts are applied. In this article we define a UML/OCL model from the Briand's framework and we relate it with the formal definition of measures upon metamodels. We describe a set of well-formed properties that a measure should verify when capturing a measurement concept (which are derived from the model). We exemplify our approach through a thorough formal definition of UML statechart diagrams measures and its well-formed constraints of size measures.

Keywords: Measurement, Metamodeling, Property-based Framework, UML, MDA.

1 Introduction

Many authors argue that many difficulties may arise when measures are defined in an unclear or imprecise way. The lack of precision of what is defined by a measure may produce that the persons who builds the measure extraction tool, makes their own decision during implementation [29]. In this way, they can arrive at incorrect values of the measure. This situation arises when measures are not repeatable (the same result would not be produced each time a measure is repeatedly applied to a same artifact by a

---

* Corresponding author.

B. Murgante et al. (Eds.): ICCSA 2013, Part III, LNCS 7973, pp. 43–58, 2013.

different person [14], [15]). Consequently, when measures are not repeatable, quality evaluators of models can take incorrect and undesirable decisions of the external quality attributes of their models. So, a complete definition of measure should include not only in natural language but also in formal language, because how well a measure is understood will influence the way the measure is implemented and used. Given the relevance of model-driven engineering [1], [16] many authors make profit of the metamodel of software artifacts that are available to formally define a measure upon a metamodel. The metamodels give the more suitable framework of the main measured concepts and relationships upon which measures may be specified. The usage of the meta modeling approach for defining model-specific measures have been introduced in [3], [4], for defining class diagram measures upon the UML metamodel. Later, Reynoso et al. formally defined measures for OCL [17] upon the OCL metamodel [19], and for BPMN models [25]. In a model-driven development process companies need to measure their models [26] to enhance its quality [27], [28] and to address safety-critical concerns [26].

A precise definition of the measure will ensure repeatability, however their formal definition could neglect the mathematical properties of the measurement concept captured by the measure (the last is known as formal or theoretical validation). Briand et al. [24] argue about the importance of a precise definition of the mathematical properties that characterize the most important measurement concepts like size, complexity, etc. regardless the specific software artifacts to which these concepts are applied. They provided a mathematical framework to define several concepts such as size, length, complexity, cohesion and coupling. The rigorous of the framework is provided by the mathematical underpinning. We were interested in defining a set of OCL properties that a formal definition of a measure should verify according to each measurement concepts of the framework. So, in this paper we show a model to define a set of constraints from Property-based Framework of Briand el al. (PbFB) and then we exemplify how to use it in the formal definition of a statechart measure. We show how the formal definition of UML Statechart Diagram (SD) measures using a metamodeling approach are defined and can be added with well-formed properties of the PbFB model. We use UML models because they become the primary artifacts, focus and products [23], [22] in recent Model-Driven Engineering (MDE) initiatives.

The advantages of a formal definition of a measure upon a metamodel are many: (1) the formal definition of measures: misunderstanding between the authors and the readers of the measures is avoided; (2) measures extraction tools can obtain the same results of a same measure; (3) experimentation is not hampered [2] due to the fact that an experiment and each of its replicas use measures which are repeatable; (4) measure value can be computed before and after a modification of design (or model transformation) is introduced, so measure can help to assess the quality of the diagram [20]; (5) they could facilitate the implementation of measures extraction tools, even more, they are used to generate a fully fledged measurement software [26].

Beside the aforementioned advantages of a formal definition of measure, a theoretical definition of measure constitutes a necessary constraint to assure that the measure captures a measurement concept [29]. In the same way the theoretical definition of a measure is a necessary conditions but not sufficient, due to the fact the measures should suffer an empirical validation, we argue that a formal definition of measure upon a metamodel is a necessary condition of its definition and it should be complemented with the add-on of well-formed properties derived of their theoretical definition.

Montperrus et al argue in [26] that a measurement software can be generated from an abstract and declarative specification of metrics, from a model of metric specifications. We believe that the PbFB model we provide in this article will be also useful for building measurement software, it helps to know and to monitoring how a measurement concepts are captured by a measure during the model-driven process.

This paper starts in the next section defining the UML/OCL model for the property-based framework of Briand (PbFB) and shows the OCL properties for size. Section 3 gives a brief presentation of the measures for UML Statechart Diagrams defined by Lemus [10]. The statechart diagrams are the software artifacts upon which we will exemplify a set of formal definitions of measures and the application of the PbFB model. In addition, section 3 presents the UML Statechart metamodel and an example of an instance of the metamodel illustrated how measures can be captured. Section 4 provides the formal specification of the measures and the application of the PbFB Model. Finally, the last section presents some concluding remarks and future work. Appendix A shows the complete definition of properties of the PbFB model.

2   The PbFB Model

In the property-based framework of Briand [24] a system is represented as a class having a set of elements and a set of binary relationship between the elements of the system. In Figure 1 we depicted the System class and two related classes, Element and Relation classes. The system verifies the property that *all the relationships link elements of the system:*

```
context System
inv: self.getRelationships()->forall( r:Relation |
self.getElements()->includes(r.source) and
self.getElements()->includes(r.target) )
```

Given a system, *a system is considered a module if and only if its elements (or relationships) are a subset of the elements (or relationships) of the system which contains it.* We model a module as a system according to its definition. The *module* attribute in System class (Boolean type) identifies whether a system is a module[1]. The modules contained in a system are modeled using a relationship which links System with itself (see Figure 1). So, a module satisfies the following property:

```
context System
inv: if module then
            self.getElements()->forall(e: Element |
            self.system.getElements()->includes(e)) and
   self. getRelationships()->forall(r: Relation |
   system.getRelationships()->includes(r))
   else true
```

---

[1] A module can also be modelled as subclass of System but practically there is no difference between a system and a system module which can be distinguished with this attribute.

The elements of a module are connected to the elements of the rest of the systems by incoming and outgoing relationships. The *InputR()* operation obtains the set of relationships from elements outside module *m* to those of module *m*. *InputR()* is defined as:

```
context System:: inputR(): Set(Relation)
inv: if module then
self.system.getRelationships()->collect(r |
self.getElements()->includes(r.target) and not
self.getElements()->includes(r.source))
```

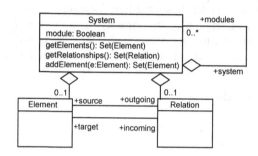

Fig. 1. A model used in the formalization of the PbFB

Similarly, it is possible to define *OutputR()*, a set of relationships from elements of a module *m* to those of the rest of the system.

Included defines an observer operations to verify that a system includes the relations and elements of another:

```
context System:: included(m:System):Boolean
inv: if self.module and m.module then
m.getElements()->forall(e |
    self.getElements()->includes(e)) and
m.getRelationships()->forall(r |
  self.getRelationships()->includes(r))
```

2.1 Properties of Size

We suppose that *size_md()* defines a measure which captures the size of the system. Size_md should also verify the following well-formed properties according to Briand et al.:

Non-negativity. The size of a system is non-negative.

```
context System:: size_md()
post non-negativity: result > 0
```

Null Value. The size of a system is null whether the system has no elements.

```
context System:: size_md()
inv null_value: self.getElements()->isEmpty()
  implies self.size_md() = 0
```

Module Additivity. The size of a system is equal to the size of its disjoint modules

```
context System:: size_md()
inv module_additivity:
self.modules->forall(m1,m2 |
  m1.getElements()->intersect(
  m2.getElements())->isEmpty() )
and
self.modules->collect(m1.getElements())->flatten() =
self.getElements()    implies
          self.size_md()= self.modules->collect(m |
          m.size_md())->sum()
```

The last property is equivalent to Size.IV property of Briand et al. Size.III is a particular case of Size.IV. Therefore, the following property is also valid:

Size Monotonocity Property. Adding elements to a system cannot decrease its size.

```
context System:: add_element(e: Element)
post monotonocity_property:
  self@pre.size_md() <= self.size_md()
```

An additional property is defined in Briand et al. which follows from the above properties. The size of a system is not greater than the sum of the sizes of its modules due to the presence of common elements:

```
context System:: merging_modules()
inv merging_modules:
if self.modules and self.getElements() = self.modules-
>collect(m: Module | m.getElements())->flatten() implies
self.size_md()= self.modules->collect(m | m.size_md())-
>sum()
```

Appendix A shows the formal definition of properties for cohesion and coupling according to the Briand et al. framework [24]. Properties for complexity, length are available in a technical report (see appendix A).

3   Measures for UML SD

The UML has now become the de facto standard software systems modeling and the UML State Diagram (SD) has become an important technique for describing the dynamic aspects of a software system [12]. An SD contributes to the behavioural specification of a type in a model. A thorough definition of a set of measures for structural properties of UML SD is presented in [10] based on the hypothesis that structural properties of an UML SD (the software artifacts measured) have an impact on the cognitive complexity of modelers (subjects), and high cognitive complexity leads the UML SD to exhibit undesirable external qualities on the final software product [14], such as less understandability or a reduced maintainability [5], [6]. These measures are supposed to be good indicators of the understandability of such diagrams. This fact was empirically validated in [8], [9], [11]. These measures were defined following a method consisting of three main steps [7]: measure definition, theoretical validation and empirical validation. But initially the measures were informally defined using natural language. So, one contribution of this paper is to thoughtfully show its formal definition whose purpose was briefly described in a short paper [30].

3.1   The UML Statechart Metamodel

A SD describes possible sequences of states and actions through which the element instances can proceed during its lifetime as a result of reacting to discrete events (for example, signals or operation invocations). The abstract syntax for state machines is expressed graphically in UML StateChart Metamodel [18], which covers all the basic concepts of state machine graphs such as states, transitions, guards, etc. In this section we will first give an overview of the Metamodel explaining its main metaclasses and relationships and then we will show the whole metamodel. Secondly, we will describe an SD as an instance of the UML Statechart metamodel.

3.2   Overview of the Main Metaclasses

Every state machine has a top state (see Figure 2), usually a composite state, that contains all the other elements of the entire state machine. The graphical rendering of this top state within an SD is optional.

The State hierarchy has a State superclass and three subclasses, CompositeState, SimpleState and FinalState. This hierarchy, in fact, is part of the StateVertex hierarchy as it is shown in Figure 3, which also includes the PseudoState, SynchState and SubState classes. The composite state may contain any state of the StateVertex hierarchy.All the classes, attributes of classes and relationships previously depicted in Figures 2 and 3 are part of the UML Statechart metamodel depicted in Figure 4 [18].

Each State in an SD may have associated actions, such as entry, exit or a do-activity actions (see in Figure 4 the relationships between the State and Actions classes having the *entry*, *exit* and *doActivity* rolenames). Nevertheless, no more than one action of a specific type is allowed for a particular state.

Transitions usually connect two states, for example two Simple States, a Simple State with a Final State, etc. These connections are described in Figure 4 through two relationships between the StateVertex and Transition classes, where each of them identifies the source and target StateVertex which is connected through the transition. So, any transition connects exactly a source to a target statevertex. From a StateVertex point of view, any state can have many incoming and outgoing transitions [13].

Within an SD, transitions may also be labeled with Guard and Events. This situation is modelled through the Guard and Event classes which are related to the Transition class (see the relationship between Transition and Guard classes with the *guard* rolename, and the relationship between Transition and Event classes with the *trigger* rolename).

The set of all transitions within an SD is modeled through a relationship between the StateMachine and the Transition classes.

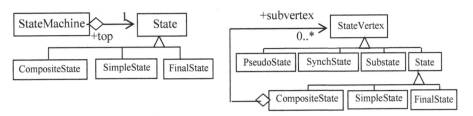

Fig. 2. The StateVertex Hierarchy   Fig. 3. The top State of a StateMachine

3.3 A Metamodel Instance Sample

In order to understand both the UML Statechart metamodel and the specification of measures for UML SD we will give in this section, a sample of how an SD is represented as a metamodel instance. The diagram we will use is shown in Figure 5 and its representation as an instance metamodel is depicted in Figure 6.

Although the SD example is rather simple because it is composed of five simple states and two composite states, its representation as a metamodel instance is quite complicated and involves more than thirty objects.

The SD shown in Figure 5 describes a simple quality control process for testing incoming raw materials. At the beginning the received material is in the NoVerification state. The raw material is tested in house (Testing state) when the supplier of the material is not a certified supplier otherwise the raw material state is changed to the Accepted state. This is modelled through two labels, check[CertifiedSupplier] and check[NoCertifiedSupplier] which represent an Event ('check') having each of them a Guard ('CertifiedSupplier' or 'NoCertifiedSupplier'). When a raw material is in the Testing state the *entry* activity of the state is triggered, and the material is tested. The test has two possible results: approved or rejected. According to each situation the raw material will change to the Accepted or Rejected state respectively. A report is filled when materials are rejected. Accepted raw materials are stored within the warehouse (and the stock is updated) whereas rejected raw materials are returned to the sender. In both cases, the process of reception for the raw material ends.

Now we will explain the diagram shown in Figure 5 as an instance of the UML Statechart Metamodel. Figure 6, an object diagram, consists of many objects which are instances of the UML statechart metamodel' metaclasses (explained in section 3.1 and 3.2). In order to easily refer to the objects of the metamodel object diagram of Figure 6 we have named a vertical and an horizontal axes (of the SD) with letters and numbers in order to use them as a simple cartesian coordinate system. For example, in the position A1, a Statemachine' object, named RawMaterialReception, is shown to represent the diagram itself. Because every state machine has a top state that contains all the other elements of the entire state machine, the RawMaterialReception' State-machine object has a top state, a composite state, depicted in the A2 position of the diagram. This composite state, which has no name, composes four states: an initial state (a PseudoState object having the 'initial' value in the kind attribute [18], see position C1 of Figure 6), two simple states (named NoVerification and WithinWarehouse, see positions F1 and D2 of Figure 6 respectively), a final state (position B2 of Figure 6) and a composite State (position A4 of Figure 6). This last composite state has three simple states: Accepted (position C4), Testing (position G4) and Rejected (position I5) states.

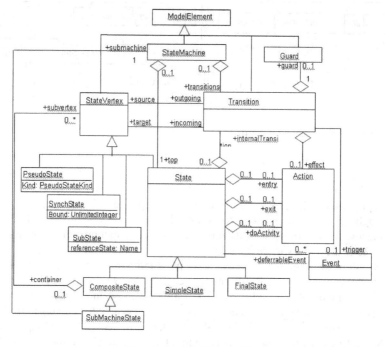

Fig. 4. The SD Metamodel

Figure 6 also shows eight transitions between states. The transition connects different kinds of states. For example:

1. Transitions between Simple states: a transition between the NoVerification and Accepted states is shown in the E3 position. Other transition between simple states are depicted in C3, D4, G3, H4 positions.
2. Transitions between Simple and Final States: a transition between the WithinWarehouse simple state and the Final state is shown in C2 position. Another transition between a simple and a final state is shown in the I1 position.
3. Transition between Initial and Simple State: a transition between the initial and the NoVerification states is shown in the E1 position.

Each of the aforementioned transitions connects a source state to a target state. The relationships between the StateMachine object and the transition objects are not shown in Figure 6 to avoid clutter the readability of the diagram.

Four transitions (shown in I1, H4, D4 and C3 positions) have associated a Guard object (shown in the I2, H5, D5 and B3 positions respectively).Similarly, two transitions (see E3 and G3 positions) have an Event (D3 and G2 positions) with a Guard associated (F3 and H3 positions) to the event.

3.4 Specification of Measures

In this section we will show the specification of SD measures using the UML Statechart metamodel [18] (described in section 3.1).

The specification of the measures relies on three query operations:

1. Alltransitions operation, defined in the StateMachine metaclass, obtains the set of transitions in an SD.
2. AllStates operation, defined in the StateMachine metaclass, selects the set of all the states within an SD.
3. AllSubStates operation, used by the two previous operations and defined in the StateVertex metaclass, obtains the set of all Subvertex included in a SD. It is recursively defined.

Their OCL definitions are shown below:

```
context StateMachine::allTransitions::Set(Transition)
body: result =  self.transitions>
union(self.allSubStates().internaltransitions)
context StateMachine::allStates:Set(State)
body : result = self.top.allSubstates()

context StateVertex::allSubstates::Set(StateVertex)
body: result =if self.oclIsKindOf(CompositeState)
then self.oclAsType(CompositeState).subvertex->union
self.oclAsType(CompositeState).subvertex
-> select (s:StateVertex| s.allSubstates())
else Set{} endif
```

For obtaining the value of each SD measure of Lemus [10] we defined in the StateMachine metaclass an operation with the same name as the measure. So, 14 operations were defined. Using the AllSubState operation we can define the value of many SD measures. This operation returns the set of all the states (of different kinds: initial, final, simples, etc.) included in a diagram, even those states which are part of composite states. For example, selecting from the allsubstate operation result, those states of an SD which have associated a doActivity action, it is possible to obtain the Number of Activity (NA) measure's value. The quantity of objects selected represents the value of the NA measure.

```
context StateMachine::NA():Integer
body: result = self.top.allSubstates()->select(s |
s.oclType(State)  and s.doActivity->notEmpty() )-> size()
```

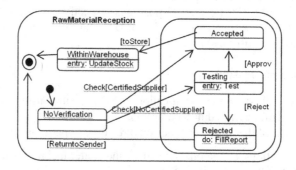

Fig. 5. A state machine called RawMaterial Reception

When the NA() operation is requested in the rawmaterialreception object of Figure 6, the obtained result is 1, due to the fact that only one state (see I5 position of Figure 6) has a *doActivity* action associated. In a similar way we can obtain the value of Number of Simple State (NSS) measure of Lemus [10]. First, we select those State-Vertex which are instance of SimpleState class, then we obtain the quantity of the object contained in this selection. When the NSS() operation is requested in the raw-materialreception object of Figure 6, the result is 5 (the five simple states). The rest of the measures of [10] are formally defined in a technical report [31].

```
context StateMachine::NSS():Integer
body: result = self.top.allSubstates()->select (s |
s.oclType(SimpleState))-> size()
```

4   Example of NSS and Its Properties

Lemus et al. [10] using the property-based framework of Briand et al. show that NSS, NA, NLCS, NT, NCS, NCT, NCTG1 are size measures. In their theoretical validation (see [10]) a UML Statechart Diagram is considered a system (i.e. is the StateMachine class in Figure 4) the elements are states (i.e. is the stateVertex class in Figure 4), the relations are transitions (the Transition class in Figure 4) and modules are considered a subset of states and transitions of the SD.

In our formal definition, the StateMachine class is added with a self-relationship to model the modules of a system. The getElements() and getRelationships are defined in the following way:

```
context StateMachine::getElements()::Set(State)
body: self.allStates()
context StateMachine::getRelationships():Set(Relation)
body : self.alltransitions()
```

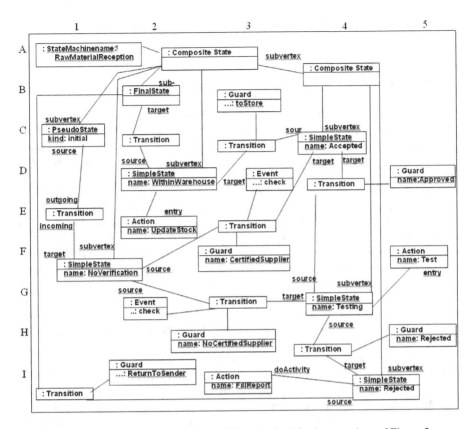

Fig. 6. An object diagram (metamodel instantiation) for the statechart of Figure 5

So we can add a set of well-formed properties to each size measure definition according to the properties of size defined in section 2.1. For example for the NSS definition we add the following properties:

```
context StateMachine::NSS():Integer
body: result = self.top.allSubstates()->select(s |
s.oclType(State)  and s.doActivity->notEmpty() )-> size()
post non-negativity: result > 0
inv null_value: self.getElements()->isEmpty() implies
self.NSS() = 0

inv module_additivity: self.modules->forall(m1,m2 |
m1.getElements()->intersect( m2.getElements())->isEmpty()
)
and
self.modules->collect(m1.getElements())->flatten() =
self.getElements() implies
self.NA()= self.modules->collect(m|m.NSS())->sum()
```

A complete list of properties for the rest of measurement concepts is included in Appendix A.

5   Conclusions

The quality of models is acquiring more relevance within the introduction of model-driven engineering. In order to assess the quality of a model, it is unavoidable the definition and application of measures for them. Many measures had been proliferated in literature but they are not correctly defined, and the worst, they are not integrated with the software artifact and with the measurement concept to which they are related (or they claim to be). Measures defined upon a metamodel not only had benefits in terms of the final product but also are ready to support ongoing model-driven development process. Their definition can be derived from model to model through model transformation. The main contribution of this paper is the definition of a UML/OCL model which captures the property-based framework of Briand et al [24]. The model includes the formal properties defined by Briand et al. for size, cohesion, length, complexity and coupling measure defining a set of OCL constraints upon a small UML model. In this paper we used the statechart measures of Lemus et al. [10], [11] to show how to connect the formal definition of measures upon the UML statechart metamodel and the OCL properties the measure should verify to capture the measure concept of size defined by Briand et al. [24].

Refactoring techniques [21] and MDA-based system for measures extraction which improve the design of UML systems can take advantage of using the formal de definition of measures and the constraints related to the measurement concepts their capture. Measure value can be computed before and after the refactoring is applied, to evaluate the change according the quality of the statecharts [12], constraints should be valid as part of assert-restrictions of the system. Using this approach, the rigorous definition of measure can be useful to build solid software measurement tools and to verify that measure constraints are verified by its metamodel, models and objects.

Acknowledgments. This research is part of the 048/12 'Hacia el Fortalecimiento de la Sociedad en el Uso y Aplicación Geoespacial y las TICS' of Patagonia San Juan Bosco University, Chubut (Argentina) and the 'Modelos y Tecnologías para Gobierno Electrónico' of Comahue University, Neuquén (Argentina)".

References

1. Atkinson, C., Kühne, T.: Model-Driven Development. A Metamodeling Foundation, IEEE Software 20(5), 36–41 (2003)
2. Baroni, A.L.: Formal Definition of Object-Oriented Design Metrics. Master of Science in Computer Science Thesis, Vrije Universiteit Brussel, Belgium (2002)
3. Baroni, A.L., Braz, S., e Abreu, F.B.: Using OCL to Formalize Object-Oriented Design Metrics Definitions. In: Proceedings of QUAOOSE 2002, Malaga, Spain (2002)
4. Baroni, A.L., e Abreu, F.B.: A Formal Library for Aiding Metrics Extraction. In: Proceedings of the Int. Workshop on Object-Oriented ReEngineering at ECOOP 2003 (2003)

5. Briand, L.C., Bunse, L.C., Daly, J.W.: A Controlled Experiment for evaluating Quality Guidelines on the Maintainability of Object-Oriented Designs. IEEE Trans. on Softw. Eng. 27(6), 513–530 (2001)
6. Briand, L.C., Wust, J., Ikonomovski, S., Lounis, H.: Investigating Quality Factors in Object Oriented Designs: An Industrial Case-Study. In: 21st Int. Conf. on Software Engineering, pp. 345–354 (1999)
7. Calero, C., Piattini, M., Genero, M.: Method for Obtaining Correct Metrics. In: Proceedings of the 3rd Int. Conference on Enterprise and Information Systems (ICEIS 2001), pp. 779–784 (2001)
8. Cruz-Lemus, J.A., Genero, M., Manso, M.E., Piattini, M.: Evaluating the Effect of Composite States on the Understandability of UML Statechart Diagrams. In: Briand, L.C., Williams, C. (eds.) MoDELS 2005. LNCS, vol. 3713, pp. 113–125. Springer, Heidelberg (2005)
9. Cruz-Lemus, J.A., Genero, M., Morasca, S., Piattini, M.: Assessing the Understandability of UML Statechart Diagrams with Composite States - A Family of Empirical Studies. Submitted to Empirical Software Engineering (2007)
10. Cruz-Lemus, J.A., Genero, M., Piattini, M.: Metrics for UML Statechart Diagrams. In: Proceedings of Metrics for Conceptual Models. Imperial College Press, UK (2005)
11. Cruz-Lemus, J.A., Genero, M., Piattini, M., Toval, A.: An Empirical Study of the Nesting Level of Composite States within UML Statechart Diagrams. In: Akoka, J., Liddle, S.W., Song, I.-Y., Bertolotto, M., Comyn-Wattiau, I., van den Heuvel, W.-J., Kolp, M., Trujillo, J., Kop, C., Mayr, H.C. (eds.) ER Workshops 2005. LNCS, vol. 3770, pp. 12–22. Springer, Heidelberg (2005)
12. Denger, C., Ciolkowski, M.: High Quality Statecharts through Tailored, Perspective-Based Inspections. In: EUROMICRO 2003, pp. 316–325 (2003)
13. Evermann, J., Wand, Y.: Toward Formalizing Domain Modeling Semantics in Language Syntax. IEEE Trans. on Soft. Eng. 31(1) (2005)
14. ISO/IEC 9126. Software Product Evaluation-Quality Characteristics and Guidelines for their Use, Geneva
15. Kitchenham, B., Pfleeger, S.L., Fenton, N.: Towards a Framework for Software Measurement Validation. IEEE Trans. on Softw. Eng. 21(12), 929–944 (1995)
16. Object Management Group. MDA The OMG Model Driven Architecture (2002) , http://www.omg.org./mda/
17. Object Management Group. UML 2.0 OCL 2nd revised submission. OMG Document, http://www.omg.org
18. Object Management Group. UML Specification Version 1.4.2, OMG Document formal04-07-02, http://www.omg.org
19. Reynoso, L., Genero, M., Piattini, M.: OCL2: Using OCL in the Formal Definition of OCL Expression Measures. In: Proceedings of the 1st Workshop on Quality in Modeling QIM co-located with the ACM/IEEE MODELs 2006 (2006)
20. Saeki, M., Kaiya, H.: Model Metrics and Metrics of Model Transformation. In: Proceedings of 1st Workshop on Quality in Modeling, Genova, Italy, October 1, pp. 31–45 (2006)
21. Sunyé, G., Pollet, D., Le Traon, Y., Jézéquel, J.-M.: Refactoring UML Models. In: Gogolla, M., Kobryn, C. (eds.) UML 2001. LNCS, vol. 2185, pp. 134–148. Springer, Heidelberg (2001)
22. Tang, M.-H., Chen, M.-H.: Measuring OO Design Metrics from UML. In: Jézéquel, J.-M., Hussmann, H., Cook, S. (eds.) UML 2002. LNCS, vol. 2460, pp. 368–382. Springer, Heidelberg (2002)

23. Vinter, R., Loomes, M., Kornbrot, R.: Applying Software Metrics to Formal Specifications: A Cognitive Approach. In: Proceedings of 5th. Int. Symposium on Softw. Metrics, March 20- 21, pp. 216–223 (1998)

24. Briand, L.C., Morasca, S., Basili, V.R.: Property-Based Software Engineering Measurement. IEEE Trans. Softw. Eng. 22(1), 68–86 (1996)

25. Reynoso, L., Rolón, E., Genero, M., García, F., Ruiz, F., Piattini, M.: Formal Definition of Measures for BPMN Models. In: Abran, A., Braungarten, R., Dumke, R.R., Cuadrado-Gallego, J.J., Brunekreef, J. (eds.) IWSM 2009. LNCS, vol. 5891, pp. 285–306. Springer, Heidelberg (2009)

26. Monperrus, M., Jézéquel, J.-M., Champeau, J., Hoeltzener, B.: A Model-Driven Measurement Approach. In: Czarnecki, K., Ober, I., Bruel, J.-M., Uhl, A., Völter, M. (eds.) MODELS 2008. LNCS, vol. 5301, pp. 505–519. Springer, Heidelberg (2008)

27. Genero, M., Piattini, M., Chaudron, M.: Quality of UML models. Information & Software Technology 51(12), 1629–1630 (2009)

28. Monperrus, M., Jézéquel, J.M., Champeau, J., Hoeltzener, B.: Measuring Models. In: Rech, J., Bunse, C. (eds.) Model-Driven Software Development: Integrating Quality Assurance. IDEA Group (2008)

29. Reynoso, L., Genero, M., Piattini, M.: Refinement and Extension of SMDM, a Method for Defining Valid Measures. J. UCS 16(21), 3210–3244 (2010)

30. Reynoso, L., Cruz-Lemus, J.A., Genero, M., Piattini, M.: Formal Definition of Measures for UML Statechart Diagrams using OCL. In: Proceedings of the, ACM Symposium on Applied Computing, SAC 2008, pp. 846–847 (2008)
    Reynoso, L., Amaolo, M., Dolz, D., Vaucheret, C., Álvarez, M.: A Briand et al.'s framework-based UML/OCL Model for Increasing the Rigorousness of Measures Definition (2013), Tech. Report UNC-UNPSJB, https://www.dropbox.com/sh/9vs6uy0re1owm8t/1tc2XIM3Yy/TechnicalReportFbPB.pdf

Appendix A: Formal Definition of Properties of Cohesion, Coupling, Complexity, Length

This appendix lists the properties of *cohesion* and *coupling* of PbFB according to the Briand et al. framework [24]. Due to space limitations the *length* and *complexity* properties are not shown in this appendix but they are available in a technical report [31].

Properties of Cohesion and Coupling

These properties are meaningful when a modular system is defined:

```
context System::is_modular_system()
inv: self.getElements()->forall(e|self.modules
  ->exists(m |m.getElements()->includes(e) )
and self.modular->forall(m1, m2 |m1.are_disjoint(m2))
context System::cohesion_md()
pre: is_modular_system()

context System::coupling_md()
pre: is_modular_system()
```

A.1 Properties of Cohesion

Non-negativity and Normalization: The cohesion of a module of a modular system (or modular system) belongs to a specified interval .

```
context System::cohesion_md()
pre: is_modular_system()
     post non-negativity_normalizacion: result >= 0 and
result <= self.maxcohesion
```

Null Value: The cohesion of a module of modular system (or modular system) is null if is empty.

```
context System::cohesion_md()
inv null_value: self.getRelationships()->isEmpty()
          implies self.cohesion_md()=0
```

Monotonicity: Adding intra-module relationships does not decrease [module|modular system] cohesion. If there is no intra-module relationship among the elements of a (all) module(s), then the module (system) cohesion is null.

```
context System:: cohesion_md ()
inv monotonicity : self. cohesion_md() <=
self.module->any(true).addRelationship().cohesion_md()
```

Cohesive Modules: The cohesion of a [module|modular system] obtained by putting together two unrelated modules is not greater than the [maximum cohesion of the two original modules |the cohesion of the original modular system].

```
context System:add_modules(m:System)
post: result.relationships = re-
sult.getRelationships()@pre-
>including(m.getRelationships())
result.elements = result.getElements()@pre-
>including(m.getElements())
```

```
context System:: cohesion_md ()
inv cohesive_modules :
self.modules->exists(m1, m2 | are_notconnected(m1, m2:
System)
m1.add_modules(m2).cohesion_md() <=
m1.cohesion_md().max(m2.cohesion_md() )
```

A.2 Properties of Coupling

Non-negativity: The coupling of a [module|modular system] is non-negativity.

```
context System::coupling_md()
         inv non-negativity: result >= 0
```

Null Value: The coupling of a [module|modular system] is null if [OuterR(m)|R-IR] is empty.

We need two auxiliary functions, InputR(m) and OutputR(m). InputR(m) are the incoming relationships of a modular system whereas Output(m) are the outcoming relationships.

```
context System:: inputR():Set(Relation)
post: result = self.getRelationships()->collect( r |
self.getElements()->includes(r.target ) and
self.getElements()->excludes(r.source) )
context System:: outputR():Set(Relation)
post: result = self.getRelationships()->collect( r |
self.getElements()->includes(r.source ) and
self.getElements()->excludes(r.target) )
```

```
context System:: coupling_md()
body null_value: self.inputR()->isEmpty() implies coupl-
ing_md() = 0 and
self.outputR()->isEmpty() implies coupling_md() = 0
```

Monotonicity: Adding inter-module relationships does not decrease coupling.

```
context System:: coupling_md()
inv monotonocity:
self.modules()->includes(m1, m2 |
self.are_not_connected(m1, m2) implies
self.coupling_md() <=
self.addRelationshipbetween_nonCC(m1, m2).coupling_md())
```

Merging Modules: The coupling of a [module|modular system] obtained by merging two modules is not greater than the [sum of the couplings of the two original modules|coupling of the original modular system], since the two modules may have common inter-module relationships.

```
context System:: coupling_md()
inv merging_modules :
    self.modules->forall(m1, m2 |
    m1.add_modules(m2).coupling_md() <=
    m1.coupling_md() + m2.coupling_md()   )
            self.getelements() = m1.getElements()
            ->including(m2.getElements())->flatten()
    implies self.length_md =
    m1.length_md().max( m2.length_md()))
```

Improving Requirements Specification
in WebREd-Tool by Using a NFR's Classification

José Alfonso Aguilar[1], Sanjay Misra[2], Anibal Zaldívar[1], and Roberto Bernal[3]

[1] Señales y Sistemas (SESIS) Facultad de Informática Mazatlán
Universidad Autónoma de Sinaloa, México
82120 Mazatlán, Mexico
[2] Department of Computer and Information Sciences
Covenant University, Nigeria
[3] Facultad de Informática Culiacán
Universidad Autónoma de Sinaloa, México
{ja.aguilar,azaldivar,roberto.bernal}@uas.edu.mx, ssopam@gmail.com
http://sesis.maz.uasnet.mx

Abstract. In Software Engineering (SE), a system has properties that emerge from the combination of its parts, these emergent properties will surely be a matter of system failure if the Non-Fuctional Requirements (NFRs), or system qualities, are not specified in advance. In Web Engineering (WE) field occurs very similar, but with some other issues related to special characteristics of the Web applications such as the navigation (with the application of the security). In this paper, we improve our Model-Driven tool, named WebREd-Tool, extending the requirements metamodel with a NFRs classification, the main idea is to help the Web application designer with the NFRs specification to make better design decisions and also to be used to validate the quality of the final Web application.

Keywords: Web Engineering, Requirements Engineering, Softgoal, GORE, i-star, A-OOH, NFRs, WebREd-Tool, MDE, MDD.

1 Introduction

Throughout the years, several methods for the development of Web applications (OOWS [1], WebML [2], NDT [3] and UWE [4], A-OOH[5]) have emerged [6], regrettably, only a few offers methodological support for the Requirements Engineering (RE) stage. Nevertheless, the complexity and continuos evolution of the Web applications demands the development of methods and tools (specially) for helping the developer's to perform the RE process [7] in order to improve the Web Engineering (WE) field. In this respect, the developer needs solutions (tool support) that take into account both Functional (FR) and Non-Fuctional (NFR) Requirements from the beginning of the Web application development process, what undoubtedly, will help to assure that the final product corresponds qualitatively to the users expectations. Functional Requirements (FRs) describes

B. Murgante et al. (Eds.): ICCSA 2013, Part III, LNCS 7973, pp. 59–69, 2013.

the system services, behavior or functions, whereas Non-Fuctional Requirements (NFRs), also known as quality requirements, specify a constraint in the application to build or in the development process [8].

An effective definition of requirements improves the quality of the final product, in this context, NFRs are critical to the successful implementation of almost every non-trivial software system, this is evidenced by the fact that many documented system failures are directly attributed to the inadequate implementation and maintenance of NFRs [9]. Unfortunately, in most of the Web Engineering approaches, a complete analysis of requirements is performed considering only FRs, thus leaving aside the NFRs until the implementation stage [10]. Following this evidency, there have been many attempts to provide techniques and methods to deal with some aspects of the RE process for the development of Web applications, but there is still a need for solutions which enable the designer to consider both FR and NFRs involved in the Web application development process [11] from the initial stage (requirements stage).

As a fact, requirements are ambiguous during elicitation process, but the introduction of the concept of *goals* helps in dealing with ambiguity and clarifying requirements. In recent years, the inclusion of Goal-oriented Requirements Engineering (GORE) in Web Engineering [7,5,12] offers a better analysis in Web application design due to the fact that requirements are explicitly specified in goal-oriented models, thus supporting developers in evaluating the implementation of certain requirements (FR and NFRs) for desigining successful software and the ability to reason about the software, the organization and the stakeholders goals in the same analysis. This has allowed the stakeholders to choose among the design decisions that can be taken to satisfy the goals and evaluate the implementation of certain requirements in particular (including NFRs). In this field, FRs are related to goals and sub-goals whereas NFRs are named softgoals, commonly used to represent objectives that miss clear-cut criteria, thus, analyzing Non-functional Requirements in terms of goals help in refining, exploring alternatives and resolving conflicts.

This paper is as extension of our recent work [12] about the importance of take into account those components from Requirements Engineering (RE) which are not considered with the necessary emphasis in Web Engineering field such as: Change Impact Analysis (CIA)[13,14], Requirements Traceability (RT) [15] and Non-Functional Requirements Optimization [16]. To this aim, we improve our Model-Driven tool named WebREd-Tool[1] adopting a NFRs classification in order to support the designer to make better design decisions and also to be used to validate the quality of the final Web application. In particular, the novelty of our ongoing work presented in this paper consists of: (i) the conduction of a literature review related to NFRs classification; (ii) the realization of an analysis of the most common Non-Fuctional Requirements used in Web and the

---

[1] The WebREd-Tool was the best demo tool and poster winner in the International Conference on Web Engineering 2012 (ICWE), developed in conjunction by the Universidad Autónoma de Sinaloa (Mexico), University of Alicante (Spain) and the Universidad Politécnica de Valencia (Spain) [17].

elaboration of a proposal for a basic classification of Non-Functional Requirements for Web Engineering, to do it, six type of NFRs have been considered due to they are the most commonly used in the Web Engineering field: *Usability, Performance, Reliability, Safety, Security* and *Efficiency*; (iii) the integration of NFRs classification in the Web requiremens metamodel used by the WebREd-Tool for requiremens specification, the integration consist in the specialization of the softgoal element from the graphic language used by the WebREd-Tool for Web requirements specification (i^* modeling language [18]).

The main benefit of our approach is that provides specific information about the different NFRs involved during the development process from the initial stage, thus allowing developers to make more informed design decisions for implementing a Web application that fully-satisfies the user expectations. Finally, it is important to mention that the WebREd-Tool is the proof of concept of our Goal-oriented Requirements Engineering (GORE) approach for requirements specification in Web Engineering [6,12].

The rest of the paper is organized as follows: Section 2 presents some related work relevant to the context of this work. Section 3 describes our GORE proposal where is found the contribution of this work. In Section 4, our Model-Driven tool, WebREd-Tool, is shortly described. The specialization of the requirements specification for the NFRs and its application is described in Section 5. Finally, the conclusion and future work is presented in Section 6.

2   Background

Recent studies with regard to Requirement Engineering techniques for the development of Web applications [19] have highlighted that most of the Web Engineering approaches focus on the analysis and design stages and do not give a comprehensive support to the requirements stage (such as WebML [2], OOHDM [20], WSDM [21] or Hera [22]). In some cases, NFRs are considered in a very general manner by almost all the approaches and only two of them, namely NDT and WebML, also provides dedicated tool support, as reviewed in [2] and [3].

Regarding approaches that consider NFRs requirements from early stages of the development process, in [23] the authors propose a metamodel for representing usability requirements for Web applications and in [10] the authors present the state-of-the-art for NFRs in Model-Driven Development (MDD), as well as an approach for considering NFRs into a MDD process from the very beginning of the development process. Chung [24] adopted a goal and process-oriented approach in NFR framework for dealing with Non-Functional Requirements using AND/OR tree. This framework was focused on quality goal satisficing where as Dardenne [25] proposed a goal-based framework with a focus on goal satisfaction. Unfortunately, these approaches are not designed to be used in the Web Engineering field. To the best of our knowledge, the only approaches that use GORE techniques in Web Engineering have been presented in [26,27]. Unfortunately, although these approaches use the i^* modeling framework [18,28] to represent requirements in Web domain, they do not benefit from every i^* feature because don't use all the expressiveness of the i^* framework to represent the special type of requirements of the Web applications such as the related with navigational issues. To overcome this situation, our previous work [12] adapts the well-known taxonomy of Web requirements presented in [29] for the i^* framework.

To sum up, there have been many attempts to provide techniques and methods to deal with some aspects of the Requirements Engineering process for Web applications. Nevertheless, there is still a need for solutions that considers NFRs from beginning of the Web application development process in order to improve the quality of the Web application perceived by users.

3   A Goal-Oriented Modeling Framework Applied in Web Engineering

This section describes our proposal to specify requirements in the context of the A-OOH (Adaptive Object-Oriented Hypermedia method) Web modeling method [30]. A-OOH is an extension of the OOH (Object-Oriented Hypermedia) [31] method with the inclusion of personalization strategies. This development method is combined with a modeling language named i^* for requirements specificication. The i^* (pronunced *eye-star*) is one of the most widespread goal-oriented frameworks, its has been applied for modeling organizations, business processes, requirements specifications and requirements analysis, among others. As a goal-oriented analysis technique, the i^* framework focuses on the description and evaluation of alternatives and their relationships to the organizational objectives [12].

We shortly describe next an excerpt of the i^* framework which is relevant for the present work. For a further explanation, we refer the reader to [18,28]. Essentially, the i^* framework consists of two models: the strategic dependency (SD) model, to describe the dependency relationships (represented as ⊃-) among various actors in an organizational context, and the strategic rationale (SR) model, used to describe actor goals and interests and how they might be achieved. Therefore, the SR model (represented as a dashed circle ⟨⟩) provides a detailed way of modeling the intentions of each actor (represented as a circle ○), i.e., internal intentional elements and their relationships:

- A goal (elipse ◯) represents an (intentional) desire of an actor. Interestingly, goals provide a rationale for requirements but they are not enough for describing how the goal will be satisfied. This can be described through means-end links (→▷) representing alternative ways for fulfilling goals.
- A task (hexagon ⬡) describes some work to be performed in a particular way. Decomposition links (—+—) are useful for representing the necessary intentional elements for a task to be performed.
- A resource (rectangle ☐) represents some physical or informational entity required for the actor.
- A softgoal (eight-shape ◯) is a goal whose satisfaction criteria is not clear-cut. How an intentional element contributes to the satisfaction or fulfillment

of a softgoal is determined via contribution links (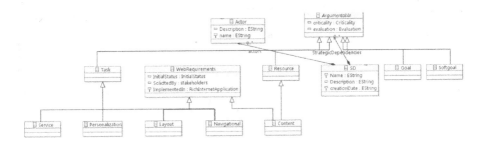). Possible labels for a contribution link are "make", "some+", "help", "hurt", "some-", "break", "unknown", indicating the (positive, negative or unknown) strength of the contribution.

Even though i^* provides good mechanisms to model actors and relationships between them, it needs to be adapted to the Web Engineering domain to reflect special Web requirements that are not taken into account in traditional requirement analysis approaches. As the A-OOH approach is UML-compliant, we have used the extension mechanisms of UML in order to adapt the i^* modeling framework to the taxonomy of Web requirements (Content, Service, Navigational, Layout, Personalization and Non-Functional Requirements) presented in [29]. To do so, (i) we defined a profile to formally represent the adaptation of each one of the i^* elements with each requirement type from the Web requirements clasification adopted [5]; and (ii) we implemented this profile in an EMF (Eclipse Modeling Framework) metamodel adding new EMF clases according to the different kind of Web requirements: the Navigational, Service, Personalization and Layout requirements extends the Task element and the Content requirement extends the Resource class. It is worth noting that NFRs, until now, can be modeled by directly using the softgoal element. In Figure 1 can be seen an extract of the EMF metamodel for Web requirements specification using the i^* framework. The metamodel has been implemented in the Eclipse [32] IDE (Integrated Development Enviroment).

Fig. 1. An overview of the original i^* metamodel implemented in Eclipse (EMF)

The development process of this method is founded in the MDA (Model-Driven Architecture) [11]. MDA is an OMG's standard and consists of a three-tier architecture with which the requirements are specified at the Computational Independent Model (CIM), from there are derived the Web application conceptual models which corresponds with the Platform Independent Model (PIM) of the MDA. Finally, the Web application conceptual models are used to generate the implementation code; this stage corresponds with the Platform Specific Model (PSM) from the MDA standard. A crucial part of MDA is the concept of transformation between models either model-to-model (M2M) or model-to-text (M2T). With the M2M transformations is possible the transformation from a model in other one. To use the advantages of MDA, our proposal supports the automatic derivation of Web conceptual models from a requirements model by means of a set of M2M transformation rules defined in [6,12].

4   The WebREd-Tool

The WebREd-Tool[2] is a set of Eclipse [32] plugins that have been developed to assist the designer in the early phases of a Web application development process. With the WebREd-Tool, the designer can specify the Web application requirements by using the *i** modelling framework. The WebREd-Tool assists the designer comparing different configurations of functional requirements, while balancing and optimizing non-functional requirements based on the Pareto efficiency [33]. The WebREd-Tool is based on the Model-Driven Development (MDD) paradigm applied in the context of the Web Engineering, this specialization of the MDD is called Model-Driven Web Engineering (MDWE) [11].

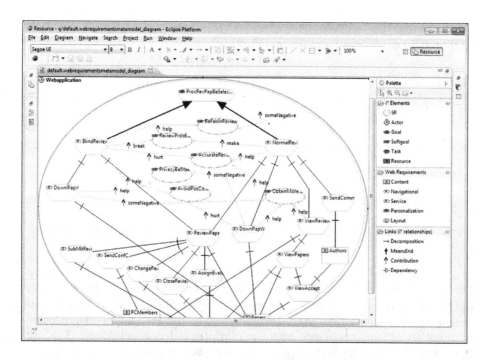

Fig. 2. WebREd-Tool implemented in Eclipse

[2] http://code.google.com/p/webred/

The tool development consists of three main parts. The first one consists on the implementation of the adapted i^* modeling framework for the Web domain. This adaptation was made by defining a EMF (Eclipse Modeling Framework) [34] metamodel and creating a specific class for each type of component of the i^* framework (see Figure 1). In the second part, the metamodel was used within the GMF (Graphical Modeling Framework) [35] to create a graphical editor (see Figure 2). With the graphical editor, the designer can specify the Web application requirements in a graphical environment using the i^* components such as goals, tasks, softgoals, decomposition, means-end and contribution links and the Web requirements types including service, navigational, content, personalization and layout. The tool-box is shown on the right side of Figure 2, including the aforementioned modeling elements. The third part is the implementation of the Pareto algorithm and, based on it, the visualization of requirement configurations.

The WebREd-Tool provides user support on issues such as Change Impact Analysis [13] and [14], Softgoal optimization [17] and [16], as well as requirements traceability [15] in a Model-Driven Web Engineering context [6]. Further explanation is available at [11] and [17].

Although this work was perceived in the context of the A-OOH method [12], it is in fact a stand-alone, independent approach that can thus be used in any Web Engineering method. Finally, this proposal supports an automatic derivation of Web conceptual models from a requirements model by means of a set of transformation rules, the derivation of the Web application source code is still in development.

5   Softgoal Specialization

It is worth noting that the development of Web applications have some particular requirements that differs from the traditional requirements, especially when it comes to Non-Functional Requirements. These type of requirements are defined and clasified in the seminal work of [36], based on the literature review performed [9], in this work, we propose the definition of six types of NFRs. An overview of each kind of NFR for Web Engineering is described below:

- Usability. Refers to the user's ability to use the Web application without requiring any special training.
- Performance. It is used to describe the best use of the resources, it is related to performance.
- Reliability. Its the capability to maintain the performance of the Web application over the time without losing throwput.
- Safety. It is used in order to ensure that the Web application will do only what it is meant to do.
- Security. Refers to protect all the information managed in the Web applications, including the session management and the user authentication.
- Efficiency. Represents the optimal use of resources, for example the server requests.

This classification of NFRs for Web Engineering is used in order to extend the
i^* framework. Specifically, this classification will be used to enrich the expressiveness of the *Softgoal* element from the i^* modeling framework. A softgoal
is an objective without clear-cut criteria [28] and can represent Non-Functional
Requirements and relations between Non-Functional Requirements in a goal-
oriented modeling context. To this aim, it was necessary to modify the original
Web requirements metamodel (see an extract of the metamodel in Fig. 1) in
order to extend the definition of the *softgoal* element in a similar form as was
done in our previous work [12] to adapt the FRs classification presented in [29]
(see Fig. 3).

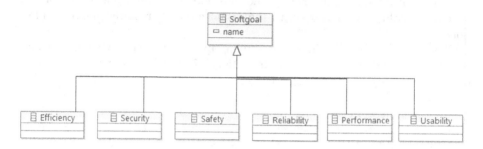

Fig. 3. An overview of the i^* metamodel modified with the NFRs taxonomy

Once the i^* framework was entended with the softgoal specialization, the
next step consisted in making a re-engineering process in order to build a new
GMF editor, thus integrating new elements to the tool-bar to be able to use the
abstract syntax (metamodel), ie the representation of each one of the elements
to represent the Non-Functional Requirements.

6   Conclusions and Future Work

In this work, we have presented an extension to our goal-oriented RE approach
for the development of Web 1.0 applications named WebREd-Tool. Seeing that
a Web application architecture is composed of a collection of design decisions,
each design decision can help or hinder certain NFRs. Thus, current tools and
methods are focus on expressing components and connectors in the Web application, therefore, design decisions and their relationships with Non-Functional
Requirements are often captured in separate design documentation. This disas-
sociation makes architecture comprehension and architecture evolution harder.
In this work, our proposal offers several advantages such as including the specification of Non-Functional Requirements from the requirements analysis stage
considering the design decisions from the initial stages of the Web application
development process. Since it is supported under a MDA-based process, it reflects the requirements captured in the final product. Future work consists in:

(i) the development of a set of model-to-model transformations test, (iii) the reengineering process in order to verify all the original functions of the WebREd-Tool (requirements traceability, change impact analysis and softgoal optimization) and (iii) the integration of our goal-oriented approach in a full-MDD solution for the development of Rich Internet Applications (RIA's) within the OOH4RIA approach [37].

Acknowledgments. This work has been partially supported by: Facultad de Informática Mazatlán from Universidad Autónoma de Sinaloa (México), PIFI (Programa de Fortalecimiento Institucional) P2012-25-126, PROFAPI 2012/002 and PROFAPI2013 with the project: "Aplicación de la Ingeniería de Requisitos Orientada a Objetivos en la Ingeniería Web Dirigida por Modelos" from Universidad Autónoma de Sinaloa (México).

References

1. Fons, J., Valderas, P., Ruiz, M., Rojas, G., Pastor, O.: Oows: A method to develop web applications from web-oriented conceptual models. In: International Workshop on Web Oriented Software Technology (IWWOST), pp. 65–70 (2003)
2. Ceri, S., Fraternali, P., Bongio, A.: Web modeling language (webml): A modeling language for designing websites. The International Journal of Computer and Telecommunications Networking 33(1-6), 137–157 (2000)
3. Escalona, M.J., Aragón, G.: Ndt. a model-driven approach for web requirements. IEEE Transactions on Software Engineering 34(3), 377–390 (2008)
4. Koch, N.: The expressive power of uml-based web engineering. In: International Workshop on Web-oriented Software Technology (IWWOST), pp. 40–41 (2002)
5. Garrigós, I., Mazón, J.-N., Trujillo, J.: A requirement analysis approach for using i* in web engineering. In: Gaedke, M., Grossniklaus, M., Díaz, O. (eds.) ICWE 2009. LNCS, vol. 5648, pp. 151–165. Springer, Heidelberg (2009)
6. Aguilar, J.A.: A Goal-oriented Approach for Managing Requirements in the Development of Web Applications. PhD thesis, University of Alicante, Spain (2011)
7. Nuseibeh, B., Easterbrook, S.M.: Requirements engineering: a roadmap. In: International Conference on Software Engineering (ICSE), pp. 35–46 (2000)
8. Gupta, C., Singh, Y., Chauhan, D.: Dependency based Process Model for Impact Analysis: A Requirement Engineering Perspective. International Journal of Computer Applications 6(6), 28–30 (2010)
9. Jureta, I.J., Faulkner, S., Schobbens, P.-Y.: A more expressive softgoal conceptualization for quality requirements analysis. In: Embley, D.W., Olivé, A., Ram, S. (eds.) ER 2006. LNCS, vol. 4215, pp. 281–295. Springer, Heidelberg (2006)
10. Ameller, D., Gutiérrez, F., Cabot, J.: Dealing with Non-functional Requirements in Model-Driven Development. In: 18th IEEE International Requirements Engineering Conference, RE (2010)
11. Aguilar, J.A.: A Goal-oriented Approach for the Development of Web Applications: Goal-oriented Requirements Engineering (GORE) and Model-Driven Architecture (MDA) in the Development of Web Applications. LAP LAMBERT. Academic Publishing (2012)

12. Aguilar, J.A., Garrigós, I., Mazón, J.N., Trujillo, J.: An MDA Approach for Goal-oriented Requirement Analysis in Web Engineering. J. UCS 16(17), 2475–2494 (2010)
13. Aguilar, J.A., Garrigós, I., Mazón, J.-N.: Impact analysis of goal-oriented requirements in web engineering. In: Murgante, B., Gervasi, O., Iglesias, A., Taniar, D., Apduhan, B.O. (eds.) ICCSA 2011, Part V. LNCS, vol. 6786, pp. 421–436. Springer, Heidelberg (2011)
14. Aguilar, J.A., Garrigós, I., Mazón, J.-N., Zaldívar, A.: Dealing with dependencies among functional and non-functional requirements for impact analysis in web engineering. In: Murgante, B., Gervasi, O., Misra, S., Nedjah, N., Rocha, A.M.A.C., Taniar, D., Apduhan, B.O. (eds.) ICCSA 2012, Part IV. LNCS, vol. 7336, pp. 116–131. Springer, Heidelberg (2012)
15. Aguilar, J.A., Garrigós, I., Mazón, J.N.: Modelos de weaving para trazabilidad de requisitos web en a-ooh. In: VII Taller sobre Desarrollo de Software Dirigido por Modelos (DSDM), JISBD, Congreso Español de Informatica (CEDI), Valencia, España, SISTEDES, pp. 146–155 (2010)
16. Aguilar, J.A., Garrigós, I., Mazón, J.-N.: A goal-oriented approach for optimizing non-functional requirements in web applications. In: De Troyer, O., Bauzer Medeiros, C., Billen, R., Hallot, P., Simitsis, A., Van Mingroot, H. (eds.) ER Workshops 2011. LNCS, vol. 6999, pp. 14–23. Springer, Heidelberg (2011)
17. Aguilar Calderon, J.A., Garrigós, I., Casteleyn, S., Mazón, J.-N.: WebREd: A model-driven tool for web requirements specification and optimization. In: Brambilla, M., Tokuda, T., Tolksdorf, R. (eds.) ICWE 2012. LNCS, vol. 7387, pp. 452–455. Springer, Heidelberg (2012)
18. Yu, E.: Modelling Strategic Relationships for Process Reenginering. PhD thesis, University of Toronto, Canada (1995)
19. Aguilar, J.A., Garrigós, I., Mazón, J.N., Trujillo, J.: Web Engineering approaches for requirement analysis- A Systematic Literature Review. In: 6th Web Information Systems and Technologies (WEBIST), vol. 2, pp. 187–190. SciTePress Digital Library, Valencia (2010)
20. Schwabe, D., Rossi, G.: The object-oriented hypermedia design model. Communications of the ACM 38(8), 45–46 (1995)
21. De Troyer, O.M.F., Leune, C.J.: Wsdm: A user centered design method for web sites. Comput. Netw. ISDN Syst. 30(1-7), 85–94 (1998)
22. Casteleyn, S., Woensel, W.V., Houben, G.J.: A semantics-based aspect-oriented approach to adaptation in web engineering. In: Hypertext, pp. 189–198 (2007)
23. Molina, F., Toval, A.: Integrating usability requirements that can be evaluated in design time into model driven engineering of web information systems. Adv. Eng. Softw. 40, 1306–1317 (2009)
24. Mylopoulos, J., Chung, L., Nixon, B.: Representing and using nonfunctional requirements: A process-oriented approach. IEEE Trans. Softw. Eng. 18(6), 483–497 (1992)
25. Dardenne, A., van Lamsweerde, A., Fickas, S.: Goal-directed requirements acquisition. In: Selected Papers of the Sixth International Workshop on Software Specification and Design, 6IWSSD, pp. 3–50. Elsevier Science Publishers B.V., Amsterdam (1993)
26. Bolchini, D., Paolini, P.: Goal-driven requirements analysis for hypermedia-intensive web applications, vol. 9, pp. 85–103. Springer (2004)
27. Molina, F., Pardillo, J., Toval, A.: Modelling web-based systems requirements using WRM. In: Hartmann, S., Zhou, X., Kirchberg, M. (eds.) WISE 2008. LNCS, vol. 5176, pp. 122–131. Springer, Heidelberg (2008)

28. Yu, E.: Towards modeling and reasoning support for early-phase requirements engineering. In: RE, pp. 226–235 (1997)
29. Escalona, M.J., Koch, N.: Requirements engineering for web applications - a comparative study. J. Web Eng. 2(3), 193–212 (2004)
30. Garrigós, I.: A-OOH: Extending Web Application Design with Dynamic Personalization. PhD thesis, University of Alicante, Spain (2008)
31. Cachero, C., Gómez, J.: Advanced conceptual modeling of web applications: Embedding operation. In: 21th International Conference on Conceptual Modeling Interfaces in Navigation Design (2002)
32. Eclipse (2012), http://www.eclipse.org/
33. Szidarovszky, F., Gershon, M., Duckstein, L.: Techniques for Multiobjective Decision Making in Systems Management. Elsevier (1986)
34. EMF, http://www.eclipse.org/emf/
35. GMF, http://www.eclipse.org/gmf/
36. Zhang, W., Yang, Y., Wang, Q., Shu, F.: An empirical study on classification of non-functional requirements. In: Proceedings of the 23rd International Conference on Software Engineering & Knowledge Engineering (SEKE 2011). Knowledge Systems Institute Graduate School, Miami Beach (2011)
37. Meliá, S., Martínez, J.J., Mira, S., Osuna, J., Gómez, J.: An eclipse plug-in for model-driven development of rich internet applications. In: Benatallah, B., Casati, F., Kappel, G., Rossi, G. (eds.) ICWE 2010. LNCS, vol. 6189, pp. 514–517. Springer, Heidelberg (2010)

Application of an Extended SysML Requirements Diagram to Model Real-Time Control Systems

Fabíola Goncalves C. Ribeiro[1], Sanjay Misra[2], and Michel S. Soares[1]

[1] Federal University of Uberlândia, Uberlândia, Brazil
[2] Covenant University, Ota Nigeria
{fgcr.ufg,ssopam,mics.soares}@gmail.com

Abstract. Most techniques for modeling requirements present many problems and limitations, including modeling requirements at a single level of abstraction, and are specific to model functional requirements. The objective of this article is to perform a study on modeling requirements of Real-Time Systems through an extension of the SysML Requirements Diagram focusing on the traceability of non-functional and functional requirements. The proposed approach has demonstrated to be effective for representing software requirements of real-time systems at multiple levels of abstraction and classification. The proposed metamodel represents concisely the traceability of requirements in a high abstraction level.

Keywords: SysML, Requirements Engineering, Modeling of Software, Traceability of Requirements.

1 Introduction

Requirements Engineering is often characterized in the literature as the most critical and complex process of the software development life cycle [1]. The process of requirements engineering has dominant impact on the capabilities of the resulting product [2]. According to Brooks [3], knowing what to build, which includes requirements elicitation and technical specification, is the most difficult phase in the design of software.

The increasing complexity of software systems makes the Requirements Engineering activities both more important and difficult. This assertion is easily verifiable when developing Real-Time Systems (RTS) which are highly dependent on restricted timing requirements. Real-time systems must respond to externally generated stimulus within finite and specifiable period. Therefore, efforts must be expent to analyze, design and validate systems for real-time targeting, providing greater reliability and security to them. In order to minimize the complexity of the development of RTS, graphical models can be applied. Modeling is an important activity in the development of RTS, because it contributes to decrease the complexity and to improve the understanding of these systems.

B. Murgante et al. (Eds.): ICCSA 2013, Part III, LNCS 7973, pp. 70–81, 2013.

UML has been frequently applied for modeling requirements in the RTS domain. For instance, in [4], UML is used to represent the modeling of design decisions and non-functional requirements by extending the language with attributes of stimulus, source of stimulus, environment, artifact, response and response from measures. These new attributes are incorporated into the class diagram for modeling non-functional requirements. In [5] a method for specifying software requirements using UML with the addition of stereotypes is presented.

The shortcomings of UML are widely described in the literature. The language is considered too informal and ambiguous. Behavior diagrams, such as the Sequence diagram, cannot represent time constraints effectively [6], as they are essentially untimed, expressing only chronological order. As discussed in [7], UML presents difficulty in expressing non-functional properties of the system, very important requirements for real-time applications.

Proposals to address the problems of UML in relation to modeling real-time software were created. These include the *profiles* SPT [8] and MARTE [9]. These profiles extend UML and add elements that model time requirements, system requirements and non-functional properties. SysML (Systems Modeling Language), a new language derived from UML, was proposed recently. SysML [10] allows the modeling of various types of applications in systems engineering, enabling the specification, analysis, design, verification and validation of complex systems. The language has been successfully applied to model requirements. The work presented in [11] refers to the application of the SysML Requirements diagram for the specification of system requirements. In another article [12], the focus is on user requirements.

The main aim of this article is to extend the metamodel of the SysML [10] Requirements diagram by adding new properties and relationships to enable the representation of requirements of real-time software at different levels of abstraction, and then demonstrating explicitly the tracing of each requirement in each level of representation.

This article is organized as follows. The proposed metamodel for representing requirements at multiple levels of abstraction and classification, and to describe tracing between requirements is presented in Section 2. Sections 3.1 and 3.2 are about the set of requirements to be modeled in the case study. Finally, the discussions are presented in Section 4 and the contributions and conclusions of this research are described in Section 5.

2   Metamodel for Requirements Modeling and Tracing

SysML is a highly customizable and extensible modeling language [10]. It allows the creation of domain-specific models through stereotypes and other extensions. Profiles may specialize language semantics, provide new graphical icons and domain-specific model libraries. When creating profiles, it is not allowed to change language semantics. The original SysML Requirements model is shown in Figure 1. The SysML Requirements diagram is a stereotype of the UML Class diagram extended with new attributes. The SysML Requirements diagram can

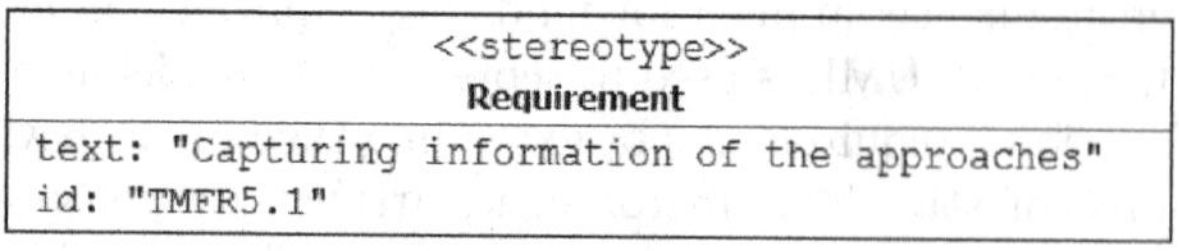

<<stereotype>>
Requirement
text: "Capturing information of the approaches" id: "TMFR5.1"

Fig. 1. Basic node for SysML requirements diagrams

be used to standardize the requirements documents with a specific pattern to be used.

The SysML Requirements diagram allows several ways to represent requirements relationships. These include relationships for defining requirements hierarchy, deriving requirements, satisfying requirements, verifying requirements and refining requirements. The relationships can improve the specification of systems, as they can be used to model requirements. Additional relationships were proposed in this article and are briefly explained as follows.

SysML allows splitting complex requirements into more simple ones, as a *hierarchy* of requirements related to each other. The advantage is that the complexity of systems is treated from the early beginning of development, by decomposing complex requirements. For instance, high-level business requirements may be gradually decomposed into more detailed software requirements, forming a hierarchy.

The derive relationship relates a derived requirement to its source requirement. New requirements are often created from previous ones during the requirements engineering activities. The derived requirement is under a source requirement in the hierarchy. In a Requirements diagram, the derive relationship is represented by the keyword *deriveReqt*.

The created attributes for the extended requirements take into account many of the specifications contained in the IEEE 830-1998 standard for describing software requirements [13]. An extended requirement (represented by the stereotype $<< ExtRequirement >>$) is proposed in this article, including additional attributes. In addition, derived from this extended requirement, an extended requirement for non-functional requirements is proposed (represented by $<< ExtRequirementNRF >>$) with additional attributes. Three types of non-functional requirements were proposed in the metamodel, as can be seen in Figure 2.

The new defined attributes for ExtRequirement are: priority, abstractLevel, constraint, scenario, creationDate, modificationDate, and versionNumber.

The priority attribute defines the relevance of a requirement in relation to the other, i.e., indicating the order that the requirements should be addressed. Values are of type String, including for instance, priority of type "must", "should", "could", and "won't". The requirement level indicates classification level in the hierarchy. The constraint attribute enables to show requirements that have some type of restriction. This attribute is of type Boolean. In case it is set to *true*, the identifier (ID) and the detailed description of this restriction are contained in a

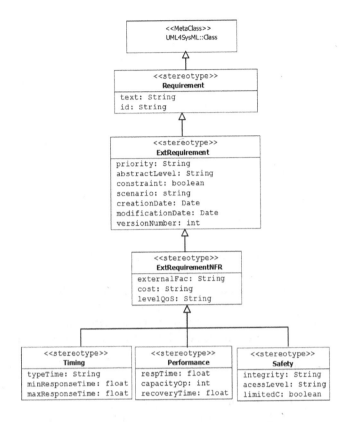

Fig. 2. Metamodel for SysML Requirements Diagram

table of restrictions. Scenario is an attribute of type String which basically identifies the scenario to which the requirement is related. The attributes creationDate and modificationDate indicate the requirements version, which is useful to keep track of multiple versions of the requirement. The attribute versionNumber allows to determine the version of creation/elaboration of a requirement.

The stereotype ExtRequirementNFR is used to describe non-functional requirements of software. The proposed attributes are externalFac, cost, and levelQoS. ExternalFac determines whether a requirement is dependent on an external factor in order to be developed. The cost attribute allows to establish criteria of costs to satisfy a requirement that infuences directly in decisions about the viability of its development. Possible values to be assigned include High, Medium, or Low. The levelQos demonstrates the level of quality required for the requirement.

The timing type of non-functional requirement relates to the description of time of a software. Its attributes are typeTime, which can assume values physical time or logical time, minResponseTime and maxResponseTime, which are used to describe timing constraints of a requirement. The performance type has three attributes. RespTime indicates the maximum response time associated

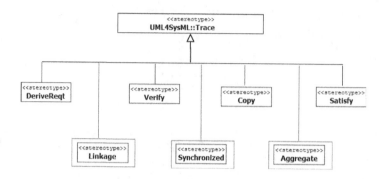

Fig. 3. Metamodel of Extended Relationship

with a requirement. Its value allows to establish which level of performance is associated or is to be guaranteed by the requirement. The capacityOp attribute indicates the possible number of simultaneous operations that are allowed in a given time period (e.g., number of reports generated for storage, operations per second, and so on). The attribute recoveryTime describes the maximum time required for recovery from a failure. The safety type of non-functional requirement has attributes integrity (level of integrity that must be guaranteed), acessLevel (establish the level of access of stakeholders to a function), and limitedC (enables to demonstrate whether communication should be limited between this requirement and other functions/modules of the system.

The extended model is able to represent requirements at different levels of abstraction, correlated requirements at the same level and, also, synchronism between requirements as shown in Figure 3. The << *linkage* >> stereotype represented by an arrow and a circle with an internal cross on both ends has the purpose of improving the activity of tracing requirements to the design models. Its graphical representation can be visualized in Figure 6.

The << *aggregate* >> stereotype (Figure 5), was elaborated in order to explicitly demonstrate correlated requirements (in the same classification level) with a requirement/function one level up, i.e., requirements that are represented with this stereotype will together provide functionality expected by a higher requirement and are strongly bonded together.

Finally, the new stereotype << *synchronized* >> should be used for non-functional requirements that in addition to performance constraints of non-functional requirements should be processed in parallel with other requirements. Its graphical representation is shown in Figure 4.

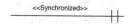

Fig. 4. The << *Synchronized* >> relationship

Fig. 5. Relationships $<< aggregate >>$

Fig. 6. The linkage relationship

3   Case Study

In this section, a subset of a list of requirements for a Road Traffic Management System (RTMS) is presented, using natural language to be further modeled and analyzed. The list of requirements given below is a subset from a document which contains 137 atomic requirements for a RTMS. The requirements were gathered through an extensive search in the literature of RTMS [14] [15].

3.1   Requirements for the Case Study

The elaborated requirements document depicts requirements for a Control System for a Road Traffic Intersection where the requirements are related to the time control at the intersection, flow control of vehicles, configuration and control of adaptive controllers, and receiving, storing and processing various data from each road approaches which interconnect the intersection, and also, controlling the green time of the signal. The configuration of the document is described as follows:

- Fourteen general purpose requirements as shown in Table 1.
- Sixty one functional requirements. Given the space of this article, only non-functional requirements modeled in this study were represented (Table 2).
- Sixty two non-functional requirements. Some of the Non-functional requirements used in the case study can be seen in Table 3.

The requirements in each one of the different levels were presented in natural language. However, the focus of this work is the use of graphical models for improved representation of system requirements of the RTMS. The SysML constructions for modeling requirements are explained in detail in the following section.

3.2   Modeling the Case Study

It can be seen from Figure 7 that requirements TM5, TM6, TM8, TM9 and TM12 are related to the requirement TM1 through the *derive* relationship using $<< deriveReqt >>$ and the *hierarchical* relationship. The *derive* relationship is justified by the fact that the high-level requirements mentioned above are all

Table 1. High Level Requirements for a Road Traffic Management System

ID	Requirement Name
TM1	The system must control the standard of vehicular traffic at the intersection.
TM2	The system must allow synchronization of semaphores.
TM3	The system must collect all kinds of information of the road approaches in order to properly evaluate these data.
TM4	The system must allow management of traffic history.
TM5	The system must possess the adaptive control of schedules to the intersection in response traffic flow.
TM6	The system must actuate in response to intersection traffic flow.
TM7	The system must have the emergency preemption mode, i.e., preferential movement of emergency vehicles.
TM8	The system must allow the control of the intersection in response to manual commands.
TM9	The system must allow control of intersection in response to replacement of remote commands.
TM10	The system must control incident management.
TM11	The system of the intersection should be able to interact with the software control panel.
TM12	The system must allow the automatic operation of semaphores.
TM13	The system must use TCP/IP SNMP interface for inter-communication system.
TM14	The system could generate statistical data to assist decision making.

Table 2. Functional Requirements for a Road Traffic Management System

ID	Requirement Name
TMFR5.1	The system must capture information of the approaches (detect volume)
TMFR5.2	The system must store information sent from the vehicle detection loop
TMFR5.3	The system must process traffic information
TMFR5.4	The system must trigger new state in sufficient time to the reprogramming of intersection controllers
TMFR5.5	The system must maintain statistics of counting vehicles

derived from the requirement that controls the traffic pattern of vehicles at the intersection.

The *hierarchical* relationship (represented by a circle with a cross inside) represents one stronger relationship between a more general requirement and the more specific requirements.

Figure 8 demonstrates the use of the new created relationship (the << *aggregate* >> relationship and its stereotype are shown in Figure 5). The relationship << *aggregate* >> shows when two or more requirements at the same level are linked to a more general requirement. In Figure 8 the associations between a requirement at the same level (in the same "branch" of the tree of modeling) with more than one requirement was suppressed due to readability.

The application of the metamodel to the case study is described in Figure 9. It clearly demonstrates through the new stereotype << *linkage* >> which requirement at any level of abstraction (at a high level as TM5, or TMRF5.3) is interlinked

Table 3. Non-Functional Requirements for a Road Traffic Management System

ID	Requirement Name
TMNFR5.1	The information of approaches (detection volume) must be captured with a maximum of 100ms.
TMNFR5.2	The storage of traffic information, sent by loop detection vehicle, must be done with maximum of 1000ms.
TMNFR5.3	The traffic information must be processed with a minimum of 150ms. and a maximum of 200ms.
TMNFR5.4	The system must trigger new state in sufficient time to program the controllers with a maximum of 100ms.
TMNFR5.5	The new state of the actuators should be modified in synchronization with the other controllers on the network.
TMNFR5.6	The controllers must receive their state safely.
TMNFR5.7	The detection volume must be performed with maximum performance and reliability.

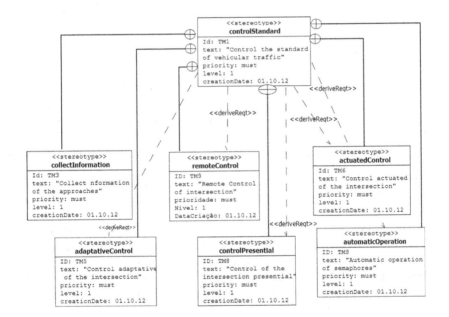

Fig. 7. High Level Requirements and their relationships in SysML

with another one at Figure 10 where the hierarchy is clearly expressed among requirements. Figure 6 shows the relationship $<<linkage>>$. Thus, the tree that demonstrates traceability of a requirement can be drawn from the $<<linkage>>$ relationship. Any requirement that has in its two ends the $<<linkage>>$ relationship indicates a point of connection between a more abstract requirement with another less abstract one. The requirements where parallel strategies should be applied are expressed by the $<<synchronized>>$ relationship.

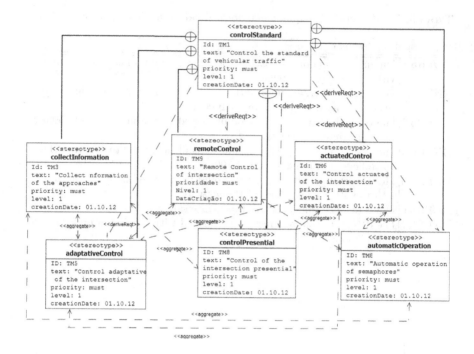

Fig. 8. High Level Requirements and their relationships with << *aggregate* >>

The linkage relationship, differently from the original SysML relationship of trace, provides well-defined semantics, as it allows toclearly observe the hierarchy levels or trace of a set of requirements. The tracing is demonstrated by hyperlinks of a basic requirement, whether functional or non-functional, with others in higher or lower levels. It is also possible to observe the classification and organization of such requirements. The obvious advantage of these new relationships and stereotypes is that complexity of the software design can be treated from the beginning of development by classification (in which the functional requirements, for example, are easily identified and can be manipulated directly) through the detailing of the various attributes of requirements relevant to the design of a real-time system.

4   Discussion

The extension of the SysML Requirements diagram to modeling critical requirements of RTS, the approach proposed in this article, is convenient to accomplish the many characteristics relevant to a document of software requirements (as suggested in [13]).

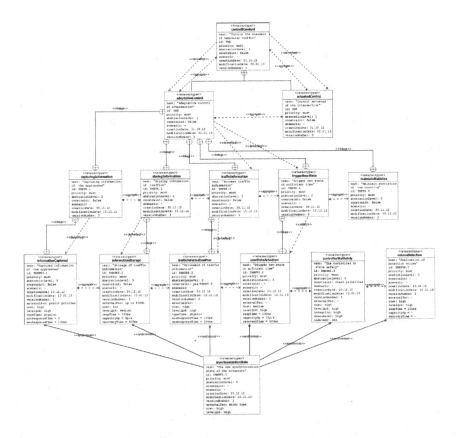

Fig. 9. Application of the metamodel

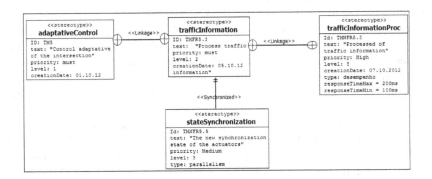

Fig. 10. Traceability between requirements of different levels

The completeness was demonstrated as all the general requirements, functional and non-functional ones may be related concisely to be further analyzed. The conflict between any requirements that are at the same level of the hierarchy or between high-level requirements with their less abstract requirements can be clearly visualized through the new relationships and the proposed modeling. The use of the new stereotype $<< aggregate >>$ also lists requirements on the same level, and with that, conflicts in the same set of requirements are easily discovered.

The ranked by importance, proposed as an attribute, is useful to define the priority attribute between requirements. This can be useful in order to clarify which requirements must be prioritized in phases of validation, testing or development. As each requirement is described separately the complexity of these changes is minimized, since a change in any requirement can be made completely and consistently maintaining structure and style of requirements. Expressing each requirement separately is highly desirable. This characteristic is addressed in this article by modeling requirements using a well-defined SysML Requirements diagram, and by organizing the relationship between requirements.

5   Conclusion

The SysML is an interesting and reasonably modeling language for designing requirements of RTS. However, the current state of the language is not complete enough to satisfy many of the needs of representation, description and manipulation of requirements for RTS. The language provides the representation of a limited number of concepts related to these systems as, for instance, elements allowing the representation of time to build computing models with interpretations of time. The proposal in this article is to create extensions to the SysML Requirements diagram and adding new relationships which can express the classification of requirements for modeling RTS at different levels of abstraction. The extensions to the basic metamodel of SysML aims for better identification of software requirements at different abstraction levels, enabling the mapping of requirements relationships among themselves. In addition, the relationships between requirements are shown together in order to make it easier to trace these requirements in the many different levels of abstraction and thus facilitating the verification and the testing of a set of requirements in any phase of the life cycle of RTS. The proposed metamodel makes it possible to trace requirements, thus enabling the identification of their origin, detailing why and when they were included in the requirements document. This allows better care for many types of conditions required for the development of RTS and also improves the support of evaluation of the impact of requested changes. As a result, an improvement in the quality of software development is expected, as well as an improvement of the development process as a whole.

References

1. Minor, O., Armarego, J.: Requirements Engineering: A Close Look at Industry Needs and Model Curricula. Australian Journal of Information Systems 13(1), 192–208 (2005)
2. Parviainen, P., Tihinen, M., van Solingen, R.: Requirements Engineering: Dealing with the Complexity of Sociotechnical Systems Development. Idea Group Inc. (2005)
3. Brooks, F.P.: No Silver Bullet: Essence and Accidents of Software Engineering. Computer 20(4), 10–19 (1987)
4. Martin, G.: UML for Embedded Systems Specification and Design: Motivation and Overview. In: Design, Automation and Test in Europe Conference and Exhibition, pp. 773–775 (2002)
5. Cote, I., Heisel, M.: A UML Profile and Tool Support for Evolutionary Requirements Engineering. In: 15th Software Maintenance and Reengineering, pp. 161–179 (2011)
6. Soares, M.S., Julia, S., Vrancken, J.: Real-time Scheduling of Batch Systems using Petri Nets and Linear Logic. Journal of Systems and Software 81(11), 1983–1996 (2008)
7. Silvestre, E.A., Soares, M.S.: Multiple View Architecture Model for Distributed Real-Time Systems Using MARTE. In: 20th International Conference on Information Systems Development, pp. 98–113 (2011)
8. OMG: UML Profile for Schedulability, Performance, and Time, Version 1.1. Technical report, OMG (2005)
9. OMG: UML Profile for MARTE: Modeling and Analysis of Real-time Embedded Systems Version, 1.1. Technical report, OMG (2011)
10. OMG, S.: Systems Modeling Language (SysML) Specification - version 1.1. (2010)
11. Soares, M.S., Vrancken, J.: Requirements Specification and Modeling through SysML. In: International Conference on Systems, Man and Cybernetics, pp. 1735–1740 (2007)
12. Soares, M.S., Vrancken, J.: Model-Driven User Requirements Specification Using SysML. Journal of Software 3, 57–69 (2008)
13. IEEE: IEEE Recommended Practice for Software Requirements Specifications (1998)
14. Laplante, P.A.: Real-Time Systems Design and Analysis, 3rd edn. John Wiley & Sons, Piscataway (2004)
15. Klein, L.A.: Traffic Detector Handbook. 3 edn. Prentice Hall, USA (Department of Transportation - Federal Highway Administration)

Frequent Statement and De-reference Elimination for Distributed Programs

Mohamed A. El-Zawawy[1,2]

[1] College of Computer and Information Sciences,
Al Imam Mohammad Ibn Saud Islamic University (IMSIU)
Riyadh, Kingdom of Saudi Arabia
[2] Department of Mathematics, Faculty of Science, Cairo University
Giza 12613, Egypt
maelzawawy@cu.edu.eg

Abstract. This paper introduces a new approach for the analysis of *frequent statement and de-reference elimination* for distributed programs run on parallel machines equipped with hierarchical memories. The address space of the language studied in the paper is globally partitioned. This language allows programmers to define data layout and threads which can write to and read from other thread memories.

Simply structured type systems are the tools of the techniques presented in this paper which presents three type systems. The first type system defines for program points of a given distributed program sets of calculated (*ready*) statements and memory accesses. The second type system uses an enriched version of types of the first type system and determines which of the specified statements and memory accesses are used later in the program. The third type system uses the information gather so far to eliminate unnecessary statement computations and memory accesses (the analysis of *frequent statement and de-reference elimination*).

Two advantages of our work over related work are the following. The hierarchical style of concurrent parallel computers is similar to the memory model used in this paper. In our approach, each analysis result is assigned a type derivation (serves as a correctness proof).

Keywords: *Ready statement and memory access* analysis, *semi-expectation* analysis, *frequent statement and de-reference elimination* analysis, certified code, distributed programs, semantics of programming languages, operational semantics, type systems.

1 Introduction

Distributed programming is about building a software that has concurrent processes co-operating in achieving some task. For a problem specification, the type, number, and the way of interaction of processes needed to solve the problem is decided beforehand. Then a supercomputer can be computationally simulated by a group of workstations to carry different processes. A group of supercomputers can in turn be combined to

B. Murgante et al. (Eds.): ICCSA 2013, Part III, LNCS 7973, pp. 82–97, 2013.

provide a computing power greater than that provided by any single machine. This enormous computing power provided by distributed systems is why the distributed-programming style [1,20,30] is quite important and attractive. Among examples of distributed programming languages (DPLs), based on machines having multi-core processors and using partitioned-global model, are Titanium which is based on Java, Unified Parallel C (UPC), Chapel, and X10.

Recomputing a non-trivial statement and re-accessing a memory location are waste of time and power if the value of the statement and the content of the location have not been changed. The purpose of *frequent statement and de-reference elimination* analysis is to save such wasted power and time. This is an interesting analysis because it involves connecting statement and de-reference calculations to program points where these calculation values may be reused. The analysis also requires doing changes to the program points at the ends of these connections. Such changes to program points have to be done carefully so that they do not destroy the compositionality. Our approach to treat this analysis is a type system [6,4,15,5,11] built on a combination of two analyses; one of them builds on the results of the other one.

For different programming languages, in previous work [14,3,6,4,15,5,8,9,7,16,10], we have proved that the type-systems style is certainly an adaptable approach for achieving many static analyses. This paper proves that this style is flexibly useful to the involved and important problem of frequent statement and de-reference elimination of distributed programs.

This paper introduces a new technique for frequent statement and de-reference elimination for distributed programs run on hierarchical memories. Simply structured type systems are the main tools of our technique. The proposed technique is presented using a language (the appendix- Figure 3) equipped with basic commands for distributed execution of programs and for pointer manipulations. The single program multiply data (*SPMD*) model is the execution archetypal used in this paper. On different data of different machines this archetypal runs the same program. The analysis of *frequent statement and de-reference elimination* for distributed programs is achieved in three steps each of which is done using a type system. The first of these steps achieves *ready statement and memory access* analysis. The second step deals with *semi-expectation* analysis and builds on the type system of the first step. The third type systems takes care of the analysis of *frequent statement and de-reference elimination* and is built on the type system of the second step.

Motivation

The left-hand-side of Figure 1 presents a motivating example of our work. We note that lines 4 and 6 de-reference $a * b$ which has already been de-referenced in line 2 with no changes to values of a and b in the path from 2 to 6. This is a waste of computational power and time (accessing a secondary storage). One objective of the research in this paper is avoid such waste by transforming the program into that in the right-hand-side of the figure. This is not all; we need to do that in a way that provides a correctness proof for each such transformation. We adopt a style (type systems) that provides these proofs (type derivations).

```
0.                                            l := c*d; k := a*b;
1.  x := a*b+c*d;                             x := k+l;
2.  x := convert (*(a*b),2);                  x := convert (*k,2);
3.  y := transmit c*d from 3;                 y := transmit l from 3;
4.  if (*(a*b) = *(c*d))                      if (*k = *l)
5.     then y := transmit *(c*d) from 2;         then y := transmit *(c*d) from 2;
6.     else x := convert (*(a*b),2);             else x := convert (*(a*b),2);
```

Fig. 1. A motivating example

Contributions

Contributions of this paper are new techniques, in the form of type systems, for:

1. The analysis of *ready statement and memory access* for distributed programs.
2. The analysis of *semi-expectation* for distributed programs.
3. The analysis of *frequent statement and de-reference elimination* of distributed programs.

Organization. Organization in the rest of the paper is as following. Section 2 presents the type system achieving the analysis of *ready statement and memory access* for distributed programs. The analysis of *semi-expectation* as an enrichment of the type system presented in Section 2 is outlined in Section 3. The main type system carrying the analysis of *frequent statement and de-reference elimination* is contained in Section 4. Related and future work are briefed in Section 5. An appendix to the paper briefly reviews our language syntax, memory model, and operational semantics which are presented in more details in [7].

2   Ready Statement and Memory Access Analysis

If the value of a statement and the content of a memory location have not been changed, then the compiler should not recompute the statement or re-access the location. The purpose of *Frequent Statement and De-reference Elimination* is to save the wasted power and time involved in these repeated computations. This is not a trivial task; compared to other program analyses, it is a bit complex. This task is done in stages. The first stage is to analyze the given program to recognize *ready* statements and memory locations.

The analysis of *ready* statements and memory locations calculates for every program point the set of statements and memory locations that are *ready* at that point in the sense of Definition 1. This section presents a type system (*ready* type system) to achieve this analysis for distributed programs.

Definition 1. *1. At a program point pt, a statement S is ready if each computational path to pt:*
   (a) contains an evaluation of S at some point (say pt') and
   (b) does not modify S (changing value of any of S's variables) between pt' and pt.

2. *At a program point pt, a memory location l is ready if each computational path to pt:*
   (a) *reads l at some point (say pt') and*
   (b) *does not modify content of l between pt' and pt.*

The *ready* analysis is a forward analysis that takes as an input a set of statements and memory locations (the *ready* set of the first program point). It is sensible to let this set be the empty set. The set of types of our *ready* type system has the form: *points-to-types* × $\mathcal{P}(Stmt^+ \cup gAddrs)$, where

1. $Stmt^+$ is the set of nontrivial statements (Figure 3 – the paper appendix),
2. *gAddrs* is the set of global addresses on our machine. This set is defined precisely in the appendix of this paper, and
3. *points-to-types* is a set of points-to types (typically have the form of maps from the union of variables and global addresses to the power set of global addresses [6,4,15,5]).

The subtyping relation has the form $\leq_p \times \supseteq$ where $\leq_p$ is the order relation on the points-to types and $\supseteq$ is the order relation on $\mathcal{P}(Stmt^+ \cup gAddrs)$. A state on an execution path is of type $rs \in \mathcal{P}(Stmt^+ \cup gAddrs)$ if all elements of *rs* are *ready* at this state according to Definition 1. Judgments of the *ready* type system have the form $S : (p, rs) \rightarrow_m (A', p', rs')$. The symbols p and p' denote the points-to types of the pre and post states of executing S. The set A' denotes the set of addresses that S may evaluate to. We assume that all such pointer information are given along with the statement S. Techniques like [6,4,15,5] are available to compute the pointer information. For a given statement along with pointer information and a *ready* pre-type *rs*, we present a type system to calculate a post *ready*-type rs' such that $S : (p, rs) \rightarrow_m (A', p', rs')$. The type derivation of this typing process is a proof for the correctness of the *ready* information. The meaning of the judgment is that if elements of *rs* are *ready* before executing S, then elements of rs' are *ready* after executing S.

The inference rules of the *ready* type system are as follows:

$$\frac{n : p \rightarrow_m (A', p')}{n : (p, rs) \rightarrow_m (A', p', rs)} \qquad \frac{S_1 : (p, rs) \rightarrow_m (A'', p'', rs'')\quad S_2 : (p'', rs'') \rightarrow_m (A', p', rs')}{S_1\ i_{op}\ S_2 : (p, rs) \rightarrow_m (\emptyset, p', rs' \cup \{S_1\ i_{op}\ S_2\})}\ (i_{op}^r) \qquad \frac{x : p \rightarrow_m (A', p')}{x : (p, rs) \rightarrow_m (A', p', rs)}$$

$$\frac{S_1 : (p, rs) \rightarrow_m (A'', p'', rs'')\quad S_2 : (p'', rs'') \rightarrow_m (A', p', rs')}{S_1\ b_{op}\ S_2 : (p, rs) \rightarrow_m (\emptyset, p', rs' \cup \{S_1\ b_{op}\ S_2\})}\ (b_{op}^r) \qquad \frac{*S : p \rightarrow_m (A', p')\quad S : (p, rs) \rightarrow_m (A'', p'', rs'')}{*S : (p, rs) \rightarrow_m \begin{cases} (A', p', rs'' \cup \{*S, g\}), & A'' = \{g\}; \\ (A', p', rs'' \cup \{*S\}), & |A''| \neq 1. \end{cases}}\ (*^r)$$

$$\frac{}{skip : (p, rs) \rightarrow_m (\emptyset, p, rs)} \qquad \frac{x := S : p \rightarrow_m (A', p')\quad S : (p, rs) \rightarrow_m (A'', p'', rs'')}{x := S : (p, rs) \rightarrow_m (A', p', rs'' \setminus \{S \in Stmt \mid x \in free(S)\})}\ (:=^r)$$

$$\frac{S_1 \leftarrow S_2 : p \rightarrow_m (A', p')\quad S_1 : (p, rs) \rightarrow_m (A_1, p_1, as_1)\quad S_2 : (p_1, as_1) \rightarrow_m (A_2, p_2, as_2)}{S_1 \leftarrow S_2 : (p, rs) \rightarrow_m (A', p', as_2 \setminus (A' \cup \{S \mid S : p \rightarrow_m (A_s, \_)\ \&\ A_s \cap A' \neq \emptyset\}))}\ (\leftarrow^r)$$

$$\frac{S_1 : (p, rs) \rightarrow_m (A'', p'', rs'')\quad S_2 : (p'', rs'') \rightarrow_m (A', p', rs')}{S_1; S_2 : (p, rs) \rightarrow_m (A', p', rs')}\ (seq^r) \qquad \frac{S : (p, rs) \rightarrow_m (A'', p'', rs'')\quad S_t : (p, rs'') \rightarrow_m (A', p', rs')\quad S_f : (p, rs'') \rightarrow_m (A', p', rs')}{if\ S\ then\ S_t\ else\ S_f : (p, rs) \rightarrow_m (A', p', rs')}\ (if^r)$$

$$\frac{S_1[S_2/x] : (p,rs) \to_m (A',p',rs')}{(\lambda x.S_1)S_2 : (p,rs) \to_m (A',p',rs')} \ (appl^r)$$

$$\frac{S : (p,rs) \to (A'',p'',rs') \quad S_t : (p'',rs') \to_m (A',p',rs)}{while \ S \ do \ S_t : (p,rs) \to_m (A',p',rs')} \ (whl^r)$$

$$\frac{S : (p,rs) \to_m (A',p',rs')}{\lambda x.S : (p,rs) \to_m (A',p',rs')} \ (abs^r)$$

$$\frac{fd(name) : (p,rs) \to_m (A',p',rs')}{name : (p,rs) \to_m (A',p',rs')} \ (name^r)$$

$$\frac{(\lambda x.S')S : (p,rs) \to_m (A',p',rs')}{letrec \ x = S \ in \ S' : (p,rs) \to_m (A',p',rs')} \ (letrec^r)$$

$$\frac{new_l : p \to_m (A',p')}{new_l : (p,rs) \to_m (A',p',rs)} \ (new^r)$$

$$\frac{convert \ (S,n) : p \to_m (A',p') \quad S : (p,rs) \to_m (A,p'',rs')}{convert \ (S,n) : (p,rs) \to_m (A',p',rs')} \ (convert^r)$$

$$\frac{transmit \ S_1 \ from \ S_2 : p \to (A',p') \quad S_2 : (p,rs) \to_m (A_2,p_2,as_2) \quad S_1 : (p_2,as_2) \to_m (A_1,p_1,rs')}{transmit \ S_1 \ from \ S_2 : (p,rs) \to_m (A',p',rs')} \ (trans^r)$$

$$\frac{(p'_1,rs'_1) \le (p_1,as_1) \quad S : (p_1,as_1) \to_m (p_2,as_2) \quad (p_2,as_2) \le (p'_2,rs'_2)}{S : (p'_1,rs'_1) \to_m (p'_2,rs'_2)} \ (csq^r)$$

$$\frac{Defs : \emptyset \curvearrowright fd \quad S : (p,rs) \to_m (A',p',rs')}{Defs : S : (p,rs) \to_m (A',p',rs')} \ (prg^r)$$

Comments on the inference rules are in order. We note that numbers, variables, and the allocating statement (*new*) do not affect the *ready* pre-type. In line with semantic rules (i^r_{op}) and (b^r_{op}) (the paper's appendix), nontrivial arithmetic and boolean statements and their nontrivial sub-statements are made *ready*. The direct assignment rule ($:=^r$) expresses that after executing the assignment the sub-statements of r.h.s. become *ready* and that all statements involving x become *unready* as the value of x may become different. The rule ($*^r$) reflects the fact that the statement $*S$ becomes *ready* after executing the de-reference. Moreover if S evaluates to a single address according to the underlying pointer analysis, then this address becomes *ready* as well. However if S evaluates to a large set of addresses (more than one), then we are not sure which of these addresses is the concerned one and hence can not conclude any readiness information about addresses. The rule ($\leftarrow^r$) adds the sub-statements of S_1 and S_2 to the *ready* pre-type. Since the content of address referenced by S_1 is possibly changed after executing the statement, all statements involving de-referencing this address are removed from the set of *ready* items. Remaining rules are self-explanatory. The Boolean statements *true* and *false* have inference rules similar to that of n.

All in all, the information provided by type derivations obtained using this and the following type systems is classified into two sorts. The first sort is about knowing the program point at which a particular statement becomes *ready*. The second sort of information is about the program point at a which a pre-computed value of a *ready* statement can be replaced with the statement.

Now we recall the assumption that our distributed system consists of $|M|$ machines. For a given statement S and a given machine m, the type system given above calculates for each program point of S, the set of *ready* items. The following rule can be used to combine the information calculated for each machine to get a new *ready* information for each program point. The new *ready* information is valid on any of the $|M|$ machines.

$$\frac{\forall m \in M. \ S : (\sup\{p,p_j \mid j \ne i\}, \sup\{rs,rs_j \mid j \ne i\}) \to_m (A_m,p_m,rs_m)}{S : (p,rs) \to_M (\cup_i A_i, \sup\{p_1,\dots,p_n\}, \sup\{rs_1,\dots,rs_n\})} \ (main\text{-}rs)$$

The rule above supposes we have a suitable notion for join of pointer types.

It is not hard to prove the soundness of the above type system:

Theorem 1. *Suppose that* $(S, \delta) \rightarrow (V, \delta')$, $S : (p, rs) \rightarrow (A', p', rs')$ *and the items of rs are ready at the point corresponding to* δ *on the execution path. Then the items of rs' are ready at the point corresponding to* δ' *on the execution path.*

3   Semi-expectation Analysis

The aim of frequent statement elimination is to introduce new variables to accommodate values of frequent statements and reusing these values rather than recomputing the statements. Analogously, the aim of frequent de-references elimination is to introduce new variables to accommodate values of frequent de-references and reusing these values rather than re-accessing the memory. The information gathered so far by the *ready* type system introduced in the previous section is not enough to achieve frequent statements and de-references elimination. We need to enrich the *ready* information, assigned to each program point, with a new information called *semi-expectable* information:

Definition 2.   *1. At a program point p, a statement S is semi-expectable if there is a computational path from p that:*
  (a) contains an evaluation of S at some point (say p'), where S is ready at p', and
  (b) does not evaluate S between p' and p.
  2. At a program point p, a memory location l is semi-expectable if each computational path to p:
  (a) reads l at some point (say p') where l is ready at p', and
  (b) does not read l between p' and p.

The *semi-expectation* analysis is a backward analysis that takes as an input a set of statements and memory locations (the *semi-expectable* set of the last program point). It is sensible to let this set be the empty set. The following example gives an intuition for the previous definition:

$$if \ (\ldots) \ then \ a := y + t \ else \ b := *r; \ c := (y + t) / *r.$$

Neither the statement $y + t$ nor the statement $*r$ is *ready* after the if statement because they are not computed in all branches. Hence it is not true to replace these statements with variables towards optimizing the last statement of the example. The job of the type system presented in this section is to provide us with this sort of information. More precisely, as the statements $y + t$ and $*r$ are not *ready* after the if statement, the second statement of the example does not make them *semi-expectable*.

The *semi-expectation* analysis assigns for each program point the set of items that are *semi-expectable*. The analysis is based on the *readiness* analysis and is backward. The set of types of the *semi-expectation* type system has the form: *points-to-types* $\times$ $\mathcal{P}(Stmt^+ \cup gAddrs) \times \mathcal{P}(Stmt^+ \cup gAddrs)$. The subtyping relation has the form $\leq_p \times \supseteq$ $\times \supseteq$. A state on an execution path is of type $se \in \mathcal{P}(Stmt^+ \cup gAddrs)$ if all elements of se are *semi-expectable* according to Definition 2. Judgments of the *semi-expectation* type system have the form $S : (p, rs, se) \rightarrow_m (A', p', rs', se')$. For a given statement along with pointer information, readiness information, and a *semi-expectation* type se', we present a type system to calculate a pre *semi-expectable*-type se such that $S : (p, rs, se) \rightarrow_m (A', p', rs', se')$. The type derivation of this typing process is a proof for the correctness of the *semi-expectable* information. The meaning of the judgment is that if elements of se' are *semi-expectable* after executing S, then elements of se must have been *semi-expectable* before executing S.

The inference rules of the *semi-expectation* type system are as follows:

$$\frac{n : p \rightarrow_m (A', p')}{n : (p, rs, se) \rightarrow_m (A', p', rs, se)} \qquad \frac{x : p \rightarrow_m (A', p')}{x : (p, rs, se) \rightarrow_m (A', p', rs, se)}$$

$$\frac{S_1 : (p, rs, se) \rightarrow_m (A'', p'', rs'', se'') \quad S_2 : (p'', rs'', se'') \rightarrow_m (A', p', rs', se')}{S_1 \ i_{op} \ S_2 : (p, rs, se \cup (rs \cap \{S_1 \ i_{op} \ S_2\})) \rightarrow_m (\emptyset, p', rs' \cup \{S_1 \ i_{op} \ S_2\}, se')} \ (i^e_{op})$$

$$\frac{S_1 : (p, rs, se) \rightarrow_m (A'', p'', rs'', se'') \quad S_2 : (p'', rs'', se'') \rightarrow_m (A', p', rs', se')}{S_1 \ b_{op} \ S_2 : (p, rs, se) \rightarrow_m (\emptyset, p', rs' \cup \{S_1 \ b_{op} \ S_2\}, se')} \ (b^e_{op})$$

$$\frac{*S : p \rightarrow_m (A', p') \quad S : (p, rs, se) \rightarrow_m (A'', p'', rs'', se'')}{*S : (p, rs, se \cup (re \cap \{*S, g\})) \rightarrow_m \begin{cases} (A', p', rs'' \cup \{*S, g\}, se''), \ A' = \{g\}; \\ (A', p', rs'' \cup \{*S\}, se''), \ |A'| \neq 1. \end{cases}} \ (*^e)$$

$$\frac{}{skip : (p, rs, se) \rightarrow_m (\emptyset, p, rs, se)} \qquad \frac{x := S : p \rightarrow_m (A', p') \quad S : (p, rs, se) \rightarrow_m (A'', p'', rs'', se'')}{x := S : (p, rs, se) \rightarrow_m (A', p', rs'' \setminus \{S \in Stmt \mid x \in free(S)\}, se'')} \ (:=^e)$$

$$\frac{S_1 \leftarrow S_2 : p \rightarrow_m (A', p') \quad S_1 : (p, rs, se) \rightarrow_m (A_1, p_1, as_1, se_1) \quad S_2 : (p_1, as_1, se_1) \rightarrow_m (A_2, p_2, as_2, se_2)}{S_1 \leftarrow S_2 : (p, rs, se) \rightarrow_m (A', p', as_2 \setminus (A' \cup \{S \mid S : p \rightarrow_m (A_s, \_) \ \& \ A_s \cap A' \neq \emptyset\}), se_2)} \ (\leftarrow^e)$$

$$\frac{S_1 : (p, rs, se) \rightarrow_m (A'', p'', rs'', se'') \quad S_2 : (p'', rs'', se'') \rightarrow_m (A', p', rs', se')}{S_1 ; S_2 : (p, rs, se) \rightarrow_m (A', p', rs', se')} \ (seq^e)$$

$$\frac{S : (p, rs, se) \rightarrow_m (A'', p'', rs'', se'') \quad S_t : (p, rs'', se'') \rightarrow_m (A', p', rs', se') \quad S_f : (p, rs'', se'') \rightarrow_m (A', p', rs', se')}{if \ S \ then \ S_t \ else \ S_f : (p, rs, se) \rightarrow_m (A', p', rs', se')} \ (if^e)$$

$$\frac{S_1[S_2/x] : (p, rs, se) \rightarrow_m (A', p', rs', se')}{(\lambda x. S_1) S_2 : (p, rs, se) \rightarrow_m (A', p', rs', se')} \ (appl^e)$$

$$\frac{S : (p, rs, se) \rightarrow (A'', p'', rs', se') \quad S_t : (p'', rs', se') \rightarrow_m (A', p', rs, se)}{while \ S \ do \ S_t : (p, rs, se) \rightarrow_m (A', p', rs', se')} \ (whl^e)$$

$$\frac{fd(name) : (p, rs, se) \rightarrow_m (A', p', rs', se')}{name : (p, rs, se) \rightarrow_m (A', p', rs', se')} \ (name^e) \qquad \frac{S : (p, rs, se) \rightarrow_m (A', p', rs', se')}{\lambda x. S : (p, rs, se) \rightarrow_m (A', p', rs', se')} \ (abs^e)$$

$$\frac{(\lambda x. S') S : (p, rs, se) \rightarrow_m (A', p', rs', se')}{letrec \ x = S \ in \ S' : (p, rs, se) \rightarrow_m (A', p', rs', se')} \ (letrec^e) \qquad \frac{new_l : p \rightarrow_m (A', p')}{new_l : (p, rs, se) \rightarrow_m (A', p', rs, se)} \ (new^e)$$

$$\frac{convert \ (S, n) : p \rightarrow_m (A', p') \quad S : (p, rs, se) \rightarrow_m (A, p'', rs', se')}{convert \ (S, n) : (p, rs, se) \rightarrow_m (A', p', rs', se')} \ (convert^e)$$

$$\frac{transmit \ S_1 \ from \ S_2 : p \rightarrow (A', p') \quad S_2 : (p, rs, se) \rightarrow_m (A_2, p_2, as_2, se_2) \quad S_1 : (p_2, as_2, se_2) \rightarrow_m (A_1, p_1, rs', se')}{transmit \ S_1 \ from \ S_2 : (p, rs, se) \rightarrow_m (A', p', rs', se')} \ (trans^e)$$

$$\dfrac{\begin{array}{c}(p_1',rs_1',se_1') \le (p_1,as_1,se_1)\\ S:(p_1,as_1,se_1) \to_m (p_2,as_2,se_2)\\ (p_2,as_2,se_2) \le (p_2',rs_2',se_2')\end{array}}{S:(p_1',rs_1',se_1) \to_m (p_2',rs_2',se_2')}\ (csq^e)
\qquad
\dfrac{Defs:\emptyset \frown fd \quad S:(p,rs,se) \to_m (A,p',rs',se')}{Defs:S:(p,rs,se) \to_m (A,p',rs',se')}\ (prg^e)$$

Some comments on the inference rules are in order. In the rule (i_{op}^e), given the post type se', we calculate the pre-type se'' for the statement S_2. Then the resulting pre-type is used as a post-type for the statement S_1 to calculate the pre-type se. In line with Definition 2, the arithmetic statement $S_1\ i_{op}\ S_2$ is added to se only if it belongs to rs. Similar explanations illustrate the rule $(*^e)$. The remaining rules mimic the rules of the *ready* type system.

Now we recall the assumption that our distributed system consists of $|M|$ machines. For a given statement S and a given machine m, the type system given above calculates for each program point of S, the set of *semi-expectable* items. Now the following rule can be used to combine the information calculated for each machine to get a new *semi-expectable* information for each program point. The new *semi-expectable* information is valid on any of the $|M|$ machines.

$$\dfrac{\forall m \in M.\ S:(\sup\{p,p_j\mid j\neq i\},\sup\{rs,rs_j\mid j\neq i\},se_m) \to_m (A_m,p_m,rs_m,\inf\{se',se_j\mid j\neq i\})}{S:(p,rs,\inf\{se_1,\ldots,se_{|M|}\}) \to_M (\cup_i A_i,\sup\{p_1,\ldots,p_{|M|}\},\sup\{rs_1,\ldots,rs_{|M|}\},se')}\ (main\text{-}rs)$$

The difference in the way that this rule treats the *semi-expectable* information and the way *ready* information is treated is explained by the fact that the *ready* analysis is forward while the the *semi-expectation* analysis is backward.

It is not hard to prove the soundness of the above type system:

Theorem 2. *Suppose that* $(S,\delta) \to (V,\delta')$, $S:(p,rs,se) \to (A',p',rs',se')$ *and the items of se' are semi-expectable at the point corresponding to δ' on the execution path. Then the items of se are semi-expectable at the point corresponding to δ on the execution path.*

4   Frequent Statement and De-reference Elimination

This section presents a type system that is an enrichment of the type system presented in the previous section. The type system of this section achieves the frequent statement and de-reference elimination. The type system uses a function $sn : S^+ \to Stmt\text{-}names$ that assigns each nontrivial statement a name. These names are meant to carry values of frequent statements and de-references. The judgments of our type system have the form $S:(p,rs,se) \to_m (A',p',rs',se') \Rightarrow (ns,S')$. The type information (p,rs,se) and (A',p',rs',se') were calculated by the previous type system. S' is the optimization of S and ns is a sequence of assignments that links optimized statements with the names of their un-optimized versions. The inference rules for the frequent statements and de-references elimination are as follows:

$$\dfrac{n:p \to_m (A',p')}{n:(p,rs,se) \to_m (A',p',rs,se) \Rightarrow (skip,n)}
\qquad
\dfrac{x:p \to_m (A',p')}{x:(p,rs,se) \to_m (A',p',rs,se) \Rightarrow (skip,x)}$$

$$\frac{*S \in as \quad *S : p \to_m (A',p') \quad S : (p,rs,se) \to_m (A'',p'',rs'',se'') \Rightarrow (ns,S')}{*S : (p,rs,se \cup (re \cap \{*S,g\})) \to_m \begin{cases} (A',p',rs'' \cup \{*S,g\},se''), & A' = \{g\}; \\ (A',p',rs'' \cup \{*S\},se''), & |A'| \neq 1. \end{cases} \Rightarrow (ns,sn(*S))} \ (*_1^f)$$

$$\frac{*S \notin as \quad *S \in se' \quad *S : p \to_m (A',p') \quad S : (p,rs,se) \to_m (A'',p'',rs'',se'') \Rightarrow (ns,S')}{*S : (p,rs,se \cup (re \cap \{*S,g\})) \to_m \begin{cases} (A',p',rs'' \cup \{*S,g\},se''), & A' = \{g\}; \\ (A',p',rs'' \cup \{*S\},se''), & |A'| \neq 1. \end{cases} \Rightarrow (ns;sn(*S) := *S',sn(*S))} \ (*_2^f)$$

$$\frac{*S \notin as \quad *S \notin se' \quad *S : p \to_m (A',p') \quad S : (p,rs,se) \to_m (A'',p'',rs'',se'') \Rightarrow (ns,S')}{*S : (p,rs,se \cup (re \cap \{*S,g\})) \to_m \begin{cases} (A',p',rs'' \cup \{*S,g\},se''), & A' = \{g\}; \\ (A',p',rs'' \cup \{*S\},se''), & |A'| \neq 1. \end{cases} \Rightarrow (ns,*S')} \ (*_3^f)$$

$$\frac{S_1 \ i_{op} \ S_2 \in re \quad \begin{array}{l} S_1 : (p,rs,se) \to_m (A'',p'',rs'',se'') \Rightarrow (ns_1,S_1') \\ S_2 : (p'',rs'',se'') \to_m (A',p',rs',se') \Rightarrow (ns_2,S_2') \end{array}}{S_1 \ i_{op} \ S_2 : (p,rs,se \cup (rs \cap \{S_1 \ i_{op} \ S_2\})) \to_m (\emptyset,p',rs' \cup \{S_1 \ i_{op} \ S_2\},se') \Rightarrow (ns_1;ns_2,sn(S_1 \ i_{op} \ S_2))} \ (i_{op(1)}^f)$$

$$\frac{\begin{array}{l} S_1 \ i_{op} \ S_2 \notin as \\ S_1 \ i_{op} \ S_2 \in se' \end{array} \quad \begin{array}{l} S_1 : (p,rs,se) \to_m (A'',p'',rs'',se'') \Rightarrow (ns_1,S_1') \\ S_2 : (p'',rs'',se'') \to_m (A',p',rs',se') \Rightarrow (ns_2,S_2') \end{array}}{S_1 \ i_{op} \ S_2 : (p,rs,se \cup (rs \cap \{S_1 \ i_{op} \ S_2\})) \to_m (\emptyset,p',rs' \cup \{S_1 \ i_{op} \ S_2\},se') \Rightarrow (ns_1;ns_2;sn(S_1 \ i_{op} \ S_2) := (S_1' \ i_{op} \ S_2'),sn(S_1 \ i_{op} \ S_2))} \ (i_{op(2)}^f)$$

$$\frac{\begin{array}{l} S_1 \ i_{op} \ S_2 \notin as \\ S_1 \ i_{op} \ S_2 \notin se' \end{array} \quad \begin{array}{l} S_1 : (p,rs,se) \to_m (A'',p'',rs'',se'') \Rightarrow (ns_1,S_1') \\ S_2 : (p'',rs'',se'') \to_m (A',p',rs',se') \Rightarrow (ns_2,S_2') \end{array}}{S_1 \ i_{op} \ S_2 : (p,rs,se \cup (rs \cap \{S_1 \ i_{op} \ S_2\})) \to_m (\emptyset,p',rs' \cup \{S_1 \ i_{op} \ S_2\},se') \Rightarrow (ns_1;ns_2,S_1' \ i_{op} \ S_2')} \ (i_{op(3)}^f)$$

$$\overline{skip : (p,rs,se) \to_m (\emptyset,p,rs,se) \Rightarrow (skip,skip)}$$

$$\frac{x := S : p \to_m (A',p') \quad S : (p,rs,se) \to_m (A'',p'',rs'',se'') \Rightarrow (sn,S')}{x := S : (p,rs,se) \to_m (A',p',rs'' \setminus \{S \in Stmt \mid x \in free(S)\},se'') \Rightarrow (skip,ns;x := S')} \ (:=^f)$$

$$\frac{\begin{array}{l} S_1 : (p,rs,se) \to_m (A_1,p_1,as_1,se_1) \Rightarrow (ns_1,S_1') \\ S_2 : (p_1,as_1,se_1) \to_m (A_2,p_2,as_2,se_2) \Rightarrow (ns_2,S_2') \end{array} \quad S_1 \leftarrow S_2 : p \to_m (A',p')}{\begin{array}{l} S_1 \leftarrow S_2 : (p,rs,se) \to_m (A',p',as_2 \setminus (A' \cup \{S \mid S : p \to_m (A_s,\_) \ \& \ A_s \cap A' \neq \emptyset\}),se_2) \\ \Rightarrow (skip,ns_1;ns_2;S_1' \leftarrow S_2') \end{array}} \ (\leftarrow^f)$$

$$\frac{\begin{array}{l} S_1 : (p,rs,se) \to_m (A'',p'',rs'',se'') \Rightarrow (ns_1,S_1') \\ S_2 : (p'',rs'',se'') \to_m (A',p',rs',se') \Rightarrow (ns_2,S_2') \end{array}}{S_1;S_2 : (p,rs,se) \to_m (A',p',rs',se') \Rightarrow (ns_1;ns_2,S_1';S_2')} \ (seq^f)$$

$$\frac{S : (p,rs,se) \to_m (A'',p'',rs'',se'') \Rightarrow (ns,S') \quad \begin{array}{l} S_t : (p,rs'',se'') \to_m (A',p',rs',se') \Rightarrow (ns_t,S_t') \\ S_f : (p,rs'',se'') \to_m (A',p',rs',se') \Rightarrow (ns_f,S_f') \end{array}}{if \ S \ then \ S_t \ else \ S_f : (p,rs,se) \to_m (A',p',rs',se') \Rightarrow (skip,ns;if \ S' \ then \ ns_t;S_t' \ else \ ns_f;S_f')} \ (if^f)$$

$$\frac{(\lambda x.S_1)[S_2/x] : (p,rs,se) \to_m (A',p',rs',se') \Rightarrow (ns,S')}{(\lambda x.S_1)S_2 : (p,rs,se) \to_m (A',p',rs',se') \Rightarrow (ns,S')} \ (appl^f)$$

$$\frac{S : (p,rs,se) \to (A'',p'',rs',se') \Rightarrow (ns,S') \quad S_t : (p'',rs',se') \to_m (A',p',rs,se) \Rightarrow (ns_t,S_t')}{while \ S \ do \ S_t : (p,rs,se) \to_m (A',p',rs',se') \Rightarrow (skip,ns;while \ S' \ do \ (ns_t;S_t';ns))} \ (whl^f)$$

$$\frac{fd(name) : (p,rs,se) \to_m (A',p',rs',se') \Rightarrow (ns,S')}{name : (p,rs,se) \to_m (A',p',rs',se') \Rightarrow (ns,S')} \ (name^f)$$

$$\frac{S : (p,rs,se) \to_m (A',p',rs',se') \Rightarrow (ns,S')}{\lambda x.S : (p,rs,se) \to_m (A',p',rs',se') \Rightarrow (skip,ns;\lambda x.S')} \ (abs^f)$$

$$\frac{new_l : p \rightarrow_m (A',p')}{new_l : (p,rs,se) \rightarrow_m (A',p',rs,se) \Rightarrow (skip,new_l)} \ (new^f)$$

$$\frac{(\lambda x.S')S : (p,rs,se) \rightarrow_m (A',p',rs',se') \Rightarrow (ns,S'')}{letrec \ x = S \ in \ S' : (p,rs,se) \rightarrow_m (A',p',rs',se') \Rightarrow (ns,S'')} \ (letrec^f)$$

$$\frac{convert \ (S,n) : p \rightarrow_m (A',p') \qquad S : (p,rs,se) \rightarrow_m (A,p'',rs',se') \Rightarrow (ns,S')}{convert \ (S,n) : (p,rs,se) \rightarrow_m (A',p',rs',se') \Rightarrow (skip,ns;convert \ (S',n))} \ (convert^f)$$

$$\frac{\begin{array}{c} transmit \ S_1 \ from \ S_2 : p \rightarrow (A',p') \\ S_2 : (p,rs,se) \rightarrow_m (A_2,p_2,as_2,se_2) \Rightarrow (ns_2,S'_2) \\ S_1 : (p_2,as_2,se_2) \rightarrow_m (A_1,p_1,rs',se') \Rightarrow (ns_1,S'_1) \end{array}}{transmit \ S_1 \ from \ S_2 : (p,rs,se) \rightarrow_m (A',p',rs',se') \Rightarrow (ns_1;ns_2,transmit \ S'_1 \ from \ S'_2)} \ (trans^f)$$

$$\frac{\begin{array}{c} (p'_1,rs'_1,se'_1) \le (p_1,as_1,pa_1) \\ (p_2,as_2,pa_2) \le (p'_2,rs'_2,se'_2) \end{array} \quad S : (p_1,as_1,pa_1) \rightarrow_m (p_2,as_2,pa_2) \Rightarrow (ns,S')}{S : (p'_1,rs'_1,pa_1) \rightarrow_m (p'_2,rs'_2,se'_2) \Rightarrow (ns,S')} \ (csq^f)$$

$$\frac{Defs : \emptyset \frown fd \quad S : (p,rs,se) \rightarrow_m (A,p',rs',se') \Rightarrow (ns,S')}{Defs : S : (p,rs,se) \rightarrow_m (A,p',rs',se') \Rightarrow Defs : ns;S'} \ (prg^f)$$

We note the following on the inference rules. A big deal of optimization is achieved by the three rules for $*S$. These rules are $(*^f_1), (*^f_2),$ and $(*^f_3)$. The rule $(*^f_1)$ takes care of the case where $*S$ is *ready* and is replaceable by its name under the function *sn*. The rule $(*^f_2)$ treats the case where $*S$ is *semi-expectable* and is not *ready* before calculating the statement. In this case, a statement name of $*S$ is used. The rule $(*^f_3)$ considers the case where $*S$ is neither *semi-expectable* at the program point after execution nor *ready* before calculating the statement. In this case, the statement $*S$ does not get changed. Similarly, the three rules $(i^f_{op(1)}), (i^f_{op(2)}),$ and $(i^f_{op(3)})$ treat different cases for arithmetic statements. The Boolean statements are treated with rules quite similar to that of arithmetic statements. The rule (whl^f) reuses frequent sub-statements of the guard. This is done via adding *ns* in the positions clarified in the rule. Remaining rules of system are self-explanatory.

For expressing the soundness, we introduce the following definition:

Definition 3. *Suppose that δ is a state defined on the set of locations, Loc (Definition 4). Suppose also that δ_* is a state defined on Loc $\cup$ Stmt-names. The expression $\delta \equiv_{se} \delta_*$ denotes the fact that δ and δ_* are equivalent with respect to the semi-expectation type se. More precisely $\delta \equiv_{se} \delta_*$ iff:*

1. $\forall j \in Loc. \ \delta(j) = \delta_*(j),$ *and*
2. $\forall S \in se. \ (S,\delta) \leadsto_m (v,\delta') \Longrightarrow \delta_*(sn(S)) = v.$

The soundness of frequent statements and de-references elimination means that the original and optimized programs are equivalent in the following sense:

- The states of the two programs coincide on the *Loc*, and
- If a statement is both *ready* and *semi-expectable*, then its semantics in the original-program state equals the value of its corresponding name in optimized-program state.

This gives an intuition to the previous definition. The following soundness theorem is proved by a structure induction.

Theorem 3. *Suppose that* $S : (p, rs, se) \to_m (A', p', rs', se') \Rightarrow (ns, S')$ *and* $\delta \equiv_{se} \delta_*$. *Then*

- $(S, \delta) \leadsto_m (v, \delta') \Longrightarrow \exists \delta'_*. \ \delta' \equiv_{se'} \delta'_*$ *and* $(S', \delta_*) \leadsto_m (v, \delta'_*)$.
- $(S', \delta_*) \leadsto_m (v, \delta'_*) \Longrightarrow \exists \delta'. \ \delta' \equiv_{se'} \delta'_*$ *and* $(S, \delta) \leadsto_m (v, \delta')$.

5 Related and Future Work

The techniques of common sub-expression elimination (CSE) [18,17] are among the most closed to our work. In [27], a type system for CSE of the *while* language is introduced. The work presented in our paper can be realized as a generalization of that presented in [27]. The generality of our work is evident in our language model which is much richer with distributed and pointer commands. Consequently, the operational semantics that we measure the soundness of our system against is much more involved than that used in [27]. Using new opportunities appearing while scheduling of control-intensive designs, the work in [25] introduces a technique that dynamically eliminates CSE. To optimize polynomial expressions (important for applications like domains, computer graphics, and signal processing), the paper [19] generalizes algebraic techniques originally designed for multilevel logic synthesis. This generalization uses factoring and eliminating common subexpressions.

The association of a correctness proof with each result of the static analysis is important and needed by applications like proof-carrying code and certified code. The work presented in this paper has the advantage over most related work of constructing these proofs. Adding to the value of using type systems, the proofs constructed in our proposed approach have the form of type derivations. The work in [6,4,15,5,8,9,7,16] presents many examples of other static analyses that are in the form of type systems.

In [21], a technique for flow-insensitive pointer analysis of programs that run on parallel and hierarchical machines and that share memory is introduced. Via a two-level hierarchy, [22] and [23] present constraint-based approaches to evaluate locality information and sharing attributes of references. Our language model is a generalization of models presented in [21,22].

Much research acclivities [21,24] was devoted to analyze distributed programs. This is motivated by the importance of distributed programming as a main stream of programming today. The examining and capturing of causal and concurrent relationships are among important issues to many distributed systems applications. In [28], an analysis that examines the source code of each process constructs an inclusive graph, POG, of the possible behaviors of systems. Data racing bugs [2] can be a side effect of the parallel access of cores of a multi-core process to a physically distributed memory. In [2] a technique, called DRARS, is proposed for avoidance and replay of this data race. Parallel programs on DSM or multi-core systems, can be debugged using DRARS. The classical problems of satisfiability decidability and algorithmic decidability are approached in [29] on the distributed-programs model of message sending. In this work, distributed programs are represented by communicating via buffers.

Mathematical domains (sets) and maps between domains mathematically represent data structures and programs in the area of denotational semantics. For future work, it is interesting to study the possibility of translating concepts of frequent statement and de-reference elimination to the side of denotational semantics [14,3]. This translation has the impact of easing achieving theoretical studies about frequent statement and de-reference elimination. Established theoretical results may then be translated back to the side of data structures and programs. Similarly, we also intend to test the gains of applying the type systems approach on the problems treated by our work in [10,12,13].

6   Conclusion

This paper introduces a new technique for the analysis of *frequent statement and de-reference elimination* for distributed programs running on parallel machines equipped with hierarchical memories. Type systems are the tools of the techniques presented in this paper which presents three type systems. The first type system defines for program points of a given distributed program sets of calculated (*ready*) statements and memory accesses. The second type system determines which of the ready statements and memory accesses are used later in the program. The third type system eliminates unnecessary statement computations and memory accesses.

References

1. Barpanda, S.S., Mohapatra, D.P.: Dynamic slicing of distributed object-oriented programs. IET Software 5(5), 425–433 (2011)
2. Chiu, Y.-C., Shieh, C.-K., Huang, T.-C., Liang, T.-Y., Chu, K.-C.: Data race avoidance and replay scheme for developing and debugging parallel programs on distributed shared memory systems. Parallel Computing 37(1), 11–25 (2011)
3. El-Zawawy, M.A.: Semantic spaces in Priestley form. PhD thesis, University of Birmingham, UK (January 2007)
4. El-Zawawy, M.A.: Flow sensitive-insensitive pointer analysis based memory safety for multithreaded programs. In: Murgante, B., Gervasi, O., Iglesias, A., Taniar, D., Apduhan, B.O. (eds.) ICCSA 2011, Part V. LNCS, vol. 6786, pp. 355–369. Springer, Heidelberg (2011)
5. El-Zawawy, M.A.: Probabilistic pointer analysis for multithreaded programs. ScienceAsia 37(4), 344–354 (2011)
6. El-Zawawy, M.A.: Program optimization based pointer analysis and live stack-heap analysis. International Journal of Computer Science Issues 8(2), 98–107 (2011)
7. El-Zawawy, M.A.: Abstraction analysis and certified flow and context sensitive points-to relation for distributed programs. In: Murgante, B., Gervasi, O., Misra, S., Nedjah, N., Rocha, A.M.A.C., Taniar, D., Apduhan, B.O. (eds.) ICCSA 2012, Part IV. LNCS, vol. 7336, pp. 83–99. Springer, Heidelberg (2012)
8. El-Zawawy, M.A.: Dead code elimination based pointer analysis for multithreaded programs. Journal of the Egyptian Mathematical Society 20(1), 28–37 (2012)
9. El-Zawawy, M.A.: Heap slicing using type systems. In: Murgante, B., Gervasi, O., Misra, S., Nedjah, N., Rocha, A.M.A.C., Taniar, D., Apduhan, B.O. (eds.) ICCSA 2012, Part III. LNCS, vol. 7335, pp. 592–606. Springer, Heidelberg (2012)
10. El-Zawawy, M.A.: Recognition of logically related regions based heap abstraction. Journal of the Egyptian Mathematical Society 20(2) (September 2012)

11. El-Zawawy, M.A.: Detection of probabilistic dangling references in multi-core programs using proof-supported tools. In: Murgante, B., et al. (eds.) ICCSA 2013, Part V. LNCS, vol. 7975, pp. 523–538. Springer, Heidelberg (2013)
12. El-Zawawy, M.A., Daoud, N.M.: Dynamic verification for file safety of multithreaded programs. IJCSNS International Journal of Computer Science and Network Security 12(5), 14–20 (2012)
13. El-Zawawy, M.A., Daoud, N.M.: New error-recovery techniques for faulty-calls of functions. Computer and Information Science 5(3), 67–75 (2012)
14. El-Zawawy, M.A., Jung, A.: Priestley duality for strong proximity lattices. Electr. Notes Theor. Comput. Sci. 158, 199–217 (2006)
15. El-Zawawy, M.A., Nayel, H.A.: Partial redundancy elimination for multi-threaded programs. IJCSNS International Journal of Computer Science and Network Security 11(10), 127–133 (2011)
16. El-Zawawy, M.A., Nayel, H.A.: Type systems based data race detector. IJCSNS International Journal of Computer Science and Network Security 5(4), 53–60 (2012)
17. Gopalakrishnan, S., Kalla, P.: Algebraic techniques to enhance common sub-expression elimination for polynomial system synthesis. In: DATE, pp. 1452–1457. IEEE (2009)
18. Ho, H., Szwarc, V., Kwasniewski, T.A.: Low complexity reconfigurable dsp circuit implementations based on common sub-expression elimination. Signal Processing Systems 61(3), 353–365 (2010)
19. Hosangadi, A., Fallah, F., Kastner, R.: Optimizing polynomial expressions by algebraic factorization and common subexpression elimination. IEEE Trans. on CAD of Integrated Circuits and Systems 25(10), 2012–2022 (2006)
20. Seragiotto Jr., C.: Thomas Fahringer. Performance analysis for distributed and parallel java programs with aksum. In: CCGRID, pp. 1024–1031. IEEE Computer Society (2005)
21. Kamil, A., Yelick, K.A.: Hierarchical pointer analysis for distributed programs. In: Riis Nielson, H., Filé, G. (eds.) SAS 2007. LNCS, vol. 4634, pp. 281–297. Springer, Heidelberg (2007)
22. Liblit, B., Aiken, A.: Type systems for distributed data structures. In: POPL, pp. 199–213 (2000)
23. Liblit, B., Aiken, A., Yelick, K.A.: Type systems for distributed data sharing. In: Cousot, R. (ed.) SAS 2003. LNCS, vol. 2694, pp. 273–294. Springer, Heidelberg (2003)
24. Lindberg, P., Leingang, J., Lysaker, D., Khan, S.U., Li, J.: Comparison and analysis of eight scheduling heuristics for the optimization of energy consumption and makespan in large-scale distributed systems. The Journal of Supercomputing 59(1), 323–360 (2012)
25. Nicolau, A., Dutt, N.D., Gupta, R., Savoiu, N., Reshadi, M., Gupta, S.: Dynamic common sub-expression elimination during scheduling in high-level synthesis. In: ISSS, pp. 261–266. IEEE Computer Society (2002)
26. Onbay, T.U., Kantarci, A.: Design and implementation of a distributed teleradiaography system: Dipacs. Computer Methods and Programs in Biomedicine 104(2), 235–242 (2011)
27. Saabas, A., Uustalu, T.: Program and proof optimizations with type systems. Journal of Logic and Algebraic Programming 77(1-2), 131–154 (2008); The 16th Nordic Workshop on the Prgramming Theory (NWPT 2006)
28. Simmons, S., Edwards, D., Kearns, P.: Communication analysis of distributed programs. Scientific Programming 14(2), 151–170 (2006)
29. Toporkov, V.V.: Dataflow analysis of distributed programs using generalized marked nets. In: DepCoS-RELCOMEX, pp. 73–80. IEEE Computer Society (2007)
30. Truong, H.L., Fahringer, T.: Soft computing approach to performance analysis of parallel and distributed programs. In: Cunha, J.C., Medeiros, P.D. (eds.) Euro-Par 2005. LNCS, vol. 3648, pp. 50–60. Springer, Heidelberg (2005)

Appendix: Memory Model, Language, and Operational Semantics

This appendix briefly reviews our language syntax, memory model, and operational semantics which are presented in more details in [7]. Hierarchical [24,26] memory models are typically used in parallel computers (Figure 2).

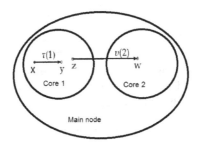

Fig. 2. Hierarchy memory

In PGAS languages, each pointer is assigned the memory-hierarchy level that the pointer may reference. Two-levels hierarchy is shown in Figure 2 in which τ is a core pointer and may point to addresses on core 1 and υ is a node pointer. These pointer domains are represented via assigning a width (number) to each pointer. Increasing hierarchy levels is the trend in hardware research. Hence benefiting from the hierarchy [24,26] is very important for software.

$$name ::= \text{'string of characters'}.$$
$$S \in Stmts ::= n \mid true \mid false \mid x \mid S_1 \; i_{op} \; S_2 \mid S_1 \; b_{op} \; S_2 \mid *S \mid skip \mid name \mid x := S \mid S_1 \leftarrow S_2 \mid$$
$$S_1 ; S_2 \mid if \; S \; then \; S_t \; else \; S_f \mid while \; S \; do \; S_t \mid \lambda x.S \mid S_1 S_2 \mid letrec \; x = S \; in \; S' \mid$$
$$new_l \mid convert \; (S,n) \mid transmit \; S_1 \; from \; S_2.$$
$$Defs ::= (name = S); Defs \mid \varepsilon.$$
$$Program ::= Defs : S.$$

where
$x \in lVar$, an infinite set of variables, $n \in \mathbb{Z}$ (integers), $i_{op} \in \mathbb{I}_{op}$ (integer-valued binary operations), and $b_{op} \in \mathbb{B}_{op}$ (Boolean-valued binary operations).

Fig. 3. The programming language $while_d$

Figure 3 presents the syntax of $While_d$ [21,22] used to present the results of this paper. $While_d$ has pointer, parallelism, and function constructions [24,26] and supports the SPMD parallelism model in which the same code is executed on all machines. The hierarchy height of memory is denoted by h. Therefore, pointer widths are in $[1,h]$. The language uses a fixed set of variables, $lVar$, each of which is machine-private. For each program, the semantics specifies a map fd of the domain *Function-defs*:

$$fd \in Function\text{-}defs = \text{'strings of characters'} \rightarrow Stmts.$$

This map is defined using the following inference rules:

$$\frac{}{\varepsilon : fd \curvearrowright fd}\,(fd_1) \qquad \frac{Defs : fd[name \mapsto S] \curvearrowright fd'}{(name = S);Defs : fd \curvearrowright fd'}\,(fd_2)$$

The semantics introduces transition relations $\leadsto_m$; between pairs of statements and states and pairs of values and states. The following definition introduces components used in the inference rules of the semantics.

Definition 4. *1. The set of global variables, denotes by $gVar$, is defined as $gVar = \{(m,x) \mid m \in M, x \in lVar\}$.*
2. The set of global addresses, denotes by $gAddrs$, is defined as $gAddrs = \{g = (l,m,a) \mid l \in L, m \in M, a \in lAddrs\}$.
3. $loc \in Loc = gAddrs \cup gVar$.
4. $v \in Values = \mathbb{Z} \cup gAddrs \cup \{true,false\} \cup \{\lambda x.S \mid S \in Stmt\}$.
5. $\delta \in States = Loc \longrightarrow Values$.

The symbols M, W, and $lAddrs$ denote the set of machines labels (integers), the set of width $\{1,\ldots,h\}$, and the set of local addresses located on each single machine, respectively. The set of labels of allocation sites is denoted by L.

The semantics judgments have the form $(S,\delta) \leadsto_m (v,\delta')$ meaning that executing S on the machine m and in the state δ produces the value v and modifies δ to δ'. We have $\delta[x \mapsto v] \Longleftrightarrow \lambda y.$ if $y = x$ then v else $\delta(y)$.

The inference rules of our semantics are as follows:

$$(n,\delta) \leadsto_m (n,\delta) \qquad (true,\delta) \leadsto_m (true,\delta) \qquad (false,\delta) \leadsto_m (false,\delta) \qquad (x,\delta) \leadsto_m (\delta(x),\delta)$$

$$(\lambda x.S,\delta) \leadsto_m (\lambda x.S,\delta)(abs) \qquad \frac{(S_1,\delta) \leadsto_m (n_1,\delta'') \quad (S_2,\delta'') \leadsto_m (n_2,\delta')}{(S_1\ iop\ S_2,\delta) \leadsto_m \begin{cases} (n_1\ iop\ n_2,\delta'),\ n_1\ iop\ n_2 \in \mathbb{Z}; \\ abort, \qquad\qquad otherwise. \end{cases}}\,(int\text{-}stmt)$$

$$\frac{(S_1,\delta) \leadsto_m (b_1,\delta'') \quad (S_2,\delta'') \leadsto_m (b_2,\delta')}{(S_1\ bop\ S_2,\delta) \leadsto_m \begin{cases} (b_1\ bop\ b_2,\delta'),\ b_1\ bop\ b_2\ \text{is a Boolean value}; \\ abort, \qquad\qquad otherwise. \end{cases}}\,(bo\text{-}stmt) \qquad (skip,\delta) \leadsto_m (0,\delta)$$

$$\frac{(S,\delta) \leadsto_m (g,\delta')}{(*S,\delta) \leadsto_m \begin{cases} (\delta'(g),\delta'),\ g \in gAddrs; \\ abort, \qquad otherwise. \end{cases}}\,(de\text{-}ref) \qquad \frac{(S,\delta) \leadsto_m abort}{(x := S,\delta) \leadsto_m abort} \quad \frac{(S,\delta) \leadsto_m (v,\delta')}{(x := S,\delta) \leadsto_m (v,\delta'[x \mapsto v])} \quad \frac{(S_1,\delta) \leadsto_m abort}{(S_1;S_2,\delta) \leadsto_m abort}$$

$$\frac{(S_1,\delta) \leadsto_m abort\ \text{or for}\ v \notin gAddrs.\ (S_1,\delta) \leadsto_m (v,\delta'')}{(S_1 \leftarrow S_2,\delta) \leadsto_m abort}\,(\leftarrow_1) \qquad \frac{(S_1,\delta) \leadsto_m (v,\delta'')}{(S_1 \leftarrow S_2,\delta) \leadsto_m abort}\,(\leftarrow_2) \qquad \frac{(S_1,\delta) \leadsto_m abort}{(S_1;S_2,\delta) \leadsto_m abort}$$

$$\frac{\begin{array}{c}(S_1,\delta) \leadsto_m (g,\delta'') \\ (S_2,\delta'') \leadsto_m (v,\delta''') \\ g \in gAddrs\end{array}}{(S_1 \leftarrow S_2,\delta) \leadsto_m (v,\delta'''[g \mapsto v])}\,(\leftarrow_3) \qquad \frac{\begin{array}{c}(S_1,\delta) \leadsto_m (v_1,\delta'') \\ (S_2,\delta'') \leadsto_m (v_2,\delta')\end{array}}{(S_1;S_2,\delta) \leadsto_m (v_2,\delta')} \quad \frac{\begin{array}{c}(S_1,\delta) \leadsto_m (v_1,\delta'') \\ (S_2,\delta'') \leadsto_m abort\end{array}}{(S_1;S_2,\delta) \leadsto_m abort} \quad \frac{\begin{array}{c}(S,\delta) \leadsto_m (true,\delta'') \\ (S_t,\delta'') \leadsto_m abort\end{array}}{(if\ S\ then\ S_t\ else\ S_f,\delta) \leadsto_m abort}$$

$$\frac{(S,\delta) \leadsto_m (true,\delta'') \quad (S_t,\delta'') \leadsto_m (v,\delta')}{(if\ S\ then\ S_t\ else\ S_f,\delta) \leadsto_m (v,\delta')}$$

$$\frac{(S,\delta) \leadsto_m (false,\delta'') \quad (S_f,\delta'') \leadsto_m abort}{(if\ S\ then\ S_t\ else\ S_f,\delta) \leadsto_m abort}$$

$$\frac{(S,\delta) \leadsto_m (false,\delta'') \quad (S_f,\delta'') \leadsto_m (v,\delta')}{(if\ S\ then\ S_t\ else\ S_f,\delta) \leadsto_m (v,\delta')}$$

$$\frac{(S,\delta) \leadsto_m abort}{(if\ S\ then\ S_t\ else\ S_f,\delta) \leadsto_m abort}$$

$$\frac{(S,\delta) \leadsto_m abort}{(while\ S\ do\ S_t,\delta) \leadsto_m abort}$$

$$\frac{(S,\delta) \leadsto_m (false,\delta'')}{(while\ S\ do\ S_t,\delta) \leadsto_m (skip,\delta)}$$

$$\frac{(S,\delta) \leadsto_m (true,\delta'') \quad (S_t,\delta'') \leadsto_m abort}{(while\ S\ do\ S_t,\delta) \leadsto_m abort}$$

$$\frac{\begin{array}{c}(S,\delta) \leadsto_m (true,\delta'')\\ (S_t,\delta'') \leadsto_m (v'',\delta'')\\ (while\ S\ do\ S_t,\delta'') \leadsto_m (v',\delta')\end{array}}{while\ S\ do\ S_t : (\delta,p) \leadsto_m (v',\delta')}$$

$$\frac{\begin{array}{c}(S,\delta) \leadsto_m (true,\delta'')\\ (S_t,\delta'') \leadsto_m (v'',\delta'')\\ (while\ S\ do\ S_t,\delta'') \leadsto_m abort\end{array}}{(while\ S\ do\ S_t,\delta) \leadsto_m abort}$$

$$\frac{(S_1,\delta) \leadsto_m (\lambda x.S_1',\delta'') \quad (S_1'[S_2/x],\delta'') \leadsto_m (v,\delta')}{(S_1 S_2,\delta) \leadsto_m (v,\delta')}\ (appl)$$

$$\frac{(S,\delta) \leadsto_m (v,\delta'') \quad (S'[v/x],\delta'') \leadsto_m (v',\delta')}{(letrec\ x = S\ in\ S',\delta) \leadsto_m (v',\delta')}\ (letrec)$$

$$\frac{a \in lAddrs \quad a\ is\ fresh\ on\ m}{(new_l,\delta) \leadsto_m ((l,m,a),\delta[(l,m,a) \mapsto null])}$$

$$\frac{(S,\delta) \leadsto_m (g = (l,m',a),\delta') \quad hdist(m,m') \le n}{(convert(S,n),\delta) \leadsto_m (g,\delta')}\ (conv)$$

$$\frac{(fd(name),\delta) \leadsto_m v,\delta')}{(name,\delta) \leadsto_m (v,\delta')}\ (name)$$

$$\frac{(S_2,\delta) \leadsto_m (m',\delta'') \quad m' \in M \quad (S_1,\delta'') \leadsto_{m'} (v,\delta')}{(transmit\ S_1\ from\ S_2,\delta) \leadsto_m (v,\delta')}\ (trans)$$

$$\frac{\theta : \{1,2,\ldots,|M|\} \to M \quad (S,\delta) \leadsto_{\theta(1)} (v_1,\delta_1) \leadsto_{\theta(2)} (v_2,\delta_2) \leadsto_{\theta(3)} \cdots \leadsto_{\theta(|M|)} (v_{|M|},\delta_{|M|})}{(Defs : S,\delta) \leadsto_M (v_{|M|},\delta_{|M|})}\ (main\text{-}sem)$$

The semantics of running a program *Defs : S* on the distributed systems is treated by the rule (*main-sem*).

Agile Software Development: It Is about Knowledge Management and Creativity

Claudio León de la Barra[1], Broderick Crawford[1,2], Ricardo Soto[1,3], Sanjay Misra[4], and Eric Monfroy[5]

[1] Pontificia Universidad Católica de Valparaíso, Chile
FirstName.Name@ucv.cl
[2] Universidad Finis Terrae, Chile
[3] Universidad Autónoma de Chile, Chile
[4] Covenant University, Nigeria
smisra@futminna.edu.ng
[5] LINA, Université de Nantes, France
FirstName.Name@univ-nantes.fr

Abstract. Software development is a knowledge intensive activity and its success depends on knowledge and creativity of the developers. In the last years the traditional perspective on software development is changing and agile methods have received considerable attention. The purpose of this paper is to provide an understanding of knowledge management and creativity in relation with new software engineering trends. The implications of these findings are considered, and some possible directions for future research are suggested.

Keywords: Knowledge Management, Software Engineering, Agile Development.

1 Introduction

Traditionally most of software engineering research is technical and deemphasizes the human and social aspects. Nowadays, it is interesting to consider the new proposals of agile methodologies for software development in order to analyse and evaluate them at the light of the existing teamwork practices. The agile principles and values have emphasized the importance of collaboration and interaction in the software development. Scrum Agile Development Process [31] and eXtreme Programming XP [3] have attained worldwide fame for their ability to increase the productivity of software teams by several magnitudes through empowering individuals, fostering a team-oriented environment, and focusing on project transparency and results. There are recent studies reporting efforts to improve agile process. Agile software development addresses software process improvement within teams. The work in [27][25] argues for the use of diagnosis and action planning to improve teamwork in agile software development. The action planning focused on improving shared leadership, team orientation and learning. Therefore, we can point out that the action planning focuses on improving

B. Murgante et al. (Eds.): ICCSA 2013, Part III, LNCS 7973, pp. 98–113, 2013.
© Springer-Verlag Berlin Heidelberg 2013

shared leadership, team orientation and learning [5][11][12]. We believe that the innovation and development of new products is an interdisciplinary issue [32][26], we are interested in the study of the potential of new concepts and techniques to foster knowledge management and creativity in software engineering.

This paper is organised as follows: Section 2 presents the background and general concepts on Creativity in Software development. Section 3 is dedicated to the presentation of Knowledge Management (KM) in Software Engineering. We include a short overview of basic concepts from the area of Knowledge Management in Section 4, presenting the two approaches to KM: Product and Process. In Section 5 we describe the Creative Process. A Background on Agile Development Approaches is given in section 6. Section 7 presents some important issues about Creativity in the context of Agile development. Section 8 briefly shows Software Reuse from KM perspective. In Section 9 we compare how knowledge sharing is handled by both Agile and Tayloristic methods in some relevant topics and activities in software development projects. Finally, in Section 10 we conclude the paper and give some perspectives for future research.

2   Creativity in Software Development

Since human creativity is thought as the source to resolve complex problem or create innovative products, one possibility to improve the software development process is to design a process which can stimulate the creativity of the developers. There are few studies reported on the importance of creativity in software development. In management and business, researchers have done much work about creativity and obtained evidence that the employees who had appropriate creativity characteristics, worked on complex, challenging jobs, and were supervised in a supportive, noncontrolling fashion, produced more creative work. Then, according to the previous ideas the use of creativity in software development is undeniable, but requirements engineering is not recognized as a creative process in all the cases [21]. In a few publications the importance of creativity has been investigated in all the phases of software development process [13][14][8][18][9] and mostly focused in the requirements engineering [28][24]. Nevertheless, the use of techniques to foster creativity in requirements engineering is still shortly investigated.

3   Knowledge Management in Software Engineering

The main argument to Knowledge Management in software engineering is that it is a human and knowledge intensive activity. Software development is a process where every person involved has to make a large number of decisions and individual knowledge has to be shared and leveraged at a project and organization level, and this is exactly what KM proposes. People in such groups must collaborate, communicate, and coordinate their work, which makes knowledge management a necessity. In software development one can identify two types of knowledge: Knowledge embedded in the products or artifacts, since they are the result of highly creative activities and Meta-knowledge, that is knowledge about the products and processes. Some of the sources of knowledge (artifacts, objects, components, patterns, templates and containers) are stored in electronic form. However, the majority of knowledge is tacit, residing in the brains of the employees. A way to address this problem can be to develop a knowledge sharing culture, as well as technology support for knowledge management. There are several reasons to believe that knowledge management for software engineering would be easier to implement than in other organizations: technology is not be intimidating to software engineers and they believe the tools will help them do a better job; all artifacts are already in electronic form and can easily be distributed and shared; and the fact that knowledge sharing between software engineers already does occur to a large degree in many successful software collaborative projects [29].

4   A Framework for Knowledge Management

Knowledge Management focuses on corporate knowledge as a crucial asset of the enterprise and aims at the optimal use and development of this asset, now and in the future. Knowledge Management has been the subject of much discussion over the past decade and different KM life-cycles and strategies have been proposed. One of the most widely accepted approaches to classifying knowledge from a KM perspective is the *Knowledge Matrix* of Nonaka and Takeuchi [26].

4.1   Two Approaches to KM: Product and Process

Traditional methods of software development use a great amount of documentation for capturing knowledge gained in the activities of a project life-cycle. In contrast, the agile methods suggest that most of the written documentation can be replaced by enhanced informal communications among team members and customers with a stronger emphasis on tacit knowledge rather than explicit knowledge. In the KM market a similar situation exists and two approaches to KM have been mainly employed; we will refer to them as the *Product* and the *Process* approaches. These approaches adopt different perspectives in relation to documentation and interactions between the stakeholders [23].

Knowledge as a Product. The product approach implies that knowledge can be located and manipulated as an independent object. Proponents of this approach claim that it is possible to capture, distribute, measure and manage knowledge. This approach mainly focuses on products and artefacts containing and representing knowledge.

Knowledge as a Process. The process approach puts emphasis on ways to promote, motivate, encourage, nurture or guide the process of learning, and abolishes the idea of trying to capture and distribute knowledge. This view mainly understands KM as a social communication process, which can be improved by collaboration and cooperation support tools. In this approach, knowledge is closely tied to the person who developed it and is shared mainly through person-to-person contacts. This approach has also been referred to as the *Collaboration* or *Personalisation* approach. Choosing one approach or other will be in relation to the characteristics of the organization, the project and the people involved in each case [2].

5   Creative Process

The creative process constitutes the central aspect of team performance, because it supposes a serie of clearly distinguishable phases that had to be realized by one or more of the team members in order to obtain a concrete creative result.

The phases - on the basis of Wallas [33] and Leonard and Swap [19] - are the following ones:

- *Initial Preparation:* the creativity will bloom when the mental ground is deep, fertile and it has a suitable preparation. Thus, the deep and relevant knowledge, and the experience precedes the creative expression.

- *Encounter:* the findings corresponding to the perception of a problematic situation. For this situation a solution does not exist. It is a new problem.

- *Final Preparation:* it corresponds to the understanding and foundation of the problem. It's the immersion in the problem and the use of knowledge and analytical abilities. It includes search for data and the detailed analysis of factors and variables.

- *Generation of Options:* referred to produce a menu of possible alternatives. It supposes the divergent thinking. It includes, on one hand, finding principles, lines or addresses, when making associations and uniting different marks of references and, on the other hand, to generate possible solutions, combinations and interpretations.

- *Incubation:* it corresponds to the required time to reflect about the elaborated alternatives, and "to test them mentally".

- *Options Choice:* it corresponds to the final evaluation and selection of the options. It supposes the convergent thinking.

- *Persuasion:* closing of the creative process and communication to other persons.

Considering the creativity as a "nonlinear" process some adjustments are necessary, redefinitions or discardings that force to return to previous phases, in a complex creative dynamic. Therefore, for each one of the defined phases it is possible to associate feedbacks whose "destiny" can be anyone of the previous phases in the mentioned sequence.

5.1   Roles in a Creative Team

Lumsdaine and Lumsdaine [20] raise the subject of the required cognitives abilities (mindsets) for creative problem resolution. Their typology is excellent for the creative team, and the different roles to consider. These roles are the following ones:

- The *Detective* is in charge of collecting the greatest quantity of information related to the problem. It has to collect data without making judgements, even when it thinks that it has already understood the problem exactly.

- The *Explorer* detects what can happen in the area of the problem and its context. It thinks on its long term effects and it anticipates certain developments that can affect the context (in this case, the team). The explorer perceives the problem in a broad sense.

- The *Artist* creates new things, transforming the information. It must be able to break his own schemes to generate eccentric ideas, with imagination and feeling.

- The *Engineer* is the one in charge of evaluating new ide-as. It must make converge the ideas, in order to clarify the concepts and to obtain practical ideas that can be implemented for the resolution of problems.

- The *Judge* must do a hierarchy of ideas and decide which of them will be implemented (and as well, which ones must be discarded). Additionally, it must detect possible faults or inconsistences, as well as raise the corresponding solutions. Its role must be critical and impartial, having to look for the best idea, evaluating the associated risks.

- The *Producer* is in charge of implementing the chosen ideas.

Leonard and Swap [19] have mentioned additional roles, possible to be integrated with the previous ones, because they try to make more fruitful the divergence and the convergence in the creative process:

The *Provoker* who takes the members of the team "to break" habitual mental and procedural schemes to allow the mentioned divergence (in the case of the "artist") or even a better convergence (in the case of the "engineer"). Think tank that it is invited to the team sessions to give a renewed vision of the problem-situation based on his/her experticia and experience.

The *Facilitator* whose function consists in helping and supporting the team work in its creative task in different stages.

The *Manager* who cares for the performance and specially for the results of the creative team trying to adjust them to the criteria and rules of the organization (use of resources, due dates).

Kelley and Littman [17], on the other hand, have raised a role typology similar to Lumsdaine and Lumsdaine [20], being interesting that they group the roles in three categories: those directed to the learning of the creative team (susceptible of corresponding with the detective, explorer, artist, provoker and think tank roles), others directed to the internal organization and success of the team (similar to the judge, facilitator and manager roles) and, finally, roles whose purpose is to construct the innovation (possibly related to the role of the engineer and judge).

6 Agile Software Methods

Agile software development is a group of software development methods based on iterative and incremental development, where requirements and solutions evolve through collaboration between self-organizing and cross-functional teams.

It promotes adaptive planning, evolutionary development and delivery, a time-boxed iterative approach, and encourages rapid and flexible response to change. It is a conceptual framework introduced in the Agile Manifesto in 2001 [4]. Between others, well-known agile software development methods include Scrum and eXtreme Programming.

6.1 Scrum

Scrum was inspired by complexity theory, systems dynamics and Nonaka and Takeuchi's theory of Knowledge creation. Scrum adapts aspects from these fields setting a project-management-oriented agile software development method [25]. Self-management is a relevant characteristic in Scrum, representing a radically new approach for planning and managing software projects. Scrum provides team members the opportunity for mutual continuous recognition of competences betwen colleagues. It is an orderly vehicle for the development of communication, collaboration, trust and cohesion. The term Scrum was first adapted as a metaphor, from [26] referring to the holistic action of an entire team going the entire distance, together. This rugby idea is attributed to a Honda project leader named Hiroo Watanabe who said: "I am always telling the team members that our work is not a relay race, in which my work starts here and yours there. Everyone should run all the way from start to finish, like rugby, all of us should run together, pass the ball left and right and reach the goal as a united body". In general, Scrum es described as a development process for small teams which includes a series of short development iterations named "sprints". A Scrum team is given significant authority and responsability in many aspects of their work.

Scrum is an iterative, incremental framework [30](See Figure 1). Scrum structures product development in cycles of work called Sprints, iterations of work which are typically 1-4 weeks in length, and which take place one after the other. The Sprints are of fixed duration they end on a specific date whether the work has been completed or not, and are never extended. At the beginning of each Sprint, a cross-functional team selects items from a prioritized list of requirements, and commits to complete them by the end of the Sprint; during the Sprint, the deliverable does not change. Each work day, the team gathers briefly to report to each other on progress, and update simple charts that orient them to the work remaining. At the end of the Sprint, the team demonstrates what they have built, and gets feedback which can then be incorporated in the next Sprint. Scrum emphasizes producing working product at the end of the Sprint is really done.

6.2 Scrum Roles

There are three primary roles: The Product Owner, The Team, and The Scrum-Master [10].

The *Product Owner* is responsible for achieving maximum business value, by taking all the inputs into what should be produced (from the customer or end-user of the product, as well as from Team Members and stakeholders) and

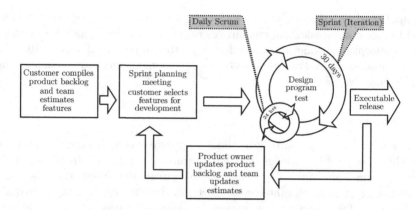

Fig. 1. Agile development with Scrum

translating this into a prioritized list. The *Team* builds the product that the customer is going to consume. The team in Scrum is cross-functional (it includes all the expertise necessary to deliver the potentially shippable product each Sprint) and it is self-managing, with a very high degree of autonomy and accountability. The team decides what to commit to, and how best to accomplish that commitment. The team in Scrum is typically five to ten people, although teams as large as 15 and as small as 3 report benefits, and for a software project the team might include analysts, developers, interface designers, and testers. The team builds the product, but they also provide input and ideas to the Product Owner about how to make the product as good as it can be. The *ScrumMaster* is one of the most important elements of Scrum success. The ScrumMaster does whatever is in their power to help the team be successful. The ScrumMaster is not the manager of the team; instead, the ScrumMaster serves the team, protects the team from outside interference, and guides the teams use of Scrum. The ScrumMaster makes sure everyone on the team (as well as those in management) understands and follows the practices of Scrum, and they help lead the organization through the often difficult change required to achieve success with Agile methods. Since Scrum makes visible many impediments and threats to the team's effectiveness, it's important to have a strong ScrumMaster working energetically to help resolve those issues, or the team will find it difficult to succeed. Scrum teams should have someone dedicated full-time to the role of Scrum-Master (often the person who previously played the role of Project Manager), although a smaller team might have a team member play this role (carrying a lighter load of regular work when they do so). Great ScrumMasters have come from all backgrounds and disciplines: Project Management, Engineering, Design, Testing. The ScrumMaster and the Product Owner should not be the same individual; at times, the ScrumMaster may be called upon to push back on the Product Owner (for example, if they try to introduce new deliverables in the middle of a Sprint). And unlike a Project Manager, the ScrumMaster does not tell people what to do or assign tasks they facilitate the process, supporting the team as it organizes and manages itself so if the ScrumMaster was previously in a position managing the team, they will need to significantly evolve their mindset and style of interaction in order for the team to be successful with Scrum. In addition to these three roles, there are other important contributors to the success of the project: Perhaps the most important of these are *Managers*. While their role evolves in Scrum, they remain critically important they support the team by respecting the rules and spirit of Scrum, they help remove impediments that the team identifies, and they make their expertise and experience available to the team.

6.3 eXtreme Programming XP

Extreme Programming is an iterative approach to software development [3], the process is shown in Figure 2. The methodology is designed to deliver the software that customer needs when it's needed. This methodology emphasizes team work. Managers, customers, and developers are all part of a team dedicated to deliver quality software. XP implements a simple, yet effective way to enable groupware style development. XP improves a software project in four essential ways; communication, simplicity, feedback, and courage.

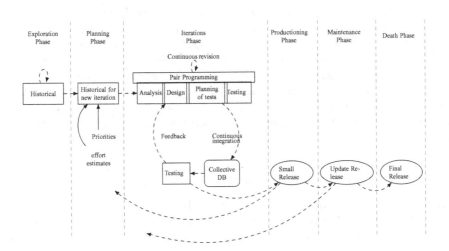

Fig. 2. Lifecycle XP Phases

6.4 Roles in XP

XP defines the following roles for a software development process [3]. The *Programmer* writes source code for the software system under development. This role is at the technical heart of every XP project because it is responsible for the main outcome of the project: the application system. The *Customer* writes user stories, which tell the programmer what to program. "The programmer knows how to program. The customer knows what to program". The *Tester* is responsible for helping customers select and write functional tests. On the other side, the tester runs all the tests again and again to create an updated picture of the project state. The *Tracker* keeps track of all the numbers in a project. This role is familiar with the estimation reliability of the team. Whoever plays this role knows the facts and records of the project and should be able to tell the team whether they will finish the next iteration as planned. The *Coach* is responsible for the development process as a whole. The coach notices when the team is getting "off track" and puts it "back on track". To do this, the coach must have experience with XP. The *Consultant*, whenever the XP team needs additional special knowledge, they "hire" a consultant who possesses this knowledge. The consultant transfers this knowledge to the team members, enabling them to solve the problem on their own. The *Big Boss or Manager* provides the resources for the process. The big boss needs to have the general picture of the project, be familiar with the current project state, and know whether any interventions are needed to en-sure the project's success.

7   Creativity in Agile Development

Regarding to the structure dimension of a new product development team (in particular software), it is possible to relate the roles in creativity to the roles defined in the agile methodology distinguishing: base roles, that is, those directly related to the creative processes and software development, and support roles, whose function is to support or lead the other roles for a better performance. In relation with the structure dimension it's important to considerate how the team can operate. In order to implement the functionality of each role, we must considerate two aspects: basic organizational conditions and the pertinent creative process.

7.1   Basic Organizational Conditions

The creative team performance is determined by the organizational conditions in which it's inserted [1][16][19]. Some conditions are necessary - although not sufficient - for the creative performance. We are interested in explore the influence of autonomy, communication, cooperation and learning, the handling of possible conflicts, pressure, formalization, performance evaluation, available resources (time) and the physical atmosphere of work. The *autonomy* refers to the capacity of the people and the team as a whole to act and make decisions. This aspect is related to the following agile practices: the actual client, since it is part of the team and, in addition, has decisional capacity delegated by its own organization; the use of metaphors, of codification standards and the existence of "right" rules really represent codes of shared thought and action, that make possible the autonomy of the team members; the small deliveries and the fact of the collective property allow that all the involved ones share official and explicit knowledge, that results in a greater independence of the members and the possibility of a minor coordination among them. The team member's *communication*, cooperation and learning are fortified since the client is present and there exist opened spaces to work together and in a pair programming mode. The work dynamics is based on a game of planning and metaphors involving all the participants from the beginning (client and equipment developer). Also, the use of codification standards, the small deliveries, the collective property of the code and the simple design, allow that the person has clear performance codes and rules about what is expected and acceptable (internal culture) in order to establish the required communication and cooperation. The *handling of possible conflicts* between the client and the development team, and internally at team level is favored by agile practices handling it (presence of the client, pairs programming, planning game, continuous integration, tests, collective property), or to reduce it and to avoid it (small deliveries, simple design, 40 hour a week and codification standard). Cooperation and trust are associated to this issue. The *pressure* (that in creativity is appraised as favorable until certain degree, favoring the performance, and detrimental if it exceeds this degree), is susceptible then to favor it through the client in situ, the programming by pairs, the planning game, the tests and continuous integration. It's possible to avoid, or at least to reduce, the pressure through the refactorization, the small deliveries, the collective property, and the fact that surpassing the 40 weekly working hours is seen as an error. The *formalization*, that gives account of all those formal aspects (norms, procedures) defined explicitly and that are known, and even shared, by the members of the team. It's assured through planning game, metaphors, continuous integration, the collective property, the 40 hours per week and the codification standards guiding the desirable conduct and performance of the team. The *evaluation of the performance* is made in XP through pair programming (self evaluation and pair evaluation), frequent tests and even through the 40 weekly hours (as a nonexceedable metric indicating limit of effectiveness), all at the light of the planning (including the standards). Finally, the presence of client constitutes the permanent and fundamental performance evaluation of the team and the products. The evaluation characteristics empower the learning processs. The *time* dedicated has fundamental importance in XP team respecting the available resources. This aspect is strongly stressed in creativity. The pair programming and the developer multifunctional role allow to optimize the partial working-times, as well as the whole project time, ensuring a positive pressure. The *physical atmosphere of work*, referred in creativity to the surroundings that favor or make difficult the creative performance (including aspects like available spaces, noise, colours, ventilation, relaxation places) are assured only partially with agilists through the open spaces, as a way to assure the interaction between members of the team.

7.2   The Creative Process

The team performance is directly determined by the creative process [19]. In practice, it's important to correlate the phases defined in a creative process with a specific agile process, we performed the comparison with XP.

- The initial preparation and "finding" defined in the creative process correspond to the exploration phase in XP, where the functionality of the prototype and familiarization with the methodology are established.
- The final stage of preparation is equivalent with the phases of exploration and planning in XP, defining more in detail the scope and limit of the development.
- The option generation phases, incubation and election of options defined in the creative process correspond to the iterations made in XP and also with the liberations of the production phase (small releases). In XP there is not a clear distinction of the mentioned creative phases, assuming that they occur to the interior of the team.
- The feedback phase (understanding this one as a final stage of the process, and not excluding that can have existed previous micro - feedbacks since the creative process is nonlinear) it could correspond in XP with the maintenance phase.
- The persuasion phase is related to the phase of death established in XP, constituting the close of the development project with the final liberation.

7.3   Creative Roles and XP Roles

As previously mentioned in the creative process there are base and supporting roles. The base roles are directly related to the creative and software development process and the supporting roles support or lead the base roles to a better performance. The following is the correlation between creative and XP roles:

- The detective function consisting in collecting information related to a problem is made by the client him-self in XP, because this one generates the first contact with the software development team.
- The function of explorer consisting in defining completely the problem is made in XP as much by the client as the manager of the team, all together they appreciate the reach of the identified problem, as well as of the possible solutions.
- The function of the artist consisting in transforming the information, creating new relations, and therefore generating interesting solutions is made by the developer, that in XP methodology is in charge of the analysis, design and programming of software.
- The function of the engineer referred to clarify and to evaluate the new ideas, in terms of its feasibility is made in XP by the tester and the tracker.
- The function of the judge, understood as the definitive selection of the solutions to implant, is made in XP by the tracker and the client.

- The function of the producer, referred to the implementation of the selected ideas (strictly speaking it is working software) is made in XP by the client in his organization, including the processes and procedures that this function implies.

The supporting roles considered are:

- The provoker; creativity demands that the divergence as well as convergence in the solutions be maximum and complete. There is not explicit reference in XP methodology about divergent thinking.
- The think tank who helps the team work "from outside" is equivalent completely to the role of the consultant.
- The facilitator whose function is helping the team, corresponds in XP to the coach role.
- The manager whose function is to lead to the team in terms of its general efficiency and its effectiveness corresponds with XP's big boss or manager.

8   Knowledge Management and Software Reuse

It would be desirable to employ Agile principles while also employing reuse techniques to improve the software quality and reduce development effort, time and cost. In [22] was introduced the term Agile Reuse to describe such an approach. They argue that in practice several inherent difficulties arise when considering the compatibility of Agile and reuse techniques due to differences, often contradictory, in their fundamental principles. The benefits of reuse-based software development are obvious, nevertheless it is unclear how reusability can be working with Agile development. Meanwhile, keeping with Agile principles, reusability is considered exclusively in relation to source code. It has been said that Knowledge Management implementations in Software Engineering can extract knowledge from its sources of knowledge: documentation, artifacts, objects, components, patterns, templates and code repositories, exploiting this knowledge in future software developments. But, software reuse is not a technology problem, nor is it a management problem. Reuse is fundamentally a Knowledge Management problem. In [15] Jim Highsmith explains how over the last ten or so years, by packaging objects into components and components into templates, we have made the problem bigger, not smaller. The greater the encapsulated knowledge, the harder it is to transfer. Additionally, the essence of problem solving, innovation, creativity, intuitive design, good analysis, and effective project management involves more tacit knowledge, the harder it is to transfer. By putting tacit knowledge in a principal role and cultivating tacit knowledge environments, KM can play an important role in application development, and particularly in reuse. An understanding of knowledge sharing and transfer issues offers important insights about Reusability and Software Engineering.

9   Knowledge Sharing Support in Agile and Tayloristic Methods

About knowledge sharing in plan-driven and agile development approaches the main different strategies are in the following dimensions [6]:

Eliciting Requirements and Documentation. Common to all software development processes is the need to capture and share knowledge about the requirements and design of the product, the development process, the business domain and the project status. In Tayloristic development approaches this knowledge is externalised in documents and artifacts to ensure all possible requirements, design, development, and management issues are addressed and captured. One advantage to this emphasis on knowledge externalisation is that it reduces the probability of knowledge loss as a result of knowledge holders leaving the organisation. Agile methods advocate lean and mean documentation. Compared to Tayloristic methods, there is significantly less documentation in agile methods. As less effort is needed to maintain fewer documents, this improves the probability that the documents can be kept up to date. To compensate for the reduction in documentation and other explicit knowledge, agile methods strongly encourage direct and frequent communication and collaboration.

Training. With regards to disseminating process and technical knowledge from experienced team members to novices in the team, Tayloristic and agile methods use different training mechanisms as well. While it is not stated, formal training sessions are commonly used in Tayloristic organizations to achieve the above objective. Agile methods, on the other hand, recommend informal practices, for example, pair programming and pair rotation in case of eXtreme Programming.

Trust and Freedom. As software development is a very social process, it is important to develop organisational and individual trust in the teams and also between the teams and the customer. Trusting other people facilitates reusability and leads to more efficient knowledge generation and knowledge sharing. Through collective code ownership, stand-up meetings, on site customer, and in case of XP, pair programming, agile methods promote and encourage mutual trust, respect and care among developers themselves and to the customer as well. The key of knowledge sharing here are the interactions among members of the teams which happen voluntarily, and not by an order from the headquarters [7].

Team Work and Roles. In Tayloristic teams different roles are grouped together as a number of role-based teams each of which contains members who share the same role. In contrast, agile teams use cross functional teams. Such a team draws together individuals performing all defined roles. In knowledge intensive software development that demands information flow from different functional sub-teams, role based teams tend to lead to islands of knowledge and difficulties in its sharing among all the teams emerge. Learning, or the internalisation of explicit knowledge, is a social process. One does not learn alone but learns mainly through tacit knowledge gained from interactions with others. Furthermore, tacit knowledge is often difficult to be externalised into a document or repository. A repository by itself does not support communication or collaboration among people either. Due to the high complexity of the software process in general, it is hard to create and even more difficult to effectively maintain the experience repository [29].

10   Conclusions

In Software Engineering many development approaches work repeating the basic linear model in every iteration (analysis-design-implementation-testing), then in a lot of cases an iterative development approach is used to provide rapid feedback and continuous learning in the development team. To facilitate learning among developers, Agile methods use daily or weekly stand up meetings, pair programming and collective ownership. Agile methods emphasis on people, communities of practice, communication, and collaboration in facilitating the practice of sharing tacit knowledge at a team level.

Knowledge Management provides methods and techniques wich help teams to process and reuse their knowledge. An important finding on KM in software engineering is the need to not focus exclusively on explicit knowledge but also on tacit knowledge. They also foster a team culture of knowledge sharing, mutual trust and care. Agile development is not defined by a small set of practices and techniques. Agile development defines a strategic capability, a capability to create and respond to change, a capability to balance flexibility and structure, a capability to draw creativity and innovation out of a development team, and a capability to lead organizations through turbulence and uncertainty. They rough out blueprints (models), but they concentrate on creating working software. They focus on individuals and their skills and on the intense interaction of development team members among themselves and with customers and management.

By other side, Creativity and innovation are essential skills in almost any teamwork. Having a team that can solve problems quickly and effectively with a little creative thinking is beneficial to everyone. The performance of a team depends not only on the competence of the team itself in doing its work, but also on the organizational context. The organizational conditions in wich the team is inserted are very important too. If workers see that their ideas are encouraged and accepted, they will be more likely to be creative, leading to potential innovation in the workplace. The creation of a collaborative work environment will foster the communication between employees and reward those that work together to solve problems. Encouraging team members to take risks, the opposite of creativity is fear, then it is necessary to create an environment that is free from fear of failure: failures are a learning tool. The Extreme Programming methodology includes implicitly central aspects of a creative teamwork. These aspects can be organized according to the structure that the team adopts and the performance that characterizes to the team. The structure that the team adopts and specially the different roles that the methodology advises to define, nearly correspond with the roles at the interior of a creative team. The performance that characterizes the team through certain advisable practices, from the perspective of creativity, constitutes the necessary basic conditions, although nonsufficient, in order to favor the group creative performance. These conditions, called practices in agile methodologies, are accompanied by concrete phases of constituent activities of an agile software development process, which is possible to correspond with the creative process, which is fundamental to the creative performance.

In spite of the previous comments, we think that agile methodologies should have a more explicit reference to the provoker role that is thoroughly described in creativity as a fundamental factor to generate innovation. This can be explained because, in general, agile methodologies do not aim, as a central element, to generate an original software, but an effective one. Secondly, it is essential the distinction and formalization of the creative phases to generate options incubation and option choices. It is assumed that they take place in the iterative and production process, although it is not focused in "originality". Thirdly, a more direct mention to the physical atmosphere of work, that in creativity are considered as highly relevant to enhance the performance. These aspects should have a greater consideration since software development is a special case of product development.

References

1. Amabile, T.: How to kill creativity. Harvard Business Review, 77–87 (September-October 1998)
2. Apostolou, D., Mentzas, G.: Experiences from knowledge management implementations in companies of the software sector. Business Process Management Journal 9(3) (2003)
3. Beck, K.: Extreme programming explained: embrace change. Addison-Wesley Longman Publishing Co., USA (2000)
4. Beck, K.: Agile alliance (2001), http://agilemanifesto.org
5. Cafer, F., Misra, S.: Effective project leadership in computer science and engineering. In: Gervasi, O., Taniar, D., Murgante, B., Laganà, A., Mun, Y., Gavrilova, M.L. (eds.) ICCSA 2009, Part II. LNCS, vol. 5593, pp. 59–69. Springer, Heidelberg (2009)
6. Chau, T., Maurer, F., Melnik, G.: Knowledge sharing: Agile methods vs tayloristic methods. In: Twelfth International Workshop on Enabling Technologies: Infrastructure for Collaborative Enterprises, WETICE, pp. 302–307. IEEE Computer Society, Los Alamitos (2003)
7. Cockburn, A., Highsmith, J.: Agile software development: The people factor. IEEE Computer 34(11), 131–133 (2001)
8. Crawford, B., de la Barra, C.L.: Enhancing creativity in agile software teams. In: Concas, G., Damiani, E., Scotto, M., Succi, G. (eds.) XP 2007. LNCS, vol. 4536, pp. 161–162. Springer, Heidelberg (2007)
9. Crawford, B., de la Barra, C.L., Soto, R., Misra, S., Monfroy, E.: Knowledge management and creativity practices in software engineering. In: Liu, K., Filipe, J. (eds.) KMIS, pp. 277–280. SciTePress (2012)
10. Deemer, P., Benefield, G.: The scrum primer. An introduction to agile project management with scrum (2007)
11. Fernández-Sanz, L., Misra, S.: Influence of human factors in software quality and productivity. In: Murgante, B., Gervasi, O., Iglesias, A., Taniar, D., Apduhan, B.O. (eds.) ICCSA 2011, Part V. LNCS, vol. 6786, pp. 257–269. Springer, Heidelberg (2011)

12. Fernandez-Sanz, L., Misra, S.: Analysis of cultural and gender influences on teamwork performance for software requirements analysis in multinational environments. IET Software 6(3), 167–175 (2012)
13. Glass, R.: Software creativity. Prentice-Hall, USA (1995)
14. Gu, M., Tong, X.: Towards hypotheses on creativity in software development. In: Bomarius, F., Iida, H. (eds.) PROFES 2004. LNCS, vol. 3009, pp. 47–61. Springer, Heidelberg (2004)
15. Highsmith, J.: Reuse as a knowledge management problem, http://www.informit.com/articles/article.aspx?p=31478
16. Isaksen, S., Lauer, K., Ekvall, G.: Situational outlook questionnaire: A measure of the climate for creativity and change. Psychological Reports, 665–674 (1999)
17. Kelley, T., Littman, J.: The Ten Faces of Innovation: IDEO's Strategies for Defeating the Devil's Advocate and Driving Creativity Throughout Your Organization. Doubleday Random House, USA (2005)
18. de la Barra, C.L., Crawford, B.: Fostering creativity thinking in agile software development. In: Holzinger, A. (ed.) USAB 2007. LNCS, vol. 4799, pp. 415–426. Springer, Heidelberg (2007)
19. Leonard, D., Swap, W.: When Sparks Fly: Igniting Creativity in Groups. Harvard Business School Press, Boston (1999)
20. Lumsdaine, E., Lumsdaine, M.: Creative Problem Solving: Thinking Skills for a Changing World. McGraw-Hill, New York (1995)
21. Maiden, N., Gizikis, A., Robertson, S.: Provoking creativity: Imagine what your requirements could be like. IEEE Software 21, 68–75 (2004)
22. McCarey, F., Cinnéide, M.Ó., Kushmerick, N.: An eclipse plugin to support agile reuse. In: Baumeister, H., Marchesi, M., Holcombe, M. (eds.) XP 2005. LNCS, vol. 3556, pp. 162–170. Springer, Heidelberg (2005)
23. Mentzas, G.: The two faces of knowledge management. International Consultant's Guide, pp. 10–11 (May 2000), http//imu.iccs.ntua.gr/Papers/O37-icg.pdf
24. Mich, L., Anesi, C., Berry, D.: Applying a pragmatics-based creativity-fostering technique to requirements elicitation. Requir. Eng. 10, 262–275 (2005)
25. Moe, N., Dingsoyr, T., Dyba, T.: A teamwork model for understanding an agile team: A case study of a scrum project. Information and Software Technology 52, 480–491 (2010)
26. Nonaka, I., Takeuchi, H.: The Knowledge Creating Company. Oxford University Press, USA (1995)
27. Ringstad, M.A., Dingsøyr, T., Brede Moe, N.: Agile process improvement: Diagnosis and planning to improve teamwork. In: O'Connor, R.V., Pries-Heje, J., Messnarz, R. (eds.) EuroSPI 2011. CCIS, vol. 172, pp. 167–178. Springer, Heidelberg (2011)
28. Robertson, J.: Requirements analysts must also be inventors. IEEE Software 22, 48–50 (2005)
29. Rus, I., Lindvall, M.: Knowledge management in software engineering. IEEE Software 19(3), 26–38 (2002), http://fc-md.umd.edu/mikli/RusLindvallKMSE.pdf
30. Schwaber, K., Beedle, M.: Agile Software Development with Scrum, 1st edn. Prentice Hall PTR, Upper Saddle River (2001)
31. Sutherland, J.: Agile can scale: Inventing and reinventing scrum in five companies. Cutter IT Journal 14, 5–11 (2001)
32. Takeuchi, H., Nonaka, I.: The new new product development game. Harvard Business Review (1986)
33. Wallas, G.: The art of thought. Harcourt Brace, New York (1926)

Formalization and Model Checking
of SysML State Machine Diagrams by CSP#

Takahiro Ando[1], Hirokazu Yatsu[1], Weiqiang Kong[1],
Kenji Hisazumi[2], and Akira Fukuda[1]

[1] Graduate School of Information Science and Electrical Engineering,
Kyushu University, Japan
{ando.takahiro,hirokazu.yatsu,fukuda}@f.ait.kyusyu-u.ac.jp,
weiqiang@qito.kyushu-u.ac.jp
[2] System LSI Research Center, Kyushu University, Japan
nel@slrc.kyushu-u.ac.jp

Abstract. SysML state machine diagrams are used to describe the behavior of blocks in the system under consideration. The work in [1] proposed a formalization of SysML state machine diagrams in which the diagrams were translated into CSP# processes that could be verified by the state-of-the-art model checker PAT. In this paper, we make several modifications and add new rules to the translation described in that work. First, we modify three translation rules, which we think are inappropriately defined according to the SysML definition of state machine diagrams. Next, we add new translation rules for two components of the diagrams – junction and choice pseudostates – which have not been dealt with previously. As the contribution of this work, we can achieve more reasonable verification results for more general SysML state machine diagrams.

Keywords: SysML state machine diagrams, formal semantics, model checking, CSP#.

1   Introduction

The OMG Systems Modeling Language (SysML) [2] is a systems modeling language that supports specification description, design, analysis, and verification of systems.

SysML is a language extension of Unified Modeling Language (UML) [3], the de facto standard software modeling language. SysML has nine types of diagrams and the state machine diagrams are one of them. In such diagrams, the life-cycle behavior of a block in a system is expressed as a state transition system. SysML diagrams, including state machine diagrams, do not have a strict formal semantics. This interferes with checking for correctness of the description and makes it difficult to verify with formal methods especially.

Model checking [4] is a well-known formal verification technique for formally analyzing state transition systems. In model checking, a target system is modeled with a formal description language and the model is then exhaustively

B. Murgante et al. (Eds.): ICCSA 2013, Part III, LNCS 7973, pp. 114–127, 2013.

explored to check whether desired properties of the system are satisfied. SPIN [5], NuSMV [6], and UPPAAL [7], etc. are state-of-the-art model checkers. In using these tools, a state transition system is modeled in their respective formal description languages, and properties to be checked are written in formal specification languages such as Linear Temporal Logic (LTL) and Computation Tree Logic (CTL), etc.

PAT [8] is a model checking tool whose performance is comparable to that of SPIN, and it uses CSP# as its model description language. CSP# is a language extension of Hoare's Communicating Sequential Process (CSP) [9] with description supports for conditional choice, shared variable and, sequential programs, etc. CSP is a kind of process algebra and it has been used in the formalization of concurrent systems. CSP and CSP# are suitable to describe processes with interruptions and parallel processes.

In this paper, we investigate formalization and model checking for SysML state machine diagrams. In our formalization, the diagrams are translated into CSP# descriptions, and then, we apply PAT model checker to check the CSP# models against desired properties. Similar translation rules from the UML state machine diagrams into CSP# have been proposed previously in [1]. In this paper, we make modification to some translation rules described in [1], which we think are inappropriate according to the definition of SysML state machine diagrams. In addition, we give new translation rules for two components of the diagrams that have not been dealt with in [1]. Moreover, we demonstrate and evaluate our translation rules with a case study for a simple user certification system.

Organization. In Section 2, we describe related work. Section 3 presents our translation rules for SysML state machine diagram. In Section 4, we demonstrate and evaluation our translation rules in a case study. Section 5 concludes this paper and mentions future work.

2   Related Work

Quite a lot research has been done for applying formal verification technique for formally analyzing state machine diagrams. For example, the work in [10] gives a comprehensive survey on researches that apply model checking to state machines, in which various model checkers including SPIN [5], SMV [11], and FDR [12], etc. are used.

In [13], state machines are translated into models written in the Promela language and then verified using SPIN. This work deals only with basic components of state machine diagrams. In [14], a tool called vUML is proposed for verification of UML models using SPIN, but detailed translation rules into Promela, the input language of SPIN, are not described.

Translation for the state machines into SMV input format has been proposed in [15]. In the work, each state (and each event) is assigned a single variable for symbolic model checking. Therefore, even for small diagrams, the state explosion problem in symbolic model checking can become more obvious due to the increase of variable numbers. In [16], the semantics and a symbolic encoding of UML state machine diagrams, which has more complex notations, are shown and the methods of verification is based on NuSMV. The symbolic encoding of state machines has also been given in [17]. The work proposes to apply SAT-based bounded model checking for analyzing state machines.

Formalizations in which state machines are translated into process algebra have been proposed in [18], [19], [20], and [1]. These works are closely related to our work. In [18], state machines are formalized with mCRL2 [21], a language based on the Algebra of Communicating Process (ACP) [22], and the model is verified using the mCRL2 toolset. In [19], [20], and [1], CSP is selected as the process algebra to formalize state machines. The formalization in [20] can handle entry and exit behaviors, but cannot handle some pseudostates which represent complex transitions. And, formalized models in [20] are verified by FDR2 [12] model checker.

[1] is a direct predecessor of this paper. The work gives a formalization in which state machines are translated into CSP#, a variant language of CSP. In this paper, we make modifications to some translation rules proposed in [1], and add several rules for components of state machine diagrams that are not dealt with in that work.

3   Formalization of State Machine Diagram

In this section, we briefly explain the elements of SysML state machine diagrams and the formal language CSP#. Then, we give the translation rules from state machine diagrams into CSP.

3.1   State Machine Diagram of SysML

When modeling a system in SysML, a state machine diagram is used for illustrating behavior of a block that is a component of the system. A state machine diagram consists of states and transitions between the states. The complete syntax and notations of state machine diagrams are defined in [2].

A state represents a situation in the life-cycle of the block. In general, the situation is expressed by invariant conditions, etc. Each state is allowed to have *entry*/*do*/*exit* behavior. The *entry* behavior is performed when a block goes into the state, and the *exit* behavior is performed when the block leaves the state. The *do* behavior is performed after the *entry* behavior, and it continues performing until it terminates or the block leaves the state.

There are several types of states such as atomic states, composite states and submachine states. An atomic state has no sub-state. A composition state has one or more orthogonal regions and each region has sub-states. A submachine state expresses a state machine that is to be reused in other diagram. The *entry*/*exit* points of a submachine state correspond to initial/final states in the diagram which the submachine state refers to.

A transition between states is allowed to have three attributes – *trigger*, *guard*, and *effect*. An event acts as a trigger that activates the transition. Only if the trigger event occurs and the guard condition is satisfied, the transition can be fired. When the transition is fired, its effects are performed and then the machine changes from the current state changes to its target state.

In addition, state machine diagrams have several types of pseudostates. Initial pseudostates, final states, junction pseudostates, and choice pseudostates, etc. are representative pseudostates. Initial pseudostates and final states represent the start and end of state machines, respectively. Junction pseudostates and choice pseudostates are notations that are used to describe common parts of multiple transitions.

3.2 CSP#

CSP [9] is a process algebra and is well-known as a formal language for describing interaction patterns of parallel systems. CSP# is the system description language of the model checking tool PAT, which adds conditional choice and shared variable, etc. to CSP.

In the following, we enumerate the elements of a subset CSP# that are used in this paper by giving their intuitive meanings.

Definition 1. *A process P is defined as follows.*

$$P ::= Stop \mid Skip \mid e\{prog\} \rightarrow P \mid P_1; P_2 \mid P_1 \square P_2$$
$$\mid P_1 \parallel P_2 \mid P_1 \parallel\mid\mid P_2 \mid [b]P \mid P_1 \triangle P_2$$
$$\mid case\{b_1 : P_1;\ b_2 : P_2;\ \cdots\ ;\ default : P\}$$

where, P_1 and P_2 are processes, e is an event name that may have optional sequential program prog, and b, b_1 and b_2 are boolean expressions.

Stop represents a dead lock process and *Skip* represents a process that terminates successfully. $e \rightarrow P$ represents a process that performs as process P after event e occurs. When e has a sequential program, it performs the program atomically, and afterwards behaves as the process P. $P_1; P_2$ behaves firstly as process $P1$ first and then as process $P2$. $P_1 \square P_2$ represents a non-deterministic choice, where the first occurred event determines whether it behaves as P_1 or P_2. $P_1 \parallel P_2$ represents a parallel composition process that synchronizes the common events of P_1 and P_2. $P_1 \parallel\mid\mid P_2$ represents an interleaved parallel composition. $[b]P$ represents a process with a guard condition and, only if the boolean expression b is satisfied, it behaves as P. $P_1 \triangle P_2$ represents a process having an interrupt process. It behaves as P_1 until the first event of P_2 occurs and, after the event occurs, it behaves as P_2. In *case* process, condition expressions are checked in turn, and it behaves as the process that corresponds to the first found satisfied condition.

3.3 Translation Rules

Next, we give translation rules from state machine diagrams into CSP#. In CSP#, states and transitions as well as the whole state machine diagrams are

Table 1. Transition Rules from State Machine Diagrams to CSP#

Elements	CSP#	Comments
State Machine sm	$tr(sm) = tr(i)$, where i is the top level initial pseudo state of sm.	Same for regions in composite state.
Initial pseudostate i	$tr(i) = tr(s)$, where s is the target state of the outgoing transition from i.	Same for initial pseudostates in any region.
Final state f	$tr(f) = Skip$;	This means a state machine terminates successfully.
Atomic State s	$tr(s) = tr(entry); ((tr(do); Stop) \triangle (tr(t_1) \square tr(t_2) \square \cdots \square tr(t_n)))$, where $t_1, t_2, \ldots, t_n$ are outgoing transitions from s.	The process of coming after $tr(do)$ is not $Skip$ but $Stop$.
Entry behavior $entry$	$tr(entry) = e_1 \rightarrow e_2 \rightarrow \cdots \rightarrow e_n \rightarrow Skip$, where $e_1, e_2, \ldots, e_n$ are actions in a sequence of entry behavior.	Same for do/exit behavior of a state and effect of a transition.
Transition t	$tr(t) = [guard](event \rightarrow tr(exit); tr(effect); tr(s_t))$, where $exit$ is a exit behavior of source state and s_t is a target state.	
Composite state cs	$tr(cs) = tr(entry); (((tr(do); Stop) \,\|\|\, tr(r_1) \,\|\|\, tr(r_2) \,\|\|\, \cdots \,\|\|\, tr(r_m)) \triangle (tr(t_1) \square tr(t_2) \square \cdots \square tr(t_n)))$, where $r_1, r_2, \ldots r_m$ are regions in cs and $t_1, t_2, \ldots, t_n$ are outgoing transitions.	

all described as processes of CSP#. Events and statements in state machine diagrams are translated into events of CSP#.

Table 1 summarizes translation rules for the elements of state machine diagrams. In the table, for brevity, translation rules are expressed by using an informal function tr, and the translation result for an element e is denoted as $tr(e)$.

Our rules in Table 1 are based on the rules proposed previously in work [1], however there are several differences between them. In the original translation rules of [1], the *entry/exit* behavior of a state and the *effect* of a transition are translated into an atomic process in the form of a sequence of events or statements. However, in our rules, atomic notations are not used for sequences of events or statements, because we believe there is a possibility that certain unsafe behavior of the target system could be hidden due to over abstraction when these sequences are treated as atomic processes in CSP#. That is, when the sequences are treated as atomic processes, the number of behavior patterns are reduced due to decrease of possible interleaving composition of the processes. As a result, certain event occurrences order that may lead to a problem/bug might not be checked when verifying the processes.

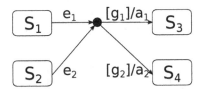

Fig. 1. An Example with Junction Pseudostate

In addition, our translation rules for states have two differences compare to the original rules. The first difference is that the $tr(entry)$ process that represents *entry* behavior is excluded from the region in which an interrupt from an outgoing transition process is caught. This is to reflect that entry behavior must be completed before other behavior and transitions become executable.

Another difference is to concatenate $tr(do)$ process and *Stop* process, which represent *do* behavior and dead lock, respectively, by using operator ";". In the original translation, a translated process is allowed to terminate successfully even if no outgoing transitions are fired after a *do* behavior of some state completes. This is judged from the following two facts.

– When $tr(do)$ process completes, the process behaves as a *Skip* process.
– In CSP#'s semantics, it is allowed that the process "$Skip \triangle P$" terminates successfully without executing process P.

Such a translation is not appropriate because it allows for state machines to terminate successfully in a state that is not a final state. From the CSP# equation, $Stop \triangle P = P$, the behavior that ignores the outgoing transitions can be prohibited by revising "$tr(do)$" as "$tr(do); Stop$".

Next, we give new translation rules for two pseudostates, i.e., junction pseudostates and choice pseudostates, which are used as representations for organizing multiple transitions.

Junction Pseudostates. A junction pseudostate is used to describe common parts of multiple transitions. Therefore, a junction pseudostate can be translated in CSP, by translating the original transitions that are unwound by using information of incoming/outgoing transitions of the pseudostate.

The unwound transition set of a junction pseudostate $Trans_{unwound}$ is given as follows, where $Trans_{in}$ is the set of incoming transitions of the pseudostate and $Trans_{out}$ is the set of outgoing transitions.

$$Trans_{unwound} = \{t \mid \exists t' \in Trans_{in}. \exists t'' \in Trans_{out}.$$
$$\big(source(t) = source(t') \wedge event(t) = event(t')$$
$$\wedge\, guard(t) = guard(t'') \wedge effect(t) = effect(t'')$$
$$\wedge\, target(t) = target(t''))\}$$

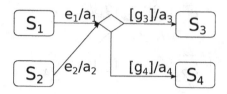

Fig. 2. An Example with Choice Pseudostate

Where the five functions *source*, *event*, *guard*, *effect* and *target* return corresponding source state, event, guard, effects and target state, respectively, of the transition given as an input of the functions. The translation for a junction pseudostate and its incoming and outgoing transitions is the result of applying the translation rule for transitions to each element of $Trans_{unwound}$. For example, the junction pseudostate and its incoming/outgoing transitions on Fig.1 are translated into the following four processes.

$$P_{13} = [g_1](e_1 \rightarrow tr(exit(S_1)); \ tr(a_1); \ tr(S_3))$$
$$P_{14} = [g_2](e_1 \rightarrow tr(exit(S_1)); \ tr(a_2); \ tr(S_4))$$
$$P_{23} = [g_1](e_2 \rightarrow tr(exit(S_2)); \ tr(a_1); \ tr(S_3))$$
$$P_{24} = [g_2](e_2 \rightarrow tr(exit(S_2)); \ tr(a_2); \ tr(S_4))$$

where, *exit* is a function that takes a state and returns *exit* behavior. The processes, P_{13}, P_{14}, are referred in $tr(S_1)$ as translation results of outgoing transitions of the state S_1 as follows.

$$tr(S_1) = tr(entry(S_1)); \ ((tr(do(S_1)); \ Stop) \triangle (P_{13} \ \square \ P_{14}))$$

P_{23}, P_{24} are referred in $tr(S_2)$ similarly.

Choice Pseudostates. A choice pseudostate has multiple incoming and outgoing transitions as a junction pseudostate does. However, the timing to evaluate the guard conditions of outgoing transitions is different from a junction pseudostate. For a choice pseudostate, the guards of outgoing transitions are not evaluated until reaching the choice pseudostate. Thus, the result of *effects* of the incoming transition influences the evaluations. Based on the above analysis, divided into an incoming part and an outgoing part, a choice pseudostate and its related transitions are translated into CSP#.

An incoming transition of a choice pseudostate is translated into CSP# in almost the same way as a transition between states. However, instead of the process of the target state, the process P_{choice}, the translation result of the outgoing part, is used.

Each outgoing transition is also translated almost like a translation between states. However, the process for the *exit* behavior of the source state is omitted. This is because the choice pseudostate is considered as the source state of these outgoing transitions. The process P_{choice}, which represents the set of outgoing transitions of the choice pseudostate, is defined as follows.

$$P_{choice} = tr(t_1) \ \square \ tr(t_2) \ \square \cdots \square \ tr(t_n)$$

where, t_1, t_2, ..., t_n are outgoing transitions of the choice pseudostate. The choice pseudostate and its related transitions on Fig. 2 are translated as follows.

$$P_1 = e_1 \rightarrow tr(exit(S_1)); \ tr(a_1); \ P_{choice}$$
$$P_2 = e_2 \rightarrow tr(exit(S_2)); \ tr(a_2); \ P_{choice}$$
$$P_3 = [g_3](tr(a_3); \ tr(S_3))$$
$$P_4 = [g_4](tr(a_4); \ tr(S_4))$$
$$P_{choice} = P_3 \ \square \ P_4$$

where, P_1, P_2 are processes that represent transitions from states S_1 and S_2, respectively, and they are also referred in $tr(S_1)$ and $tr(S_2)$, respectively. In addition, P_3, P_4 are processes that represent transitions toward states S_3 and S_4, respectively.

4   A Case Study

In this section, we demonstrate our translation rules by translating the state machine diagram of a simple user certification system illustrated in Fig. 3 into CSP#. In the following, a process translated from a state, e.g., labeled with "name", is written as "$name()$".

First, let $System()$ be the process that represents the whole behavior of the state machine in the diagram. The behavior of the state machine is the behavior that follows the top level initial pseudostate, so the process is written as follows.

$$System() = TopLevelInit();$$

where, $TopLevelInit()$ represents the process translated from the top level initial pseudostate. According to Table 1, the initial pseudostate and the final state at the top level are translated as follows.

$$TopLevelInit() = idle();$$
$$TopLevelFinal() = Skip;$$

Next, we show translations for the atomic states "idle", "initializing" and "diagnosing". The idle and diagnosing state do not have $entry/do/exit$ behavior. So, the processes that represent these behaviors of the two states are $Skip$. In addition, since "$Skip; P = P$" is defined in CSP#, the translation results of them can be simply expressed as follows.

$$idle() = (StartUp \rightarrow initializing()) \ \square \ (TurnOff \rightarrow TopLevelFinal());$$
$$diagnosing() = (SystemOK \rightarrow operating()) \ \square \ (SystemNG \rightarrow idle());$$

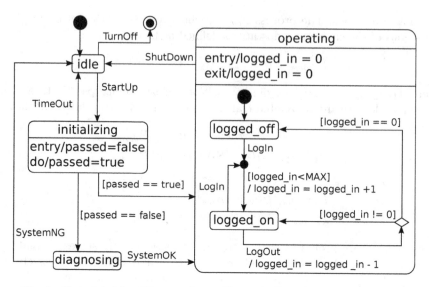

Fig. 3. State Machine Diagram for an Example User-Certification System

On the other hand, the initializing state has *entry/do* behavior that change the value of the variable *passed*. In our translation rules, these behavior is considered as action sequences and is translated to event sequences of CSP#. An event in CSP# is allowed to be attached with a sequential program. For this initializing state, its *entry* behavior is translated into "{*passed* = *false*} → *Skip*", and its *exit* behavior is translated into "{*passed* = *true*} → *Skip*". In addition, the variable *passed* used in these behaviors should also be declared as parts of the translation. Finally, the initializing state (i.e., process) is translated as follows.

> *// Declaration of Shared Variable*
> *var passed = false;*
>
> $initializing() = \{passed = false\} \rightarrow (\ (\{passed = true\} \rightarrow Stop)$
> $\triangle\ (\ (TimeOut \rightarrow idle())\ \square\ ([passed == true]\ operating())$
> $\square\ ([passed == false]\ diagnosing())\ )\ );$

However, when applying previous translation rules proposed in [1], this initializing state is translated as follows.

> *// When applying previous translation rules*
> $initializing() = (\ \{passed = false\} \rightarrow \{passed = true\} \rightarrow Skip)$
> $\triangle\ (\ (TimeOut \rightarrow idle())\ \square\ ([passed == true]\ operating())$
> $\square\ ([passed == false]\ diagnosing())\ );$

In this translation by previous rules in [1], the entry behavior process of initializing state, {*passed* = *true*}, is included in the region where an interrupt from an outgoing transition process is caught, and *Stop* in the translation by our rules is replaced with *Skip*. Adopting this previous translation, the process *System()* which represents the whole system's behavior can receive the following event sequences and then terminate successfully.

$$StartUp \rightarrow TimeOut \rightarrow TurnOff \rightarrow Skip$$
$$StartUp \rightarrow \{passed = false\} \rightarrow \{passed = true\} \rightarrow Skip$$

The first sequence means that the user certification system can receive the *Time-Out* event before the entry behavior of the initializing state is executed. On the other hand, the second sequence means that the system can be terminated successfully in the initializing state without firing no outgoing transition. However, we think that these behaviours are inappropriate as the system behaviors according to the definition of the SysML state machine diagrams. That is, we believe that the translated process of the system should not receive the event sequences. Using our translation, there is no possible that they are received.

From now on, we describe the translation of the composite state "operating", which has a region with junction and choice pseudostates. The composite state has an *entry* behavior, and this behavior must be executed before the internal state transition. After the *entry* behavior completes, the state transition started from the initial pseudostate in the internal region is performed. The composite state also has an *exit* behavior, and the behavior is executed when the outgoing transition to the idle state is fired by event *ShutDown*. This composite state is translated as follows, where *operatingSubInit()* is the process translated from the internal initial pseudostate.

```
// Composite State
var logged_in = 0;
operating() = {logged_in = 0}
    → ( (operatingSubInit() ||| Stop)
         △ (ShutDown → {logged_in = 0} → idle()) );
```

The two sub-states "logged_off" and "logged_on" are associated with the transitions with junction and choice pseudostates. No translation rule for states and transitions associated with these pseudostates is proposed in [1]. However, we can translate them by using our translation rules added in this paper.

The junction pseudostate in the composite state groups two transitions, from logged_off to logged_on and from logged_on to logged_on. These two transitions have *LogIn* event as a trigger and the boolean expression "$logged\_in < MAX$" as a guard condition. Both of the original unwound transitions are translated as follows.

```
// for junction pseudostate
[logged_in < MAX] (LogIn → {logged_in = logged_in + 1} → logged_on())
```

The choice pseudostate has an incoming transition whose effects decrement the value of variable *logged_in* by 1, and two transitions whose guard condition compares *logged_in* with 0. For a choice pseudostate, as mentioned earlier, the guard condition of an outgoing transition is evaluated after the effect of the incoming transition is performed. Therefore, the incoming and outgoing parts of the choice pseudostate are translated as follows, respectively.

// incoming part of choice pseudostate
LogOut → {*logged_in* = *logged_in* − 1} → *Choice()*

// outgoing part of choice pseudostate
Choice() = *([logged_in* == *0] logged_off())* □ *([logged_in* != *0] logged_on())*

From the above, the internal state transition in the composite state operating is translated as follows.

// for the region in "operating" composite state
operatingSubInit() = *logged_off();*

logged_off() =
    [logged_in < *MAX] (LogIn* → {*logged_in* = *logged_in* + 1} → *logged_on());*

logged_on() =
    ([logged_in < *MAX] (LogIn* → {*logged_in* = *logged_in* + 1} → *logged_on()))*
    □ *(LogOut* → {*logged_in* = *logged_in* − 1} → *Choice());*

// outgoing part of choice pseudostate
Choice() = *([logged_in* == *0] logged_off())* □ *([logged_in* != *0] logged_on());*

The CSP# descriptions obtained as above can be used as an input model for the model checking tool PAT [8]. In the rest of this section, we describe about model checking of the translated descriptions (model) with PAT.

We consider the following three properties as requirements that should be satisfied by the state machine of Fig. 3.

1. The state machine is deadlock free.
2. The state machine can reach the state in which the condition, "*logged_in* > 0", is satisfied.
3. The value of the variable *logged_in* always satisfies "$0 \leq logged\_in \leq MAX$".

PAT has built-in functions for checking deadlock-freeness and reachability for a CSP# process. For properties 1 and 2, they are written as simple assertions of CSP# as follows, and they can be checked by evaluating these assertions with PAT.

#assert System() deadlockfree;   (for Property 1)

#define logged_on_prop (logged_in > 0);* ⎫
#assert System() reaches logged_on_prop; ⎬   (for Property 2)

Property 3 is expressed in LTL as follows.

$$\Box((logged\_in \geq 0) \wedge (logged\_in \leq MAX))$$

In order to check whether the state machine satisfies this LTL property, the following assertion is defined and evaluated.

> #define user_num_prop ((logged_in >= 0) && (logged_in <= MAX));
> #assert System() | = []user_num_prop;

When these three assertions are evaluated with PAT, all properties are evaluated as valid. Finally, we consider to check whether the system satisfies the following property 4.

4. When the event *LogIn* is received, the value of the variable *passed* is inevitably true.

Property 4 is defined as an assertion as follows and evaluated with PAT.

> #define passed_prop (passed == true);
> #assert System() | = [](LogIn → passed_prop);

When the assertion is evaluated with PAT, the property is evaluated as "NOT valid", and the following event sequence is given as a counter example.

$$StartUp \rightarrow \tau \rightarrow SystemOK \rightarrow \{logged\_in = 0\} \rightarrow LogIn$$

where τ represents an event not labeled. This sequence represents a path to reach the logged_off sub-state through the diagnosing state from the initializing state and to receive the event *LogIn* at the logged_off sub-state. If the system behaves along the path, the value of *passed* is false until *LogIn* event is received. Therefore we can make sure that the system does not satisfy the property 4.

The processes that can be model-checked with PAT are not limited to the top level process *System()* that represents behavior of the whole state machine. For instance, if the process *System()* in the above assertions is replaced with the process *operatingSubInit()*, the internal state transition of the composite state could be model checked. Since the initial pseudostate in each region is translated into a CSP# process in our translation rules, such operations can be performed easier. When invariant properties (i.e., properties that should be satisfied through the whole system) are to be checked, checking them from the internal state transitions could help achieve early detection of defects that may be possibly lurk in deep nests. The translation rules given in this paper have the advantage that such hierarchical checking can be handled flexibly.

5   Conclusions and Future Work

In this paper, we described translation rules from SysML state machine diagrams to CSP# processes. This translation allows formal verification of the correctness of SysML state machine diagrams. We modified some translation rules proposed in [1], which we believe are not appropriate according to the formal definitions of SysML state machines. As a result, model checking of SysML state machine diagrams has become more accurate against the definition of the diagrams.

In addition, we added translation rules for two elements – junction pseudostates and choice pseudostates, which are not handled in [1]. These rules have contributed to increasing component types which are targets of model checking, and then,the coverage of verifiable diagrams has extended. Moreover, we conducted a case study to demonstrate the actual translation, and then, we showed some model checking results with PAT. These model checking results showed that, compared with the case where the previous translation rules in [1], verification of SysML state machine diagrams are more accurate when our translation rules are used.

Future Work. Regarding future work, we consider that it is important to deal with the followings. In order to expand the coverage of our methods, the behavior with message communication across multiple state machine diagrams should be handled. In addition, it is necessary to develop translation rules for other types of SysML diagrams such as block diagrams, parametric diagrams, and sequence diagrams, etc. Moreover, we plan to implement our translation rules and integrate the implementation with a web-based model driven development (MDD) tool Clooca [23], by which SysML diagrams including state machine diagrams can be graphically developed. Clooca is used currently for education purpose mainly. We have a vision that we will be able to perform a series of tasks like the followings on the Web: 1) draw SysML diagrams with Clooca, 2) translate the diagrams and generate processes in CSP#, and 3) apply model checking to check correctness of the diagrams. We expect that we can offer higher quality education for the MDD formalization and model checking by this combination of our transition with Clooca.

Acknowledgment. This research is conducted as a program for the "Regional Innovation Strategy Support Program 2012" by Ministry of Education, Culture, Sports, Science and Technology (MEXT), Japan.

References

1. Zhang, S.J., Liu, Y.: An Automatic Approach to Model Checking UML State Machines. In: IEEE International Conference on Secure Software Integration and Reliability Improvement Companion, pp. 1–6 (2010)
2. OMG: OMG Systems Modeling Language Version 1.3 (June 2012), http://www.omg.org/spec/SysML/1.3/PDF
3. OMG: OMG Unified Modeling Language Superstructure Version 2.4.1 (August 2011), http://www.omg.org/spec/UML/2.4.1/Superstructure/PDF
4. Edmund, M., Clarke, J., Grumberg, O., Peled, D.A.: Model Checking. The MIT Press (1999)
5. Holzmann, G.: The Model Checker SPIN. IEEE Trans. 23(5), 279–295 (1997)
6. Cimatti, A., Clarke, E., Giunchiglia, F., Roveri, M.: NUSMV: A New Symbolic Model Verifier. In: Halbwachs, N., Peled, D.A. (eds.) CAV 1999. LNCS, vol. 1633, pp. 495–499. Springer, Heidelberg (1999)
7. Larsen, K.G., Pettersson, P., Yi, W.: Model-Checking for Real-Time Systems. In: Reichel, H. (ed.) FCT 1995. LNCS, vol. 965, pp. 62–88. Springer, Heidelberg (1995)

8. Sun, J., Liu, Y., Dong, J.S.: Model Checking CSP Revisited: Introducing a Process Analysis Toolkit. In: Margaria, T., Steffen, B. (eds.) ISoLA 2008. CCIS, vol. 17, pp. 307–322. Springer, Heidelberg (2008)
9. Hoare, C.A.R.: Communicating Sequential Processes. Commun. ACM 21(8), 666–677 (1978)
10. Bhaduri, P., Ramesh, S.: Model Checking of Statechart Models: Survey and Research Directions. CoRR cs.SE/0407038 (2004)
11. McMillan, K.L.: Symbolic Model Checking: An approach to the state explosion problem. PhD thesis, Pittsburgh, PA, USA (1992)
12. The Formal Systems website: FDR2.91 (November 2010), http://www.fsel.com/
13. Latella, D., Majzik, I., Massink, M.: Automatic Verification of a Behavioural Subset of UML Statechart Diagrams Using the SPIN Model-checker. Formal Asp. Comput. 11(6), 637–664 (1999)
14. Lilius, J., Paltor, I.P.: vUML: A Tool for Verifying UML Models, 255–258 (1999)
15. Clarke, E.M., Heinle, W.: Modular Translation of Statecharts to SMV. Technical report (2000)
16. Dubrovin, J., Junttila, T., Högskolan, T., Dubrovin, J., Junttila, T., Dubrovin, C.J., Junttila, T.: Symbolic Model Checking of Hierarchical UML State Machines. Technical report, Helsinki University of Technology Laboratory (2007)
17. Niewiadomski, A., Penczek, W., Szreter, M.: A New Approach to Model Checking of UML State Machines. Fundam. Inf. 93(1-3), 289–303 (2009)
18. Hansen, H.H., Ketema, J., Luttik, B., Mousavi, M.R., van de Pol, J.C.: Towards Model Checking Executable UML Specifications In MCRL2. Innovations in Systems and Software Engineering 6, 83–90 (2010)
19. Rasch, H., Wehrheim, H.: Checking Consistency in UML Diagrams: Classes and State Machines. In: Najm, E., Nestmann, U., Stevens, P. (eds.) FMOODS 2003. LNCS, vol. 2884, pp. 229–243. Springer, Heidelberg (2003)
20. Ng, M.Y., Butler, M.: Towards Formalizing UML State Diagrams in CSP. In: 1st International Conference on Software Engineering and Formal Methods, pp. 138–147. IEEE Computer Society (2003)
21. Groote, J.F., Mathijssen, A., Reniers, M., Usenko, Y., Weerdenburg, M.V.: The Formal Specification Language mCRL2. In: Proceedings of the Dagstuhl Seminar. MIT Press (2007)
22. Baeten, J.C.M., Weijland, W.P.: Process Algebra. Cambridge University Press (1990)
23. Technical Rockstars: clooca, http://www.clooca.com

From Arrows to Netlists Describing Hardware

Matthias Brettschneider and Tobias Häberlein

University of Applied Science Albstadt-Sigmaringen
{brettschneider,haeberlein}@hs-albsig.de

Abstract. This paper describes how to transform a functional domain specific langauge (DSL) into hardware represented by a netlist. In earlier papers we proposed the usage of an algebraic structure called "arrows" (basically an abstraction of Haskell's higher-order type ($\rightarrow$) for describing DSLs. This structure forms the basis of a novel concept that gives the developer a tool at hand to describe hardware functionally in a natural way. Aside of that an arrow provides not only a tool to synthesize, but to verify and reason about the input DSL. We have taken this concept to the next stage, from a static size length arrow into a fixed size length vector arrows fitting much better to logic gates with a fixed number of in- and output pins. There are many sound possibilities to use the algebraic arrow data structure to model hardware. This paper presents some of them which showed to be most useful. A simple example, the implementation of a cyclic redundancy check (CRC) algorithm, is used to illustrate the presented techniques.

1   Introduction

As hardware becomes more and more like software, it is not surprising that there are various approaches to bring software development concepts into the hardware development process. Until now, the way hardware is defined is mostly influenced by the imperative/procedural paradigm. Our approach is mainly influenced by the functional programming paradigm.

Modeling hardware in functional languages is not new, there have been various approaches in the past. The beginning dates back to the early 80s where a group around Mary Sheeran came up with a functional hardware description language (HDL) muFP[20]. In the same decade John O'Donnell presented his functional HDL HDRE[13,14]. These two HDL's gave birth to numerous later approaches that are base more or less on them.

One of Sheeran's students, Coen Claessen, later came up with a monadic solution called Lava[1]. Ingo Sanders et al introduced an approach, ForSyDe [17,18] that is based on meta programming techniques. Lava, ForSyDe and also Hydra [14] represent HDL's embedded around the functional programming language Haskell. There is HML[7], a simple HDL which is embedded inside ML, another functional programming language.

There are functional languages that are exclusively designed to describe hardware, like the language SAFL (=Statically Allocated Functional

B. Murgante et al. (Eds.): ICCSA 2013, Part III, LNCS 7973, pp. 128–143, 2013.
© Springer-Verlag Berlin Heidelberg 2013

Language)[19,10,11]. This one compiles direct into a netlist; it does not support dynamic allocated storage like heaps or stacks since these kind of software features do not map well on silicium.

Another candidate for the functional meta programming languages section is the language reFL^{ct}[4]. It's one that is developed at Intel and tailored for hardware design and theorem proving.

There are good reasons why meta programming is beneficial for the use as HDLs. Meta programming allows a program to have a representation of itself, usually by providing a data structure of the abstract syntax tree. With this tool at hand the developer is able to *compute* parts of their program rather than *write* them. This leads to a direct and natural implementation of program transformations.

With cryptol[2] another class of functional HDLs enters the scene. While all the other HDL's are general purpose, cryptol is designed to model cryptographic hardware. A sub-language of cryptol is designed by Galois, explicit to generate FPGA implementations [3]. Cryptol has been adopted by an American agency and is not public developed anymore.

2   Arrows for Hardware Development

Arrows are lately developed concept in functional programming [5]. As a generalization of Monads, Arrows fit the task of hardware development much better. Arrows enable a data flow view onto problems. This fits the hardware developers view which has always been a data flow view, as in circuit board layouts or in state flow diagrams.

The typical problem one faces while describing hardware through functional programming languages is, a functional function is much more feature rich than the hardware pendant. In the functional programming world functions are so called "first class citizens", which means that they can be passed around like ordinary values. Therefore another function might not only expect values as parameters, but also functions. This is similar to C++'s template mechanism, where for example a sorting algorithm is type independently defined. In the functional paradigm such a behavior is achieved by passing an ordering function to the sorting mechanism.

The other effect of functions being first class is called currying. This means that it is perfectly correct for any function to not return a value but to return a function. Currying is something that is present in every function that is supplied with less parameters than expected. Such function calls result in yet another function which now expects the not supplied parameters as arguments.

It is clear that these concepts can not be translated into hardware where a function is symbolized by a circuit. A good HDL must come up with a mechanism to even out these differences. Arrows are a method for this, including their mathematical basis. Arrows offer an interface to the feature set of functions. Similar to a construction kit an arrow enables the developer to choose the features of a function. This is done by providing instances to the arrow type classes for arbitrary date types. We therefore are able to define arrows that match the feature set of a hardware circuit.

Another reason for the use of arrow is that a concept which is less feature rich can be modeled more abstract. Typical HDL's do not take this into account. With the arrow approach we gain a maximum level of abstraction while not permitting features that can not translate into silicium. Arrows are a good compromise between a shallow embedded approach, where the host language is only used to construct the netlist and a deep embedded approach, where a data type inside the host language is the netlist.

The arrow *proc*-notation [15] also transports the data flow view into the textual source. The generation of a netlist from an arrow is done by a "pretty print" function, that is straight forward to implement. This also holds for the generation of typical HDL code like VHDL. They can be used to simulate behavior or proof properties of the circuit. This section gives a brief overview of the Grid-Arrow that can be used to generate VHDL code or to simulate a circuit. The name Grid is analogously chosen to the snap circuit toys from childhood, where one also could combine hardware elements arbitrary.

2.1 Grid-Arrow

The Grid-Arrow is a combination of an executable arrow and a descriptive hardware component part. In a first approach one can imagine the descriptive part to live inside the arrow. But that does not work out, so they are stored equal to each other. The mathematical structure of a tuple represents this conjunction. The arrow resides in the first, the graph occupies the second part of the tuple.

```
newtype Grid a b c = GR (a b c, CircuitDescriptor)
```

Alongside with the arrow execution, the generated Netlist is stored inside the graph structure called a CircuitDescriptor. The Grid type is a surface where hardware elements can be combined to compose new structures and analyze them. It is not only the surface to combine hardware components, it's also recursive. So it is used as hardware component itself which means a Grid is an arrow on its own.

2.2 Circuit-Descriptor

The descriptive part of the arrow is called a CircuitDescriptor, and describes a circuit in detail. The combination sequence of the hardware parts, the pin layout, their cabling structure and also recursively the subcomponents are represented inside the CircuitDescriptor. A circuit with such a descriptor alongside has enough details to be expressed as a Netlist.

There are three distinct forms of circuits:

- *Combinatorial:* are the parts of a circuit that are describable without a clock or buffered data. Boolean logic gates or deterministic finite automata are in this category.

- *Register:* A register is used to describe the storage of data on the one hand and is used to clock computations on the other hand. The implementation of a register highly depends on the target platform. There is no need to know the details at this point.
- *Loop:* Hardware components that are referring onto them self are handled in the looping section.

```
data CircuitDescriptor
  = MkCombinatorial
    { nodeDesc :: NodeDescriptor
    , nodes    :: [CircuitDescriptor]
    , edges    :: [Edge]
    , cycles   :: Tick
    , space    :: Area
    }
  | MkRegister ...
  | MkLoop ...
  | NoDescriptor
    deriving (Eq)
type Netlist = CircuitDescriptor
```

The MkCombinatorial and the MkLoop parts reveal a functional graph structure that is build from two lists. One containing the edges and another one containing nodes. Additionally there are equal information for every circuit. They are stored inside the NodeDescriptor with the graph. At last, information about the hardware's performance (space and time) is stored alongside with the graph.

```
data NodeDescriptor
  = MkNode
    { label   :: String
    , nodeId  :: ID
    , sinks   :: Pins
    , sources :: Pins
    }
    deriving (Eq)
```

To avoid corrupted data types, smart constructors are used. They ensure correct formed graphs by checking the incoming data against given constraints and then generate only valid graphs. For example the combinatorial section only accepts non looping connections.

A wire is a connection between Pins of one or two hardware components. The output is called the source and the input is called the sink of a component.

2.3 Lists through Tuples

The Haskell arrow interface is build from tuples. Circuits come often with a list like interface. There are operators like in binary logic that perfectly match the tuple structure, but in most cases the list interface is needed. Circuits ofte come with a list like interface and therefore we modell this throug tuples. The process of restructuring the output of one function into the input to another function is called retuple'ing. A function that does such tuple'ing is `aFloatR`. This one takes an argument of the structure $(x, (y, z))$ to a different structure $(y, (x, z))$.

```
aFloatR :: (Arrow a) => Grid a (my, (b, rst)) (b, (my, rst))
aFloatR
  =   aDistr
  >>> first aSnd
```

There are two arrows involved here, `aDistr` and `aSnd`. With `aDistr` the first element is distributed over the second one and with `first aSnd` the first occurrence is dropped. The `aFloatR` arrow moves the first element one position to the right. Of curse this arrow assumes a structure where the first element of a tuple is a value, the second is another tuple: $(v_1, (v_2, (v_3, (v_4, v_5, (\dots)))))$

A common task over the list like tuples is to apply an arrow to the front and than concentrate the remaining calculation on the inner structure. For this purpose the operator `>:>` helps to clarify the code that describes such computations.

```
aA >:> aB = aA >>> (second aB)
```

The operator `>:>` is a special case of the `>>>` operator. The first supplied arrow is executed over the whole input, while the second one is applied to only the second part. Such an operator comes handy as long tuple structures are used and are processed from the left.

While working with tuple lists, it is needed to reorder the parenthesis around the data so that selector functions like `first` process the right data. This is done by the two arrows `a_xXX2XXs` and `a_XXx2xXX` that exploit the associative law of arrows. The names mirror the changing paranthesis structure, upper case letters show the grouped position.

$$a\_xXX2XXx : (a, (b, c)) \rightarrow ((a, b), c)$$

$$a\_XXx2xXX : ((a, b), c) \rightarrow (a, (b, c))$$

3   Hardware Specification with (Classical) Arrows

Within this section we present a cycle redundancy check (CRC) [16] algorithm, implemented with arrows. Additionally it's shown how netlists are generated. The section concludes by highlighting some of the problems with this approach.

3.1   CRC – Implementation

A CRC - cycle redundancy check - is a method to calculate checksums of data. The calculation of a CRC is based upon a polynomial division, where the reminder is the CRC. Our approach calculates the CRC with a functional galois type shift register.

The shift register is modeled from a list and the positions of where to feedback. The list is simulated by tuples and therefore the feedback value is feed back into the list by the aFloatR arrow.

Different feedback positions are captured in multiple arrows called aMoveBy$_n$.

```
aMoveBy5
  =   aFloatR
  >:> aFloatR
  >:> aFloatR
  >:> aFloatR
  >:> aFlip
  where
     (>:>) aA aB = aA >>> (second aB)
```

The principle behind every CRC is captured in the arrow aCRC8

```
aCRC8 polynom polySkip padding start reminder
  =   (padding &&& aId)
  >>> aMoveBy8

  >>> (start &&& reminder)
  >>> first polynom

  >>> step
  >>> step
  >>> step
  >>> step
  >>> step
  >>> step

  >>> aFlip
  >>> polySkip
  >>> polynom

  where step =  a_xXX2XXx
            >>> first
                (   aFlip
                >>> polySkip
                >>> polynom
                )
```

The CRC input data are alignet and the padding bits are floated to the right end of the tuple list.

$$((p_0, p_1), (b_0, (b_1, (b_2, (b_3, b_4))))) \rightarrow (b_0, (b_1, (b_2, (b_3, (b_4, (p_0, p_1))))))$$

After that, the arrow is an abstract representation of the CRC calculation. By parameterising over aCRC8, every concrete CRC algorithm can be constructed. Due to Haskells inability to dynamicaly change the count of certain expressions in the function body, our example is limited to 8Bit.

3.2   Example

A CRC-Code uses a polynomial that especially fits the needs of the domain. This example shows the CRC-Code for USB. With the polynomial:

$$USB : x^5 + x^2 + 1$$

The polynomial describes which positions are feed back and are xor'ed with the bit that has traveled through the shift register. For the USB CRC bit_0 and bit_2 are xor'ed with bit_5.

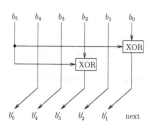

Fig. 1. USB CRC

To summarize this for the USB CRC, the highest exponent is 5. Therefore:

- the polySkip needs to be aMoveBy5
- there are 5 padding bits
  aConst (False, (False, (False, (False, False))))
- the split between start and reminder is after bit 5

3.3   Building Netlist-Graphs

A Netlist is a data structure, that describes how an electronic component is build up. The supporting data structure is a graph that consists of multiple lists. One list contains all the singleton components where another list holds the connecting wires.

A wire connects two components, starting from the output pin of one component and ending at the input pin of another component. Every component is uniquely identified by an integer value. Besides the identifier, every component also has a descriptive name. This one is used to generate hardware describing code like VHDL which is a little easier to debug later on. The last essential describing part about the component are the list of in- and outgoing pins.

All this information are sub summed under the "NodeDescriptor" data type. Our NodeDescriptor distinguishes four types of components. One of them are the combinatorial components. They are defined without any form of clock in mind. It is possible to compute the result of such a component. Also one could substitute the conglomerate of combinatorial components by a single function, called f.

Multiple combinatorial components are combined to bigger components by a clocked register. The register has an input for a clock signal and therefore takes care of the clock synchronization. Between every two registers must be at least one combinatorial component.

A register on its own is separate component that is also listed separate. The buildup of a register is not handled here. The register is defined per architecture so that a solution is a multi platform solution.

The last type of components are the ones that refer onto themselves. They have a combinatorial core from which some wires point right back into the same component. So these wires have the same in- and output Id's. Of course such looping components are synchronized by a register in the loop.

A wire is identified by 4 integer value Id's. Every Pin has one and also every component. Wires that come from the "outside" or that go "out" do not have a component there and therefore don't have a sink- and source Id. To also work with such cases, the component Id's are not only integers but Maybe Integer. This has a direct translation into the real world, where components exist, that aren't built in yet. Exactly this components do have pin's which are connected to the gates but aren't soldered in yet.

3.4 Translate the Netlist into VHDL

With a netlist being an universal description of the hardware circuits it is our goal to convert the hardware describing arrow into one. Every HDL (hardware description language) comes with quirks that must be respected. To translate a netlist into a specific hdl like VHDL one considers the idiosyncrasy of VHDL. That are things like the header of a VHDL file, the components which are described within Entity's or the wires that are called portmaps in VHDL. To the outside, the translation process hides behind a pretty print function, called "showCircuit".

Functions that name single elements are named starting with "name" in the function name. VHDL quirks are represented by the functions that start with "vhdl_". At last there are the auxiliary functions for the remaining tasks.

The "name"-Functions are needed to assign the different elements of the netlist a desired name. The information used to generate such names are extracted from the component description called the CircuitDescriptor. These information are then collected into a temporary representation called a dictionary. The Id's of the components are bound together with the names of the components by ziping them. Consider the In- and Outputs: First of all, the PinIDs are combined with the default-names for input-wires; the same is done with output pins and output-wire names. This chunk of data is further combined with the component id's. The result is a component dictionary that contains for every pin a corresponding name.

Specific VHDL quirks are taken care of by the "vhdl_" functions. This functions generate correct VHDL source code by looking at CircuitDescriptors.

Nearly every "vhdl." function takes one of the dictionary's as input to generate more reasonable names. Every element that is used in the VHDL context to depict a netlist, is produced e.g. by one of the functions "Entity", "Component", "Signal" or "Portmap". A component in the VHDL sense is build from entity's. To generate a VHDL netlist, all the subnodes from the graph-view are taken. An entity is generated for every one of the subnodes. If all entity's are generated the VHDL component header is the big bracket around them.

The execution of the described functions produce a `Grid` arrow as result. As stated in Section 2.1 a Grid consists of the arrow representing the behavior and the graph representing the structure that causes this behavior. Every component comes with a trivial graph containing only one node. That node describes the component in detail.

The hardware components are combined with each other by arrow operators to generate new components with different behavior. Alongside with the combinations of the hardware a new describing graph is generated. This graph describes now, from what components the new one is build up. And it also describes how the single parts are interconnected.

To interconnect two circuits into a new one, a function `splice` is used. This function is abstract in the sense that it can splice two circuits together in different ways. They correspond to the arrow operations. For example graphs that are combined by (>>>) are spliced with the `seqRewire` method. The function connects the sources of circuit 1 with the sinks of circuit 2 one by one. After that step all the remaining sinks of the two circuits are extracted into one list and all the remaining sources are extracted into another list. These two lists become the new sinks and sources of the generated component. To generate a proper graph of circuits one has to make sure, that all the identifiers inside the components are again unique.

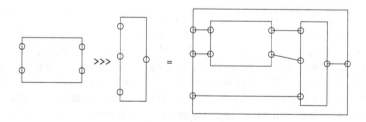

Fig. 2. Combine two components

It turns out, that this graph in fact is a form of a netlist. Once the graph is created, it is straight forward to translate it into another netlist format (e.g. VHDL).

3.5 Drawbacks

This approach has problems, that are visible in the CRC example. The input data of these hardware components must fit into tuples. Logic gates fit well into this scheme, but for hardware components with more than two pins data structures like lists are needed. Lists can be simulated with tuples as it is done in the above example. But the simulation comes at a cost. A big part of the code only restructures the tuples. This process becomes tedious, confusing and error prone. At least if bigger real world hardware components are be modeled. Think about 32 or even 64-Bit ALU's.

The tedious tuple code could be avoided by using Patterson's proc-notation for arrows[15]. The proc-notation abstracts the tuples to named variables. The hardware can be described more "visually" in the source. But this is just syntactic sugar, there are still tuples in use. The following code shows how the USB CRC is modeled with the proc-notation:

```
usbPolynom
  = proc (x5, (x4, (x3, (x2, (x1, x0))))) →do
      o1 <- aXor -< (x5, x0)
      o2 <- aId -< (x1)
      o3 <- aXor -< (x5, x2)
      o4 <- aId -< (x3)
      o5 <- aId -< (x4)
      returnA   -< (o5, (o4, (o3, (o2, o1))))
```

The second problem affects the resulting auto generated arrow code. Patterson's notation heavily uses the arr function. Because it is possible to "lift" any function passed to arr into an arrow, it is impossible to keep track of that outcoming function type. The type information of the passed function is needed to generate a proper component. This makes it possible to generate component arrows ad-hoc. Nonetheless a proper name to that function is missing in every case.

4 Coping with the Tupel-Problem – "Vector-Arrows"

Arrows match binary hardware pretty good. In this section we show howto overcome the problems with n-ary ($n > 2$) hardware components.

The reason for the arrows preference of binary circuits is it's tuple structure. Also the tuples allow the in- and output to be of any kind. In the digital world, hardware transmits nearly always Boolean values and also integers are transmitted via bundled Boolean values. And even the developers of Sensors try to digitize pretty early in their design process. [8]

So arrows are an adequate abstract form to describe hardware, they are modular and they come with an enhanced type system that helps to reduce errors. The part which does not match the hardware perfect are the tuples that prefer binary circuits. Therefore the type a b c is far too general for a generic gates: the type of the in going data should match the type of the out coming values.

A set of wires is not a tuple and for instance, the tuple `(a,(b,(c,d)))` represents the same set of wires as the tuple `(a,((b,c),d))`. The number of possible – different with respect to syntax and type – representations of a set of wires exponentially explodes with a growing number of wires that we are trying to model.

The ideal type for hardware development is the one that is static in their data types and dynamic in the length. Arrows in contrast are dynamic in the data type and are static in their amount of the values. So arrows are opposing the hardware's nature here. It is possible to transform arrows so that they match hardware more naturally, without loosing the beneficial properties for hardware design.

A vector fulfills the desired data type needs of the hardware behavior. The vector Vec_n^b is one of n values over the type b. The arrow class with vectors instead of tuples is described by the following class definition.

```
class VArrow a where
    arr    :: ((Vec_n^b) → (Vec_m^b))  → a (Vec_n^b) (Vec_m^b)
    firsts :: a (Vec_n^b) (Vec_m^b) → a (Vec_(n+k)^b) (Vec_(m+k)^b)
    seconds :: a (Vec_n^b) (Vec_m^b) → a (Vec_(k+n)^b) (Vec_(k+m)^b)
    (>^>) :: a (Vec_n^b) (Vec_m^b) → a (Vec_m^b) (Vec_k^b) → a (Vec_n^b) (Vec_k^b)
    (&^&) :: a (Vec_n^b) (Vec_m^b) → a (Vec_n^b) (Vec_k^b) → a (Vec_n^b) (Vec_(m+k)^b)
    (*^*) :: a (Vec_n^b) (Vec_m^b) → a (Vec_k^b) (Vec_j^b) → a (Vec_(n+k)^b) (Vec_(m+j)^b)
```

In the following sections we sketch different ways to express such "Vector-Arrows" and present different approaches to work with them.

4.1   Solution Using Horn Clauses

Vectors are a data structure which can not be implemented inside haskells type system. In fact, dependent types are needed to express vectores with the right abstraction. Within haskell's type system, lists match vectors in a way. Vectors do have a defined length; lists are of infinite length. Our first approach would be to constrain a list in it's length. Therefore it is necessary to introduce a counter into the data type. We choose type level natural numbers to restrict the list in length.

```
data Vec n a where
    T    :: Vec VZero a
    (:.) :: (VNat n) => a → (Vec n a) → (Vec (VSucc n) a)
```

The type level numbers are introduced as peano numbers. This is similar to the mechanism shown by Kiselyov et al [6].

The `Arrow`-Class of this GADT vector is defined similar to the classical arrow. `firsts` translates an arrow into one that takes it's the first n inputs from the front of the vector. `second` takes the input in a analog way from the end. The Problem is to decide, where the "end" starts.

As Haskell is statically typed, everything in the data type is declared. All variables that hold type level numbers must be defined in the data type context.

To divide a vector in a `first` and a `second` half, one must calculate the peano number of the split position. This is done by type classes like `VAdd`. They must be listed in the type context of the Vector. Working with this contexts reminds of Prolog's Horn Clauses programming style.

It is possible to divide the list just at the "right spot", once the consuming arrow is known. `first f` splits the input into exact the length that `f` expects as input and a reminder vector. So with `aXor` being an arrow that takes two Booleans `first` applied to `aXor` must be an arrow from at least two Booleans to at least one.

```
aXor :: a (Vec (VSucc (VSucc VZero)) Bool) (Vec (VSucc VZero) Bool)
```

```
first (aXor)
 :: (VNat k, VAdd (VSucc (VSucc VZero)) k nk, VAdd (VSucc VZero) k mk)
 => a (Vec ((VSucc (VSucc VZero)) + k) Bool) (Vec ((VSucc VZero) + k) Bool
```

This unsugared syntax is actually a major drawback. It is not possible to write just $2 + k$. Instead one has to add `VAdd (VSucc (VSucc VZero)) k nk` as context of the data type. This might only be tedious for easy arrows but it becomes an major error source for larger ones.

Another drawback with this solution is the way the `VAdd` type class is defined.

4.2 Solution Using Type Families

The next approach is to replace the Horn like clauses by actual calculations. Haskell has lately come up with a type family extension [21] that comes in handy for this task.

A type family in a way enables to calculate the type constructor of a data type. A feature the horn-clause solution lacks. The length of the vector is stored in a peano number. As peano numbers are expressed as a bunch of type constructors, type families are used to calculate the constructor amount. The Haskell-Wiki describes this vividly: A type family is to a data type, what the method of a type class is to a function.

```
type family (n::VNat) :+ (m::VNat) :: VNat
type instance VZero  :+ m   = m
type instance VSucc n :+ m   = VSucc (n :+ m)
```

The first Line defines the type family, with an operator `:+` that operates on peano numbers. With the addition type family the arrow could be defined more similar to the original type class.

```
class Category a => VArrow a where
   arr   :: (Vec n b → Vec m b)
        → a (Vec n b)      (Vec m b)

   firsts :: a (Vec n b)      (Vec m b)
        → a (Vec (n :+ k) b) (Vec (m :+ k) b)
```

```
seconds :: ((m :+ k) ~ (k :+ m))
       => a (Vec n b)        (Vec m b)
        → a (Vec (k :+ n) b) (Vec (k :+ m) b)

(***)  :: a (Vec n b)        (Vec m b)
        → a (Vec n' b)        (Vec m' b)
        → a (Vec (n :+ n') b) (Vec (m :+ m') b)

(&&&)  :: a (Vec n b)        (Vec m b)
        → a (Vec n b)         (Vec m' b)
        → a (Vec n b)         (Vec (m :+ m') b)
```

The problem with this approach is the compiler. It is not possible for GHC to deduce arithmetics in the type context. This is also valid for simpler calculations down to even the equality of $n + m = m + n$. It's possible to supply the compiler with these information by hand. But this are only hints into a direction. At this point it gets tedious to supply hints for easy deductions and gets unpractical for complicated arithmetic's.

4.3   Solution Using Dependent Types

A dependently typed programming language is a language, that mixes expressions with types. It therefore allows the calculation of types within that same language. There are multiple languages with dependent types out there. The one with a pretty good connection to Haskell is Agda[9] [12]. The following describes the arrow class in Agda with dependent types.

In summary the previous solutions all try to fix a unitary problem. And this problem depends on the fact that vectors aren't purely functional. At least it's not practicable in Haskell to define and work with them. A vectors length must show up in it's data type so that a function can match against it. The problem here is to calculate values at the type level. The transfer of the inner tuples into the vectors changes the inner data structure of arrows. Therefore the projection maps from that data type must translate to a vector version. first and second exactly match the tuple projection maps in Haskell, namely fst and snd.

With vectors the first statement is straight forward defined. The second expression is the head-scratcher. The implementation of first constraints it's inputs to only those, that are equal or larger than the ones f expects. The second case adds the right alignment on top. This can be achieved by calculating the size of the sub vector, ignoring him and applying the function to the reminder. It can also be done via reversing the vector multiple times and than applying the function to the front of the vector.

While aSwitch is of type Vec_n^{Int}, the expression seconds aSwitch must be of type Vec_{k+n}^{Int}, $k \geq 0$. This is hard to express in a functional data type. In a dependent language this is easily expressible by the data type.

The arrow type class form Haskell must be translated into a vector arrow data structure inside Agda, where typically records do the job. Therefore the vector arrow record in Agda is defined like so:

```
record VArrow (_~>_ : Set →Set → Set) : Set1 where
 field
  arr    : forall {B n m}
    → (Vec B n → Vec B m) → (Vec B n) ~> (Vec B m)

  firsts : forall {B n m k}
    → Vec B n ~> Vec B m → Vec B (n + k) ~> Vec B (m+k)

  seconds : forall {B n m k}
    → Vec B n ~> Vec B m → Vec B (n + k) ~> Vec B (m+k)

  _>>>_   : forall {B n m k} → Vec B n ~> Vec B m
    → Vec B m ~> Vec B k → Vec B n ~> Vec B k

  _***_   : forall {B n m k j} → Vec B n ~> Vec B m
    → Vec B k ~> Vec B j → Vec B (n+k) ~> Vec B (m+j)

  _&&&_   : forall {B n m k} → Vec B n ~> Vec B m
    → Vec B n ~> Vec B k → Vec B n ~> Vec B (m+k)

 infixr 2 _>>>_
 infixr 2 _***_
 infixr 2 _&&&_
```

This specifies a record over a function like type _~>_. With the **forall** definition one matches the context of a Haskell type definition. The Vec denotes a vector type defined in Agda that is used to specify the arrow record. A close look at the **firsts** and **seconds** statements reveals, that they have the same type. If **firsts** comes with a type $(n + k)$ then it would be reasonable to use $(k + n)$ in the type of the **seconds** statement. This can not be done because of the way the addition is defined.

With this vector record at hand it is possible to apply the concepts shown earlier. The Grid-type is defined in a following way inside agda.

```
data Grid (A : Set → Set → Set) : Set1 where
  GR : forall {B C} → (( A B C ) , Nat) → Grid A
```

It is now possible to define the graph constructing function on top of the Grid-type. This will also generate a connection graph of the circuit structure, that can later be translated into all forms of hardware describing code.

5 Conclusion

With this paper we showed how to model hardware with arrow descriptions and generate circuits from them. Arrows bring just the right amount of abstraction into the hardware design process. This offers on the one hand a precise tool for the developer. On the other hand it stays as flexible as it is usual in the software design process.

The classical arrows come with a tuple inner structure which is the foundation of our problems. Although we are able to simulate list-like behavior it is neither practical to work with nor is it correct in a data type sense. To take arrows from tuples to arrows with vectors is therefore mandatory. We showed different ways, how to get as close as possible to vector arrows.

Our concept is powerful enough to generate graphs and therefore netlists from only the arrow definitions. We think of netlists as a form of graphs. On the other way round the graph is a mathematical description of what a netlist is. We also showed how to get from a graph to a netlist by a straight forward "pretty-print" function.

A proof of the superiority of the functional approach is the modeled CRC circuit. Although we used the classical arrows for modeling, it wasn't much pain to model the CRC algorithm with them. It is up to the developer to use a decent data layout so that the tuple structure stays out of the way. The problems that come from the tuple arrows are theoretically gone with a vector arrow. It is just not possible to define vectors in such a way inside Haskell that blend with the arrows.

6   Future Work

We are going to follow the vector approach inside Agda. The graph generation is something that can be done completely inside a dependent typed language. And there could be a connection between Agda and Haskell due to the IO Monad, but this connection is still work in progress. To fully utilize Agda, we have to define the Agda data types much more in detail. Also the combination and connection algorithms need to be translated from Haskell into Agda.

It might be possible to completely switch over to an Agda solution. But at the moment this is just an idea and needs further research. Also Agda is a language that is still in development and there are major changes from time to time. So it is not clear how reasonable a solution completely in agda is.

References

1. Bjesse, P., Claessen, K., Sheeran, M., Singh, S.: Lava: Hardware design in haskell. ACM Press (1998)
2. Galois, Inc., Portland, Oregon. Cryptol Programming Guide (October 2008), http://www.galois.com/files/Cryptol/Cryptol_Programming_Guide.pdf
3. Galois, Inc., Portland, Oregon. From Cryptol to FPGA (October 2008), http://www.galois.com/files/Cryptol/Cryptol_Tutorial.pdf
4. Grundy, J., Melham, T., O'Leary, J.: A reflective functional language for hardware design and theorem proving. Journal on Functional Programming 16(2), 157–196 (2006)
5. Hughes, J.: Generalising monads to arrows. Science of Computer Programming 37, 67–111 (1998)
6. Kiselyov, O., Lmmel, R., Schupke, K.: Strongly typed heterogeneous collections. In: Proceedings of the Haskell Workshop, Snowbird, Utah (2004)

7. Li, Y., Lesser, M.: HML, a novel hardware description language and its translation to VHDL. IEEE Transactions on VLSI 8(1), 1–8 (2000)
8. Makinwa, K.: Smart sensors - no signal to small. In: DCIS (2011)
9. Martin-Lf, P.: Intuitionistic type theory. Bibliopolis, Napoli (1984)
10. Mycroft, A., Sharp, R.: The FLaSH compiler: Efficient ciruits from functional specifications. Technical report, University of Cambridge, Computing Laboratory (2000)
11. Mycroft, A., Sharp, R.: Hardware/Software co-design using functional languages. In: Margaria, T., Yi, W. (eds.) TACAS 2001. LNCS, vol. 2031, pp. 236–251. Springer, Heidelberg (2001)
12. Norell, U.: Towards a practical programming language based on dependent type theory (2007)
13. O'Donnell, J.: Hardware description with recursion equations. In: Proceedings of the IFIP 8th International Symposium on Computer Hardware Description Languages and their Applications, pp. 363–382. North-Holland (April 1987)
14. O'Donnell, J.T.: The Hydra Computer Hardware Description Language. University of Glasgow (2003)
15. Paterson, R.: A new notation for arrows. In: ICFP, pp. 229–240 (2001)
16. Peterson, W.W.: Cyclic codes for error detection. In: Proceedings of the Institute of Radio Engineers, Gainesville, Florida (1961)
17. Sander, I.: System Modeling and Design Refinement in ForSyDe. PhD thesis, Royal Institute of Technology, Stockholm, Sweden (April 2003)
18. Sander, I., Jantsch, A.: System Modeling and Transformational Design Refinement in ForSyDe. IEEE Transactions on Computer-Aided Design of Integrated Circuits and Systems 23(1), 17–32 (2004)
19. Sharp, R. (ed.): Higher-Level Hardware Synthesis. LNCS, vol. 2963. Springer, Heidelberg (2004)
20. Sheeran, M.: μFP: A language for VLSI design. In: LISP and Functional Programming, pp. 104–112 (1984)
21. Yorgey, B.A., Weirich, S., Cretin, J., Jones, S.P., Vytiniotis, D., Magalhães, J.P.: Giving haskell a promotion. In: Proceedings of the 8th ACM SIGPLAN Workshop on Types in Language Design and Implementation, TLDI 2012, pp. 53–66. ACM, New York (2012)

Evaluation of Process Architecture Design Methods

Mery Pesantes, Hugo A. Mitre, and Cuauhtémoc Lemus

Center for Mathematical Research (CIMAT),
Av. Universidad 222, Fracc. La Loma, 98068 Zacatecas, Mexico
{mpesantes,hmitre,clemola}@cimat.mx
http://www.ingsoft.mx

Abstract. Process Architecture has become a recently emerged discipline. Although several methods have been established to design process architectures, there is a lack of available studies comparing these methods. There is also little consensus on the technical and no technical issues that a method must address and which is most suitable for a particular situation. Four process architectures design methods of different authors and disciplines were selected. This paper provides an overview and comparative analysis of these methods according to a framework devised as part of our analysis. The framework considers the following criteria: context, stakeholder, contents, quality and validation. Comparison revealed different ideologies between the methods. Some methods considered the attribute of reuse at the product quality level. Furthermore, several methods have been validated in the academy and they are considered in this paper as a starting point for validation.

Keywords: Process Architecture, Process Architecture Design, Evaluation Framework.

1 Introduction

Process Architecture (PA) has been attracting attention from researchers and practitioners since the last two decades. Disciplines such as Process Management (PM), Business Process Management (BPM), Enterprise Architecture (EA) and Software Process Engineering (SPE) have their own published methods to design a PA. However, few works have been made to show similarities and differences between these methods. .

Based on our review, a work has been found to compare the PA design methods in the BPM discipline. Green and Ould [12] propose a framework for evaluating and classifying PA methods. Their framework compares PA methods by textual descriptions obtained in the evaluation of each of its facets.

Our work has selected the following methods because they clearly specify the integration of business and process requirements in the PA design and besides they are the most representative and complete by discipline: a) Ould's method [30], b) Maldonado and Velázquez's method [24], c) Borsoi & Becerra's method

B. Murgante et al. (Eds.): ICCSA 2013, Part III, LNCS 7973, pp. 144–159, 2013.
© Springer-Verlag Berlin Heidelberg 2013

[7], and d) Dai & others' method [9]. Other methods beyond the scope of this document are: Jeston and Nelis [19] propose a framework to implement BPM programs and projects. Li and others [11] present a PA using Aspect Oriented Programming (AOP) methodology. And, Boehm and Wolf [6], [5] introduce one open PA and examine some architectural element interfaces.

The purpose of our work is to understand, evaluate and compare four PA design methods supported by an evaluation framework. The intention of this paper is not to provide an exhaustive survey on each discipline but provide a state of the art of PA current practices and help to contrast alternative approaches to design PAs.

This document is organized as follows: Section 2 shows the importance of PA design. Section 3 introduces a comparison framework for understanding and evaluating PA design methods. Section 4, three PA design methods are briefly described. Section 5 provides a comparison of PA methods against the framework. And, finally a summary of our work is presented in section 6.

2   Process Architecture Design

PA adopts different definitions from the point of view of PM [4], BPM [28,19], EA [13,14] and SPE [10,23]. As a result of this multidisciplinary approach, it is defined as "a conceptual framework for designing and maintaining processes and its relationships, which must be aligned with business objectives and strategy and enterprise architecture" [35]. Moreover, a "process is composed of process elements" [29,2] and a "process element is composed of activities, input and output interfaces, pre-conditions and post-conditions and knowledge Base" [3].

Many researchers have developed specifics methods to design a PA. We can also find efforts generating awareness, ideas and approaches related to this theme such as: software process asset [34], properties and views [27,33], the process architect role [20,19,16], tool using process metamodels [32] and harmonization frameworks [37,38,36]. However, there is no explicit agreement or recommended approach that provides further insight on the suitability of a method for a particular situation. Some of the ideas that emerged around of PA design follow:

– Mutafelija addressed the subject of PA views and properties [27]. It refers three views: functional, behavioral, and organizational. Borsoi and others [7] represent a PA as a set of views. They identify four views: structural, behavioral, organizational, and automation.
– Carr and others [8] provide a process interface description language named Mini-interact that is a simple approach to describe process interfaces that is accessible to non-programmers as well as to programmers. Boehm and Wolf [6] provide a framework through the identification of architectural elements and the specification of element interfaces.
– Kasser [20] identifies a new role named process architect. This role will cover the gap in the functions performed by the three organizational roles (systems architecting, systems engineering and project management) when viewed from the perspective of planning and implementing the development of a

system. Jeston and Nelis [19] and Harmon [16] mention that a useful way to embed PA in the organization is the establishment of a "business process architecture committee". This committee has the responsibility for maintaining an overview of the entire company. They should know what business process support what goals and produce plans for processes redesign efforts.

3   Evaluation Framework

The framework introduced is used as an analysis tool. This framework has its underpinnings on the following methods:

1. The NIMSAD (Normative Information Model-based Systems Analysis and Design) evaluation framework [18]. NIMSAD framework uses the entire problem solving process as the basis of evaluation and it can be used to evaluate methods on any category. According to NIMSAD, there are four essential elements for method evaluation. Firstly, the method is evaluated from the element of the problem situation, i.e. the method context. The second element is the intended problem solver, i.e. the user of the method. The third element is the problem solving process, i.e. the method itself. The fourth element is the self-evaluation of the previous elements. Because methods rarely consider evaluation of the context, or user or contents, our evaluation method considers validation of the method and its outputs. In addition, the quality element from two perspectives: product and process is considered [17].
2. The definition of a method and its ingredients has influenced the third element of the framework, i.e. the method contents. Kronlöf [22] defines method ingredients as follows: 1) an underlying model, 2) a language, 3) defined steps and ordering of these steps and 4) guidance for applying the method. Additionally, tools are also considered because help in execution of the methods.
3. The IEEE 1471 standard supports the process architect with requirements on architectural descriptions. This standard composes a unification of best practice solutions from a variety of architectural concepts. The core of IEEE 1471 is its focus on architectural support and guidance in the domain of architectural descriptions [1].
4. The analysis on approaches for comparing models will consider the characterization based on classes of comparison proposed by Halvorsen and Conradi named comparison based on characteristics [15].
5. For identifying the factors that are relevant to compare PA methods, we looked at existing PA works[23,10,30,31,16,26,12,20,24,19,3,25,9,37,7].

Therefore, our characterization proposal is denoted in Figure 1. The goal of our evaluation framework was not to rate the methods but to provide an overview of current PA methods, identifying similarities and differences. A neutral, common and quite extensive NIMSAD framework for method evaluation was utilized to derive the fundamental categories and elements in the framework.

The framework identifies five categories to ease readability and comprehension. We have added the category of quality to the NIMSAD framework. We

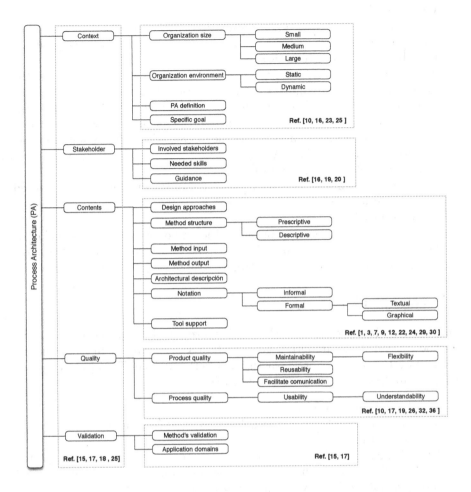

Fig. 1. Evaluation framework to compare process architecture methods

believe that "quality" is an important indicator to users and it will support the selection of a method. The following subsections explain briefly each category and its elements. Each category is divided into five parts: context, stakeholder, contents, quality and evaluation, as denoted in Figure 1. Also, we show in Table 1 various questions treating to answer elements e.g. specific goal, method structure, product and process quality, etc.

3.1   Context Category

This category describes the attributes of the organization and environment where the method may be used. Usually, the context is the organization whose problems are to be solved by the software being developed. The elements of the context category are based on [10,23,16,25] and they are as follows:

– **Organization Size**

This element indicates the organization size for which the method is suitable, i.e. some methods are easier to use for large, medium or small companies.

– **Organizational Environment**

This element is related with the context of organizations. This means the method considers the context of an organizations twofold: static or dynamic. A dynamic environment is when the organization requires changing its business. These changes may lead to change to the set of processes. Static environment considers no changes within the organization. This assumption may lead to build an inflexible PA.

– **Process Architecture Definition**

This element identifies an explicit definition of PA provided by the method or it should help define what it means.

– **Specific Goal**

The goal of the method is its primary objective or end result. The goal must be clearly defined, for instance, identify processes, prioritize processes, align the processes with the business strategy, elaborate processes, etc.

3.2 Stakeholder Category

This category identifies the stakeholders involved in the implementation of the method. Stakeholders are either people who have an interest in the proposed PA, architecture artifacts or who actually uses the method. The elements of this category are based on [16,19,20] and include:

– **Involved Stakeholders**

This element identifies the stakeholders required to execute the method.

– **Needed Skills**

This element considers skills that are needed to get full benefit from the method. The use of the method requires some skills. The stakeholders should meet skills needed for executing the method.

– **Guidance**

This element is a special guideline that the method provides for using it. The guidance supports the stakeholders in the execution of the method.

3.3 Contents Category

The contents category is concerned with a method itself. It includes those elements, which help gain a better understanding of the more technical aspects of method. This category considers the elements that any method should contain based on [22,29,30,1,12,3,24,9,7]. This consists of seven elements.

– **Design Approaches**

The approach to design a PA is the theoretical background or model on which it is based. Some methods have originated from previous frameworks, while others are based on mathematical rules.

- **Method Structure**
  This element is defined as the detailed description of the design steps, their sequence to be followed, requirements of input and output, roles and artifacts provide by the method. This category classifies the types of method structure as *descriptive* or *prescriptive*. *Descriptive method* includes general steps without detailing requirements of input and output, artifacts and roles that must participate. *Prescriptive method* includes detail steps, their sequence to follow, requirements of input and output, artifacts and roles.

- **Method Input**
  This element is the information that is needed as input for the method.

- **Method Output**
  This element describes the intended output artifacts of the method, i.e. what kind of information is produced by the method.

- **Architectural Description**
  This element is a recommended practice by IEEE 1471-2000 standard [1] considered as a collection of products to document an architecture. The architectural description is organized by *views*. Each view addresses one or more of the concerns of the stakeholders. The views are built using a pattern or template named *viewpoint*. A viewpoint is a specification of the conventions for constructing and using a view.

- **Notation**
  This element refers to the language for expressing PA design. A well-defined and expressive notation makes possible to communicate unambiguously those design decisions to different stakeholders. The types of notation include the following: *Informal notation* is a technique that does not have complete sets of rules to constrain the models that can be created. Natural language (written text) and unstructured pictures are typical instances. *Formal notation* is a technique that has rigorously defined syntax and semantics. There is an underlying theoretical model against which a description expressed in a mathematical notation can be verified. Languages based on predicate logic are typical instances. These notations can also be classified as either *textual* or *graphical*. *Textual* allows a lot of fine grained detail to be specified in form of text. *Graphical* is one whose elements are visual or in a graphic form.

- **Tool Support**
  This last element is used to support different activities of the method such as documentation of PA design, user assistance in preventing human mistakes in the process. There is a lack of tools that support the entire design process.

3.4 Quality Category

Identifying desired system qualities before a system is built allows system designers to frame a solution (starting with its architecture) to match the desired needs of the system within the context of constraints (available resources, interface with legacy systems, etc.). When a designer understands the desired qualities before a system is built, then the likelihood of selecting or creating the right architecture is improved. In order to know if quality has been achieved or degraded, it has to be measured, but determining what and how to measure is difficult. The quality attributes are features or characteristics that affect process architecture's quality. This category considers two fundamental focus of quality assurance based on [17,10,19,26,32,37]:

- **Product Quality**
  This element is defined as a set of quality attributes for final product evaluation. The quality attributes considered for final PA are: *maintainability* including *flexibility, facilitate communication* and *reusability*. *Maintainability* is the ease with which a PA can be modified to correct faults, improve some attributes, or adapt to a changed environment. Within this attribute considers the quality factor named flexibility. *Flexibility* the ease with which a PA can be modified for use in environments other than those for which it was specifically designed (support planned or unplanned changes). *Facilitate communication* measures how the PA makes easier the communication between stakeholder, such as experts and non-experts. *Reusability* defines the capability for components of PA to be suitable for use in other domains and in other scenarios. Reusability minimizes the duplication of components and also the implementation time.
- **Process Quality**
  This element is defined as a set of quality attributes for method process evaluation. The quality attribute considered in the process of the method is *usability* and this includes *understandability*. *Usability* is the ease with which a user can learn to operate, prepare inputs for, and interpret outputs of a method. This attribute considers the quality factor named understandability. *Understandability* considers that a method has a clear purpose and its description contains adequate comments.

3.5   Validation Category

Validation is needed to confirm that a product meets the goals and needs of its users. The validation category has only two elements based on [15,17]:

- **Method's Validation**
  This element refers the techniques used to validate the method. This information may encourage or discourage the evaluators to select one particular method over the other. Also, the method should show some kind of evidence regarding the maturity of the method, such as evidence of its use, applicability, benefits and costs. Some methods of validation can be experiment, real application examples or industrial case studies, etc.
- **Application Domains**
  It is related with those domains where the method may be applied. The method proves its suitability in the domain for which was designed, so the method has to be validated. All the methods have been validated already or are under validation on various domains (education domain).

Table 1. PA method classification and comparison framework

Category	Elements	Questions	Possible Values
Context	Organization size	Which is the size of the organization more adequate to implement the method?	Small, medium or large
	Organizational environment	How the method considers the context of business organization?	Static or dynamic
	PA Definition	Does the method explicitly consider a particular definition of PA?	it is Based in process models, related with the enterprise, etc.
	Specific goal	What is the specific goal of the method?	align the processes with the business strategy, elaborate processes, etc.
Stakeholder	Involved stakeholders	Who are the stakeholders addressed by the method?	SPEG, process engineers, process architects etc.
	Needed skills	What skills does the user need to accomplish the tasks required by the method?	
	Guidance	How does the method guide the user while applying the method?	Manuals, methodologies, user guidance, etc.
Contents	Design approaches	What types of design approaches are included in the method?	Value chain, Zachmans framework, object oriented methodology, etc.
	Method structure	What are the design steps that are used to accomplish the method's specific goal?	
	Method inputs	What is the starting point for the method?	Requirement of process models, knowledge of the organization, etc.
	Method outputs	What are the results of the method?	Guidelines, documentation of process architecture, etc.
	Architectural description	What are the architectural viewpoints the method applies?	viewpoints or views (behavioral, functional, organizational).
	Notation	Does the method define a language or notation to represent the models, diagrams and other artifacts it produces?	Informal, formal, textual or graphic.
	Tool support	What are the tools supporting the method?	BPMN, UML, etc.
Quality	Product Quality	Which quality attributes are covered on final process architecture?	Maintainability, facilitate communication, reusability, etc.
	Process Quality	Which quality attributes are covered by the method process?	Usability
Validation	Methods validation	Has the method been validated? How has it been validated?	Experiments, small examples, case studies, etc.
	Application domains	What are the domains the method is validated?	Education, etc.

4 Overview of Process Architecture Methods

- **Ould's Method**
  In the BPM discipline, Ould developed this method [32,30]. The fundamental concept of the Riva approach is that an organization's PA can be abstracted from the essential business entities(EBEs). Riva steps are: 1) Brainstorm the subject matter of the organization to identify EBEs. 2) Identify those EBEs that have a lifetime, these are called units of work (UOWs). 3) Create a UOW diagram that shows the relationships between them. 4) Transform the UOW diagram into corresponding first-cut PA and apply heuristics on it in order to produce a second-cut architecture.

- **Maldonado & Velázquez's Method**
  This method has been developed in the EA discipline. It defines systematically the PA of any organization using the Zachman's framework [24]. This method consists of four steps: 1) Capturing the organizational structure. 2) Analyzing the flow of information between the different organizational units, customers, and providers. 3) Identifying and modeling the configuration of value. 4) Given the specified value configuration, identifying, defining, and interrelating the different essential processes.

- **Borsoi & Becerra's Method**
  This method has been developed in the BPM discipline. PA is defined as a set of views represented by models and relationships between them [7]. This method has three phases: Conceptual architecture phase determines the process views, models and elements, and defines how to compose the metamodels and to group these metamodels in the defined views. Reference architecture phase aims to define the standard process models. Operational architecture phase aims to define a kind of guidance to process implementation.

- **Dai & Others' Method**
  This method has been developed in the SPE discipline. It defines PA as an evolution of process components and connectors. PA describes the overall structure of a software evolution process and it is used as the blueprint for EPC (Evolution of Process Component) composition [9]. The method includes five steps: 1) Process requirement determines what the customer requires. 2) Process architecture describes the overall structure of a software evolution process.3) All evolution process components defined in PA are constructed. 4) Process composition composes all constructed evolution process components into a complete software evolution process. 5) Process simulation, software evolution processes will be executed by process engine.

5 Evaluation of Process Architecture Methods

Table 2 summarizes the evaluation results of four PA design methods. The evaluation results are divided into five categories: Context, Stakeholder, Contents, Quality and Validation. The evaluation was done on the basis of publications found in a literature review based on [21]. The most important conclusion was that more research is needed on the quality attributes of a PA.

5.1   Context

The evaluation results of this category are divided into four elements:

- **Organization Size**
  Only Maldonado & Velázquez's method mentions that it can be used by any organization. Other methods do not specify whether they were created for specific size organizations (i.e. small, medium or large).
- **Organizational Environment**
  This element is considered by Ould as dynamic. It assumes that Strategy Process (CSP) analyzes all the factors both inside and outside the organization, which could affect the UOW and any other changes that occur. CSP also assesses the performance of the UOW and how improvements could be made. But there is still an unexplored activity, which consists of viewing the whole organization strategically in the long-term. Maldonado & Velázquez's method assumes a static organizational environment because it does not specify mechanisms of change. For example, if the organizational chart is changed or processes are added. The others methods do not make explicit the organizational environment in which they can be used.
- **PA Definition**
  All methods have the same general definition about PA. It is processes and its relationships. But some methods have different scope. Maldonado & Velázquez's method defines a PA as set of key processes in the organization, showing relationships among them and with customers and suppliers. Borsoi & Becerra's method defines a PA as set of views represented by models and its relationships. Finally, Dai& others' method proposes the concept of PA made up of software evolution process components and connectors.

- **Specific Goal**
  All the methods have the same overall goal, i.e. to design process architectures. However, to find a difference, a specific goal denotes what points are emphasized by the method in the development of PA and what phases of process cycle are supported by PA. Each method has different scope. Riva Method [30] aims to elicitation, analysis, modeling and design processes. Maldonado & Velázquez's method [24] is focused on identifying key processes of an organization. Borsoi & Becerra's method [7] defines process models in distinct abstraction and detail levels. Dai & others' method [9] aims to design processes from existing software process components. It also emphasizes on software process reuse.

5.2   Stakeholder

As seen in Table2, the evaluation results of this category are:

- **Involved Stakeholders**
  The stakeholders are either people who actually use the method. Each evaluated method identifies a different stakeholder (i.e. process designer, planner

Table 2. Using the framework to understand and evaluate Process Architecture Methods

Evaluation Framework		Software Process Architecture Methods			
Category	Elements	Ould [30]	Maldonado & Velázquez [24]	Borsoi & Becerra [7]	Dai & others [9]
Context	Organization size	Not specified	Any organization	Not specified	Not specified
	Organizational environment	Dynamic	Static	Not specified	Not specified
	PA definition	Processes and relationships	Processes and relationships among them and with customers and suppliers	Object process and relationships	Evolution process components (EPC's) and connectors
	Specific goal	Elicitation, modeling, analysis and design of processes	Identifying of essential processes	Defining process models	Designing and composing process
Stakeholder	Involved stakeholders	Process designer	Planner	Not specified	Process modeler
	Needed skills	Not specified	Not specified	Not specified	Not specified
	Guidance	Heuristics: UOW diagrams to PAD	Not specified	Orientations (process models and views)	Not specified
Contents	Design approaches	Role-based process modeling	Product-based process modeling	Object-oriented activity-based process modeling	Component-oriented activity-based process modeling
	Method structure	Prescriptive (detailed steps)	Prescriptive (detailed steps)	Prescriptive (phases, activities, input, output)	Dscriptive (general steps)
	Method input	Essential business entities	Organization chart	Information of process domain and context, basic process concepts	Process requirements
	Method output	PAD and RAD diagrams	Key processes Standard process models Specialized process models	Process metamodels	Complete software evolution process
	Architectural description	Not specified	Not specified	Views	Not specified
	Notation	Graphical	Graphical-informal	Graphical-formal	Formal
	Tool support	Inter actor, RADmodeller Visio stencil	Not specified	Not specified	Not specified
Quality	Product quality	Reuse	Not specified	Reuse	Reuse
	Process quality	Not specified	Not specified	Not specified	Not specified
Validation	Method's validation	Case studies	Case studies	Experiments and Case studies	Not specified
	Application domains	Pharmaceutic, high education, digital libraries	High education, financial services, telecommunication	High education Small Software Factories	Not specified

or process modeler) except the Borsoi & Becerra's method that does not specify the role. This is clear evidence that the role responsible for developing a PA is not standardized.

- **Needed Skills**
  Considering the question of what are the skills the method users need when applying the method, the following issues were concluded. None method explicitly considers skills that stakeholders should have, but one of the essential method properties is the method notation. Two of the methods have a special notation to learn (RAD, PAD and UOW [30] and EPDCL [9]) and another method specifies that can complement its base notation with any language (such as UML or BPMN [7]). Tools that support each method are other challenges of skills that the user must have to implement this method. Furthermore, each method has its own ideology needed to learn.

- **Guidance**
  Most of the methods suffer lack of method documentation. Some methods (i.e. Riva method and Maldonado & Velázquez's method) provide case studies that could help to stakeholders as guidelines. In addition, Riva method provides heuristics to transform UOW diagrams to PAD diagrams.

5.3   Contents

The contents category is concerned with a PA method itself. It includes those elements, which help gain a better understanding of the technical aspects of a design method. Table 2 summarizes evaluation results in seven elements:

- **Design Approaches**
  Each method has different process design approach. Riva method is based in Role based process modeling approach. The focus is on the entity that performs a process element. Roles represent the sequences of activities carried out by roles engaged in a cooperative behavior. Maldonado & Velázquez's method is based in product-based process modeling approach. Borsoi & Becerra's method is based in object-oriented activity based process modeling approach. It considers object orientation concepts to design processes. So, this approach describes a process as a set of ordered activities. Dai & Others' method considers that PA can play a very important role in software process reuse. Evolution process reuse is a special type of software reuse which is focused on composing software evolution processes from existing software evolution process components rather than constructing them directly. It is based in component-oriented activity based process modeling approach.

- **Method Structure**
  PA methods can be classified by considering their structure. It is necessary to know how it supports the problem-solving process in terms of guidance on the activities to be followed and their sequence. Three of the methods were identified as prescriptive. Riva method and Maldonado & Velázquez's method include detailed steps with a sequence to follow and artifacts defined. Borsoi & Becerra's method divides its structure in sequential phases with

detailed activities and requirement and results clearly identified. On the other hand, Dai & Others' method was identified as descriptive because it only mentions general steps of its development process. Moreover, most of the activities of studied methods vary in number, complexity and granularity. We can identify at least one common complex activity in all methods, i.e. process requirements. For example, Riva method needs to identify essential business entities of the organizations but what would be appropriate.

– **Method Input and Output**
  All methods identify inputs and outputs in their development process. Riva method requires as input EBEs and outputs UOW, Process architecture diagram (PAD) and Role activity diagram (RAD). Maldonado & Velázquez's method requires as input the organizational chart and output the key processes in a PA diagram. Borsoi & Becerra's method requires as input general information of process domain and context and notational and basic process concepts. It also requires as output process metamodels, standard process models and specialized process models. Dai & Others' method mentions of implicit manner that its inputs are process requirements and its outputs are complete software evolution processes.

– **Architectural Description**
  Communicating PA to its stakeholders is one of critical factor of a successful PA design. The role of architectural view is also considered vital in communicating architectural information. The importance of multiple views as an effective means of separating the concerns during architectural design has been considered explicitly by the Borsoi & Becerra's method. It mentions that a view is represented by means of process models. This method considers three views: process metamodels (conceptual), standard process model (reference) and specialized process model (operational).

– **Notation**
  Riva method uses as a graphical-formal notation the following diagrams: UOW, PAD and RAD. Borsoi & Becerra's method has a syntactic and semantic base is complemented by the syntax and the semantic of the language used to depict the metamodels and the models. This method uses a graphical-formal notation. Dai & others' method uses a formal notation. This method has two notations: Evolution process component description language (EPCDL) and Evolution process description language (EPDL). EPCDL to describe the overall structure of a software evolution process and the relations, it is a high-level specification language. And EPDL is a computer language that is used to transform EPCDL specification to evolution process descriptions, it is low level description language. Only the Maldonado & Velázquez's method uses a graphical-informal notation. It does not specify in its graphic the meaning of each element.

– **Tool Support**
  Most of methods do not define tool support. Instead, Riva method provides two tools such as RAD modeler and Visio stencil plug-in. Theses tools support the modeling of UOW, PAD and RAD diagrams.

5.4   Quality

Table 2 presents the summary of main findings about the quality category, it is divided into two elements:

- **Product Quality**
  One of the most significant features of method differentiation and classification is the number of quality attributes that a method deals. Most of the surveyed methods focus mainly on the quality attribute of reuse. Riva method mentions that the PA can be reused by others organizations in the same business domain. Borsoi & Becerra conducted experiments of their method. They verified that the PA promotes the reuse of artifacts. Dai & others' method emphasizes to compose software process from existing software process components. There are four forms to build EPC's. First, an EPC can inherit characteristics from an existing EPC. Second, an EPC can be built by means of EPF tailor. Third, an EPC can be built by means of EPF composition. Fourth, an EPC can build a new EPC by means of evolution process metamodel (EPMM). Only Maldonado & Velázquez's method does not specify the quality attribute that supports.
- **Process Quality**
  The methods do not specify the quality attributes that support their development process.

5.5   Validation

Table 2 summarizes evaluation results of this category, this is as follows:

- **Methods Validation**
  Two methods, Riva method and Maldonado & Velázquez's method, have been validated in practical industrial and academic case studies. Another method, Borsoi & Becerra's method, has been validated with experiments.
- **Application Domains**
  Riva method has been applied in the domains of Pharmaceutical, high education and digital libraries. Maldonado & Velázquez's method has been applied in financial services, high education and telecommunication. Finally, Borsoi & Becerra's method was applied in the domain of education.

6   Conclusions

The main contribution of this study is to provide a comparative analysis and overview on four methods for designing PA (methods of Ould, Maldonado & Velázquez, Borsoi & Becerra, and Dai & Others) according to specially developed question framework. We have also demonstrated how the proposed framework can be used to identify the essential features of a good design method.

The comparison of existing methods revealed several features supported by most of the methods. All methods have the same general goal but with different scope in their specific goal. Process modeling approaches, structure and notation are clearly defined in each method. The most popular domain for applying the methods has been education. Another important feature is the product quality; three methods were created considered the quality attributes of reuse. Issues that are not addressed by existing methods were also identified. Only one method, Riva, provides a tool support for designing PA. None method considered attributes of its development process and neither identified needed skills of its stakeholders.

We do not suggest that our framework is complete in its existing form. Rather, we expect this framework to be modified and extended in the future as a result of our ongoing research. In particular the elements of product and process quality need more work. It is identifying what quality attributes should be considered in the design of a kind of PA and how these quality attributes can be implement in a PA.

Finally, we are contemplating other issues, such as to evaluate the completeness of PAs verification method, identify the utility of deriving processes from a PA, the suitability of a PA for small, medium and large organizations and the development of a methodology to create PA in a multimodel environment.

References

1. IEEE Recommended practice for architectural description of software-intensive systems, IEEE Std 1471-2000. IEEE Computer Society (September 2000)
2. Bass, L., Clements, P., Kazman, R. (eds.): Software Architecture in Practice, 2nd edn. Addison-Wesley Professional (2003)
3. Bhuta, J., Boehm, B., Meyers, S.: Process elements: Components of software process architectures. In: Li, M., Boehm, B., Osterweil, L.J. (eds.) SPW 2005. LNCS, vol. 3840, pp. 332–346. Springer, Heidelberg (2006)
4. Biazzo, S., Bernardi, G.: Process management practices and quality systems standards: Risks and opportunities of the ISO 9001 certification. Bussiness Process Management Journal 9(2), 149–169 (2003)
5. Boehm, B.: Anchoring the software process. IEEE Software 13(4), 73 (1996)
6. Boehm, B., Wolf, S.: An open architecture for software process asset reuse. In: 10th ISPW (1996)
7. Borsoi, B., Risco, J.L.: A method to define an object oriented software process architecture. In: 19th ASWEC, Santa Barbara, California, USA (October 2008)
8. Carr, D.C., Dandekar, A., Perry, D.E.: Experiments in process interface descriptions, visualizations and analyses. In: Schäfer, W. (ed.) EWSPT 1995. LNCS, vol. 913, pp. 119–137. Springer, Heidelberg (1995)
9. Dai, F., Li, T., Zhao, N., Yu, Y., Huang, B.: Evolution process component composition based on process architecture. In: IITAW (2008)
10. Feiler, P., Humphrey, W.: Software process development and enactment: Concepts and definitions. In: ICSP, pp. 28–40 (February 1993)
11. Fu, Z., Li, T., Hu, Y.: An approach to aspect-oriented software evolution process architecture. In: ICICTA, Changsha, Hunan (October 2009)
12. Green, S., Ould, M.: A framework for classifying and evaluating process architecture methods. In: Software Process Improvement and Practice (2005)
13. T.O. Group: The open group architecture framework (TOGAF)-version 9.0 enterprise edition (2013), http://www.opengroup.org/togaf/

14. Hafner, M., Winter, R.: Processes for enterprise application architecture management. In: 41st HICSS, pp. 7–10 (January 2008)
15. Halvorsen, C.P., Conradi, R.: A taxonomy to compare SPI frameworks. In: Ambriola, V. (ed.) EWSPT 2001. LNCS, vol. 2077, pp. 217–235. Springer, Heidelberg (2001)
16. Harmon, P.: Business Process Change. A Mananger's Guide to Improving, Redesigning and Automating Processes. Morgan Kaufmann (2003)
17. ISO. Software Engineering - Product Quality - Part 1: Quality model. ISO/IEC 9126-1 (June 2001)
18. Jayaratna, N.: Understanding and Evaluating Methodologies: NIMSAD, a Systematic Framework. McGraw-Hill, London (1994)
19. Jeston, J., Nelis, J.: Business Process Management: Practical Guidelines to Successful Implementations. Elsevier (2006)
20. Kasser, J.E.: Introducing the role of process architecting. In: 15th INCOSE (2005)
21. Kitchenham, B., Brereton, O.P., Budgen, D., Turner, M., Bailey, J., Linkman, S.: Systematic literature reviews in software engineering - a systematic literature review. Information and Software Technology 51(2), 7–15 (2009)
22. Kronlöf, K. (ed.): Method integration: concepts and case studies. John Wiley & Sons, Inc., New York (1993)
23. Lonchamp, J.: A structured conceptual and terminological framework for software process engineering, Berlin, Germany (1993)
24. Maldonado, A., Velázquez, A.: A method to define the process architecture. In: Twelfth Conference on Information Systems, Acapulco, México (August 2006)
25. Minoli, D.: Enterprise Architecture A to Z: Frameworks, Business Process Modeling, SOA and Infrastructure Technology, Auerbach (2008)
26. Mutafelija, B., Stromberg, H.: Systematic Process Improvement Using ISO 9001:2000 and CMMI. Artech House (2003)
27. Mutafelija, B., Stromberg, H.: Architecting standard processes with SWEBOK and CMMI. In: SEPG Conference (2006)
28. Osterle, H.: Business in the Information Age: Heading for New Processes. Springer, Berlin (1995)
29. Osterweil, L.: Software processes are software too. In: ICSE, pp. 2–13 (1987)
30. Ould, M.: Designing a re-engineering proof process architecture. Business Process Management Journal 3(3), 232–247 (1997)
31. Ould, M.: Basing an information systems strategy on the organization's process architecture. In: Process Think: Winning Perspectives for Business Change in the Information Age (2000)
32. Ould, M.: Business Process Management: A Rigorous Approach. British Computer Society, UK (2005)
33. Patel, N.: Deferred action: Theoretical model of process architecture design for emergent business processes. International Journal of Business Science and Applied Management 2, 4–21 (2007)
34. Paulk, M., Weber, C., Garcia, S., Chrissis, B., Bush, M.: Key practices of the capability maturity modelsm, version 1.1 (CMU/SEI-93-TR-025). SEI (February 1993)
35. Pesantes, M., Lemus, C., Mitre, H.A., Mejia, J.: Software process architecture: Roadmap. In: 9th CERMA, Cuernavaca, Mexico (November 2012)
36. Siviy, J., Kirwan, P., Marino, L., Morley, J.: Maximizing your process improvement roi through harmonization (2008)
37. Siviy, J., Kirwan, P., Marino, L., Morley, J.: Process architecture in a multimodel enviroment (2008)
38. Siviy, J., Kirwan, P., Marino, L., Morley, J.: Strategic technology selection and classification in multimodel environment (2008)

Multi Back-Ends
for a Model Library Abstraction Layer

Ngoc Viet Tran, Andreas Ganser, and Horst Lichter

RWTH Aachen University, Ahornstr. 55, 52074 Aachen, Germany
viet.tran.ngoc@rwth-aachen.de, {ganser,lichter}@cs.rwth-aachen.de
http://www.swc.rwth-aachen.de

Abstract. Software development is moving in the direction of modeling as do quite a lot of other IT related tasks. This means, models become more and more important either as a means of communication or as parts of realizations. Unfortunately, these models are rarely reused which might be due to poor tool support.

A model recommender system is one possible way out, but it bases on high quality data which is most likely stored in a database and needs to blend into an environment. Hence, approaching model recommendations in a model driven way and generating the underlying data store which makes do with an existing infrastructure is desirable. In this paper we describe the underlying model and the obstacles we had to overcome to make this approach work for relational and non relational databases.

Keywords: Model Library, Multi Database, Model Reuse, Eclipse Modeling Framework (EMF), Model Recommender.

1 Introduction

Software development has been becoming more complex and due to that it has been facing lots of challenges. In order to overcome these challenges, many approaches were developed, aiming at data, processing, or complexity issues. Addressing complexity issues, Model Driven Development (MDD) puts forward models as abstractions from the real world. They map objects to concepts on a higher level of abstraction in order to give way to the important considerations omitting technical details. But developing and designing models is challenging enough and takes a while.

Since time and quality are very limited in most development projects, model reuse is regarded as a promising approach to increase quality while reducing development time. This is because, reuse enables to leverage existing knowledge, which was preserved from previous projects as experience. For example, one can consider a reusable model is stored in a model library. Taking this model and continuing with it means, that no more than the finishing touches need to be applied. This could either mean that some parts would need to be deleted or that some parts would need to be added. Either way, a reviewed model was the starting point.

B. Murgante et al. (Eds.): ICCSA 2013, Part III, LNCS 7973, pp. 160–174, 2013.

Looking at programming, reuse has been successful for decades already. No matter if functions and procedures, modules, libraries, frameworks, knowledge bases, or best practices, there are plenty of examples which show how to make reuse work in programming. However, modeling appeared to be a game changer, though design patterns and basic building blocks are one possibility of reuse. This holds true for various modeling languages which might be graphical. For example UML, BPMN, SysML, or even autosar are examples for such languages. Still, model reuse barely exceeds more than this basic reuse although MDD claims to have model reuse as one of its cornerstones. We believe, that the reasons why model reuse is lacking behind lie in weak conceptual foundations and tool support which need to blend into existing environments. Hence, reuse approaches need repositories or libraries which are not just easy accessible, but seamlessly integrate into existing environments.

A seamless integration of a model repository or library starts with embedding it into the existing infrastructure. Surely, every development company has a database management system installed. But there is a wide variety of databases available. It might be a simple textual database, a fully fledged relational system, or a kind of noSQL database. In each case, the existing installations should serve as a foundation for the persistence strategy of the model library. Still the API for each should stay the same, so the power of the model library remains the same. To the best of our knowledge, there is no solution to this problem.

Hence, we first present related work from other groups in section 2. Then we provide our results from our domain analysis from an existing model library in section 3. After that, we discuss the modeled library and its conceptual foundations in section 4.1. These foundations are mapped to entities in section 4.2 and discussed in further details regarding loading and saving. This is enhanced by discussing graph cuts in section 4.3. Finally, we explain the abstraction layer we achieved altogether in section 5 and conclude in section 6.

2   Related Work

Model libraries and model reuse are under research for decades already. First of which holds models primarily for versioning purposes in order to enable collaborative work. The second focuses on concepts to make models more reusable in several ways and with respect to different aspects. We intend to bring these two areas together to enable model recommendations in the long run.

Research related to model reuse is sometimes linked to other aspects as well. For example, Mens et. al, enhanced the UML metamodel with a "reuse contract" formalism [1]. Another example is MOOGLE by Lucrdio et. al. [2], which is a model search engine. It focuses on usability and uses a model extractor to get the relevant information indexed. Last, but not least, ReMoDD by France et. al. is a web based portal which should offer models to a larger community and provide a platform for discussions [3]. All these approaches do not take into account persistence issues to foster in house acceptance as we do; they look at model reuse at a global or community level.

The field of model repositories addresses mostly collaborative work on models and model versioning. First of which, e.g. is covered by Koegel et. al with a project called EMFStore [4]. It allows simultaneous editing of models which propagates changes as they are persisted. Those are forwarded immediately to other developers who are working on the same model. Purely focusing on versioning and its semantics and conflicts on models, a project called AMOR by Altmanniger [5], enhances versioning support for models proceeding ideas from SoMVer and modelCVS [6], [7]. These projects use very specific back ends which could not fit our needs, since we need to make information we store valuable for a model recommender. A project, which goes in the direction of querying model repositories, is the REBUILDER. Gomez et. al investigate Wordnet as a supporter and reasoning mechanism for retrieving UML models.

The "Model Repository" by University of Leipzig builds a bridge to related projects from model repositories to technologies [8]. It works with a graph database (Neo4j) and bursts models as their structure is into a property graph. We use very similar technologies, but for a totally different purpose and allow more than this strategy for persisting data.

Regarding technologies, it is important to note, that our persistence is either relational or graph structured. Hence, Blueprints is an API, which serves as an abstraction layer for graph databases [9]. It allows different graph databases or servers as back ends and can be connected with ideas from Hillairet to EMF [10]. But these ideas support immediate mappings from EMF [11] generated code to Blueprints only. For a library, there is more sophisticated support required as we will discuss below in terms of efficiency and in terms of data structures.

These data structures are motivated by and related to two domains. First, an "object relational mapping" (ORM) builds a foundation [12], but it does not support "object graph mapping" (OGM) out of the box. Some projects were active on these mappings but did not proof beneficial. For example, Java Objects for Neo [13] is a project that simplifies Java object mappings to a Neo4j graph database [14]. It uses annotations and does not require XML bindings. Hence, it is simple and obvious for developers. However, it is limited to some data types and Neo4j. Another project is OX [15]. It uses the terms defined by the OGM but persistence is through XDI data [16]. We aim at constructing a mapping mechanism between plain programming objects and graph vertices for our use case.

3   Model Library Recommender

Modeling became amazingly popular in recent years, so lots of models are designed and built. But support for reuse is still behind its possibilities compared to other domains. Although some model libraries and repositories exist they are not used often. Hence, we developed a model completion tool which supports, but is not limited to, class diagrams as depicted in figure 1.

Fig. 1. Model Recommendation: Drop Down List

The example shows a class diagram with three classes, namely `Person`, `Bartender` and `Party`. As known from code completion, a search box opens in the class diagram editor as `Ctrl + Space` is pressed waiting for queries. The modeler can type and hereby "ask" for recommendations which are produced by an underlying recommender. As soon as reported, the produced recommendations are shown in a drop down list, which a modeler can step through. Lingering on an entry for a moment in the drop down list opens a graphical preview including additional information.

These visualizations should help modelers to decide which model to pick and by that inserting into the existing diagram. This means, after a model is picked, the content of that recommendation is requested, rendered and combined with the current diagram. Additionally, if relationships are known with already existing classes, these are recovered as depicted in figure 2 between `Bartender` and `Cocktail`. Finally, the search box disappears and waits for the next invocation, so modelers can continue modeling as usual.

Altogether, by executing only a few simple actions, diagrams are designed and completed in shorter time. Even expected quality is supposed to be higher, since several recommender strategies can support modelers on different grounds. Starting from simple translations, context analysis, or synonyms all the way to complete structures as shown above, is available. Hence, the tool prevents modelers from making "grammatical" mistakes or typos and provides best practices in terms of correct, reviewed, and standardized models.

Please note, that this tool is not limited to class diagrams. It can work with any type of UML model and its graphical representation in a data store. In fact, the underlying enhanced knowledge graph provides a meta structure which allows to produce recommendations for any kind of model. All that needs to be adjusted is the indexing mechanism and the recommender strategies which are out of scope here.

The underlying enhanced knowledge graph was the foundation for our initial realization and works well. It is backed up by a graph database which enabled us to avoid conceptual breaks and to map data to a very natural format. I.e., each model is mapped to a vertex with additional properties and two vertices

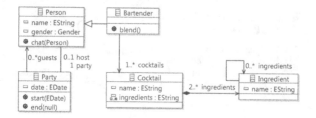

Fig. 2. Model Recommendation: Model Placed

are linked with each other via edges with additional properties. Altogether, this enhanced knowledge graph forms a property graph in terms of graph theory (cf. figure 3). As can be seen, the property graph distinguishes itself from a weighted graph since it has no weights or costs attributed to vertices and edges. Eventually, we do so for the edges so we can foster recommender strategies.

Mind that vertices and edges could be used interchangeably due to their properties; yet, their distinction is just in terms of semantics, not structure. This means, as we map a model to an edge, it could be seen as a vertex and we do something very similar. We store the outtakes between the two models which are related to the vertices. That means, if there was a relationship between one class in the model of one vertex and another class in the other vertex, this relationship is stored in the edge. Due to that, the vertices hold valid models, while edges hold the dangling outtakes. We call this concept "cross links" and it allows us to merge the models later in situations as depicted in figure 2.

A closer look at our enhanced knowledge graph unveiled, that we can use the stored information for a variety of recommender strategies. Each of these might produce different recommendations. Hence, we developed a recommender framework which comprises of an extendable core. First, different user interface strategies might handle the inputs and outputs in different editors, either textual or graphical. Second, several contexts might support different context analysis and tailoring of queries. Last, the recommender strategies produce the actual recommendations. They query the data back ends and know how to produce the most appropriate recommendation for a certain context. To that end, they wrap a data source entirely which means, for a user it would be totally transparent,

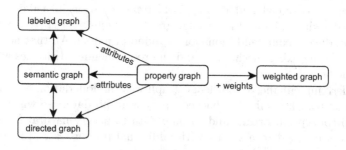

Fig. 3. Graph Types

if a recommendation comes from our system or from systems like MOOGLE or ReMoDD [2] [3].

Still, different deployment scenarios might require additional data sources. For examples, one project might have a neo4j database running and is fine using strategies working with it. Another project might have relational databases only, and might not want to maintain another type of database. Hence, it needs a different persistence strategy. Finally, there might be a project without any databases and lack of interest to install one, but eager to work with model recommendations. As a consequence, a file persisted data set might be most appropriate. Luckily we modeled the domain as depicted in figure 4 and could use a model driven approach and generate most of the libraries for each of the required back ends. Only the persistence strategies would be needed to be tailored.

4   Model Library Foundations

The model libraries mentioned in the related work lead to the conclusion, that these systems see and handle models isolated rather than interlinked. But relating models with each other makes libraries powerful for producing recommendations. This is why, we developed our own prototype with a goal to produce recommendations for reuse eventually.

4.1   A Modeled Model Library

From our existing model library, we reengineered a model for the purpose to apply a model driven approach. The model is depicted in figure 4. It shows how each model is mapped to a `LibraryElement` holding three properties. They are `name`, `owner`, and a list of `files`. Last of which is a list of URIs which point to resources. These could be .ecore files or URLs on servers. Moreover, we store .png files for preview purposes, since we did not want to burden our server with creating previews on the fly and because it proofed reasonable to change the previews manually every now and then. The preview image is used as shown in figure 1. Mind, that the preview functionality is a capability of the `Models`. We did so, because we anticipated other extensions of `LibraryElement`s which we do not discuss here.

We modeled two concepts as means to order and indirectly relate `Library-Elements`. This means, models might be either grouped or categorized which are orthogonal concepts. For example, a class diagram which models a *person* could belong to the same groups as *party* and *company*, since a *person* could attend a *party* or be an employee at a *company*. While the *party* model might fall into the category *leisure* and the *company* might fit into the category *business*, the *person* model belongs to neither of those categories. Still there is a strong relationship between *person* and *party* on the one side and *person* and *company* on the other side. This is why these two pairs are held in one group respectively. In other words, the grouping can establish bridges for models between categories or sub-categories. In terms of graph databases they are light weight hyper edges.

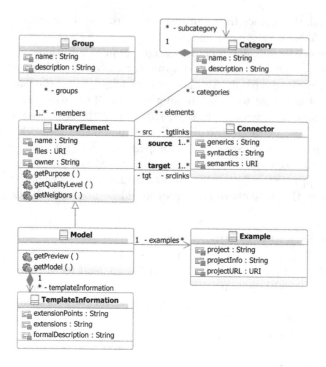

Fig. 4. Model Library Domain Model [17]

Relationships, which are named `Connectors`, are even stronger than categories and groups and directly relate `LibraryElements`. They represent immediate connections between `LibraryElements` on three different levels. First and most low level, there is syntactic information on connectors. These connectors are the cross links mentioned above. Hence, they keep syntactical information so placed models can be reestablish known relationships as shown in figure 2. Second, semantic information holds details about design rationales which are regarded beneficial for other modelers. Last, generic information serves as smallest common ground for describing links if no further information but a previous usage is available.

The modeled library as shown in figure 4 describes in technical terms how the underlying concepts are related and how the structure of the library looks like. But it does not offer a mapping of these concepts to relational or graph based databases though the model was reengineered from a graph database. Fortunately, the mapping is simple at first glance and objects instantiated from `LibraryElements`, `Models`, or `Connectors` can be mapped to relational databases using object relational mappings. Unfortunately, things are a bit more difficult for graph databases. This is due to concepts from graph databases which should be leveraged as much as possible. So, we map `Models` to vertices and we map `Connectors` to edges.

Looking at the meta connectors `Group` and `Category`, the mapping approach needs to change for the graph based database since they are indirect connectors. In contrast, for `Connectors` the mapping is natural and done to edges between elements. The reason to do so is motivated by advantages of property graphs. Since edges can hold properties as vertices can do, edges should store and provide data related to them. Still, the question is, how the mapping works for graph based databases in detail. For object relation mapping, approaches work fine except for some generated overhead.

4.2   Mapped Library Entities

The modeled library from figure 4 served as an input for a code generator. Since EMF was the supporting framework, the question arose how a runtime model could be persisted to a graph databases. This is why we looked at an unrelated example in order to demonstrate the detailed mappings. Those explain how key-value pairs are used in detail, so the property characteristics of the graph could be leveraged to the most extend.

The most fundamental starting points for the mappings are the attributes of an object which can be seen as attribute-value pairs. These can be simply mapped to properties in the graph database, becoming key-value (property-value) pairs. This means, the attributes of objects become keys. For example, a class `Person` is modeled comprising three attributes namely *firstName*, *lastName* and *age*. Consequently, a vertex in the graph database would hold these three values as keys or properties. As an unrelated example a vertex could be ("David", "Beckham", 38) or ("Michael", "Owen", 33).

Since attributes can be mapped to properties, the next requirement is to map the type information from an object to an element in the graph database. The most natural candidates are vertices, but graphs are not supposed to distinguish between types of vertices. Luckily, property graphs allow placing the type information in another property which can be evaluated by graph walkers while traversing a graph.

Moreover, we need to map relationships to the graph database. If two objects are related by a reference, the most natural way to map this to a graph database is an edge which connects these two corresponding vertices. For example, two `Person` objects could be related to each other with a `knows` relationship. After we store the vertices in the graph database, we create an edge connecting these two vertices, and put a property with a value *knows* on the edge. In case the relationships are directed, we simply add starting and ending points. Fortunately, we do not need to go into further details for our model library, since we do not need a stronger concept for relationships, yet (i.e. hyper edges).

Now, we can take our model from figure 4 and map the entities to the graph database. All the concepts, except for the `Connector` become vertices and the `Connectors` are mapped to edges. Moreover, the relationships are mapped to edges which form an enhanced knowledge graph altogether.

The mapping of the `Connectors` had two alternatives. They could have been mapped to a vertex or they could have been mapped to an edge. Either of these choices would have major impact on the realization parts. Hence, we discuss the alternative in more detail.

Mapping a `Connector` to a vertex implies that the semantics of the edges in the graph changes. This is due to edges which do not leverage the properties which are available for edges any more. Hence, edges deteriorate to simple links without any more meaning than just connecting two vertices. As a result, the graph traversal needs to take into account, that `LibraryElements` are never reachable with one hop which requires a self made traversal mechanism. Moreover, this mapping would result in more difficult traversal mechanism for relational data bases. The upside of `Connectors` as intermediate vertices would be that they allow hyper edges as cardinalities are not limited.

We opted for mapping `Connectors` to edges instead which we did for the following reasons. First, we make use of the full capabilities of the graph database which offers properties for edges. Hence, we could store the information we needed in the edges. Moreover, the graph traversal stays as easy as it is and does not require adjustments. Last, the relational mapping carried out by the ORM produces less complex and easier to traverse tables. The downside of `Connectors` as edges is that this does not immediately support hyper edges. Fortunately, our `Groups` make up for that if the underlying graph database does not support hyper edges out of the box.

The established graph structure leads to some runtime aspects like load and save strategies which need to be addressed. This is due to atomicity which is not necessarily assured in graph databases. But when we want to retrieve data, we do not want to convert all the data to objects and keep them in memory. Instead, we could use an indexing mechanism which provides a single vertex. Starting from there, the loading process could walk the neighbors via the connecting edges as necessary. Depending on the scenario depth-first search (DFS) and breadth-first search (BFS) are the candidates for the traversal.

As vertices are loaded into the memory adhering the above mentioned approach, they should be loaded without redundancy. For the EMF part this means that one object is loaded into a resource exactly once. By using traversal algorithms, we walk through the databases avoiding duplicates. All in all, the loading is simpler than the saving process.

As the required parts are loaded, it can be manipulated as a resource. After working on the data, the current content of the resource will need to be forwarded to the database. This is the usual saving process. Basically, it does the opposite of what the loading does. We simply insert, update or delete data in the databases. If the resource is in sequential order, we start with the first element and continue. From one object, we get the attribute names and values and its references to others. After that we have all the data and information of the object. Therefore, we can save it in a single vertex of the database. Based on object references, we will list the neighbors and continue. In order to traverse the entire content of the resource, a backtracking algorithm should be applied. This avoids duplicate saving or missing objects.

Concerning the extent of the saving, the easiest approach is to delete the whole database and recreate its contents from the resource. This avoids a lot of trouble, but is not practical for numerous reasons. A better approach is inserting new objects, delete old ones, and update the changed; the remainder, we leave untouched. All we need, is a mechanism for keeping the states of objects. Fortunately, a simple map or a list will do. It keeps track of whether the content of a vertex or edge changed or not, relying on object identifiers or indices. One question which remains is, what happens if the size of the resource is huge?

4.3   A Disassembled Knowledge Graph

A graph database offers a very natural means of persisting and accessing a lot of data avoiding conceptual breaks. This is due to the graph nature which lots of problems have inherent. But graph databases come around with a few obstacles as well. For example, graph databases rarely scale well in regard of ACID [18]. Hence, users can have pending transactions if another user is already saving or updating. Considering a system which has a lot of simultaneous transactions, this might lead to extensive latency, conflicting data or memory overrun.

One possible solution is to introduce sub-graphs which just cut or split a graph database into smaller and independent graphs. The basic idea of a graph cut is to make a cut-set which is a set of edges whose two adjacent vertices belong to a different subset of the graph. This leads to two disjoint subsets of the graph after cutting the entire graph. An example of this is shown in figure 5. Here, the cut crosses five edges with black ending vertices and divides the graph into a left and a right set. We follow these ideas and apply them to our system. Doing so, we need to bear in mind the following questions:

- How do we keep data of two subsets after cutting?
- How do we decide on a suitable cut-set?
- How do we keep the information of the cut-set?
- How do we store the direction of cut-set in a directed graph?

As a cut will subdivides a graph into two sets, one set holds the data (vertices and edges) of interest and the other set holds the reminder of the graph. For simplicity sake, we call the first part the user set. For the reminder set, we almost do not need to alter it because no actions will be performed on it. We just need to track the ending points of the cut-set. Regarding the user set, we need to store a list of vertices and their edges with the data and a record of the changes. Therefore, if a vertex is changed, the new data is kept on a temporary memory. This applies to property changes regarding values on edges and vertices.

If an operation deletes an edge or a vertex, it is handled differently. When an edge is deleted, the connection between two vertices is removed including the data which is inherent to the edge. As a vertex is deleted, itself and all the adjacent edges need to be removed as well. Hence, a vertex deletion almost always triggers an edges deletion.

As a vertex is changed in terms of attributes, we can forward the changes based on the vertex IDs. Regarding edges, they can be tracked using a tuple of

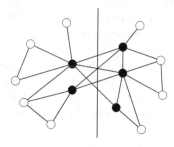

Fig. 5. Graph Cut

the vertex IDs it is adjacent to. In case of directed edges, we get the direction immediately since we store a tuple and in case we have multiple edges between vertices, we distinguish between edges in terms of another identifier.

Since all the operations from above are easily traceable, the record of changes only needs to be forwarded to the graph database. A closer look at the user set uncovers two types of vertices and edges. Either they were changed or not. In order to update the database with these changes, two approaches are possible. First, solely the changes could be forwarded to the database, or the whole user set could be forwarded to the database. The second approach is certainly not optimal for large data which comprise of just minor changes. But the upside is a less complicated algorithm since the tracing of the changes is not necessary.

The first approach, which forwards just the changes, needs to trace those which is more complex. But, based on the traces, the outdated data can be updated by new values. Certainly, the size of the user set determines the operation time and for a large user set the updating can take a considerable amount of time. Therefore, the size of the user set should be limited as much as possible. This leads to the question of a suitable cut-set.

To answer the question what the user really needs, we should look at an example. If the user set is related to one vertex A, the user set could include the vertex A and its neighbors. But, if we know that the user works just with some neighbors of A, we could only create a cut-set holding these neighbors. In case, the user performs an operation that relates to extra data, we could update the current cut-set, and then load a new cut-set that extends from the old cut-set. In our system we create cut-sets following this approach, i.e. implementing a proxy, lazy loading mechanism.

As a result, every operation has its own cut-set. If it could not finished since it needs more data, the system will load another cut-set with the current data on its old cut-set. As a minor tweak, the API allows to set the level of neighbors for each operation. For example, if a user needs only one vertex, the level is set zero. If the user needs a vertex and its next neighbors, the level is set one. If the user also needs the neighbors of its neighbors, the level is set two, and so on. Through this, we keep the information of the cut-set simple.

Looking at how the cut-set expands, we could imagine that the vertex is a center of the ball, since its neighbors can be pictured as the surface of a ball. Hence, each level has one surface and the utmost of it is the cut-set. We depicted an example in figure 6. It shows a black vertex in the center, i.e. at level zero. The white vertices are added to be the cut-set of the level one and the cut-set at level two is depicted as black boxes. Mind, that the size of the cut-set might grow exponentially. But that is not an issue in our case, since elements in a model library barely need to be considered in a context larger than two. In fact, we switch to a walker approach if the level is larger than two. Our lazy loading approach supports this out of the box.

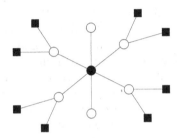

Fig. 6. Cut-Set Ball

We also load piece-cuts which are not fully cut-sets. It means that, from one vertex, we just load some of its neighbors. These piece-cuts are used for loading and saving vertices. In the knowledge graph, vertices of `LibraryElements` may connect with `Groups`, `Categories`, `Examples` or other `LibraryElements`. With a piece-cut, from one element, users could retrieve its only first level neighbors which are also `LibraryElement`. Moreover, the loading API allows querying for exactly one certain neighbor.

5   Model Library Abstraction

The domain model from figure 4 serves as an input for code generation as mentioned above and establishes an abstraction layer for model libraries in general. This is due to the extendability of the EMF framework and its persistence concept. Hence, the model and the generated code are flexibly extendable with relational, graph databases or file storages. These extensions form strategies for the model library and base on other abstraction layers to avoid a vendor lock in.

For example, one popular API which avoids vendor lock in and supports a lot of property graph databases is Blueprints [9]. It is an open source project which can be used on top of property graph databases like as MongoDB, Neo4J, OrientDB, Sail, and Tinker. They can be hosted by a Rexster server [19] which can serve as a hosting environment for multiple of these data bases. All in all,

Rexster and Blueprints serve as a mirror on implementation level for our abstraction layer which offers all the access which is necessary for elements of graph databases and traversing the graph.

Taking a closer look at one implementation of a graph database, Neo4j is accessible through its own API and in our realization it is hidden behind Blueprints. Moreover, it supports full but single transaction management at a time. As a consequence, the ACID properties are not fully supported [18] [14]. Therefore, Neo4j is safer than most other graph databases in terms of ACID, and insert and selection operations are quite simple. This means, Blueprints and Neo4j allow loading either the whole system or vertices one by one identified by ID. Unfortunately, the query mechanism does not support ranges or groups of vertices.

Moreover, graph databases serve as a basis for separation of concerns in terms of meta data and file data. This is due to the fact that graph databases are meant to store property-value pairs and no other data. As a consequence, files as our .png previews and the actual models cannot be stored in the databases. This is why we store files as URIs and do the indexing on the fly.

In contrast, relational databases break the concept of a knowledge graph while storing information. This is due to the object relational mappings. On the one side, this is a well know and researched subject. On the other side, the break of concepts forces us to emulate graph traversals for the relational data base strategies. This is time consuming and performance issues arise quickly when queries includes deep join statements. One upside is related to CRUD or ACID [20]. They are obviously well handled in relational databases. Moreover, these databases could store the files as binary large objects (BLOB) but at this end, the indexing of the content of the .ecore files would get lost and the separation of concern would break.

Altogether, we achieved an abstraction layer which we could use for code generation. This code, which was generated by EMF, uses resource strategies for the actual storing which we could change at runtime with different strategies (cf. figure 7). This means, the facade to our clients remains stable and we are able to adjust the generated code for the different data strategies. Moreover, we could

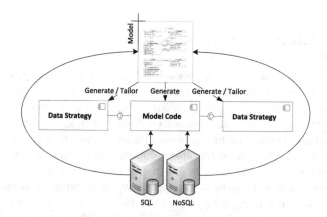

Fig. 7. Model Library Architecture

change our model domain and regenerate the model code since the strategies were mostly generated or based on external frameworks as well.

6   Conclusion

Starting with an existing implementation of a model library we took a domain model and went a model driven approach. This led to an enhanced knowledge graph which is a property graph in terms of graph theory and can serve as a foundation for a model recommender. Since the generated source code allows several ways to persist the enhanced knowledge graph, we investigated further on the ups and downs of several back ends and went into further detail discussing graph databases.

These graph databases allow a very natural way to structure knowledge and offer a very intuitive means of accessing data. But these databases do not come without obstacles. Hence, we looked into loading and storing information as well as into intersecting graphs so performance would not become an issue working with the knowledge graphs. In brief, this is a lazy loading strategy which extends a cut-set concentrically as requested. The exponential space complexity is not an issue in our context because we switch to a walker approach if the level is deeper than two. The saving strategy tracks changes so the data to be updated is kept at a minimum. The realization was done in a model reuse project which is available for download as an Eclipse Updatesite [21].

All in all, we presented an abstraction layer for a model library which outsources persistence related parts into strategies. These were relational or graph based in our case. This outsourcing was one big upside since it allows different deployment scenarios. Additionally, the opportunity to extend the library model easily and regenerate the source code offers a great deal of flexibility. Now, new features are easily added at model level and the source code can be generated and tailored from there.

Further research is now directed in enhancing the content of the model library, so better model recommendations can be produced by the model recommender. This is reasonable, since the model recommender uses strategies which might need different sets of information.

Acknowledgements. We would like to thank all our reviewers for their comments. Moreover, we would also like to thank Felix Bohuschke, Stefan Dollase, Christian Fuchs, Junior Lekane Nimpa, Ruslan Ragimov, and David Mularski for their contributions.

References

1. Mens, T., Cadenhead, T.: Supporting disciplined reuse and evolution of UML models. In: Bézivin, J., Muller, P.-A. (eds.) UML 1998. LNCS, vol. 1618, pp. 378–392. Springer, Heidelberg (1999)
2. Lucrdio, D., de M. Fortes, R., Whittle, J.: MOOGLE: a metamodel-based model search engine. Software and Systems Modeling 11, 183–208 (2010)

3. France, R., Bieman, J., Cheng, B.H.C.: Repository for Model Driven Development (ReMoDD). In: Kühne, T. (ed.) MoDELS 2006. LNCS, vol. 4364, pp. 311–317. Springer, Heidelberg (2007)
4. Koegel, M., Helming, J.: EMFStore: A model repository for EMF models. In: ACM/IEEE 32nd International Conference on Software Engineering (2010)
5. Altmanninger, K., Kusel, A., Retschitzegger, W., Seidl, M., Wimmer, M.: AMOR Towards Adaptable Model Versioning (2008), http://www.modelversioning.org
6. Altmanninger, K.: Models in Conflict – Towards a Semantically Enhanced Version Control System for Models. In: Giese, H. (ed.) MODELS 2008. LNCS, vol. 5002, pp. 293–304. Springer, Heidelberg (2008)
7. Kappel, G., G.K.: ModelCVS - A Semantic Infrastructure for Model-based Tool Integration. Technical Report, Johannes Kepler University of Linz and Vienna University of Technology. (2005)
8. Uni-Leipzig: Eclipse Model Repository., http://modelrepository.sourceforge.net/
9. Tinkerpop: Blueprints API (2013), https://github.com/tinkerpop/blueprints/
10. Hillairet, G.: Blueprints EMF (2013), https://github.com/ghillairet/blueprints-emf
11. Steinberg, D., Budinsky, F., Paternostro, M., Merks, E.: EMF: Eclipse Modeling Framework 2.0, 2nd edn. Addison-Wesley Professional (2009)
12. Barry, D., Stanienda, T.: Solving the java object storage problem. Computer 31(11), 33–40 (1998)
13. Jo4neo: Jo4neo (2009), http://code.google.com/p/jo4neo/
14. Neo Technology: Neo4J (2013), http://neo4j.org/
15. OX:OX (May 2009), http://ox.gluu.org/doku.php
16. OASIS: Oasis members collaborate on 'dataweb' standard for exchange of machine-readable information (2004)
17. Ganser, A., Lichter, H.: Engineering model recommender foundations - from class completion to model recommendations. In: Proceedings of the 1st International Conference on Model-Driven Engineering and Software Development, Modelsward 2013. SCITEPRESS (2013)
18. Haerder, T., Reuter, A.: Principles of transaction-oriented database recovery. ACM Comput. Surv. 15(4), 287–317 (1983)
19. Tinkerpop: Rexster (2013), https://github.com/tinkerpop/rexster
20. Martin, J.: Managing the data-base environment. Prentice-Hall (1983)
21. Ganser, A.: Reusing Domain Engineered Artifacts for Code Generation, https://www2.swc.rwth-aachen.de/?p=35

Explicit Untainting to Reduce Shadow Memory Usage and Access Frequency in Taint Analysis

Jae-Won Min[1], Young-Hyun Choi[1], Jung-Ho Eom[2], and Tai-Myoung Chung[1,*]

[1] Dept. of Computer Engineering, Sungkyunkwan University
[2] 300 Cheoncheon-dong, Jangan-gu, Suwon-si, Gyeonggi-do, 440-746, Korea
{jwmin,yhchoi}@imtl.skku.ac.kr, tmchung@ece.skku.ac.kr
[3] Military Studies, Daejeon University
62 Daehakro, Dong-Gu, Daejeon
eomhum@gmail.com

Abstract. Software vulnerabilities weaken the security of a system increasing possibility of being attacked by exploits in the wild. There are a lot of researches being done on efficiently finding software vulnerabilities to eliminate them. General program testing method for finding flaws in software can be categorized into whitebox and blackbox testing. In whitebox testing, tester examines the internals of the target program such as source codes while in blackbox testing, tester is not aware of the internal structure. Taint analysis is a blackbox testing method efficient for prioritizing exploitable crashes by tracking external input to the program. However due to memory usage and large amount of computation, taint analysis is slow to be used for commercial programs. There has been heuristic approaches to speed up the analysis process but it is not in state of practical use yet. In this paper, we propose a method to reduce resource usage by selectively not tracking certain memories and registers which we call *untainting*. Our evaluation results show that by untainting we can reduce number of taint operation by considerable amount.

Keywords: Security, Taint Analysis, Data Flow Analysis.

1 Introduction

Malware propagates across computer systems leaking private information and harming usability. They use software vulnerabilities inside the target system to do malicious activities. According to the annual report from Panda Security, total number of newly introduced malware in 2011 reached 26 million[1]. Moreover, exploits which triggers vulnerabilities inside programs to gain privilege or execute custom codes, are being sold for up to hundred thousands of dollars. For example, prices of iOS exploits range from 100 thousand dollars to 250 thousand dollars[2]. Security firms like Vupen[3] professionally researches software vulnerabilities and develop exploits. It is becoming more important to efficiently find software vulnerabilities for both offensive and defensive purposes.

---

* Corresponding author.

B. Murgante et al. (Eds.): ICCSA 2013, Part III, LNCS 7973, pp. 175–186, 2013.

Blackbox testing and whitebox testing is the two main category of general program testing technique. In whitebox testing, tester examines the internals of the target program such as source codes while in blackbox testing, tester is not aware of the internal structure. Because most commercial programs do not release the source code, blackbox testing is usually done on the binary of the commercial programs. Fuzzing is a blackbox testing technique which inserts random inputs to the target program and examines the output of each input. It is effective because it does not care about the internal implementation. However, among many crashes that fuzzing resulted by inserting random inputs, it is hard to find crashes that are exploitable [4]. Taint analysis tracks the input to the program which guides the analyst to focus on crashes that have higher chance of being exploitable. There are some taint analysis frameworks developed in the academia[7,8] but these frameworks are not suitable for practical use due to performance.

In this paper, we propose a method to reduce usage and access frequency of shadow memory in taint analysis. We only track and keep information of memories and registers that have higher possibility of being exploitable. Therefore we do not allocate space in shadow memory for memories and registers that are not being tracked. Moreover because number of tracking memory locations and registers are less than that of normal taint analysis, frequency of accessing taint information is lower. These characteristics lead to a more faster analysis.

Organization of this paper is as follows. In section 2 we explain the concept of taint analysis by using an example intermediate language presented in the paper. Section 3 discusses our proposal of reducing overhead in taint analysis and evaluates it in section 4. Finally we present future works and conclude this paper in section 5.

2   Taint Analysis

Taint analysis is a dynamic analysis method which tracks arbitrary input given to the program and examines how the input propagates as the program executes. Taint analysis has been applied to various fields, especially the field of computer security. There are a lot of researches being done on developing analysis frameworks to efficiently analyze programs [6,8,7]. Taint analysis can be divided into three main factors, which are taint introduction, taint propagation and taint checking[5]. Details about each factor is described in the following subsections. To explain taint analysis using formal methods, we present a assembly-like intermediate language in Fig. 1 and its operational semantics in Fig. 2. Operational semantics is written in $[\dfrac{premise}{state_1 \to state_2}$ $condition$ $]$ format. When premise is satisfied, program at $state_1$ will be changed to $state_2$. Functions with double brackets are semantic functions which returns semantic value of the given syntax and examples of them are shown in Fig. 3. $s[x \mapsto v]$ means x is substituted by v in state s and $\mu[x_1] \leftarrow x_2$ means value in the index x_1 of μ is replaced by x_2. For example, operational semantics of unary operation means that when expression e is evaluated into v in initial state s then the result of $unop$ e is $U[\![v]\!]$.

Program ::= *stmt**	
stmt s ::= if *b* then s_1 else s_2 | goto *exp* | while *b* do *s* | *type* store(*exp*, *exp*)	
exp e ::= *type* load(*exp*) | *exp* binop *exp* | unop *exp* | *v*	
boolean b ::= true | false | *exp* = *exp* | *exp* ≠ *exp* | *exp* ≤ *exp* | *exp* < *exp* | *exp* ≥ *exp* | *exp* > *exp* |	
binop ::= binary operation	
unop ::= unary operation	
value v ::= *type* immediate value	
type ::= DWORD | WORD | BYTE	

Fig. 1. Example intermediate language

2.1   Taint Introduction

To conduct taint analysis, we need some kind of initial input to be set as a seed of taint analysis. Input can be a basic keyboard input, network packet, normal files and etc. Usually all the shadow memory and registers are initialized as untainted before the start of the analysis. When the input is inserted to the program, memory or the register that saves the input is marked as tainted. Taint information is expressed in (value, taint) format. Every value that is computed by an instruction has it's own taint value. When other instruction uses the previously computed value as an operand, taint information of that value should be accessed to propagate the taint information to other memory locations or registers.

2.2   Taint Propagation

From the initial input, taint information is propagated to other memories and registers after each execution of instructions in the program. Taint propagation rules define how the taint information is propagated from one place to another according to different instructions. We call these propagation rules as taint policy. Taint policy can differ according to the specific purpose of the taint analysis. In special cases, taint information can be sanitized when the result of some instruction is independent of whether its operands are tainted or not. For example, exclusive OR on same register always yields 0, which is independent of the value in the register.

2.3   Taint Checking

Taint information can be used to determine the program state. For vulnerability detection, one might access the taint information of the program counter register to examine whether control flow of the program can be hijacked. In the case of

$$\text{INPUT}\frac{v \text{ is input value}}{< get\_input(),\ s >\to s[get\_input() \mapsto v]}$$

$$\text{UNOP}\frac{< e\ ,s >\to s[e \mapsto v],< unop\ v,\ s[e \mapsto v] >\to s[e \mapsto v][unop\ v \mapsto U[\![v]\!]]}{< unop\ e,s >\to s[e \mapsto v][unop\ v \mapsto U[\![v]\!]]}$$

$$\text{BINOP}\frac{\begin{array}{c}< e_1,\ s >\to s[e_1 \mapsto v_1],< e_2,s[e_1 \mapsto v_1] >\to s[e_1 \mapsto v_1][e_2 \mapsto v_2] \to s',\\ < v_1\ binop\ v_2,s' >\to s'[v_1\ binop\ v_2 \mapsto Bin[\![v_1,v_2]\!]]\end{array}}{< e_1\ binop\ e_2,\ s >\to s[e_1 \mapsto v_1][e_2 \mapsto v_2][v_1\ binop\ v_2 \mapsto Bin[\![v_1,v_2]\!]]}$$

$$\text{LOAD}\frac{< e,\ s >\to s[e \mapsto v],< load\ v,s[e \mapsto v] >\to s[e \mapsto v][load\ v \mapsto \mu[\![v]\!]]}{< load\ e,s >\to s[e \mapsto v][load\ v \mapsto \mu[\![v]\!]]}$$

$$\text{STORE}\frac{\begin{array}{c}< e_1,\ s >\to s[e_1 \mapsto v_1],\ < e_2,s[e_1 \mapsto v_1] >\to s[e_1 \mapsto v_1][e_2 \mapsto v_2] \to s',\\ < store(v_1,v_2),s' >\to< \mu[\![v_1]\!] \leftarrow v_2,s' >\to s''\end{array}}{< store(e_1,e_2),\ s >\to s''}$$

$$\text{GOTO}\frac{< e,\ s >\to s[e \mapsto\ v],\ < goto\ v,\ s[e \mapsto\ v] >\to< pc \leftarrow v,s[e \mapsto\ v] >\to s'}{< goto\ e,\ s >\to s'}$$

$$\text{TCOND}\frac{< s_1,s >\to s'}{< if\ b\ then\ s_1\ else\ s_2,s >\to s'} \quad \text{if B}[\![b]\!] \text{ is true}$$

$$\text{FCOND}\frac{< s_2,s >\to s'}{< if\ b\ then\ s_1\ else\ s_2,s >\to s'} \quad \text{if B}[\![b]\!] \text{ is false}$$

$$\text{TWHILE}\frac{< s_1,s >\to s',< while\ b\ do\ s_1,s' >\to s''}{< while\ b\ do\ s_1,s >\to s''} \quad \text{if B}[\![b]\!] \text{ is true}$$

$$\text{FWHILE}\frac{}{< while\ b\ do\ s_1,s >\to s} \quad \text{if B}[\![b]\!] \text{ is false}$$

$$\text{CONST}\frac{}{< v,\ s >\to s[v \mapsto N[\![v]\!]]}$$

Fig. 2. Operational semantics for example IL

U[ ] : Returns result of the unary operation
Bin[ ] : Returns result of the binary operation
B [ ] : Returns true or false of the boolean expression
μ[ ] : Maps value of memory in the address given as parameter

Fig. 3. Semantic functions

malware analysis, we can taint sensitive data we want to protect and examine if those data are sent to external servers. In Table 1. we present a basic taint policy for propagating taint information. Each function is called by appropriate operations as stated in the operational semantics in Fig. 4.

Table 1. Taint policy[5]

Functions	Policy Check
$P_{input}()$	True
$P_{const}()$	False
$P_{unop}(t)$, $P_{assign}(t)$	t
$P_{binop}(t_1, t_2)$	$t_1 \vee t_2$
$P_{mem}(t_{address})$	$t_{address}$
$P_{cond}(t)$	t
$P_{goto}(t)$	t

3   Reducing Existing Overheads in Taint Analysis

In this section, we discuss overhead issues present in taint analysis and propose a mechanism to reduce them for efficient use of the analysis.

3.1   Discussion about Tracking Memories and Registers

As explained in the previous section, taint analysis tracks the tainted input by keeping a shadow memory of taint information structures which are constantly updated while instructions are executed. Taint analysis is effective on discovering control hijacking vulnerabilities in an executable. For instance, if the program counter register stores a value which is tainted, control flow of the program can be hijacked by an external input injected by an attacker. However, the problem arise: how much of the register or memory must be tainted in order to be used to hijack the control flow? Controlling all 4 bytes (in the case of 32-bit machines) of the program counter will be the perfect circumstances for the attacker to hijack the control flow of the program because the program counter can be made to point to anywhere in the 4GB memory space. As the percentage of tainted bytes in the register or memory decreases, possibility of hijacking the control flow to a meaningfull memory address also decreases as well. Fig. 5 describes the memory range that the tainted program counter register can point to in the taint percentage perspective. T means that the specific byte is tainted and F means that it is not tainted. When all 4 bytes of the program counter is tainted, logically it can be overwritten to point at anywhere in the 4GB memory space. However when only 1 byte is tainted, memory address that program counter can point to becomes very limited. Therefore if an attacker has injected shellcode in the process memory, it is less likely that program counter will be overwritten to

$$\text{T-Input} \frac{v \ is \ input \ value}{< get\_input(), \ s > \rightarrow s[get\_input() \mapsto (v, P_{input}())]}$$

$$\text{T-Unop} \frac{< e , s > \rightarrow s[e \mapsto (v,t)], \quad < unop \ v, \ s[e \mapsto (v,t)] > \rightarrow s[e \mapsto (v,t)][unop \ v \mapsto (U[\![v]\!], P_{unop}(t))]]}{< unop \ e, s > \rightarrow s[e \mapsto (v,t)][unop \ v \mapsto (U[\![v]\!], P_{unop}(t))]}$$

$$\text{T-Binop} \frac{< e_1, \ s > \rightarrow s[e_1 \mapsto (v_1,t_1)], \quad < e_2, s[e_1 \mapsto (v_1,t_1)] > \rightarrow s[e_1 \mapsto (v_1,t_1)][e_2 \mapsto (v_2,t_2)] \rightarrow s', \quad < v_1 \ binop \ v_2, s' > \rightarrow s'[v_1 \ binop \ v_2 \mapsto (Bin[\![v_1,v_2]\!], P_{binop}(t_1,t_2))}{< e_1 \ binop \ e_2, \ s > \rightarrow s'[v_1 \ binop \ v_2 \mapsto (Bin[\![v_1,v_2]\!], P_{binop}(t_1,t_2))]}$$

$$\text{T-Load} \frac{< e, \ s > \rightarrow s[e \mapsto (v,t)], \quad < load \ v, s[e \mapsto (v,t)] > \rightarrow s[e \mapsto (v,t)][load \ v \mapsto (\mu[\![v]\!], P_{mem}(t))]}{< load \ e, s > \rightarrow s[e \mapsto v][load \ v \mapsto (\mu[\![v]\!], P_{mem}(t))]}$$

$$\text{T-Store} \frac{< e_1, \ s > \rightarrow s[e_1 \mapsto (v_1,t_1)], \quad < e_2, s[e_1 \mapsto (v_1,t_1)] > \rightarrow s[e_1 \mapsto (v_1,t_1)][e_2 \mapsto (v_2,t_2)] \rightarrow s', \quad < store(v_1,v_2), s' > \rightarrow < \mu[\![v_1]\!] \leftarrow (v_2, P_{assign}(t_2)), s' > \rightarrow s''}{< store(e_1,e_2), \ s > \rightarrow s''}$$

$$\text{T-Goto} \frac{< e, \ s > \rightarrow s[e \mapsto (v,t)], \quad < goto \ v, s[e \mapsto (v,t)] > \rightarrow < pc \leftarrow (v, P_{goto}(t)), s[e \mapsto (v,t)] > \rightarrow s'}{< goto \ e, \ s > \rightarrow s'}$$

$$\text{T-Tcond} \frac{< s_1, s > \rightarrow s'}{< if \ b \ then \ s_1 \ else \ s_2, s > \rightarrow s'} \quad \text{if } B[\![b]\!] \text{ is true}$$

$$\text{T-Fcond} \frac{< s_2, s > \rightarrow s'}{< if \ b \ then \ s_1 \ else \ s_2, s > \rightarrow s'} \quad \text{if } B[\![b]\!] \text{ is false}$$

$$\text{T-TWhile} \frac{< s_1, s > \rightarrow s', < while \ b \ do \ s_1, s' > \rightarrow s''}{< while \ b \ do \ s_1, s > \rightarrow s''} \quad \text{if } B[\![b]\!] \text{ is true}$$

$$\text{T-FWhile} \frac{}{< while \ b \ do \ s_1, s > \rightarrow s} \quad \text{if } B[\![b]\!] \text{ is false}$$

$$\text{T-Const} \frac{}{< v, \ s > \rightarrow s[v \mapsto (N[\![v]\!], P_{const}())]}$$

Fig. 4. Operational semantics with taint policy

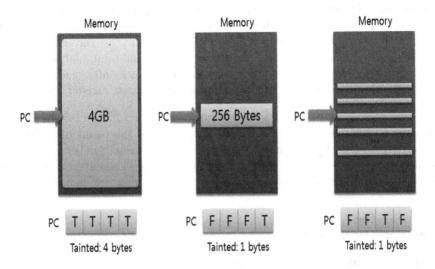

Fig. 5. Range of memory that can be changed to according to the taint ratio

point to the shellcode for execution. This means in general, lower percentage of tainted bytes has lower possibility of successful exploitation.

Is it necessary to track memory locations and registers that have low possibility of exploitation? Since taint analysis itself is a process that requires huge amount of resources especially memory and time, eliminating unnecessary computation and resource usage will accelerate the analysis. Moreover, this will help prioritize exploitable crash cases for further inspection because tainted memories and registers which has low possibility of exploitation are not tracked during the analysis. Therefore analyzers can focus on the cases with higher chances of being exploitable.

3.2 Explicitly Untainting Memory and Register

To make the taint analysis more efficient, we propose a heuristic method of explicitly *untainting* memories and registers that have low percentage of tainted bytes in respect to the access type. By changing the upper boundary of the percentage, output cases will differ. Explicit *untainting* takes place when the instruction has tainted memory or register as operands. When the instruction is executed, it fetches taint information of the operands and update them according to the specific operational semantic of the instruction. If the tainted-byte percentage of the output is below the boundary or predefined threshold, it is untracked from that point of execution. The percentage of tainted bytes is calculated by [total number of tainted bytes / size of the access type]. Taint policy of our proposal is stated in Table 2.

Binary and Unary Operations. Unary operations are done on a single operand such as a memory address or a register. Therefore the status of taint information does not change which means it behaves same as conventional taint analysis. Binary operations is the case where conventional taint analysis and our approach differ. As explained earlier, our approach checks after each taint operation whether it is necessary to track certain memory or register and update the taint information.

Boolean Expressions. Similar to unary operations, the result of evaluating boolean expressions does not have effect on the taint information because the expression itself does not update any memory or a register. Therefore our approach does not do anything different.

```
DWORD store(mem1, get_input())
DWORD store(mem2, get_input())

DWORD store(mem2, load(mem2) AND 0x000000FF)

...

DWORD store(reg1, load(mem1) + load(mem2))
goto reg1
```

Fig. 6. Example code for explicit untainting

Branch Operations. Fig.6 shows an example intermediate language statements for explicit untainting. Assume that predefined threshold is 25 percent. First two lines get 4-byte user inputs and store them in memory addresses specified by *mem1*, *mem2* which become tainted by 4 bytes. Then, the next instruction does bitwise AND operation of the value stored in memory address *mem2* and the hex value 0x000000FF leaving only lower 1 byte of the memory tainted. Assuming the predefined taint threshold is 25 percent, we mark memory address *mem2* as untainted instead of updating the taint information. Finally, target address of goto instruction is sum of values stored in *mem1* and *mem2* which becomes tainted by 4 bytes because value stored in *mem1* is tainted by 4 bytes. Therefore even if we did not track the memory address *mem2* along following instructions, target of the goto instruction is still tainted. Note that in this condition, control flow can be hijacked because destination address of the goto instruction is dependent on the user input.

String Operations. In the case of user input string being copied to the memory, Intel x86 architecture copies 4 bytes each time instead of 1 byte to be efficient. However when accessing the string, it is accessed by byte instructions. Loading 1 byte from the memory with byte instruction will exceed the threshold because the percentage of tainted bytes is 100 percent. Therefore tainted string bytes will continue to be tainted.

Others. There are other expressions or statements that are not discussed above. For example while, if-then-else, and etc. The basic building block of these statements are cases that are mentioned, therefore our heuristic analysis behaves according to the result of the basic blocks.

Table 2. Proposed taint policy

Functions	Policy Check
$P_{input}()$	True
$P_{const}()$	False
$P_{unop}(t),\ P_{assign}(t)$	t
$P_{binop}(t_1, t_2)$	$\begin{cases} false\ if\ (\%\ of\ tainted\ bytes) \leq threshold \\ true\ if\ (\%\ of\ tainted\ bytes) > threshold \end{cases}$
$P_{mem}(t_{address})$	$t_{address}$
$P_{cond}(t)$	t
$P_{goto}(t)$	t

3.3 Issues

Normally when the instruction manipulates the tainted operands, percentage of tainted bytes does not increase. Therefore if a certain tainted memory location's percentage drops below the threshold, it does not go above the threshold in the remaining of the program in normal cases. In other words, memories or registers that we considered as unimportant remain at that state until the end of the execution. However, special cases arise when binary operation has tainted and explicitly untainted operands or two explicitly untainted operands. Because when two tainted operands have tainted bytes that do not collide with each other's position in the 4-byte range, ratio of tainted bytes in the output of the operation can increase. Therefore if our policy decided to untaint one of the operand in prior, output will have lower percentage then the original which can lead to a false negative result.

Fig.7 explains the situation. Memory location *mem1* and *mem2* store user inputs which are tainted. After third and fourth instruction, memory location *mem1* and *mem2* are tainted by only 1 byte with different offset. In the original taint analysis, after calculating the sum of values stored in *mem1* and *mem2*, *reg1* becomes tainted by 2 bytes. However in proposed taint policy, memory address *mem1* and *mem2* are explicitly untainted since they are tainted by 1 byte , resulting *reg1* as untainted. Goto instruction with *reg1* as operand therefore, does not seem to be hijackable causing a false negative.

However this is an design issue. There is a tradeoff between accuracy and speed. Tracking every single tainted memory and register will provide accurate results but will be slow. On the other hand, tracking only a subset of tainted memory will result in information loss but will provide faster result. It is becoming more important to filter out crashes that are security critical because

```
DWORD store(mem1, get_input())
DWORD store(mem2, get_input())

DWORD store(mem1, load(mem1) AND 0x0000FF00)
DWORD store(mem2, load(mem2) AND 0x000000FF)

DWORD store(reg1, load(mem1) + load(mem2))
...
goto reg1
```

Fig. 7. False negative example

amount of cases to examine is large[4]. Assuming set of memories and registers that are tainted is U, our approach will track the set A, which is a subset of U. False negative cases such as Fig. 7 will be included in set U-A which is not deniable. Although false negative cases are not deniable, we must work to reduce the number of cases. If set E exists such that its elements are exploitable memories and registers, our goal will be analyzing set A which includes all the elements in E, thus saving resources of tracking U-A.

4   Evaluation

In this section, we evaluate our proposal by tracing an example program and examining number of memories and registers that are untainted. Below figure is the source code of the program that we traced for the evaluation. It is a simple program that has two local character buffers to save user input strings. There is no boundary checking for the buffers, therefore it can cause a stack buffer overflow. The threshold is assumed to be 25 percent, which is as same as previous examples.

We inserted a long input string that triggers the overflow vulnerability in the program when trying to return to the caller of the main function and traced the execution. Out of 533,134 instructions traced from the execution, 6632 instructions had at least one tainted operand. Out of 6632 tainted instructions, 1263 instructions had two operands with number of tainted bytes below the threshold and 2760 instructions had one tainted operand below the threshold and one untainted operand. As a result, 40322 instructions can be explicitly untainted which means that number of access to the shadow memory is decreased by 4032. Over 60 percent of tainted instructions can be eliminated. High percentage of untainted instructions are due to long loop of instructions that manipulates the string. There are cases that need further consideration regarding untainting, for example byte operations, string operations etc. We present these statistics to address that there is a lot of room for optimization of tainted instructions.

```
#include <stdio.h>

int main(void)
{
    char user[100];
    char pw[100];
    char* pointer=NULL;

    gets(user);
    gets(pw);

    printf("login credentials");
    printf(user);

    if(!strncmp(user,"smba",4) && !strncmp(pw,"pass",4))
        pointer = "Login success";

    printf("Result: %s",pointer);
}
```

Fig. 8. Traced program code

Table 3. Summary of analysis results

Description	Number
Total number of traced instructions	533134
Number of tainted instructions	6632
Number of explicitly untainted instructions	4032

5   Conclusion and Future Works

In this paper, we explained about the concept of taint analysis and its application to the field of computer security. Taint analysis is a data flow analysis method used to track the input to the program, which can be applied to vulnerability detection, malware analysis and so on. However, because taint propagation is done at runtime and there are many memories and registers to track, it is yet too slow for commercial uses. Therefore instead of tracking all the tainted memories and registers, we proposed a method to track only those that have more higher possibility of being exploitable. Decision of stop tracking a certain memory or register is made by examining the percentage of tainted bytes. We have addressed that as the percentage of tainted bytes gets lower, it is hard for an attacker to change the running program to a vulnerable state. Our evaluation results show that by explicit untainting we achieve performance improvement. However, we have not shown how many exploitable bugs we can find comparing to the full taint analysis, which we leave as a future work.

Acknowledgements. This work was supported by Priority Research Centers Program through the National Research Foundation of Korea(NRF) funded by the Ministry of Education, Science and Technology(2012-0005861).

References

1. 2011 Annual Report PandaLabs,
   http://press.pandasecurity.com/press-room/reports
2. Greenberg, A.: Shopping For Zero-Days: A Price List For Hacker's Secret Software Exploits (2012), http://www.forbes.com
3. Vupen Security (February 2013), http://www.vupen.com
4. Miller, C., et al.: Crash Analysis with Bitblaze In: Blackhat USA (2010)
5. Schwartz, E.J., Avgerinos, T., Brumley, D.: All You Ever Wanted to Know About Dynamic Taint Analysis and Forward Symbolic Execution (but might have been afraid to ask). In: IEEE Symposium on Security and Privacy (2010)
6. Brumley, D., Jager, I., Avgerinos, T., Schwartz, E.J.: BAP: A Binary Analysis Platform. In: Gopalakrishnan, G., Qadeer, S. (eds.) CAV 2011. LNCS, vol. 6806, pp. 463–469. Springer, Heidelberg (2011)
7. Song, D., Brumley, D., Yin, H., Caballero, J., Jager, I., Kang, M.G., Liang, Z., Newsome, J., Poosankam, P., Saxena, P.: BitBlaze: A New Approach to Computer Security via Binary Analysis. In: Information Systems Security (2008)
8. Clause, J., Li, W., Orso, A.: Dytan: A Generic Dynamic Taint Analysis Framework. In: Proceedings of the 2007 International Symposium on Software Testing and Analysis. ACM (2007)
9. Avgerinos, T., Cha, S.K., Hao, B.L.T., Brumley, D.: AEG: Automatic Exploit Generation In. In: Proceedings of the Network and Distributed System Security Symposium (2011)
10. Miller, C., et al.: Crash Analysis with Bitblaze In: Blackhat USA (2010)

A Framework for Security Testing

Daya Gupta, Kakali Chatterjee, and Shruti Jaiswal

Department of Computer Engineering, Delhi Technological University, India
dce.shruti@gmail.com

Abstract. The goal of security testing is to verify and validate the potentiality of different vulnerabilities. For identified threats ensure that security mechanism deployed during design really mitigate the threats at vulnerable points. This requires checking that during functionality execution the threats to the assets really get mitigated. In this paper we propose a Framework for Security Testing that involves identifying different attacks that are possible by different stakeholders or intruders for each functionality offered by the system. Next we validate that the design decision taken to implement the security requirement associated with that functionality is appropriate to mitigate identified threats and risks on assets involved. Finally a test report template is designed which can be used to review the deployed security mechanism.

Keywords: Security Testing, Vulnerability Point, Vulnerability Nullification, Threat Mitigation.

1 Introduction

The common security measures (security goals) are to secure the assets of the organization. Assets are generally defined as anything that has value to the organisation and needs protection from intruder or accidental detriment by recognised actors of the system. Traditionally it is known as information security and is expressed as confidentiality, integrity and availability of information in an organisation which is called CIA triad [1].

To develop an efficient, cost effective and secure system, security engineering process consisting of activities like i) Security requirements elicitation, analysis & prioritization, specification and management; (ii) Appropriate design structure; (iii) Implementation of all functionalities incorporating design decision; (iv) Testing of security modules must be present. Our previous research [3, 4] presents a framework for secure software development where one should follow the security engineering activities. Then in [5] a design template is presented that guide the security engineer to take appropriate design decisions after eliciting and prioritizing requirements. In continuation to our previous research now our focus is to present a framework for security testing.

Eliciting security requirements using viewpoint oriented approach [3] and prioritizing this security requirement using techniques of CRAMM [10] are discussed in [4].

B. Murgante et al. (Eds.): ICCSA 2013, Part III, LNCS 7973, pp. 187–198, 2013.

Once security requirement engineering is through, one needs to take proper design decision for security requirement analogues to conventional design process for functional and nonfunctional requirements. In [11] we have developed a methodology, to analyze available cryptographic techniques for different threats and proposed a design template that identifies the environmental constraints and design constraints. Based on this, appropriate cryptographic techniques may be specified that optimally meets the security requirements. After the design decision, the system needs security testing to avoid all possible loopholes and weaknesses in the system which might result into loss of information in the hands of the stakeholders or outsiders/ intruders of the system.

Recently there are some proposals on model based security testing as in [6] they have combined system modeling languages with modeling languages for security and formalize access control requirements. A brief survey on models that can be used for model based security testing also found in [7]. In [8] they are creating threat traces to test the security concept at the time of testing by matching the execution trace with threat trace created with the help of sequence diagrams. Security testing as presented in [9] is a novel scenario based method that deals with security testing during the design phase. Continuing our earlier work to develop a cost effective security engineering framework which is already proposed in [5], now our aim in this paper is to extend this framework by developing a process for Security Testing.

Our approach involves identifying vulnerable points and mapping identified threats to these vulnerable points to mitigate the identified security threats. In this paper we have proposed a framework for security testing by which we check whether the design decision taken to implement the security requirements is appropriate for the system or not.

The rest of the paper is organized as follows:

In Section 2, Proposed Security Engineering framework is discussed; Section 3 provides Proposed Framework for Security testing; Section 4 provides Implementation Example; finally Section 5 concludes the paper.

2   Proposed Security Engineering Framework

To design, build and deploy secure systems, one must integrate security into SDLC and adapt current software engineering practices and methodologies to include specific security-related activities. Security-related activities [5] include identifying security requirements, prioritizing & management of security requirements, security design, implementing security mechanisms, security testing. Our proposed framework for Security Engineering Process (SEP) is shown in Fig. 1. Now the brief explanation of the different activities performed in different phases of Security Engineering Process (SEP) are as follows:

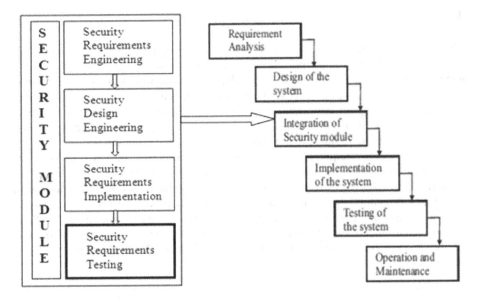

Fig. 1. Framework for Security Engineering Process

Requirement Engineering – This process is based on well known process of View Point Oriented Requirement Definition [3]. It involves discovering functional requirements. It then identifies the assets and associated threats involved with the functional requirements to finally elicit security requirement such as Privacy, Authentication, Integrity, Non- Repudiation, etc as defined by Firesmith [2]. Then elicited security requirements are analyzed and prioritized.

Design Decisions – In this phase security design decisions are taken which include mapping of security requirements with cryptographic services such as authentication, confidentiality, etc., and then mapping attacks to prioritized threats. After that design decisions are taken which consider environmental constraints such as memory, encryption speed, energy, etc and security design attributes such as throughput, target platform, cost, etc. It then generates Security Design Template (SDT) that guides security engineer to finalize design decisions that specify which cryptographic technique is best suited for particular environment and design constraints.

Implementation – This includes implementing specific techniques that are suggested in the design phase of the Security Engineering Process.

Testing – It involves evaluating the system security and determining the adequacy of cryptographic algorithms chosen during design.

3   Proposed Security Testing Framework

When the term testing comes, the two sub topics arise. These are verification and validation. Verification is the process of evaluating a system or component to determine whether the products of a given development phase satisfy the conditions imposed in the beginning of that phase. Validation is the process of evaluating a system or component during the development process or at the end of it to determine whether it satisfies the specified requirements. Verification is a static technique in which execution of code is not required hence process of verification is chosen here.

After the security requirements are prioritized, appropriate design decisions are taken and further activities arise. Process for making design decisions is already explained in [5]. After security design implementation security testing comes into picture. Early detection of security related concerns will be cost effective, as it costs more to remove any error during later phases. Security testing is done to test whether security requirements are implemented properly and are protecting the system against all the attacks/ vulnerable points. Our proposed framework for Security testing is shown in Fig. 2.

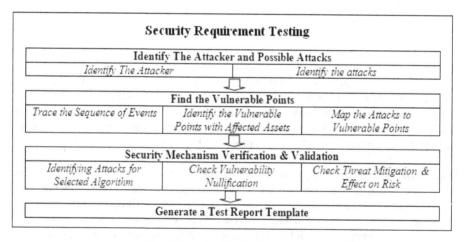

Fig. 2. Security Testing Framework

Step 1: Identify the Attacker and Possible Attacks
Identification of attacker and the attacks that he can execute is very helpful in assessing the system security.
The process consists of following sub processes:

i)   Identify the Attacker. Identify whether the attacker is an insider or an outsider. Insider means anybody who is an actor related to system directly or indirectly and outsider means an intruder who wants to intentionally attack the system. Identify the attacker for each functionality offered by the system.

ii)   Identify the Attacks. Once the attackers are identified, identify the possible attacks executed by attackers to breach the security. This is required to test whether the security mechanism is protecting against the attacks.

Step2: Find Vulnerable Points

Vulnerable point shows the weakness of the system, from where the attackers can penetrate the system and gain access to the valuables for their use. They may also cause harm to the organizational assets, so identify all these vulnerable points.

The process consists of following sub processes:

i)   Trace the Sequence of Events. To identify the vulnerable points that exist in the system, create the sequence of events that occurs to execute functionality.

ii) Identify the Vulnerable Points with Affected Asset. From the above sequence trace out the vulnerable points of system and the assets that can be affected if an attack occurs at the identified vulnerable point. Asset can be anything that has value to the organization.

iii) Map the Attacks to Vulnerable Points. Map the attacks identified in step 1 to all the identified vulnerable points, so that it can be checked easily whether the system is able to protect these points from identified attacks or not.

Step 3: Security Mechanism Verification & Validation

Now test whether the cryptographic algorithms identified during the design phase are able to protect the system against various attacks and threats possible due to the presence of vulnerable points for particular functionality

The process consists of following sub processes:

i)   Identifying Attacks for Selected Algorithm. First take the cryptographic algorithm identified during the security design decisions activity of security design engineering and then identify the attacks which are resisted by the algorithm.

ii) Check the Vulnerability Nullification. Map the attacks to the vulnerable points identified with the set of attacks that the given algorithm is able to protect against. Then will measure the level to which the vulnerability points are protected against the attack. If out of N total attacks M are mitigated then will say that the vulnerability point is protected by value M/N. Following this, calculate the Security Vulnerability Index (SVI) using the formula given below:

$$\text{Security Vulnerability Index (SVI)} = (\text{Extent of Vulnerable Points Protected}/ \text{Total Vulnerable Points Considered}) * 100$$

iii) Check Threat Mitigation & Effect on Risk. Map the attacks to the corresponding threats and check how many threats are mitigated by selected cryptographic algorithm. If algorithm is able to defend against attack, conclusion can be extracted that threats corresponding to attacks are also mitigated. If the threat is mitigated it means the value of threat rating becomes zero and hence the risk will also become zero as the risk is dependent on threat rating in our risk value computation [4]. Then calculate the Security Risk Index (SRI) using the formula given below:

$$\text{Security Risk Index (SRI)} = (\text{Total Risk Evaluated as Zero} / \text{Initially Considered Threats}) * 100$$

Step 4:   Generate Test Report Template

Create a test report template that contains all the important information related to the activities. The template will help the developer in deciding further activities as it contains all the security related information of this phase. Template has the fields like Test Case ID, Test Case Name, Attacker and Modeled Attacks, Vulnerable Points, Security Mechanism, Result, Recommendation/ Remark. The sample test template is shown below in Fig. 3.

Test Case ID	Unique ID for each Template so that it can be distinguished	Test Case Name	Unique Name must be same as functionality under test
Attacker Type	Type of attacker (insider or outsider)		
Modelled Attacks	List of attacks that attacker can execute		
Vulnerability Points	List of all points on which attack can occur during the execution of functionality		
Security Algorithm Applied	Algorithm that is identified and chosen during design for protection of the system.		
Result	The value of Security Vulnerability Index and Security Risk Index.		
Remarks	Any suggestion or recommendation if required		

Fig. 3. Reference Test Report Template

4   Case Study of Web Based Banking

In this section, apply the framework to a Web Based Banking system for verifying the security mechanisms as per requirements and constraints. In web based banking system, the actors involved can be Customer, Bank Employee, Bank Database, System Administrator, Maintenance Manager, etc. But the actors/stakeholder that comes under our consideration is only the direct actors such as Customer, Bank Employee and Bank Database. Their functional requirement, non functional requirements, security requirements, attacks, security mechanism and all other details of elicitation and design can be found in [5].

The output of the activities of first phase of security engineering, that is, security requirements engineering is shown in Table 1.

Before applying the proposed framework of security testing we have taken the security decision template as shown in Table 2. For detailed process refer [5].

Table 1. Security Requirement Prioritization for Web Based Banking System

Security Requirement	Threat	Threat rating	Vulnerability	Assets Affected	Asset Value	Risk	SR Prioritize
Authentication	1. T.PIN_Violated	3.5	.1	User login info(1,2)	4	3,4	7
	2. T.Denial_of_service	.1	.5	Smart Card info(1,2,3)	9	3,4,5	12
	3. T.Impersonate	.34	.5	Account info(1,3)	6	4,5	9
Authorization	1. T.Change_Data	3.35	.5	Certificate info(2)	9	6	6
	2.T.Certificate_steal	1	.5	Account info(1,2)	6	6,4	10
Privacy	1. T.PIN_Violated	3.5	.5	User login info(1,2)	4	3,4	7
	2.T.Unauthorize_access	.1	.5	Account info(1,2)	6	4,5	9
Integrity	1. T_replace	1	.1	Customers info(1,2)	5	5,6	11
	2. T.Data_Theft	1	.1	Transaction info(1,2)	8	5,6	
Non-Repudiation	1.T.Certificate_steal	3.33	.5	Customers info(1)	5	4	4

Table 2. Security Design Engineering Web Based Banking System

Environment	Design Constraints	Design Attribute	Priority			ECC operation on GF(p)		
			High (Value 1.0)	Medium (Value 0.5)	Low (Value 0.1)	Operation	Iterations	Total Time
LAN ☐	Encryption ☐	Throughput	✓	☐	☐	168 bit Encryption	368	30.1
WLAN ✓	Speed ✓	Storage	☐	✓	☐			
WPAN ☐	Range ☐	Target platform	✓	☐	☐	168 bit Decryption	736	30
WWAN ☐	Channel Capacity ☐	Cost	☐	☐	✓	168 bit Signature	735	30
WMN ☐	Frequency ☐	Power consumption	☐	☐	✓	168 bit Verification	606	30
WSN ☐	Bandwidth ☐	Security	✓	☐	☐	155 bit Encryption	417	30
MANET ☐		Scalability	✓	☐	☐	155 bit Decryption	822	30
		Flexibility	✓	☐	☐	155 bit Signature	835	30
		Algorithm agility	☐	✓	☐	155 bit Verification	689	30
		Complexity	☐	✓	☐			
		Interoperability	✓	☐	☐			

Now apply our proposed framework on Web Based Banking system. As our framework has four main steps, we progress through all of them one by one.

Step1: Identify the Attacker and Possible Attacks

i)   Identify the Attacker. To identify the attacker, start with the functionality *balance enquiry*. This functionality can be attacked by an insider or by an outsider or by both. Here we take the attacker as an outsider for explaining the rest of the process.

ii)   Identify the Attacks. Now identify the possible attacks. Various possible attacks are:

    a)   Replay attack
    b)   Man – in – the – Middle Attack
    c)   Dictionary Attack
    d)   Impersonation Attack
    e)   Insertion attack

Step2: Find Vulnerable Points

i)   Trace the Sequence of Events. The sequence of event for the selected functionality is as follows

- a)   Provide login details at customer end (Request)
- b)   Verification of login details at server side (Response)
- c)   Request for balance enquiry (Request)
- d)   Retrieve data from database (Response)

ii)   Identify the Vulnerable Points with Affected Assets. From the above sequence trace extract out the vulnerable points of attack and assets that get affected. Result shown in Table 3.

Table 3. Vulnerable points with Assets

S. No	Vulnerable Point Extracted	Assets Involved
1	At the time of Login	User Login Information
2	Verification of Login details	Account Information, Customer Information
3	Request for Balance	Account Information
4	Retrieving Data	Account Information, Customer Information
5	During Data Transfer	User Login Information Account Information, Customer Information

iii)   Map the Attacks to Vulnerable Points. Now map the attacks identified in step 1 to the identified vulnerable points and also map the corresponding threats. As shown in Table 4.

Table 4. Mapping of Threats Involved

S. No.	Vulnerable Point	Attack	Threats Involved
1	At the time of login	Replay Attack, Impersonation Attack	T. Impersonate, T.Data_Theft
2	Verification of Login Details	Denial of Service Attack	T.Change_Data
3	Request for Balance	Impersonation Attack	T.Impersonate
4	Retrieving Data	Denial of Service Attack	T.Change_Data
5	During Data Transfer	Man – in – the – middle Attack, Dictionary Attack, Insertion Attack	T. Impersonate, T.Data_Theft T.Disclose_Data

Step 3: Security Mechanism Verification & Validation

i)   Identifying Attacks for Selected Algorithm. Take the cryptographic algorithm selected in design phase and identify the attacks protected by the algorithm

> Security Mechanism selected: ECDSA
> List of Attacks that are resisted by ECDSA are given in Table 5

Table 5. Attacks Resisted by ECDSA

S. No.	Attacks
1	Impersonation
2	Stolen Verifier
3	Smart Card Loss
4	Denial of Service
5	Insertion

ii)   Check Vulnerability Nullification. Next check whether the identified attacks are resisted and corresponding vulnerable points are protected by the selected algorithm. Table 6 shows the vulnerability point nullification.

Functionality under Test: Balance Enquiry

Table 6. Nullification of Vulnerable Points

S. No.	Vulnerable Point	Attacks	Nullified or Not
1	At the time of login	Replay Attack, Imperso-nation Attack	1/2
2	Verification of Login Details	Denial of Service Attack	1/1 = 1
3	Request for Balance	Impersonation Attack	1/1 = 1
4	Retrieving Data	Denial of Service Attack	1/1 = 1
5	During Data Transfer	Man – in – the – middle Attack, Dictionary Attack, Insertion Attack	1/3

So out of five vulnerable points identified, selected algorithm is able to protect three vulnerable points completely, one as half and one as one third.

$$\text{Security Vulnerability Index} = ((1/2 + 1 + 1 + 1 + 1/3)/5) * 100 = 76.67\ \%$$

iii)   Check Threat Mitigation & Effect on Risk. Now check the effect of attack removal on threat and risk shown in Table 7.

Table 7. Threat Mitigation and Risk Value

S. No.	Vulnerable Point	Attack	Threats Involved	Threats Mitigated	Risk Value
1	At the time of login	Impersonation Attack, Replay Attack	T. Impersonate, T.Data_Theft	T.Impersonate	Y (1) N (0)
2	Verification of Login Details	Denial of Service Attack	T.Change_Data	T.Change_Data	Y (1)
3	Request for Balance	Impersonation Attack	T.Impersonate	T.Impersonate	Y (1)
4	Retrieving Data	Denial of Service Attack	T.Change_Data	T.Change_Data	Y (1)
5	During Data Transfer	Man – in – the – middle Attack, Dictionary Attack, Insertion Attack	T. Impersonate, T.Data_Theft T.Disclose_Data	T.Data_Theft	N (0) Y (1) N (0)

In Table 7 Risk value Y (1) corresponding to a threat represents it is being mitigated by mapped attack and value N (0) means corresponding threat is not mitigated by attack .

In the above table, most of the threats are mitigated as the corresponding attacks are resisted. Therefore, it is validated that most of the risk values become zero, as the threats are mitigated which in turn make threat rating as zero. And in CRAMM [10] if the threat rating is 0 the corresponding risk will be 0.

$$\text{Security Risk Index} = (5/\ 8) * 100 = 62.5 \ \%$$

Step 4: Generate Test Report Template

Test report template is generated to provide a brief overview of the overall process of security testing. Test Template for Balance Enquiry is shown in Fig. 4. It contains all the important information related to the activity. The template will help the developer in deciding further activities as it contains all the security related information of this phase.

Test Case Id	TC_01	Test Case Name	Balance Enquiry
Attacker Type	Outsider		
Modelled Attacks	Replay attack, Man – in – the – Middle Attack, Dictionary Attack, Impersonation Attack, Insertion attack		
Vulnerability Points	At the time of login, Verification of Login Details, Request for Balance, Retrieving Data, During Data Transfer		
Security Mechanism Applied	ECDSA		
Result	SVI = 76.67 % SRI = 62.5 %		
Remarks	None		

Fig. 4. Test Template for Balance Enquiry

5   Conclusion

In this paper framework for security testing is presented that tries to verify all the security mechanism identified during design phase are appropriate for system under study. In existing research proposed, security testing is carried out after system implementation. In [9] researchers are performing security testing during design phase but this is based only on security requirements. In our proposal we first elicit security requirements, and then take effective design decision by choosing the most optimal mechanism to implement security and then perform testing. Security testing phase will be the part of Security Engineering Process. The framework is very helpful in information system development process as here it tries to deal with all security related concerns in early stages. It will help cut down the overall budget of system development as early detection and correction of security concern incur less cost and time. Security testing will become an integral part of security engineering framework and in turn part of the conventional software engineering process.

References

1. Benjamin, F., Seda, G., Marittal, H., Thomas, S., Holger, S.: A comparison of security requirements engineering methods. Requirements Engineering 15(1), 7–40 (2010)
2. Firesmith, D.G.: Engineering Security Requirements. Journal of Object Technology 2(1), 53–68 (2003)
3. Agarwal, A., Gupta, D.: Security Requirement Elicitation Using View Points for online System. In: Emerging Trends in Engineering and Technology, ICETET 2008, pp. 1238–1243. IEEE Computer Society (2008)
4. Jaiswal, S., Gupta, D.: Security Requirement Prioritization. In: The Proceeding of SERP 2009, pp. 673–679 (2009)

5. Chatterjee, K., Gupta, D., De, A.: A Framework for Security Design Engineering Process. In: Venugopal, K.R., Patnaik, L.M. (eds.) ICIP 2011. CCIS, vol. 157, pp. 287–293. Springer, Heidelberg (2011)
6. David, B., Jurgen, D., Lodderstedt, T.: Model Driven Security: From UML Models to Access Control Infrastructures. ACM Transactions on Software Engineering and Methodology 15(1), 39–91 (2006)
7. Schieferdecker, I., Grossmann, J., Schneider, M.: Model Based Security Testing. In: Workshop on Model- Based Testing 2012 (MBT 2012). EPTCS, vol. 80, pp. 1–12 (2012)
8. Wang, L., Wong, E., Xu, D.: A Threat Model Driven Approach for Security Testing. In: Third International Workshop on Software Engineering for Secure Systems (SESS 2007). IEEE Computer Society (2007)
9. Mouratidis, H., Giorgini, P.: Security Attack Testing (SAT) – testing the security of information systems at design time. Journal of Information Systems 32(8), 1166–1183 (2007)
10. The Logic behind CRAMM's Assessment of Measures of Risk and Determination of Appropriate Countermeasures, http://www.cramm.com
11. Gupta, D., Chatterjee, K., De, A.: A Framework for Development of Secure Software. CSI Transaction on ICT (2013)

Comparing Software Architecture Descriptions and Raw Source-Code: A Statistical Analysis of Maintainability Metrics

Eudisley Anjos, Fernando Castor, and Mário Zenha-Rela

CISUC, Centre for Informatics and Systems,
University of Coimbra, Portugal
CI, Informatic Center,
Federal University of Paraiba, Brazil
Cin, Informatic Center,
Federal University of Pernambuco, Brazil
eudisley@ci.ufpb.br, fjclf@cin.ufpe.br, {eudisley,mzrela}@dei.uc.pt

Abstract. The software systems have been exposed to constant changes in a short period of time. It requires high maintainable systems and makes maintainability one of the most important quality attributes. In this work we performed a statistical analysis of maintainability metrics in three mainstream open-source applications, Tomcat (webserver), Jedit (text editor) and Vuze (a peer to peer client). The metrics are applied to source-code and to derived similar architectural metrics using scatter plot, Pearson's correlation coefficient and significance tests. The observations contradict the common assumption that software quality attributes (aka non-functional requirements) are mostly determined at the architectural level and raise new issues for future works in this field.

1 Introduction

The growth in complexity and size of object-oriented systems has lead to reasoning about the quality attributes and the possibilities to compute measures automatically. Various individuals and standardization bodies have defined several software quality attributes such as functionality, portability, usability, efficiency, and maintainability. The IEEE Standard Glossary of Software Engineering defines maintainability as: the ease with which a software system or component can be modified to correct faults, improve performance or other attributes, or adapt to a changed environment [12].

In the categorization of maintenance effort, it is clear that software maintenance takes a huge amount of effort consuming between 60 and 80% of the overall software budget [17,8]. On the other hand, maintainability might also be among the quality attributes most difficult to estimate, because this inherently involves making predictions about future activities. Despite the fact that software maintenance is an expensive and challenging task, it is often poorly managed. One reason is the lack of proven metrics for software maintainability [3].

B. Murgante et al. (Eds.): ICCSA 2013, Part III, LNCS 7973, pp. 199–213, 2013.
© Springer-Verlag Berlin Heidelberg 2013

Applying maintainability metrics over a system various parameters can be assessed and it is possible to gather numerical data related to the evaluation. The information gathered can be used to analysis and comparison with past data directing us in intriguing managerial and technical decisions. These metrics are used during every phase of software development life cycle.

Although there are maintainability metrics for any development phase, it is necessary to know if the same metrics evaluate the same features. This analysis is important to understand whether the metrics being used are not redundant. In the work of [22], a statistical analysis is performed between modularity metrics for source-code assessing their correlation. It approaches potential redundancies in the use of these metrics. Nevertheless, there is still a lack in the literature regarding metrics for evaluation of maintainability in software architectures and comparing them with code metrics.

The work presented here compares maintainability evaluation relating code metrics with architectural metrics. These metrics assess the core properties of maintainability: complexity, coupling and cohesion. The case studies use different releases of three object-oriented open-source systems: Tomcat, jEdit and Vuze. Firstly, the assessment is performed in the source-code. Afterwards, UML package diagrams are generated for each software version and the architectural metrics are applied to them. The comparisons among the metrics are presented in this paper using scatter plots and Pearson's product-moment coefficient. The results show strong similarities for some metrics and raise new issues for future works.

2   Definitions

According to IEEE 1471-2000 [18], the software architecture is *"the fundamental organization of a system embodied in its components, their relationships to each other, and to the environment, and the principles guiding its design and evolution"*.

The relevance of system evolution has been approached in many works [11,30]. There is a need to change software on a constant basis, with major enhancements being made within a short period of time in order to keep up with new business opportunities. So, the approach system evolution is a fundamental element for making strategic decisions, improving system features, and adding economic value to the software.

The IEEE standard computer dictionary [33] defines maintainability as the ease with which a software system or component can be modified in order to correct faults, improve the quality attributes, or adapt to a changed environment. The modifications may include corrections, improvements or adaptation of the software to changes in environment, requirements and functional specifications.

In ISO/IEC 9126 [21] and Khairudin [20] maintainability is described as a set of sub-characteristics that bear on the effort needed to make specified modifications such as: modularity, flexibility, adaptability, portability and so forth. These sub-characteristics are refined to attributes used to evaluate the level of maintainability in a system. Among these attributes we can mention complexity, composability, cohesion, coupling, decomposability, etc.

A system with high level of maintainability has low cost for its maintenance activities. It reduces the total maintenance cost or improves the system by performing more maintenance activities at the same total cost. The level of maintainability can be measured applying metrics over its attributes.

3 Related Work and Objectives

The main focus of existing literature for architectural evaluation methods is on scenario-based methods [15,23,6] . Most of them based on surveys, meetings and interviews, and not in the evaluation through automatic metrics application. Other works [9,6,14] provide more extent surveys but do not include architecture-level metrics either.

To predict any quality attribute from an architectural description we need to define which information we have available. For software code metrics, the lexical rules and the grammar of the language already takes care of this and there is a large of study in this field [5]. But software architecture has no standard language or standard definition and we should be careful to not use data that is not clearly part of software architecture [7]. These differences reduce the correlation between the type of information obtained by code metrics and architectural metrics.

Some work has been developed in a way to define metrics for more abstract designs, e.g. package diagrams [5]. However, reviews of architecture-level for maintainability metrics are difficult to be found in literature, as related studies in [29]. The metrics focus on class-level OO metrics (e.g., Chidamber [13], Halstead [19], McCabe [26]) and neglect metrics for higher-level code structures [24]. Moreover, none of these works realizes statistical analyses between architectural metrics and code metrics.

The work presented in [22] correlates code metrics to evaluate the level of similarities. They compare software metrics that affect modularity in open-source systems. However, there are no architectural metrics in their approach. Only code metrics are compared to assess the Pearsons correlation between them. The problem arises when we decide to compare the architectural metrics to code metrics.

The common assumption says that the quality attributes are defined previously in the design level and inserted in the architectural design [25]. This work contributes to assess the relationship between metrics for code and architectural metrics. We evaluate 73 versions for three mainstream open-source object-oriented applications: Tomcat (webserver), Jedit (text editor) and Vuze (a peer to peer client). The statistical analysis approaches scatter plot, Pearson's product-moment correlation coefficient and significance t-tests. We assume that the results obtained here support the definition of new maintainability metrics for software architectures.

4   Maintainability Metrics

Software Metrics are increasingly playing a central role in the planning and control of software development projects [32,27] The true value of software metrics comes from their association with important external software attributes such as maintainability,security, availiability, etc. [4]. Metrics are then used to assess quality attributes for software applications [28].

There are several metrics used in the evaluation of maintainability. One of the problems about these metrics is that most of them use the code as source of evaluation.

In our previous work, a set of metrics used for code evaluation was adapted to be performed on software architectures models using package diagrams. The metrics approached are: McCabe Cyclomatic Complexity (CC), Afferent Coupling (AC), Efferent Coupling (EC) and Lack of Cohesion (LOCM). The formula for metrics application is architectural designs are mentioned below. Further information can be found in [16]. Here we correlate both architectural and code metrics to approach the viability of use architectural metrics for quality assessment.

A set of metrics used for code evaluation was adapted to be performed on software architectures models using package diagrams. The metrics approached here are: McCabe Cyclomatic Complexity (CC), Afferent Coupling (AC), Efferent Coupling (EC) and Lack of Cohesion (LOCM). The next subsections contain a description of these metrics and the formulas applied for code and architecture evaluation.

$$CC = P - D + G (Package)$$
Where: P : Number of packages
D : number of dependency links (vertices)
G number of independent graphs (usually P = 1)

$$Cp = \sum_{i=1}^{R} \frac{Di}{Pi} (Package)$$
Where: R : graphs
Di : number of efferent/afferent dependency links to/from package i
Pi number of nodes (packages) in the graph i

$$LCOM = \sum_{i=1}^{R} \frac{(\frac{1}{S} * \sum \mu Li) - nP}{1 - nP} (Package)$$
Where: R : graphs
S : number of packages with dependency connections
μLi : number of dependency connections in the graph i
nP number of nodes (packages) in the graph i

5 Case Studies

In way to check the maintainability relationship between code and architecture we highlight three verification case studies: Tomcat [10], jEdit [1] and Vuze [2]. A brief description of these systems are presented below. All these case studies are object-oriented and open-source, allowing access to their change-logs and codes. They have different purposes and were chosen based on the size, use and popularity (high number of downloads).

Initially, the tests were applied on Apache Tomcat. We selected this system due to the support available in its website like change-logs, high number of versions and easiness in build it to apply the metrics. The results were satisfactory but revealed the necessity of new tests with jEdit and Vuze. The table 1 shows the main system characteristics such as: number of system versions, initial version, final version, average of packages and average of lines of code. The system evolution was considered as an essential parameter to verify the behavior of the metrics applied over code and architecture.

- **Apache Tomcat:** is an open-source software implementation of a web server and servlet container. Tomcat implements the Java Servlet and the JavaServer Pages (JSP) specifications from Oracle Corporation, and provides a "pure Java" HTTP web server environment for Java code to run.
- **jEdit:** is a programmer's text editor written in Java. It uses the Swing toolkit for the GUI and can be configured as a rather powerful IDE through the use of its plugin architecture.
- **Vuze:** is a BitTorrent client used to transfer files via the BitTorrent protocol. Vuze is written in Java, and uses the Azureus Engine. It has a lot of functionalities such as: player, meta-search, fast download, etc.

Table 1. Metrics and scope of application

System	Versions	Initial	Final	AVG Packages	AVG LOC
Tomcat	39	6.0.0	7.0.26	106	165K
jEdit	19	3.0.0	4.5.1	28	79K
Vuze	15	4.4.0	4.7.3	478	490K

6 Metrics Correlation

The purpose of this study is to assess whether there is a correlation between code and architectural metrics for maintainability evaluation in Object-Oriented systems. Since the experiment is complex to be performed, we divided the process in different steps. Each step was defined based on the expected results for the tests. The results in a step guided new decisions to determine the next one.

Once the first experiment was performed in a system with few lines of code, as in [16], we had defined 5 main steps to be follow to get the correlation. These main steps are: applications of code metrics, generation of software architecture designs, application of architectural metrics, unification of the results to analysis files and execution of statistical analysis. Each step is described below and the Figure 1 shows a diagram with the flow of the tests.

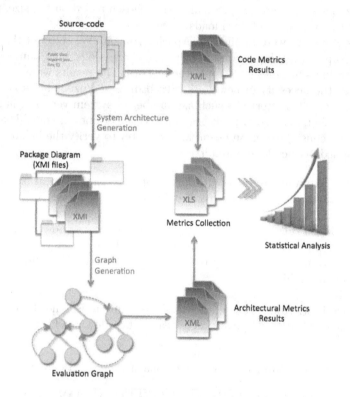

Fig. 1. Steps conducted in this work

1. *Application of the Metrics on the Source-Code*

In this step several metrics are applied on source code for all defined versions of the systems (e.g. Tomcat: 39 versions of source code). The metrics contains information about number of packages, number of lines-of-code, average of complexity, coupling, instability, number of interfaces, cohesion and so forth.

All the results were obtained using a plug-in for Eclipse IDE called Metrics 1.3.6. The results produce a XML file which contains values per class, per package, average, maximum and standard deviation. We did not used all the metrics in the comparison but for considerations of future improvements we hold all the results in the XML files. For each version of software a different XML was generated, for example for the version 6.0.35 of tomcat the file: *"tomcat-6.0.35-code.xml"* was created.

2. System Architecture Generation

The software used as case studies do not have architectural descriptions for every release available in their web pages. The system evolves but the maintenance of the architecture does not follow this evolution. Hence, possible software architecture for each version is created, for instance using UML representation.

The UML package diagram is built based on the initial source-code using a software called Visual Paradigm 9.0. The output of the generation process is the creation of two files: a PNG and a XMI. The PNG file contains an image of the package diagram for future analysis and comparison with the source code and, the XMI file contains the meta-data about the package diagram such as package names, link types and names, etc.

Two main type of links are approached here: dependency and containment. The containment depicts the hierarchical relationship between packages and contained packages. It can be used to distinguish packages and sub-packages. The dependency links are a direct relationship which occurs when the functioning of a package element requires the use of other elements in different packages. The file names follow the same pattern mentioned before, e.g. "tomcat-6.0.35-package.eps" or "tomcat-6.0.35-package.xmi".

3. Application of the Metrics on the System Architecture

The package links mentioned in the last step are used here to define an information flow directed graph. The containment links allow a definition of hierarchy trees of packages dictating the edges between the vertices (packages). The dependency links connect the elements of the packages indicating the information flow. This flow is used to determine the weight of the vertices. The direction of the flow is considered.

Using the graph, the architectural metrics proposed in the previous section are applied on the XMI file. This step produces two files: a PNG file containing the final graph with vertices and edges named "tomcat-6.0.35-graph.eps", and a XLS file with the results of architectural metrics named "tomcat-6.0.35-architecture.xls" All the process presented in this step is performed using the system described later in this section.

4. Metrics Collection

In this step the results obtained in the step 1 and 3 are joined in one XLS file. This file organizes the metrics values for all versions of the system both for code and architecture and inserts only the metrics of interest. The file created here is named: "tomcat-6.0.35-metrics.eps" and support the next step.

5. Statistical Analysis

The last step is concerned to the statistical analysis of the metric results. The statistical analysis comprises techniques to evaluate the correlation between the two metrics values: code and architecture. They are used to validate our initial hypothesis. All the process performed here use the SPSS software. The next section contains detailed information about the statistical analysis.

All the process mentioned before was managed by a system developed in this work. The system uses Python as programming language and automatize the whole process. It takes the output of all steps and merge the data in a final XLS file. It improved the results and avoid bias. Besides the use of this system allows the application of the tests in other case studies or adapt the code for different analysis.

7   Statistical Analysis and Results

7.1   Statistical Analyses

When planning experiments it is essential also to know that the results can be analysed. The statistical analysis is therefore an integral part of the experiment. This work uses statistical analysis to evaluate the relation among metrics applied on several system versions. These analysis demonstrate the correctness of our hypothesis. This work uses three main statistical analysis techniques: scatter plot, Pearson's product-moment correlation coefficient and significance t-tests (one tailed or two tailed).

The scatter plot (or scatter graph) is a mathematical diagram that shows how much one variable is affected by another. The closer the data points come when plotted to making a line, the higher the correlation between the two variables, or the stronger the correlation.

The significance indicates how sure the statistic analysis is reliable. A one or two-tailed t-test is determined by whether the total area of the curve is placed in one tail or divided equally between the two tails.

The Pearsons correlation is a measure of the strength of a linear association between two variables. The product-moment correlation draws a line over the data of two variables and the coefficient indicates how far away all these data points are to this line. The coefficient can take values in the range of -1 and 1. The negative values indicate a correlation perfectly inversely proportional. 0 means no correlation and positive values a perfectly proportional correlation. Table 2 presents a classification of the Pearson correlation coefficient.

7.2   Results

This section shows the results achieved after the statistical analysis. The information obtained was divided in two kind of data: graphs and tables. After each result table is described the valued obtained for each case study.

The graphs or scatter plots map the correlation between two metrics with the same purpose, applied on code and architecture. The curve show how close is the relationship between these two metrics. The scatter plots are created for each metric: complexity, afferent coupling, efferent coupling and cohesion. The tests conducted result in 12 graphs. The Figures 2 to Figure 5 present the graphs for Tomcat. Due the space only Tomcat graphs are showed here.

Table 2. Metrics and scope of application

Pearson's Classification (r)	Rangers	Interpretation
Low	$0 \leq r < 0.5$	Very small or no correlation between the two variables
Med	$0.5 \leq r < 0.7$	There are possibility a indirect correlation between the two variables
High	$0.7 \leq r < 1$	There are correlation which can be assumed as direct correlation between the two variables

In the graphs is possible to notice the closeness between the curve and the points obtained through the correlation applied to tomcat metrics (package and code). Although this closeness exists we have to compare the final Pearson correlation numbers to be sure.

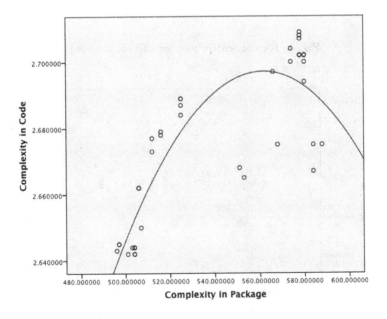

Fig. 2. Tomcat scatter plot for complexity

The following tables contains the Pearsons product-moment correlation and the significance for each pair of metrics. The same trends obtained in the scatter plot are presented in the Pearson's correlation coefficient. The results are presented below in the tables 3, for Tomcat, 4, for jEdit and 5, for Vuze.

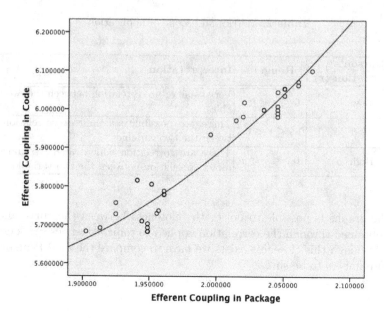

Fig. 3. Tomcat scatter plot for efferent coupling

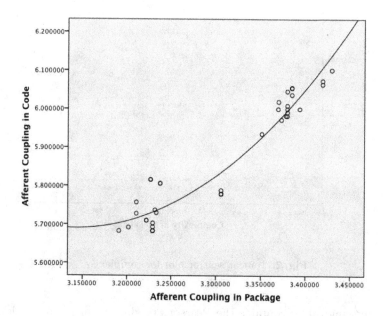

Fig. 4. Tomcat scatter plot for afferent coupling

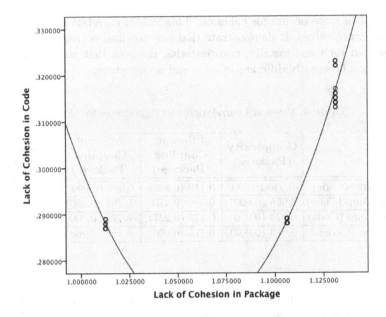

Fig. 5. Tomcat scatter plot for lack of cohesion

The values obtained for Tomcat are closely related. Almost all metrics have a high Pearson correlation coefficient. The architectural complexity is strongly correlated to code coupling, even more than to code complexity. This observation seems to indicate that an architectural description has no significant relation to code complexity. Its complexity is actually the degree of relationship between packets and therefore closely related to coupling. The significance values are zero, indicating that the coefficients are highly reliable.

Table 3. Pearson's correlation and significance for Tomcat

	Complexity (Pack)	Efferent Coupling (Package)	Afferent Coupling (Package)	Cohesion (Package)
Complexity (Code)	0.793 (0.000)	0.710 (0.000)	0.748 (0.000)	0.534 (0.000)
Efferent Coup (Code)	0.973 (0.000)	0.957 (0.000)	0.951 (0.000)	-0.727 (0.000)
efferent Coup (Code)	0.973 (0.000)	0.957 (0.000)	0.951 (0.000)	-0.727 (0.000)
Cohesion (Code)	- 0.967 (0.000)	- 0.919 (0.000)	- 0.940 (0.000)	0.719 (0.000)

jEdit results, as in Table 4, presents high correlation values to the analysis between code complexity and architecture complexity. It also keeps a strong correlation between package complexity and the code coupling (afferent and efferent). However coupling in code has a weak correlation with architectural coupling. The same occurs for cohesion. This lack of correlation can be seen in the significance values. It demonstrate that the previous correlation results did not repeated for some metrics, corroborating the idea that some metrics, like cohesion, are inherently different in code and architecture.

Table 4. Pearson's correlation and significance for jEdit

	Complexity (Package)	Efferent Coupling (Package)	Afferent Coupling (Package)	Cohesion (Package)
Complexity (Code)	-0.940 (0.000)	0.110 (0.328)	0.946 (0.000)	0.314 (0.095)
Efferent Coup (Code)	0.955 (0.000)	0.056 (0.410)	-0.798 (0.000)	-0.248 (0.153)
efferent Coup (Code)	0.974 (0.000)	0.175 (0.237)	-0.777 (0.000)	-0.225 (0.177)
Cohesion (Code)	0.613 (0.003)	0.775 (0.000)	-0.151 (0.268)	0.015 (0.475)

Vuze, the biggest system of our case study, has few similarities between the metrics. The only significant results are the comparison between package complexity with coupling (afferent and efferent). The other comparisons are not meaningful since they show a low coefficient or a high significance value. These results are presented in the Table 5.

Table 5. Pearson's correlation and significance for Vuze

	Complexity (Package)	Efferent Coupling (Package)	Afferent Coupling (Package)	Cohesion (Package)
Complexity (Code)	-0.028 (0.461)	0.207 (0.220)	-0.155 (0.290)	0.215 (0.221)
Efferent Coup (Code)	0.930 (0.000)	0.572 (0.013)	-0.530 (0.021)	0.547 (0.017)
efferent Coup (Code)	0.378 (0.001)	0.415 (0.062)	-0.363 (0.090)	0.384 (0.079)
Cohesion (Code)	0.469 (0.039)	0.282 (0.154)	-0.224 (0.211)	0.253 (0.182)

Analyzing the results make us to conclude that the complexity in the software architecture is close to the coupling in code for both case studies. It can prove that the complexity is related to how modular this architecture is. So the more coupled is the architecture, more difficult will be to change/maintain, increasing the complexity and reducing the modularity. It was proved with the correlation done for this paper.

8   Conclusions

The difficulty and cost to modify code (evolutive maintenance) and evaluate the quality for a system evolution creates the need for architectural properties evaluation. As it happens in metrics for specific kinds of code, architectural metrics must be general enough to work in different kinds of systems.

In the work described here we defined some maintainability metrics for software architectures inspired on metrics applied to source code. The metrics are applied to a set of real-world applications to assess their correlation. The observations lead us to conclude that the architectural descriptions do not seem to embody structural properties, namely maintainability, as happens at the code level. This raises a significant research issue: there are quality attributes that are not apparent at the (more abstract) software architecture level. This contradicts the commonly accepted premise that quality attributes are determined at the architectural level.

This work has raised interesting issues for future works. One relevant approach can be compare this metrics using other programming language paradigm, e.g. aspect-oriented. Some studies has approached the relation between software architectures and aspect-oriented software development as in [31]. Other study can use other architectural description languages, different of UML, to use as architectural design.

One future work which seems to be relevant and arise with the results obtained here is to understand how the modularity level of an architecture can increase. Or, how loosed-coupled is the system architecture. This would enable to define the point where the software reach the worst modularity level, requiring refactoring.

References

1. jedit: Programmer's text editor (June 2012)
2. Vuze: A bittorrent application (June 2012)
3. e Abreu, F.B., Pereira, G., Sousa, P.: A coupling-guided cluster analysis approach to reengineer the modularity of object-oriented systems. In: Proceedings of the Conference on Software Maintenance and Reengineering, CSMR 2000, p. 13. IEEE Computer Society, Washington, DC (2000)
4. Shaik, B.M.A., Reddy, C.R.K.: Empirically investigating the effect of design metrics on fault proneness in object oriented systems. International Journal of Computer Science & Engineering Technology (IJCSET) 2(4), 97–101
5. Anjos, E., Zenha-Rela, M.: A framework for classifying and comparing software architecture tools for quality evaluation. In: Murgante, B., Gervasi, O., Iglesias, A., Taniar, D., Apduhan, B.O. (eds.) ICCSA 2011, Part V. LNCS, vol. 6786, pp. 270–282. Springer, Heidelberg (2011)
6. Barcelos, Travassos: Evaluation approaches for software architectural documents: A systematic review. In: Ibero-American Workshop on Requirements Engineering and Software Environments, IDEAS (2006)
7. Bengtsson, P.: Towards maintainability metrics on software architecture: An adaptation of object-oriented metrics. In: First Nordic Workshop on Software Architecture, NOSA 1998 (1998)
8. Bode, S.: On the role of evolvability for architectural design. In: Fischer, S., Maehle, E., Reischuk, R. (eds.) GI Jahrestagung. LNI, vol. 154, pp. 3256–3263. GI (2009)
9. Breivold, H.P., Crnkovic, I.: A Systematic Review on Architecting for Software Evolvability, pp. 13–22. IEEE (2010)

10. Brittain, J., Darwin, I.F.: Tomcat: The definitive guide (2003)
11. Cai, Y., Huynh, S.: An evolution model for software modularity assessment. In: Proceedings of the 5th International Workshop on Software Quality, WoSQ 2007, p. 3. IEEE Computer Society, Washington, DC (2007)
12. Chae, H.S., Kwon, Y.R., Bae, D.H.: A cohesion measure for object-oriented classes. Softw. Pract. Exper. 30, 1405–1431 (2000)
13. Chidamber, S.R., Kemerer, C.F.: A metrics suite for object oriented design. IEEE Trans. Softw. Eng. 20(6), 476–493 (1994)
14. Clements, P., Kazman, R., Klein, M.: Evaluating Software Architectures: Methods and Case Studies. Addison-Wesley (2001)
15. Dobrica, L., Niemela, E.: A Survey on Software Architecture Analysis Methods. IEEE Transactions on Software Engineering 28(7), 638–653 (2002)
16. dos Anjos, E.G., Gomes, R.D., Zenha-Rela, M.: Assessing maintainability metrics in software architectures using COSMIC and UML. In: Murgante, B., Gervasi, O., Misra, S., Nedjah, N., Rocha, A.M.A.C., Taniar, D., Apduhan, B.O. (eds.) ICCSA 2012, Part IV. LNCS, vol. 7336, pp. 132–146. Springer, Heidelberg (2012)
17. Glass, R.L.: Facts and Fallacies of Software Engineering. Addison-Wesley (2002)
18. IEEE Architecture Working Group. Ieee std 1471-2000, recommended practice for architectural description of software-intensive systems. Technical report, IEEE (2000)
19. Halstead, M.H.: Elements of software science (Operating and programming systems series). Elsevier (1977)
20. Hashim, K., Key, E.: A software maintainability attributes model. Malaysian Journal of Computer Science 9(2) (1996)
21. Iso. International standard - iso/iec 14764 ieee std 14764-2006. ISO/IEC 14764:2006 (E) IEEE Std 14764-2006 Revision of IEEE Std 1219-1998, pp. 1–46 (2006)
22. Mustofa, A.W., Rahardjo, R., Wardoyo, J.E., Instiyanto, K.: Statistical analysis on software metrics affecting modularity in open-source software. Academy & Industry Research Collaboration Center (AIRCC) 3(3), 105–118 (2011)
23. Kazman, R., Bass, L., Klein, M., Lattanze, T., Northrop, L.: A basis for analyzing software architecture analysis methods. Software Quality Control 13(4), 329–355 (2005)
24. Koziolek, H.: Sustainability evaluation of software architectures: A systematic review. In: Proceedings of the Joint ACM SIGSOFT Conference QoSA and ACM SIGSOFT Symposium, QoSA-ISARCS 2011, pp. 3–12. ACM (2011)
25. Lundberg, L., Bosch, J., Hggander, D., Bengtsson, P.-O.: Quality attributes in software architecture design. In: Proceedings of the IASTED 3rd International Conference on Software Engineering and Applications, pp. 353–362 (1999)
26. McCabe, T.J.: A complexity measure. In: Proceedings of the 2nd International Conference on Software Engineering, ICSE 1976, p. 407. IEEE Computer Society Press, Los Alamitos (1976)
27. Babu, S., Parvathi, R.M.S.: Development of dynamic coupling measurement of distributed object oriented software based on trace events. International Journal of Software Engineering and Applications 3(1), 165–179 (2012)
28. Pressman, R., Pressman, R.: Software Engineering: A Practitioner's Approach, 6th edn. McGraw-Hill Science/Engineering/Math (2004)
29. Riaz, M., Mendes, E., Tempero, E.: A systematic review of software maintainability prediction and metrics, pp. 367–377 (October 2009)

30. Rowe, D., Leaney, J.: Evaluating evolvability of computer based systems architectures - an ontological approach. In: Proceedings of the 1997 International Conference on Engineering of Computer-based Systems, ECBS 1997, pp. 360–367. IEEE Computer Society, Washington, DC (1997)
31. Saraiva, J., Soares, S., Castor, O.: A metrics suite to evaluate the impact of aosd on layered software architectures. In: 2nd Workshop on Empirical Evaluation o Software Composition Techniques (ESCOT 2011), Lancaster, UK (2011)
32. Shaik, A., Reddy, C.R.K., Manda, B., Prakashini, C., Deepthi, K.: An empirical validation of object oriented design metrics in object oriented systems. Applied Sciences 1(2), 216–224 (2010)
33. The Institute of Electrical and Eletronics Engineers. Ieee standard glossary of software engineering terminology. IEEE Standard (1990)

Systematic Mapping of Architectures for Telemedicine Systems

Glauco de Sousa e Silva[1,*], Ana Paula Nunes Guimarães[1], Hugo Neves de Oliveira[1], Tatiana Aires Tavares[1], and Eudisley Gomes dos Anjos[1,2]

[1] Informatics Center, UFPB
João Pessoa, Brazil
{glauco.sousa6,apng89,tatianaires}@gmail.com,
{hugoneves,eudisley}@ci.ufpb.br
[2] CISUC – Centre for Informatics and Systems,
Universidade de Coimbra – Portugal

Abstract. The use of telemedicine systems is becoming increasingly common these days. Telemedicine systems exist for the purposes of education, improving the accuracy of medical diagnoses through the provision of a second opinion, and remote patient monitoring. Accordingly, the number of software solutions is increasing. Software Architecture is a subarea of Software Engineering whose goal is to study the system components, their external properties, and their relationships with other software. A good architecture can allow a system meets the mainly requirements of a project, such as performance, reliability, portability, easy maintenance, interoperability, etc. Aiming to find out what architectural styles that proposes a better performance in systems for telemedicine, a systematic mapping was done. With this mapping, it was possible to find taxonomies related to telemedicine systems, architectural styles commonly used in these systems and technologies relevant to the area.

Keywords: telemedicine, mapping study, software architecture.

1 Introduction

Aiming to solve the growing social, economic and medical needs to keep ill patients in their home, telemedicine emerges as provider for remote medical information and an efficient distance health care manager. There are several advantages in using the homecare medical monitoring technologies, such as preserving the patient's comfort, quality of life and mental condition and reducing the hospitalization costs. For these reasons, strategies for developing telemedical software must be implanted in order to make the application safe and effective enough so that it can be used in a commercial context, and not only for academic purposes.

Telemedicine systems have several purposes, such as access and update medical information in data centers, monitoring patients with cardiovascular or chronic

---

* Corresponding author.

B. Murgante et al. (Eds.): ICCSA 2013, Part III, LNCS 7973, pp. 214–229, 2013.

disease, diagnosis and treatment of illness, remote consultations, or situations where the patient's life is in risk. In order to successfully achieve the specific requirements for the software, some critical aspects must be taken into consideration, such as the location of the physical branches connected with the system and the response time for extreme situations. These requirements give the software an additional assignment of dealing with human lives.

Thus, the development of hardware and software solutions focused on telemedicine adds complexity to the processes, particularly the Software Engineering ones that accompany from specification until the implementation of these services. A crucial step is the choice of the software architecture for the telemedicine system. The architectural solution will print the dynamics of evolution and adaptation of changing the system development. This feature is especially useful in telemedicine systems, because the addition of new features or the evolution of old ones is a need. This is due to the incorporation of technologies related to the medical field and new functionalities which require an adjustment of these systems to different realities.

In [48], software architecture is seen as the study of the overall structure of the software system, especially the relations among subsystems and components. These relationships and modularization define how the software will behave and fit in the context for which it is developed. According to [47], the rate of increase in the size and complexity of software, the problem goes beyond the design of algorithms and data structures of computation. Because telemedicine systems often deal with more than one physical location and data transmission by any means of communication, it should be considered the aspects of highest level to creating the application, having a better idea of the software to be coded, which makes development more organized and less susceptible to logic or architectural errors which can be costly to be corrected in the encoding step. This design step involves the definition of architectural styles to be used in the software, as well as their modifications due to suit the problem presented and considered in the requirements specification document.

Among the various types of software architectures being used in the development of telemedicine systems, some are more appropriate to a particular type of application. This paper presents the results of a systematic mapping on software architectures in telemedicine systems. This study seeks to present and explain the most used architectural styles, characteristics, benefits and its limitations. This paper's analysis was guided through taxonomies created to group the investigated systems, in order to regulate the mapping study to the most seen topics in telemedicine systems.

2   Telemedicine Systems

According to [41], Telemedicine is the delivery of health services, where distance is a critical factor, using all information technologies and communication for the exchange of valid information for diagnosis, treatment and prevention of diseases, research and evaluation, and for the continuing education of health professionals, all with the aim of promoting the health of individuals and their communities. This definition is extremely broad and starts relating the medical art - involving patient contact, diagnosis, treatment or even surgery - with any means of communication that can unite two or more physically distant points from letters written to electronic medical records, interconnected by networks of high-speed communication.

It is important to understand that telemedicine is a process, not a technology [42]. The case study of telemedicine has become possible in recent years as a result of technological advances and continuous cost reduction [49]. Telemedicine has obvious advantages in remote or rural areas where there is a shortage of health professionals. Even in urban areas, however, telemedicine has shown accelerating the process of health care, reducing unnecessary referrals, and increasing the consistency and quality of the health system. The increased contact between the team of professionals has originated educational benefits (continuing medical education) for them and has reduced professional isolation [50].

3   Systematic Maps

Literature review of a specific area of knowledge is essential to continue the study of a particular point within the area. Thus, it is important to know what has been widely studied, developed and built in order to find a "gap" (gap to be filled) - to start a new search, for example. A literature review is basically a critical, meticulous and extensive analysis of the current publications in a particular field of knowledge. In general, the literature review is performed as part of an initial scientific study. There are methods to accomplish this task: systematic literature review and systematic mapping.

3.1   Systematic Map

"A systematic map is a defined method of constructing a scheme of classification and structure in a field of interest "[44]. It is indicated as a first step to be taken by doctoral students so that their area of interest be mapped, the main vehicles for publication are known, the major papers and their types can be identified and "gaps" found.

3.2   Systematic Mapping Study

A systematic mapping (mapping study) is a specific type of systematic review [40] which uses the same basic methodology of the systematic review: methodology of rigorous review, reliable and auditable. The difference, again according to [40], is that the systematic review focuses on well-defined research questions, more specific, which can be answered by experimental research, while the systematic mapping focuses on broader issues, exploratory in nature, whose purpose is to give an overview of a research area and find evidence of the area.

The result of the mapping itself is considered a contribution to the literature, because through analysis of the systematic map produced one can obtain information about the research questions defined, which prevents future researchers conducting a study that already has been documented.

3.3   Process of Systematic Mapping

Using the concepts achieved in [40], [44], [45], it is possible to compile the ideas and describe the essential steps of a systematic mapping study. These steps are: defining research questions that must be answered, leading the search for relevant documents, sorting of documents; elucidation of the search string based on keywords summaries (abstracts), data extraction, and mapping results. Each process step has an outcome, the end result of the process is the systematic map.

The first step is to define the subject to be mapped, ie define the scope about what you want to obtain relevant and specific data already existing in the literature and also justify the study.

Then, one should define research questions related to the subject, such as: What are the main means of publication? You can sort the papers selected following taxonomy? The research questions will be the basis for the analysis of selected papers, and their responses will form the systematic map.

The next step is to define a search protocol, which must contain all the detailed planning of the study, to guide the participants. It must include an introduction to the field of knowledge that will be mapped, a justification for the study, identifying research questions and finally, planning how the search will be performed (automatic and/or manual), the analysis (inclusion and exclusion criteria of papers and synthesis of data) and the construction of the systematic map.

The search process is the next step and is usually divided into automatic and/or manual. The manual search allows you to define the main conference area, in which the participants of the study must manually access related papers to the topic that have been published and/or presented in these conferences. The automatic search is performed after defining a search string that should contain the keywords of the topic that is being mapped, usually integrated by Boolean operators such as OR and AND. This string will be used in the search to be performed in automatic engines for obtaining related papers. These engines are able to index and locate content stored on computer systems automatically. Each search engine has a way to create strings with Boolean operators, and this must be observed to ensure the results returned.

After finalizing the automatic process of search, you should begin to examine the papers. Before starting the analysis, it should be defined the inclusion and exclusion criteria to be applied on the results returned by search engines, and it is these criteria that define which papers will be included or excluded from the mapping. These criteria must be well defined so there is no discrepancy in the evaluation of the participants, or at least that can reduce this possibility. Initially there should be some sort of triage of all returned results in all search engines used to eliminate repetition. Then, the study participants can leave for detailed analysis of the papers returned, identifying and selecting the most relevant to answer the research questions. After the selection of papers, participants will analyze them in detail in order to extract data to construct the systematic map.

Finally, it must be decided whether the papers will be read in full or will be analyzed only their abstracts and also unpublished scientific conferences or events should be analyzed. The results of the systematic mapping are usually presented in the form of graphs.

Figure 1 shows a summary of the steps needed to obtain the results in a systematic mapping:

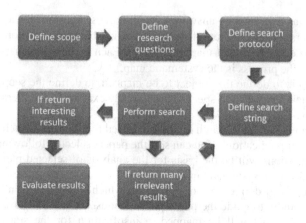

Fig. 1. Steps for conducting a systematic mapping

4   Architecture Styles

Software architecture it's necessary when the size and the complexity of software systems increase. Thus, the problem of constructing systems goes beyond the choice of algorithms and data structures. This issue also involves decisions about the structures that will form the system, the overall control structures, communications protocols, data access and synchronization, allocation of functionality to system elements, or, yet, the distribution of the physical elements of the system. The architecture exposes certain properties and encapsulates others. Ideally this representation provides a guide for an overview of the system and allows architectural designers to better analyze the system's ability to meet certain requirements and suggests a project's construction and composition of the system [47].

One of the most important developments in current software architecture is the study of architectural styles. An architectural style, also called architectural pattern, promotes a design vocabulary, this style contains the rules that define the how the elements in that vocabulary can be combined. In essence, an architectural style provides a specialized design language for a specific class of systems, that is a set of principles, properties and behavior that provides a framework for building systems.

The use of architectural styles brings several benefits:

- Reuse of design, in which common solutions with well understood properties can be reapplied in new problems [39];
- The use of architectural styles supports interoperability, ie the ability of a system to communicate transparently with another one [38].
- Code reuse, allowing, occasionally, invariant aspects of an architectural style can be "borrowed" to share implementations [39].
- The simplification of the design process and implementation of the system, which reduces the costs of implementation through reusable infrastructure, improving the integrity of the system with checks and analyzes of specific styles [38].

In a complex system a critical issue is the choice of an architecture influencing in their design and construction. A good architecture can help ensure that a system will satisfy fundamental requirements in areas such as performance, reliability, portability, scalability and interoperability. The development of a bad architecture can be disastrous for system development.

There are several architectural patterns, such as Layered, Client-server, Object-Oriented, Event based, Service-oriented, etc [39]. The architecture of a software system is almost never limited to a single architectural style, but the combination of styles that together complete the system. Many factors influence the choice of a style. These factors include: the ability of the organization to design and implementation, capabilities and experience of the developers and its constraints and organizational infrastructure [39].

5   Methodology

The development of the methodology applied in the mapping study was guided by the protocol described earlier. The main goal was to build a classification scheme and structure related to the architectures of telemedicine systems

5.1   Scope and Search Questions

Telemedicine systems make possible monitoring of patients with accurate medical data that enable the action of physicians with stability and safety even at a distance. The development of such system is increasingly common in the medical field.

Problems of machine performance, sensor or network connection between the computers of telemedicine systems, especially considering sending real-time data for monitoring, can endanger patient's life. Therefore, it's important understand the architecture of this systems so the implementation can evolve and patters can be found.

The goal of this mapping study is identify patterns of architectural styles that are commonly used in telemedicine systems. A previous study of software architectures related to telehealth systems commonly seen in the literature, along with their advantages and flaws, is a good way to start designing a solid framework for the modularization of a robust system for this type of application.

We try to answer as conclusion of this paper the research questions that guided the creation of the mapping study, applying on it some rules to get a general explanation about the mapped topic, in this case, the data streaming used in telemedicine. The questions defined at the beginning of the study were:

- Question 1: Is possible distribute the papers following taxonomy?
- Question 2: Which technologies are more often used in telemedicine systems?
- Question 3: Which architectural styles are seen more frequently in the researched studies?
- Question 4: How are the architectural styles in telemedicine systems distributed chronologically?

5.2   Source and Search Strategies

After we define the research questions, the next step was the automatic search of papers in the engines: IEEEXplore, Scopus and Science Direct.

The search string used on the digital libraries was built with the following strategy:

- Search for keywords in relevant papers already seen on an informal revision;
- Identify synonymous and alternative terms of the keywords;
- Consult experts in the field;
- Use the Boolean connector OR to incorporate alternative words and synonyms;
- Use the Boolean AND connector to connect keywords;
- Check the search string constructed performing pilot searches.

Using this steps, the first string search was built: "(health care OR telemedicine OR telehealth) AND (software engineering OR software architecture OR system architecture)". This string has proved inadequate to the needs of the research, since many returned results papers containing only asynchronous data exchange instead of the proposed analysis to process articles that addressed the architecture of e-health systems in real time. The string search was adjusted to "(health care OR telemedicine OR telehealth) AND (software engineering OR software architecture OR system architecture) AND (video OR streaming)". Even at this stage, it was decided that the search would be made in the title, keywords and abstract of the articles sought in each search engine.

5.3   Inclusion and Exclusion Criteria

To regulate the choice of papers found in search engines, we define criteria for inclusion and exclusion in the mapping, as shown in Tables 1 and Table 2.

Table 1. Inclusion Criteria

1	*Papers must present architectures for telemedicine systems*
2	*Papers must present architectures to telehealth systems with real-time data transmission*
3	*Papers must present logic descriptions of the software's architecture*
4	*Papers must be in English*

Table 2. Inclusion Criteria

1	*Telehealth systems that do not contains real-time data transmission*
2	*Papers about software architecture that have real-time data transmission but disregards telemedical context*
3	*Papers that do not describes the system's architecture*
4	*Papers that wasn't written in English*

5.4   Methods for Data Extraction

As a primary extraction method, were read only the titles and abstracts, so that the potentially relevant articles were found for the search. The articles that met with success at least part of the inclusion criteria were moved to a later stage, which has a more detailed analysis of the content of each article. Finally, we read the full papers and understood the descriptions of their architectures, created a systematic map, which divides the resulting articles by category. This enables the analysis of the results through a graphical display, facilitating the understanding.

Disagreements found by researchers acceptance and classification of articles are discussed by the whole team together and resolved.

5.5   Methods for Data Synthesis

The systematic map supplied metadata about publications, which will be analyzed and discussed in the next session of this work. The patterns found in architectures, as well as the main similarities and technologies used in the systems, are highlighted with the presentation and study of taxonomies found.

6   Discussion

In this session, the results obtained at the main landmarks of the mapping's methodology are shown and the main topics of the interest area are discussed and analyzed.

6.1   Quantitative Analysis of the Results

At the first step of the mapping, two hundred and fifty articles were found through automatic search. After the second stage, the results were refined and fifty seven articles were selected as relevant, according to their titles and abstracts. At the last step of the mapping some irrelevant articles were removed and the taxonomy was defined, through which the articles were classified. After this last stage, thirty seven articles were classified and arranged at the references from [1] to [37]. The Table III shows this steps and the number of articles divided by search engine.

6.2   Qualitative Analysis of the Results

This session discusses the qualitative results about the data obtained in the mapping study. For that, the research questions were prepared, aiming to highlight the analysis parameters of the research area.

6.2.1   Used Technologies

Another perspective observed was the commonly used technologies in telemedicine systems lately. As the articles were read, the predominant technologies in the telemedical field were observed. These numbers are shown in Fig. 2.

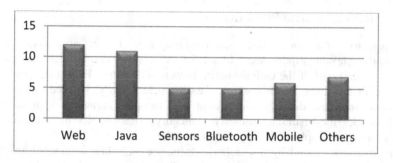

Fig. 2. Commonly used technologies in Telemedicine Systems

6.2.2 Architectural Styles

The architectural styles found in this study indicate the adoption of architectural patterns for telemedicine systems. Through a research question (Question 3: which architectural styles are more frequently seen in the researched studies?) it was possible to show the architectural styles currently in use in telemedical software.

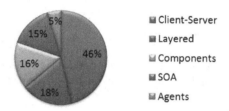

Fig. 3. Architectural Styles found in the mapping results

6.2.3 Occurrence of Taxonomies

Another approach adopted was the observation of the taxonomies more frequently used in the papers found by the research. It aims to show the distribution of the taxonomies in the types of telemedicine systems. The taxonomies were sorted according to its use in the telemedicine software which papers were accepted by the inclusion criteria of the study. The amounts of results obtained for each element in each taxonomy are seen in Fig. 4 and Fig 5.

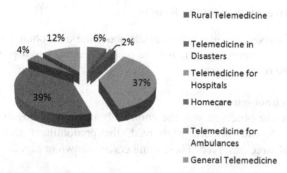

Fig. 4. Percentage of the types of systems found in the mapping

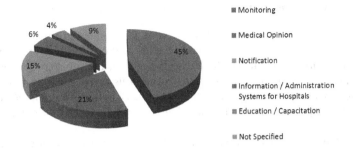

Fig. 5. Percentage of the usage categories found in the mapping

6.2.4 Publications over Time

Another view analyzed was the chronology of the publications about software architecture and telemedicine. This analysis was guided by the research question: "How the papers about data transmission systems in telemedicine software are chronologically distributed?"

The years of the publications were counted and sorted, resulting in a two-dimensional graphic, presented in Fig. 6 and Fig. 7.

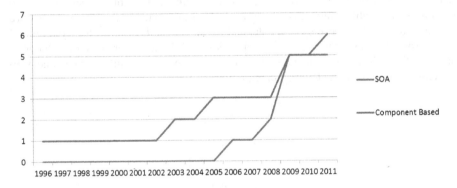

Fig. 6. Chronological distribution of SOA and Component Based Architectures

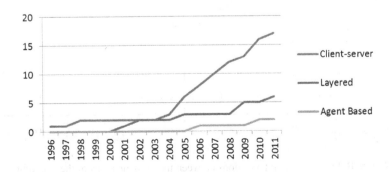

Fig. 7. Chronological distribution of Client-Server, Layered and Agent Based Architectures

6.3   Discussion of the Results

A mapping study offers the possibility to build taxonomies in order to classify a research area. Therefore, the mapping study for telemedicine systems presents a great contribution to this important health care software area. For a better view about this area, a graphic containing the intersections between usage, types of systems and the architectural styles found in the research was built as can be seen in Fig. 9. It aims to emphasize the gaps and saturated areas of the intersection between computing and medicine.

As is shown in Fig. 4 and Fig. 8, there is a great concentration of telemedical systems based in Homecare, but there is a gap in the research and development of telemedicine technology for mobile ambulatories and telemedicine for catastrophes. The left sector of the bubble graphic shows a saturation in papers describing architectures for monitoring, notification and medical opinion. Still in this scope, there are gaps in educational medicine and administration/information systems for hospitals. This highlights the interest for the development of telemedicine systems which help the medical professionals. Otherwise, the telemedicine administrative and pedagogical aspects are lacking in research.

Monitoring, as seen in Fig. 5, is a common type of telemedicine system, either in Homecare applications, where the health situation of the patients can be relatively stable; or in critical cases, where the patients can be unstable or susceptible to variations in health status that may lead to death. There are diverse software architectures used in these systems, depending on the application of the system, and range from the most basic architectural styles, as Client-Server; to more elaborate ones, as SOA. The use of technologies related to sensors, mobile devices and web systems is common in the telemedicine domain.

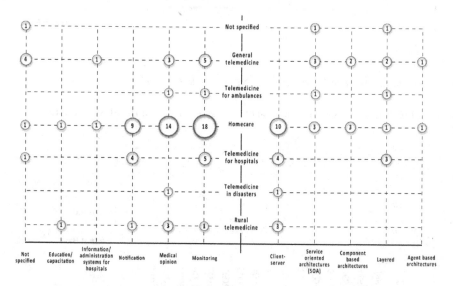

Fig. 8. Bubble graphic: intersections between the classifications of the mapping

The right session of the graphic presents the distribution of software architectural styles by type of telemedical application. There is a predominance of architectures based on the Client-Server model, since this architectural style is inherently distributed, which supplies the needs of most telemedicine applications involving multi-client data exchange or remote communication. The prevalence of this architectural style occurred because the systems described in the papers included in the mapping should necessarily have synchronous or asynchronous multi-client data exchange, that is, real time transmission systems. All these features are best implemented in a Client-Server style. The vast number of types of architectures found in Homecare systems shows the versatility of this kind of application, but the number of Client-Server software still prevailed.

The software modularization proposed by the SOA and Layers styles also had a good acceptance by all kinds of telemedical applications, and had a great variability, as can be seen in the graphic related to this development styles. The use of service oriented architectures also had a good acceptance in the researched papers, since they are focused on WEB applications, thereby easing the transmission of data. The software sub-modules created with these kinds of architecture also have a good disconnection, allowing the developers to insert new services without drastic architectural changes. Systems based in agents are not common in the telemedicine domain, despite of its modularization and natural bias towards the use of signal capturing, used often in telemedical applications.

Component based architectures did not show versatile results in telemedicine systems, despite of its propensity to subprogram independence. The use of this architectural style in Homecare systems is a result of the integration with the signal capture hardware, which can be distributed all over the patient's home, providing a base for the communication between the modules at the residence and the data receiving module at the hospital.

Fig. 6 and Fig. 7 show the continuous use of Client-Server architectures over the years. That is because it is already a consolidated style, and its properties match the needs of most telemedicine systems. Recently there was a growth in the use of service and component oriented architectures, showing that these architectural styles adapted themselves to the needs of the telemedicine scope.

The technologies presented in Fig. 2 were the most visible ones in the articles considered relevant to the mapping. JAVA [43] technologies were the most used ones, allowing the construction of multi-platform and WEB systems easily. Sensors stayed in second place in usage, as they are responsible for capturing vital signals, a common feature in telemedical applications, mainly in Homecare and Monitoring. Mobile and Bluetooth [46] technologies are gaining ground on the telemedicine area as the price of mobile phones and smartphones is getting lower, many of them containing Bluetooth communication interfaces, and easing their use in areas beyond telecommunication. WEB technologies are also being widespread in telehealth systems due to its accessibility and ease of use by the general public, which represents a significant portion of the telemedicine users.

7   Conclusion

In this paper, we described a systematic mapping that was developed about the area of system architecture focused on telemedicine. We also introduced the concepts of telemedicine and styles of software architectures. Furthermore, we presented a model of systematic mapping. With this model, a systematic search was being built to be able to analyze how architectural models are arranged in telemedical systems of all kinds and uses.

As a result series of data representing the systematic mapping of the area were collected and organized, so an analysis was made of the distribution of the papers according to the types of systems, use categories, architectures and technologies.

It can be seen, through the systematic mapping presented, that quantitatively the Client-Server architectures have presented results in functional scope of telemedical applications, since its inherent characteristic of distribution is a significant proportion of network failures by nature. The Client-Server architectural style has, yet, a high dependence on central services (usually located in hospitals), which provides the necessary assistance in case of an emergency, especially in systems monitoring. Wide acceptance was noticed with Service-Oriented architectures (SOA) and components, showing that the use of these types of styles underwent a major expansion.

It is also noteworthy the high concentration systems for Homecare and monitoring of medical conditions of patients, which represents a saturation of these types of work.

As future work, we propose studies about architectural evolution in telemedicine systems, for example, the Arthron [51], in order to analyze the architectural behavior along with the growth of the software itself, its new features and requirements. Also, we intend deepen to study deeply the hybrid architectures and their characteristics in these types of systems. And, finally, evaluate the impact of dynamic architectures in telehealth systems.

References

1. Catarinucci, L., Colella, R., Esposito, A., Tarricone, L., Zappatore, M.: A Context-Aware Smart Infrastructure based on RFID Sensor-Tags and its Application to the Health-Care Domain. In: Emerging Technologies & Factory Automation, Espanha (2009)
2. Yang, C., Chang, Y., Chu, C.: A Gateway Design for Message Passing on The SOA Healthcare Platform. In: International Symposium on Service-Oriented System Engineering, Taiwan (2008)
3. Cheng, P., Shyu, M., Chen, S., Lai, F., Luh, J., Chen, H., Lai, J.: A Healthcare Pattern Collection for Rural Telemedicine Services. In: 8th International Conference on e-Health Networking, Applications and Services (2006)
4. Deng, M., Petkovic, M., Nalin, M., Baroni, I.: A home healthcare system in the cloud – addressing security and privacy challenges. In: IEEE International Conference on Cloud Computing, USA, (2011)
5. Braga, F., Forlani, C., Signorini, M.G.: A Knowledge Based Home Monitoring System for Management and Rehabilitation of Cardiovascular Patients. In: Computers in Cardiology, France (2005)
6. Protogerakis, M., Gramatke, A., Henning, K.: A System Architecture for a Telematic Support System in Emergency Medical Services. In: 3rd International Conference on Bioinformatics and Biomedical Engineering, China, (2009)

7. Celler, G.B., Basilakis, J., Budge, M., Lovell, N.H.: A Clinical Monitoring and Management System for Residential Aged Care Facilities. In: 28th Annual International Conference of the IEEE Engineering in Medicine and Biology Society, USA, (2006)

8. Sung, M., Marci, C., Pentland, A.: Wearable feedback systems for rehabilitation. Journal of Neuroengeneering and Rehabilitation (2005)

9. Wu, W.H., Bui, A.A.T., Batalin, M.A., Au, L.K., Binney, J.D., Kaiser, W.J.: Medic: Medical embedded device for individualized care. Artificial Intelligence in Medicine 42, 137–152 (2008)

10. Bruun-Rasmussen, M., Bernstein, K.E., Chronaki, C.: Collaboration - a new IT-service in the next generation of regional health care networks. International Journal of Medical Informatics 81, 205–214 (2003)

11. Shuicai, W., Peijie, J., Chunlan, Y., Haomin, L., Yanping, B.: The development of a tele-monitoring system for physiological parameters based on the B/S model. Computers in Biology and Medicine 40, 883–888 (2010)

12. Kuntalp, M.E., Akar, O.: A simple and low-cost Internet-based teleconsultation system that could effectively solve the health care access problems in underserved areas of developing countries. Computer Methods and Programs in Biomedicine 75, 117–126 (2004)

13. Hein, A., Eichelberg, M., Nee, O., Schulz, A., Helmer, A., Lipprandt, M.: A Service Oriented Platform for Health Services and Ambient Assisted Living. In: International Conference on Advanced Information Networking and Applications Workshops, United Kingdom, (2009)

14. Griffin, L., Foley, C., Leastar, E.: A Hybrid Architectural Style for Complex Healthcare Scenarios. In: International Conference on Communications Workshops (2009)

15. Ray, S., Dogra, D., Bhattacharya, S., Saha, B., Biswas, A., Majumdar, A., Mukherjee, J., Majumdar, B., Singh, A., Paria, A., Mukherjee, S., Bhattacharya, S.D.: A Web Enabled Health Information System for the Neonatal Intensive Care Unit (NICU). In: IEEE World Congress on Services, USA, (2011)

16. Chimiak, W.J., Rainer, R.E., Cook, J.: An Adaptive Multi-Disciplinary Telemedicine System. In: Proceedings of the Fourth International Conference on Image Management and Communication, USA (1996)

17. Capozzi, D., Lanzola, G.: An Agent-Based Architecture for Home Care Monitoring and Education of Chronic Patients. In: Complexity in Engineering, Italy (2010)

18. Katehakis, D.G., Tsiknakis, M., Orphanoudakis, S.C.: An architecture for integrated regional health telematics networks. In: Proceedings of the 23rd Annual EMBS International Conference, Turkey (2001)

19. Sannino, G., De Pietro, G.: An Evolved eHealth Monitoring System for a Nuclear Medicine Department. In: Fourth International Conference on Developments in eSystems Engineering, United Arab Emirates (2011)

20. King, A., Procter, S., Andresen, D., Hatcliff, J., Warren, S., Spees, W., Jetley, R., Jones, P., Weininger, S.: An Open Test Bed for Medical Device Integration and Coordination. In: 31st International Conference on Software Engineering, Canada (2009)

21. Matsumoto, T., Ogata, S., Kawaji, S.: Designing and Implementation of Mobile Terminal for Telehealth Care Life Support System. In: The 7th International Conference on Computer Supported Cooperative Work in Design, Brazil (2002)

22. Zhang, Y., Bai, J., Lingfeng, W.: Development of a home ECG and Blood Pressure Telemonitoring Center. In: 22nd Annual EMBS International Conference, USA (2000)

23. Ahn, H., Lee, S., Lee, S.: Development of a Ubiquitous Healthcare System Implementing Real-time Connectivity between Cardiac Patients and Medical Doctors. In: 7th International Workshop on Enterprise Networking and Computing in Healthcare Industry (2005)

24. Omar, W.M., Taleb-Bendiab, A.: E-Health Support Services Based on Service-Oriented Architecture. IT Professional 8(2), 35–41 (2006)

25. Quero, J.M., Tarrida, C.L., Santana, J.J., Ermolov, V., Jantunen, I., Laine, H., Eichholz, J.: Health Care Applications Based on Mobile Phone Centric Smart Sensor Network. In: 29th Annual International Conference of the IEEE EMBS, France (2007)

26. Pour, G.: Prospects for Expanding Telehealth: Multi-Agent Autonomic Architectures. In: International Conference on Computational Intelligence for Modelling Control and Automation, and International Conference on Intelligent Agents, Web Technologies and Internet Commerce, Australia (2006)

27. Dhouib, M.A., Bougueroua, L., Istrate, D., Pino, M., Bernard, C.: HoCoS: Home Companion Software. A service oriented solution for elderly home accompanying and remote healthcare monitoring. In: 33rd Annual International Conference of the IEEE EMBS, Boston, USA (2011)

28. Barro, S., Presedo, J., Castro, D., Fernandez-Delgado, M., Fraga, S., Lama, M., Vila, J.: Intelligent telemonitoring of critical-care patients. In: IEEE Engineering in Medicine and Biology (1999)

29. Schacht, A., Wierschke, R., Wolf, M., Von Löwis, M., Polze, A.: Live Streaming of Medical Data - The Fontane Architecture for Remote Patient Monitoring and Its Experimental Evaluation. In: 14th IEEE International Symposium on Object/Component/Service-Oriented Real-Time Distributed Computing Workshops, Los Alamitos, Calif. (2011)

30. Chowdhury, A., Chien, H., Khire, S., Fan, S., Tang, X., Jayant, N., Chang, G.: Next-generation E-health communication infrastructure using converged super-broadband optical and wireless access system. In: IEEE International Symposium on a World of Wireless, Mobile and Multimedia Networks, Montreal, Canada (2010)

31. Valeanul, V., Vizitiu, A., Chiru, M.: Primary health-care system for human security in hazardous situations". In: Recent Advances in Space Technologies, Istanbul, Turkey (2007)

32. Liu, Q., Lu, S., Hong, Y., Wang, L., Dssouli, R.: Securing Telehealth Applications in a Web-Based e-Health Portal. In: The Third International Conference on Availability, Reliability and Security (2008)

33. Maji, A.K., Mukhoty, A., Majumdar, A.K., Mukhopadhyay, J., Sural, S., Paul, S., Majumdar, B.: Security analysis and implementation of web-based telemedicine services with a four-tier architecture. In: Pervasive Computing Technologies for Healthcare, Tampere, Finland (2006)

34. Lu, X.: System design and development for a CSCW based remote oral medical diagnosis system. In: Fourth International Conference on Machine Learning and Cybernetics, China (2005)

35. Petcu, V., Petrescu, A.: Systems of systems applications for telemedicine. In: 9th RoEduNet IEEE International Conference (2010)

36. Shaikh, A., Memon, M., Memon, N., Misbahuddin, M.: The Role of Service Oriented Architecture in Telemedicine Healthcare System. In: Computational Intelligence in Security for Information Systems, Burgos, Spain (2009)

37. Ku, H., Huang, C.: Web2ohs: A Web2.0-Based Omnibearing Homecare System. IEEE Transactions on Information Technology in Biomedicine 14 (2010)

38. Monroe, R.T., Kompanek, A., Melton, R., Garlan, D.: Architectural Styles, Design Patterns, and Objects. In: Computing & Processing (Hardware/Software), pp. 43–52 (1996)
39. Microsoft. Chapter 3: Architectural Patterns and Styles, http://msdn.microsoft.com/en-us/library/ee658117.aspx
40. Kitchenham, B., Budgen, O.D., Brereton, P.: The value of mapping studies - A participant-observer case study. In: Evaluation Assessment in Software Engeneering, United Kingdom, (2008)
41. Indian Space Research Organisation.: Telemedicine - Healing Touch Through Space, http://www.isro.org/publications/pdf/Telemedicine.pdf
42. Bashshur, R.L.: Telemedicine and health care. Telemedicine Journal and E-Health, [S.l.] 8 (2002)
43. Java, http://www.java.com/pt_BR/download/faq/whatis_java.xml.
44. Petersen. K, Feldt R., Mujtaba S., Mattsson M.: Systematic Mapping Studies in Software Engineering, University of Bari, Italy (2008)
45. Ramos, E.S., Brasil, M.M.A.: Um Mapeamento Sistemático sobre Padrões de Software para Reengenharia de Sistemas, http://www.infobrasil.inf.br/userfiles/ 16-S2-2-97124-Um%20Mapeamento%20Sistem%C3%A1tico.pdf
46. Bluetooth, http://www.bluetooth.com/Pages/Bluetooth-Home.aspx
47. Garlan, D., Shaw, M.: An Introduction to Software Architecture (1994)
48. Shaw, M.: The Coming-of-Age of Software Architecture Research (2001)
49. Wooton, R.: Telemedicine, http://www.ncbi.nlm.nih.gov/pmc/articles/ PMC1121135/pdf/557.pdf.
50. Wooton, R.: Telemedicine: a cautious welcome, http://www.ncbi.nlm.nih.gov/ pmc/articles/PMC2352879/pdf/bmj00570-0037.pdf
51. Vieira, E.S.F., et al.: Uma Ferramenta para Gerenciamento e Transmissão de Fluxos de Vídeo em Alta Definição para Telemedicina. Anais do Salão de Ferramentas do Simpósio Brasileiro de Redes de Computadores (2012)

Architectural Model for Generating User Interfaces Based on Class Metadata

Luiz Azevedo[1], Clovis Torres Fernandes[1], and Eduardo Martins Guerra[2]

[1] Aeronautical Institute of Technology (ITA), Praça Marechal Eduardo Gomes, 50
São José dos Campos – SP, Brazil
[2] National Institute for Space Research (INPE)
São José dos Campos – SP, Brazil
{luizfva,guerraem}@gmail.com,
clovistf@uol.com.br

Abstract. Source code duplication is the origin of several problems in a software development project. Even been aware of this situation, application developers tend to ignore them, once it takes a lot of time and effort for duplicated pieces of code to be found and eliminated. To address this issue, the present work presents a new model for source code generation for user interface development. The generation process happens at runtime, each time a page is requested, uses the resulting content of the request processing and a set of templates and is based on class metadata. As result, application developers have new tool to avoid inconsistencies that can be originated by code duplication.

Keywords: user interface, source code generator, metadata, framework, inconsistency, software architecture.

1 Introduction

Source code duplication is an evil we all must fight. There are times when we feel we don't have a choice, the application environment seems to require duplication, or we simply don't realize we are duplicating information [1] and that can negatively affect the software development in many ways.

Code duplication results in increased code size and complexity, making program maintenance more difficult. Even if programmers could find and edit each copy of a piece of code, it is impossible to ensure that all the changes were made consistently – that the common regions are identical and the differences are retained – without manually comparing each clone's body, word-by-word, and hoping that no important details were missed [2].

Whatever the reason, there are ways to avoid replicating source code. Source code generation is a well-known tool to prevent a piece of information to be spread out to several places of an application [1,3].

Structures in multiple languages can be built from a common metadata representation using a simple code generator each time the software is built [1]. Particularly, when dealing with User Interfaces (UI), another approach would be to use a single source of metadata in order to generate the source code at runtime.

B. Murgante et al. (Eds.): ICCSA 2013, Part III, LNCS 7973, pp. 230–245, 2013.

The purpose of this paper is to present the model for source code generation for user interface development. It is divided in the following manner: Section 2 introduces inconsistencies in user interfaces, Section 3 presents the categories of source code generators, Section 4 introduces the proposed model, Section 5 presents a framework created as a proof of concept, Section 6 reports the validation of this work and Section 7 presents the conclusions.

2   Inconsistency in User Interfaces

At a survey conducted in early 1990s, Myers and Rosson [4] found that up to 48% of software application source code is devoted to the user interface portion. Though there are no recent studies to verify whether this relation stays the same, a considerable part of the software development time is still dedicated to the user interface development. On that occasion, interviewees reported that achieving consistency was one of the most common problems faced when developing user interface, especially when there are multiple developers involved [4].

The lack of consistency might be a symptom of potential source code duplication but, when it comes to user interfaces, it is rarely treated as such. Focusing on a web application, three different kinds of inconsistencies, involving user interfaces, were identified.

Inconsistencies between HTML Code and Application Code

When listing the good habits of a pragmatic programmer, Hunt and Thomas [1] highlight the Don't Repeat Yourself (DRY) principle as: *Every piece of knowledge must have a single, unambiguous, authoritative representation within a system.*

While developing a user interface, a common, yet bad, practice is to explicitly express in it every piece of information about the content that will be presented to the final user. By doing so, the DRY principle is been violated, i.e. not only the Student class knows that a student has a name, but every page that presents information about a student also does. Thereby, when it comes the time a student's name is no longer stored in a single field, instead it must be stored by the student's first-name and last-name, all those pages that displayed the student name will have to be changed. If, by any chance, at least one page is not updated, an inconsistent state arises. This kind of inconsistency is usually found in applications that separate user interface code from application logic, like MVC (Model-View-Controller) applications.

This is the simplest kind of inconsistency. The support provided by integrated development environments (IDE) usually helps to avoid, or at least helps to identify this kind of inconsistency. When they don't, the first time someone navigates into an inconsistent page, an error message will be printed out.

Inconsistencies in the Way a Content Type Is Presented

Sometimes, it is desirable that a type of information always be presented the same way (consistently) to the final user. For example, a java.util.Calendar object usually holds more data than the user would want to see, thus it is customary to format its date in a way the user could read its information and the date formatting pattern should be the same throughout the whole application.

By violating that pattern, the application usability may be compromised. Besides, it is evidence that the formatting code is been duplicated among the pages of the application.

Behavioral Inconsistencies among Different User Interfaces

When a graphical element has some events associated with it, like an animation or client-side content validation, it should trigger these events in the same manner, independently from what content it holds or by what page it is been used.

Suppose that all mandatory fields for text entry have a minimum length to be considered valid. For example, in a form for adding students an input value like "a" for the student name field should not be considered valid. If this validation is not performed in a mandatory text field, inconsistent data can be informed and persisted to the application.

3   Source Code Generation

Source code generators are great tools to reduce the odds that fragments of source code be copied to several places. Source code generation is about writing programs that write programs [3]. According to Herrington [3] the benefits of the use of a source code generator are:

- Quality: large volumes of manually written source code tend to have inconsistent quality, because new and better approaches are found as the application is been developed, however, by several reasons, they cannot be applied to every portion of the application that has already been developed.
- Consistency: The source code that is built by a generator is consistent in the design decisions and conventions. That is, one can't be certain that design decisions and source code conventions will be respected when the source code is written manually. However, when a generator acts as an interface between two distinct contexts of an application, like between a database and the data access layer or between the model layer and the presentation layer, consistency is always preserved, for one context is built in according to the other.
- Abstraction: There is the possibility of creating new templates to translate the logic into other languages, onto other platforms, or programming paradigms.
- Productivity: The work time of a software engineer is highly valuable and should not be wasted with repetitive and predictable activities. The generator can treat those repetitive activities, and then the engineer can apply his efforts to a more noble activity.

According to Czarnecki [5], there are two fundamental processes of generating instances of a given concept: composition and transformation. The composition process consists on the "collage" of several components; whilst on the transformation process some transformations are made until the desired content is achieved [5]. Figure 1 presents an example, in which each process generates an instance of a star. The composition process takes some components, which are bound in an organized manner to compose the resultant instance. In Figure 1, the composition process uses one circle and four arms, previously created, to assemble a star. The transformation process, however, starts from an initial state, from which some operations are performed until the desired content is achieved.

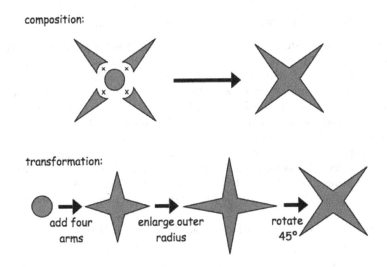

Fig. 1. Generating instances of a star by composition and by transformation [5]

Voelter [6] classifies source code generators by its applicability in the application development lifecycle, as illustrated by Figure 2.

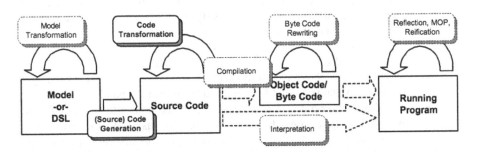

Fig. 2. Different source code applications [6]

- Model Transformation: generates new models from available models. Typically, these generated models are specialized with regards to some property [6].
- Source Code Generation: describes the generation of executable code. This usually involves some kind of de-abstraction or concretization of the model [6].
- Source Code Transformation: denotes the creation from code based on other code. This is typically used when different code artifacts are somehow merged or modified [6].
- Byte Code Rewriting: is a technique that has been introduced in the context of bytecode interpreters or virtual machines. The code that has been created by the compiling source is directly created or modified [6].
- Reflection: a program can modify itself while running [6].

Source code generators are classified into two categories as for its influence over the code built: active and passive [3]. Passive generators build source code that can be freely changed, that is, developers have full access to the content created and when necessary, it can be directly modified. Hence, when the generation is performed again, all manually made modifications will be lost [7]. This type of code generators is usually used by IDEs for creating new files with a starting content.

Active source code generators, on the other hand, are responsible by the code generated. It is important to highlight that active code generators use templates as basis of the creation process [8]. As the need to modify the code arises, the changes are made directly to the templates and are reflected on the created content. Active source code generators can be used along with the build process or at runtime as the proposed model.

There is also a classification as for the output produced. Herrington [3] describes them as follows:

- Code Munger Generator: this kind of generator takes an input file, usually source code, and searches it for patterns. When it finds those patterns, it processes their contents and generates a set of one or more output files. A well-known example of a Code Munger Generator is the JavaDoc documentation generated from a set of java source files.
- Inline-code Expansion Generator: simplifies the source code by adding a specialized syntax, in witch you specify the requirements for the code to be generated. This syntax is parsed by the generator, which then implements code based on requirements. The source code is "expanded" based on some kind of markup, which will be replaced by the generated code. The output file will have all the original source code, except for the special markup that is going to be replaced by the built content.
- Mixed-code Generator: The generator reads the input file and modifies it, overwriting the original file with the changes made. As the Inline-code Expansion Generators, the Mixed-code Generators also use special markups that indicate where the generated code will be placed. The main difference is that the Mixed-code Generators keep the original markup that will denote where the generate code was placed.

- Partial-class Generator: The generator uses an abstract definition of the source code to be created as input. Instead of filtering, or replacing code fragments, this generator takes a description of the code to be created and builds a full set of implementation code.
- Tier Generator: The generator builds all the code for one layer or section of an application. The most common example is the constructor of a database access layer of a web/client-server application.

4   User Interface Source Code Generation

Several source code generators were studied, aiming to identify the best approach in order to keep the application free of inconsistencies. The main characteristics of some of the studied source code generators will be presented below:

Some passive code generators were studied, like the scaffold generator of the Rails Framework. From an abstract definition of an entity, Rails generates the source code necessary to perform CRUD operations over an entity. However, the generated code is not production code, it is still essential to implement some customization and add business rules etc. We concluded that the less metadata was yielded to the generator the more changes were required to the controllers, views or to any other generated artifacts to achieve their desired state.

The approach that proved to be the best in keeping the application free of inconsistencies was an active tier-generator like OpenXava [9]. OpenXava uses Java object's metadata in order to compose the user interfaces. Thus, a page will be a faithful representation of the corresponding entity. Nevertheless, this is a great approach for generating CRUD user interfaces, but it restricts the possibilities of customizing the generated content.

Other generators, like SwingBean [10], provide means to adapt the user interfaces through the use of templates; besides, the developer can manually add graphical components to the user interfaces. This flexibility comes with a price, the user interfaces of an application wont be free of the three types of inconsistencies.

5   The Proposed Model

This research proposes a new model of source code generators for user interface development, called MAGIU – Architectural Model for Generating User Interfaces (in Portuguese, *Modelo Arquitetural de Geração de Interfaces com o Usuário*), it can be implemented in any object oriented language and guarantees the consistency of an application user interfaces. The proposed model suits the characteristics of active, inline-code expansion generator [3,8,7], its parameterization and generations are made at runtime.

It has been chosen to use an active source code paradigm by the possibility to customize the resulting content by acting directly over the generation process, once the templates used by the generator are passive to changes. On the other hand, the inline-code expansion paradigm was chosen so that the special markup that indicates where the generated content will be placed is deleted and will not impact the resulting content.

For the sake of simplicity, from now on, all the given examples will be for an HTML code generator and the server-side language used will be Java.

The generated content corresponds to the HTML code necessary to present or to edit server-side objects' state, and it will be placed at an HTML page each time the user requests it. Those server side objects are usually the resulting content of a calculation or a query to the database, whose metadata are read and then used to build HTML code. The resulting HTML content corresponds to an editor (an HTML form) or a display for server-side object.

The proposed model has three main sources of information in the process of building the user interface:

- Metadata: the generator reads metadata information from an instance object. This object contains the primary source of metadata consumed by the generator besides the information that will be show to the user or edited by him.
- Templates: define the organization of the HTML code generated, which will compose the user interface. Each template is directly related to a specific data type (as templates for texts, for boolean data, for numbers or user defined data). Application developers can create new templates as well as edit the existing ones in order to change the resulting content.
- Metadata Repository: stores the metadata read from various sources, except for that from the instance object given to the generator. It is usually implemented as an in memory object, used to avoid repetitive file reading or database queries. The metadata repository avoids unnecessary metadata readings [11].

Figure 3 presents the main elements of the proposed model.

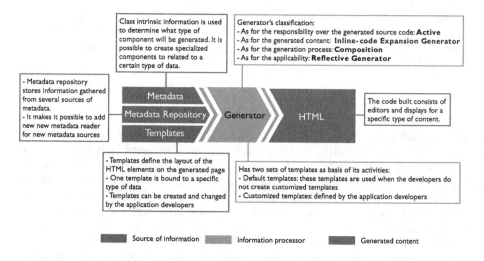

Fig. 3. Proposed Model for generation of source code for user interfaces

The metadata reading and metadata processing are two of the most important stages of the generation. Characteristics of the generated content vary from application to application and what determine these variations are the different forms in which the data is structured. By processing the metadata, the generator is capable of obtaining any information regarding the structure of all necessary data to conclude the generation.

MAGIU model supports the generation of HTML code for two different contexts: editing content and displaying content. For this, the generator must have access to an object whose content will be displayed or edited. Usually, this object is an instance of an application entity. In practice, any type of object that stores information can be used as a source of knowledge for generating a page.

The model enables the addition of new metadata sources that makes it possible for the generator to read metadata from any source, even those not foreseen by the moment of its creation.

The generator will search for a user-defined template related to the provided object. The lack of a customized template implies in the use of the default template for that corresponding type. If no template is found so far, the generator will search for the template of the immediate supertype of that object. The search continues until it reaches the `Object` class. The set of default templates holds templates for primitive types (such as integers, booleans, floating numbers etc.), text types (string), collection of objects (like the Java `java.util.Collection` interface and arrays) and the `Object` class.

It is also possible that one type of data can represent more than one type of information, i.e. a string object can represent a plain text, a password, an email address, a URL, an image path, and so on. Therefore, the application developer should be able to add a special markup to add semantics to those types and create a new template for each of these semantic types. At runtime, the generator perceives that object has a semantic type associated with it and a specialized template will be used instead.

The generator iterates over the fields of the given object repeating the generation processing for each of them. This divide and conquer approach is repeated until a type of interest is found. On the context of this work, we define a type of interest, as been an atomic type of data. Once divided, a type of interest has no meaningful information to the final user. A common example is a `java.util.Calendar` object, whose fields might not bring useful information besides the actual date that the final user would be interested in. In Java, some examples of types of interest are `java.lang.String`, `java.util.Calendar`, `java.lang.Number`, `java.lang.Character` and any of their subclasses plus the java primitive types.

Generating the source code for a type of interest is the base case of this recursive algorithm. Every type of interest has a correspondent template for display's generations and other for editor's generations. Once the HTML code is built, the generation process goes up a level.

Figure 4 illustrates the recursive generation algorithm defined by the proposed model.

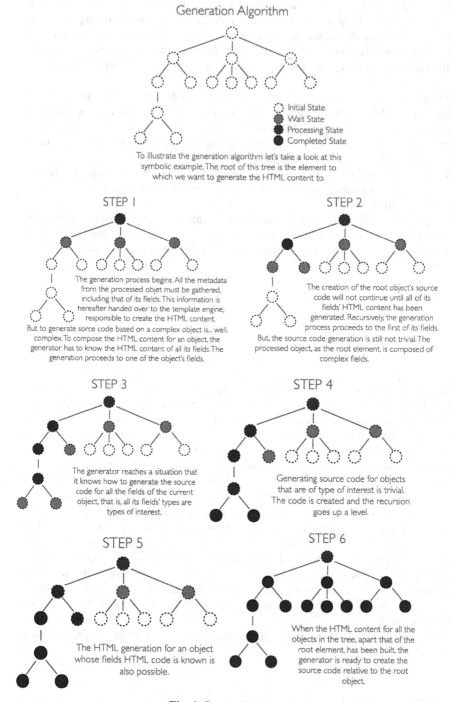

Fig. 4. Generation Algorithm

The MAGIU model helps to avoid the three kinds of inconsistencies introduced previously:

1. Inconsistencies between HTML code and application code: the HTML code is always generated from the application code;
2. Inconsistencies in the way the content is presented: a given type of application code is always presented the same way, that is, their generation process and their template are the same;
3. Behavioral inconsistencies among different user interfaces: the event calls originated by a kind of HTML element is consistent throughout the application, once that element is always generated from the same template.

However, the developers have full access to the templates, so they can introduce inconsistencies into them. It has been chosen to make customization possible, so that developers could achieve the desired user interface, but as a drawback, it is not possible to check whether the templates are consistent with each other.

6   Development Framework

A Java development framework was built, as a proof of concept, to show the real capacity of the proposed model of source code generation. A framework is a set of classes that embodies an abstract design for solutions to a family of related problems [12]. It consists of a structure that can be reused as a whole on the development of a new system [13].

All interaction with the framework is done through its façade. The client applications make calls to two methods, one to generate an editor one to generate a display for a given content.

MAGIU public interface

```
public class MAGIU {
        public String editorFor(Object model) {…}
        public String displayFor(Object model) {…}
}
```

The metadata read is encapsulated in a Metadata Container object [11] and given to a template engine. The templates have some small regions that will be filled with information, provided in the metadata container.

The following example illustrates a template used to edit the content of an object of type Object. The Object template is considered to be the default template for complex objects. The Properties collections represents the collection of metadata gathered from all the instance variables that object has, accessible through the Metadata Container objects. This collection will be iterated, generating HTML content for every field.

Example of template for edit an Object's content.

```
<div id="${DisplayName}">
  <#list Properties as prop>
    <p>
      <label class="inFieldLabel"
             for="${prop.nameAttribute}">
        ${prop.displayName}
      </label>
      ${prop.html}
    </p>
  </#list>
</div>
```

Accessing the `prop` variable, the template engine can print every information hold by the related Metadata Container, by using one of its fields:

- `nameAttribute`: metadata usually used by templates of types of interest. It is used in the "name" attribute of some HTML tags. This way, the HTML form submitted back to the server can be easily bound to an object.
- `displayName`: a display name that the current object/field has in a form or in a display page.
- `templateName`: the name of a template, used to specify which semantic type an object is. The template name is used to find the template that corresponds to current object/field.
- `value`: a reference to the value of the current object/field. When used in a template, its `toString` method is called.
- `type`: a reference to the class of the current object/field.
- `html`: the generated HTML content for the current field.

A common example of usage of a user interface generator is to build CRUD pages. Figure 5 illustrates a form for adding new users generated by the framework.

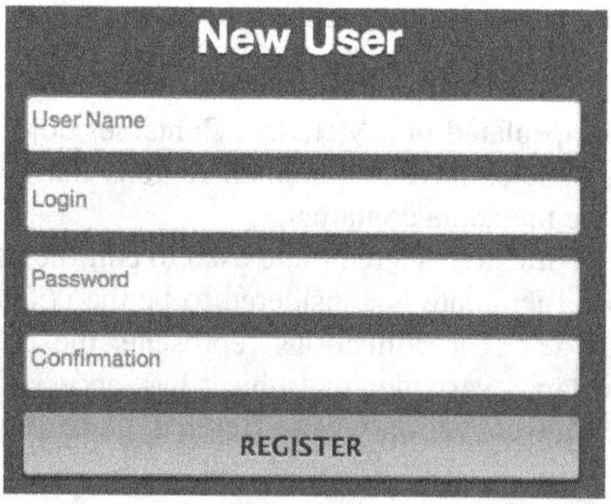

Fig. 5. Example of New User Form, created by a MAGIU generator

The following example shows an example of how a MAGIU generator can be used in a JSP page. The `user` object is attached in the request by the servlet and handled to the generator. The `editFor()` call will then return a String object containing the resulting HTML content.

Example of the usage of the generator in a JSP page.

```
<%@page import="atarefado.view.Html"%>
<%@ page language="java" contentType="text/html; charset=ISO-8859-1"
  pageEncoding="ISO-8859-1"%>
<%@ taglib prefix="my" tagdir="/WEB-INF/tags"%>
<%@ taglib prefix="html" uri="/WEB-INF/functions.tld"%>
<jsp:useBean id="user"
  type="atarefado.model.presentationmodel.UserPM" scope="request" />
<my:base title="NewUser">
  <form action="NewUser" method="post">
    <div class="centralized">
      <h1>New User</h1>
      ${html:editorFor(user)}
      <input type="submit" value="Register"
             class="goldButton" />
    </div>
  </form>
</my:base>
```

The UserPM class declaration defines the metadata processed while the form presented in Figure 5 was generated.

UserPM class declaration.

```
public class UserPM {
  @DisplayName("User Name")
  private String name;
  private String login;
  @TemplateName("Password")
  private String password;
  @TemplateName("Password")
  @DisplayName("Confirmation")
  private String passwordConfirmation;
}
```

The templates used in this generation are the Object Template, the String Template and the Password Template.

String Template

```
<input id="${model.nameAttribute}" type="text" class="text"
       name="${model.nameAttribute}">
  ${model.html}
</input>
```

Password Template

```
<input id="${model.nameAttribute}" type="password" class="text"
       name="${model.nameAttribute}">
  ${model.html}
</input>
```

7   Experiment

Bosh [13] highlights that tests must be made to verify whether the framework provides the wanted functionality and to assess its usability. Hence, an experimental study was elaborated to verify if the MAGIU model, implemented in the presented framework, really removes the potential inconsistencies in the user interfaces of an application, as proposed. It is not in the scope of this paper to assess characteristics such as usability, scalability, maintainability or developer productivity when adopting a MAGIU generator.

The study was conducted with undergraduate students of the Aeronautical Institute of Technology (ITA – Instituto Tecnológico de Aeronáutica). The main goal of the experiment was to verify which kind of inconsistency could be inserted in applications with and without the use of the proposed model. In this context, the students were responsible to perform a modification over a sample application in order to try to intentionally cause inconsistencies.

The artifacts used on the study were written in Java and, even though some participants had a limited experience in developing software, all of them had solid knowledge in Java and were capable to read and comprehend the application logic. Most of the participants had only had developed software at class and their knowledge was strictly academic. Besides, a few students had already had some previous experience involving source code generators. To those students that participated in the studies, the experiment activities came to replace one of the discipline's practical evaluations. Figure 6 illustrates the participants' previous experience with software development.

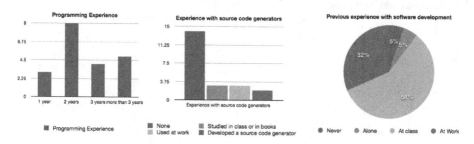

Fig. 6. Participants previous experience with software development

The experiment was divided in three phases: training, implementation and analysis. In the first one, the students went through a training process at which they were presented to the types of source code inconsistencies, its causes, and the usage of the MAGIU generator framework. Additionally, they were also exposed to two versions of a sample application called *Atarefado*, where one used the generator and the other did not.

In the second phase, the students were divided into ten groups of two or three people and each group chose a piece functionality to be implemented. Their activities consisted of trying to cause one or more inconsistency cases, stressing the MAGIU generator in every imaginable way. Initially, just the version of the *Atarefado* sample application that did not use the generator should be used (V1). If an inconsistency was found, the group should try to cause that same inconsistency to the version of *Atarefado*, supported by the generator (V2), and report the results. At the end of the experiment, the students answered a survey questionnaire to register their experience.

In the third phase, the reported information, the questionnaire answers and the group's resulting implementation were qualitatively analyzed in the attempt to identify evidences that proved that the generator eliminated the provoked inconsistencies. The main advantage of performing a qualitative analysis is that it demands that the researcher unveil the problem complexity instead of abstracting it [14]. Hence, it is possible to analyze the elements responsible for causing the inconsistencies found by the experiment participants.

Six groups reported inconsistencies in version V1 along the implementation of the chosen piece of functionality as ilustrated in Figure 7. Three of them reported that inconsistencies were also found in version V2 of the *Atarefado* sample application. After the analysis was completed, it was found that all reported inconsistencies were directly inserted on the templates. Those inconsistencies emerged when two or more templates were inconsistent with each other, the CSS classes or JavaScript functions used by one template were not necessarily used by the other.

Thus, inconsistencies found on the templates were replicated to all the pages that used them. However, to solve it, the application developer needs only to correct the inconsistent template, therefore modifying only one file. It is noteworthy that the reported of inconsistency can only happen within the templates created by the application developers. By using a consistent set of templates, the generator will always create inconsistency free content.

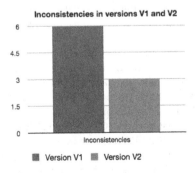

Fig. 7. Reported inconsistencies

8   Conclusions

At the beginning of this research, it was defined that consistency between two objects is the fulfillment of all the rules/relations that bind them. By breaking one of those rules, an inconsistent state is characterized. When the value of a metadata is changed, some rule is also changing, and a reflective source code generator will always perceive that. With no support of a generator like MAGIU generators, it is the application developers' responsibility to know and follow those rules.

To develop the generation model, there was a great concern not to limit the way in which the developers build their application pages. The use of the generator is not mandatory, if the developers know better approaches to build the desired page, they can make use of it.

On the other hand, when a MAGIU generator is adopted, the rules that bind entities to user interfaces can be abstracted, once they will be followed by the generator.

After concluding the analysis of the experiment data, we feel that our hypothesis gained strength. The usage of a MAGIU generator can help the application developers to avoid source code duplication and, consequently, avoid inconsistencies among the pages of an application.

Acknowledgments. We thank for the essential support of FAPESP (Fundação de Amparo à Pesquisa do Estado de São Paulo) to this research.

References

[1] Hunt, A., Thomas, D.: The Pragmatic Programmer: From Journeyman to Master. Addison-Wesley (1999)
[2] Toomim, M., Begel, A., Graham, S.L.: Managing Duplicated Code with Linked Editing. In: 2004 IEEE Symposium on Visual Languages and Human Centric Computing, Rome, pp. 173–180 (2004)
[3] Herrington, J.: Code Generation in Action. Manning Publications Co., Greenwich (2003)
[4] Myers, B., Rosson, M.: Survey on User Interface Programming. In: Proceedings of the SIGCHI Conference on Human Factors in Computing Systems, pp. 195–202. ACM Press (1992)
[5] Czarnecki, K.: Generative Programming: Principles and Techniques of Software Engineering Based on Automated Configuration and Fragment-Based Component Models. Department of Computer Science and Automation, Technical University of Ilmenau (1998)
[6] Voelter, M.: A Catalog of Patterns for Program Generation (Abril 2003),
    http://www.voelter.de/data/pub/ProgramGeneration.pdf
[7] Wilkins, A., Smith, C.: CodeSmith Generator 6.x (Julho 2011),
    http://docs.codesmithtools.com/display/Generator/
    Active+vs.+Passive+Generation
[8] Montero, M.: The Input Ouput Toolkit - IOTK. (Maio 2009),
    http://iotkfw.com/2009/05/21/active-passive-code-generation
[9] Paniza, J.: OpenXava (2011),
    http://openxava.wikispaces.com/overview_en

[10] Guerra, E.M.: SwingBean: Aplicações Swing a Jato! (2007),
     http://swingbean.sourceforge.net/
[11] Guerra, E.: A Conceptual Model for Metadata-based Frameworks. Aeronautics Institute
     of Technology, São José dos Campos (2010)
[12] Foote, B., Johnson, R.: Designing Reusable Classes. Journal of Object-Oriented Pro-
     gramming I e II, 22–35 (1988)
[13] Bosh, J., Molin, P., Mattson, M., Bengtsson, P.: Object-Oriented Framework - Problems
     and Experiences. Department of Computer Science and Business Administration, Univer-
     sity of Karlskrona, Ronneby, Sweeden (1997)
[14] Shull, F., Singer, J., Sjøberg, D.: Guide to Advanced Empirical Software Engineering.
     Springer, New York (2007)
[15] Tarr, P., Ossher, H., Harrison, W., Sutton, S.M.: N degrees of separation: multi-
     dimensional separation of concerns. In: Proceedings of the 21st International Conference
     on Software Engineering, New York, NY (1999)
[16] Hürsch, W.L., Lopes, C.V.: Separation of Concerns (February 1995)

A Novel Fuzzy Co-occurrence Matrix
for Texture Feature Extraction

Yutthana Munklang[1], Sansanee Auephanwiriyakul[1], and Nipon Theera-Umpon[2]

[1] Department of Computer Engineering,
Faculty of Engineering, Chiang Mai University,
Chiang Mai, Thailand
sansanee@ieee.org
[2] Department of Electrical Engineering,
Faculty of Engineering, Chiang Mai University,
Chiang Mai, Thailand
nipon@ieee.org

Abstract. Texture analysis is one of the important steps in many computer vision applications. One of the important parts in texture analysis is texture classification. This classification is not an easy problem since texture can be non-uniform due to many reasons, e.g., rotation, scale, and etc. To help in this process, a good feature extraction method is needed. In this paper, we incorporate the fuzzy C-means (FCM) into the gray level co-occurrence matrix (GLCM). In particular, we utilize the result from FCM to compute eight fuzzy co-occurrence matrices for each direction. There are four features, i.e., contrast, correlation, energy, and homogeneity, computed from each fuzzy co-occurrence matrix. We then test our features with the multiclass support vector machine (one-versus-all strategy) on the UIUC, UMD, Kylberg, and the Brodatz data sets. We also compare the classification result using the same set of feature extracted from the GLCM. The experimental results show that the features extracted from our fuzzy co-occurrence matrix yields a better classification performance than that from the regular GLCM. The best results on validation set using the features computed from our fuzzy co-occurrence are 77%, 95%, 99.11%, and 98.44% on the UIUC, UMD, Kylberg, and Brodatz, respectively, whereas those on the same data sets using the features from the gray level co-occurrence are 53%, 85%, 82.81%, and 95.31%, respectively. The best result on the blind test set of Brodatz data set using our fuzzy co-occurrence is 92.19%, whereas that from the gray level co-occurrence is 85.74%. Since the blind test data set is a rotated version of the training data set, we may conclude from the experiment that our features provide better property of rotation invariance.

Keywords: Fuzzy C-Means Clustering, Gray Level Co-occurrence Matrix, Fuzzy Co-occurrence Matrix, Multiclass Support Vector Machine, Texture Feature Extraction.

1 Introduction

Texture classification is one of the important areas in texture analysis that are used in many computer vision applications. Classification is not an easy problem in

B. Murgante et al. (Eds.): ICCSA 2013, Part III, LNCS 7973, pp. 246–257, 2013.
© Springer-Verlag Berlin Heidelberg 2013

real- world application because of non-uniform texture resulting from orientation, scale or other visual appearance. An important part in texture classification is the problem of texture feature extraction. One of the popular methods in texture feature extraction is using a statistical method. The first introduced method is using gray tone spatial dependence co-occurrence statistic [1]. There are several research works involving with either 1-D, 2-D, or 3-D gray level co-occurrence matrix (GLCM) [2–21]. Although the GLCM is very popular, this method may not cover the uncertainties of the gray level in the image. Since, in order to compute GLCM, the gray level in the image has to be quantized into 8, 16, or 32 integer number. However, there are a few works involving with a fuzzy co-occurrence matrix [22–24]. The membership functions in these cases have to be somehow created. This fuzzy-based method has an uncertainty in itself since the shape and the domain of the universe of discourse are not known.

In this paper, we incorporate the fuzzy C-means (FCM) clustering with the GLCM for the texture feature extraction. In particular, we utilize the FCM in the quantization part. Then, we create eight fuzzy gray level co-occurrence planes in each of the orientations. Finally, the contrast, correlation, energy, and homogeneity features are calculated from each plane. To show the performance of these features, we utilize a multi-class support vector machine. We also compare the classification results with those using the features computed from the GLCM.

2   Background Theory

We briefly review the fuzzy C-means (FCM) algorithm [25–26] here. Let $X = \{ x_1, x_2,..., x_N \}$ be a set of vectors, where each vector is a p-dimensional vector. The update equation for FCM is as follows [26]

$$u_{ij} = \frac{1}{\sum_{k=1}^{C} \left[\frac{\|x_i - c_j\|}{\|x_i - c_k\|} \right]^{\frac{2}{m-1}}} \tag{1}$$

$$c_j = \frac{\sum_{i=1}^{N} u_{ij}^m x_i}{\sum_{i=1}^{N} u_{ij}^m} \tag{2}$$

where, u_{ij} is the membership value of vector x_j belonging to cluster j, c_j, is the center of cluster j, and m is the fuzzifier. The following is the summarization of the FCM algorithm:

```
Fix the number of clusters C
Initiate prototypes
Do {
          Update membership using (1)
          Update prototypes using (2)
} Until prototypes stabilize
```

We will now briefly describe a gray level co-occurrence matrix (GLCM) [2] that is a second-order statistics of an image. The joint probability of occurrence of two gray level values with a particular distance (d) and orientation (θ) (shown in figure 1) is counted as a $P(i, j: d, \theta)$. Suppose the size of an image is $N_x \times N_y$. Let $L_x = \{1, 2, ..., N_x\}$ and $L_y = \{1, 2, ..., N_y\}$ be the horizontal and vertical spatial domain, respectively and $G = \{1, 2, ..., N_g\}$ be the set of N_g quantized gray tones. The image I can be represented as a function which assigns some gray tone in G to each resolution cell or a pair of coordinates in $L_y \times L_x$.

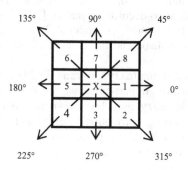

Fig. 1. GLCM orientation assignment

The joint probability of occurrence of two gray level values in each distance and direction is calculated as:

$$P(i, j, d, 0) = \#\{((k, l), (m, n)) \in (L_y \times L_x) \times (L_y \times L_x)$$

$$|k - m| = 0, |l - n| = d, I(k, l) = i, I(m, n) = j\}$$

$$P(i, j, d, 45) = \#\{((k, l), (m, n)) \in (L_y \times L_x) \times (L_y \times L_x)$$

$$|k - m| = d, |l - n| = -d, I(k, l) = i, I(m, n) = j\}$$

$$P(i, j, d, 90) = \#\{((k, l), (m, n)) \in (L_y \times L_x) \times (L_y \times L_x)$$

$$|k - m| = d, |l - n| = 0, I(k, l) = i, I(m, n) = j\}$$

$$P(i, j, d, 135) = \#\{((k, l), (m, n)) \in (L_y \times L_x) \times (L_y \times L_x)$$

$$|k - m| = d, |l - n| = d, I(k, l) = i, I(m, n) = j\}$$

$$P(i, j, d, 180) = \#\{((k, l), (m, n)) \in (L_y \times L_x) \times (L_y \times L_x)$$

$$|k - m| = 0, |l - n| = -d, I(k, l) = i, I(m, n) = j\}$$

$$P(i, j, d, 225) = \#\{((k, l), (m, n)) \in (L_y \times L_x) \times (L_y \times L_x)$$

$$|k - m| = d, |l - n| = -d, I(k, l) = i, I(m, n) = j\}$$

$$P(i, j, d, 270) = \#\{((k, l), (m, n)) \in (L_y \times L_x) \times (L_y \times L_x)$$

$$|k - m| = d, \; |l - n| = 0, \; I(k, l) = i, \; I(m, n) = j\}$$

$$P(i, j, d, 315) = \#\{((k, l), (m, n)) \in (L_y \times L_x) \times (L_y \times L_x)$$

$$|k - m| = d, \; |l - n| = d, \; I(k, l) = i, \; I(m, n) = j\} \tag{3}$$

where # denotes the number of elements in the set. In the experiment, we set $d = 1$ with $\theta = 0, 45, 90$, and 135 degrees. Hence, we have 4 GLCM in total. Then the contrast, correlation, energy, and homogeneity are computed from each GLCM. Therefore, there are 16 dimensions in total. We call this type of feature as the GLCM. Then, we computed an average of each feature from all directions to produce $\mu_{contrast}$, $\mu_{correlation}$, μ_{energy}, and $\mu_{homogeneity}$. Hence, this will be a 4 dimensional vector and is called the GLCM1. In addition, we create the GLCM2 that is an 8-dimensional vector with those aforementioned averages and the standard deviation of each feature from all directions ($\sigma_{contrast}$, $\sigma_{correlation}$, σ_{energy}, $\sigma_{homogeneity}$).

Finally, we will briefly explain the multi-class support vector machine (MSVM) [27] which is a method that assigns a class label to a vector into one of the several classes. In this research, we use one-versus-all strategy. We now briefly explain the MSVM. Suppose we have an optimum discriminant function ($D_i(\mathbf{x})$ for $i=1, \ldots, C$). From the support vector machine (SVM) [28], we have an optimum hyperplane at $D_i(\mathbf{x}) = 0$ that will separate class i from all the others. Hence, each SVM classifier gives $D_i(\mathbf{x}) > 0$ for vectors in class I, and $D_i(\mathbf{x}) < 0$ for those in all other classes. Then the classification rule is

$$\mathbf{x} \text{ is assigned to class } i \text{ if } \quad i = \arg \max_{j=1,..,n} D_j(x). \tag{4}$$

The SVM used in the experiment is the one with hard margin optimization. The radial basis function used in each SVM is

$$K(\mathbf{x}_i, \mathbf{x}_j) = \exp\left(\frac{-\|\mathbf{x}_i - \mathbf{x}_j\|^2}{2\sigma^2} \right). \tag{5}$$

3   System Description

In order to create a fuzzy co-occurrence matrix (FCOM), we first implement the FCM on an original gray scale image I with $m = 2$ and $C = 8$. Then we assign each pixel to the cluster where it has a maximum membership value. Finally, in each direction ($\theta = 0, 45, 90$, and 135), we create eight FCOM planes. In the experiment, we also set $d = 1$. The fuzzy co-occurrence matrix algorithm is as follows:

```
For each direction (θ)
  For each pixel (p)
      Find a pixel (q) that is d apart from p
      Find the assigned cluster of pixel q
      Suppose assigned cluster of p is k and assigned
        cluster of q is l
      Set all eight FCOMs to zero
      For each cluster (i)
```

```
        FCOM(i,k,1) = FCOM(i,k,1) + u_pi + u_qi
        (where u_pi and u_qi are the membership values of
        pixels p and q in cluster i)
     End For i
   End For p
 End For θ
```

Since, we implement the FCM with $C = 8$, each FCOM will have the size of 8×8. Hence, there are 32 FCOM in total. Similar to those computed from the GLCM, in each FCOM, we compute the contrast, correlation, energy, and homogeneity, as follows:

$$contrast_k = \sum_{i,j} |i-j|^2 FCOM(k,i,j), \quad for\ 1 \le k \le C \tag{6}$$

$$correlation_k = \sum_{i,j} \frac{(i-\mu_i^k)(j-\mu_j^k)FCOM(k,i,j)}{\sigma_i^k \sigma_j^k}, \quad for\ 1 \le k \le C \tag{7}$$

$$energy_k = \sum_{i,j} FCOM(k,i,j)^2, \quad for\ 1 \le k \le C \tag{8}$$

$$homogeneity_k = \sum_{i,j} \frac{FCOM(k,i,j)}{1+|i-j|}, \quad for\ 1 \le k \le C \tag{9}$$

where

$$\mu_i^k = \sum_i i \sum_j FCOM(k,i,j), \quad for\ 1 \le k \le C$$

$$\mu_j^k = \sum_j j \sum_i FCOM(k,i,j), \quad for\ 1 \le k \le C$$

$$\sigma_i^k = \sum_i (i-\mu_i^k)^2 \sum_j FCOM(k,i,j), \quad for\ 1 \le k \le C$$

and

$$\sigma_j^k = \sum_j (j-\mu_j^k)^2 \sum_i FCOM(k,i,j), \quad for\ 1 \le k \le C.$$

This will be a 128-dimensional vector and is called the FzCM. Again, similar to the GLCM, we also create the FzCM1, i.e., we compute the average of each feature from all directions. Hence, we have eight of $\mu_{contrast}$, $\mu_{correlation}$, μ_{energy}, and $\mu_{homogeneity}$. Therefore, we will have a 32-dimensional vector in this case. Again, we compute the standard deviation of each feature from all directions to produce eight of $\sigma_{contrast}$, $\sigma_{correlation}$, σ_{energy}, $\sigma_{homogeneity}$. In this case, we will have a vector of 64 dimensions, called the FzCM2, (those average values and standard deviation).

4   Experimental Results

We compute features using the FzCM, FzCM1, and FzCM2 with the data sets from four well-known benchmark texture datasets, the so-called UIUC [29], UMD [30],

Kylberg [31], and Brodatz [32]. In each experiment, we run the SVM with σ = 0.01, 0.05, 0.1, 0.25, 0.5, 0.75, 1.0, 1.5, 2.0, 2.5, 3.0, 3.5, 4.0, 4.5, and 5.0 on those features. We also set $\varepsilon = 1 \times 10^{-3}$. For the comparison, we also implement the GLCM, GLCM1, and GLCM2 with the same setting on the same data sets.

In the UIUC texture data set [29], there are 25 texture classes, each with 40 images with the size of 640×480. Hence there are 1000 images in total. An example from 25 classes is shown in figure 2. For this data set, 10-fold cross validation is implemented.

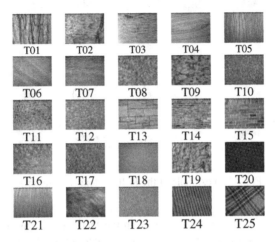

Fig. 2. An example of the UIUC texture data set

There are 1000 uncalibrated unregistered images in the UMD texture data set [30]. It can be downloaded at http://www.cfar.umd.edu/~fer/website-texture/texture.htm. In this data set, there are 25 texture classes, each with 40 images (with the size of 1280×900). An example of each texture is shown in figure 3. Again, 10-fold cross validation is implemented for this data set.

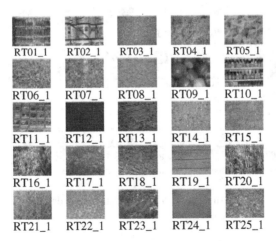

Fig. 3. An example of the UMD texture data set

The third texture data set is the Kylberg texture surfaces data set [31]. This data set is suitable for evaluating and comparing texture measures, noise sensitivity and geometric transform. There are 28 texture classes with 160 images per class. The size of each image is 576×576. Figure 4 shows some sample images from each class of this data set. As before, 10-fold cross validation is implemented.

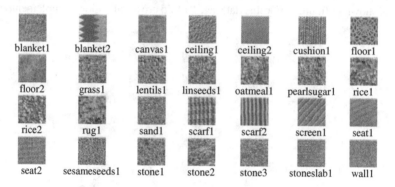

Fig. 4. An example of the Kylberg texture data set

The Brodatz texture data set [32] is used in the last experiment. The data set consists of 32 texture classes, each with 64 images (16 original, 16 rotated, 16 scaled, and 16 rotated and scaled). The size of each image is 64×64. However, in this experiment, we utilize original form images as our training data set. We also implement 8-fold cross validation. Then the blind test data set consists of the rotated images. Hence there are 512 images for the training data set and another 512 images for the blind test data set. Some examples from the Brodatz data set are shown in figure 5.

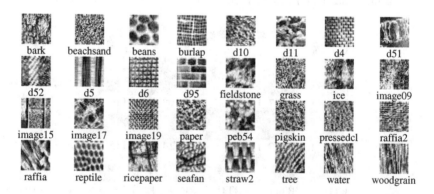

Fig. 5. An example of the Brodatz texture data set

Figures 6, 7, 8, and 9 show the best validation set results on the UIUC, UMD, Kylberg and Brodatz training data sets with all σ values. From the figures, we can see that the highest correct classification from the FzCM, FzCM1, and FzCM2 always higher than that from the GLCM, GLCM1, and GLCM2. There is only one case from the Brodatz data set that the GLCM outperforms the FzCM1 and FzCM2. However, the FzCM result is still better than the GLCM. Tables 1, 2, and 3 summarize the best correct classification and the average results from the validation sets for UIUC, UMD, and Kylberg data sets. The summarized result for the best classification on the validation set and the blind test data set from the Brodatz data set is shown in table 4. From all tables, we can see that the average classification result on the validation set from the FzCM is always better than those from the GLCM, GLCM1, and GLCM2. Although, the GLCM seems to do better than the FzCM1 and FzCM2 in the Brodatz data set, the result of the best GLCM on the blind test data set is worse than all FzCM's. The FzCM2 yields 92.19% on the blind test data set in this case.

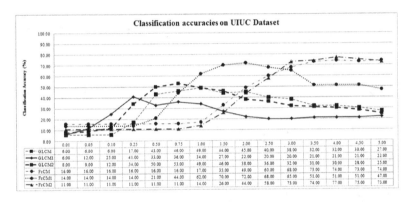

	0.01	0.05	0.10	0.25	0.50	0.75	1.00	1.50	2.00	2.50	3.00	3.50	4.00	4.50	5.00
GLCM	6.00	6.00	6.00	17.00	43.00	46.00	49.00	44.00	45.00	40.00	38.00	32.00	32.00	30.00	27.00
GLCM1	6.00	12.00	25.00	41.00	33.00	38.00	34.00	27.00	22.00	20.00	20.00	21.00	21.00	21.00	22.00
GLCM2	8.00	9.00	12.00	34.00	50.00	53.00	49.00	46.00	36.00	36.00	32.00	31.00	30.00	28.00	25.00
FzCM	16.00	16.00	16.00	16.00	16.00	16.00	17.00	33.00	49.00	60.00	68.00	73.00	74.00	73.00	74.00
FzCM1	14.00	14.00	14.00	14.00	21.00	44.00	70.00	72.00	68.00	65.00	51.00	51.00	51.00	51.00	47.00
FzCM2	11.00	11.00	11.00	11.00	11.00	11.00	14.00	26.00	44.00	58.00	73.00	74.00	77.00	75.00	73.00

Fig. 6. The best validation set classification results from the UIUC data set

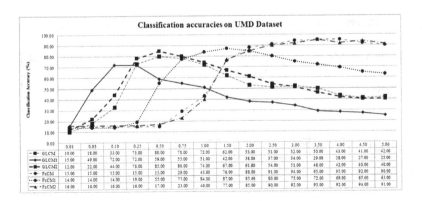

	0.01	0.05	0.10	0.25	0.50	0.75	1.00	1.50	2.00	2.50	3.00	3.50	4.00	4.50	5.00
GLCM	10.00	18.00	33.00	73.00	80.00	78.00	72.00	62.00	53.00	51.00	52.00	50.00	43.00	41.00	42.00
GLCM1	15.00	49.00	72.00	72.00	59.00	55.00	51.00	42.00	38.00	37.00	34.00	29.00	28.00	27.00	25.00
GLCM2	12.00	22.00	44.00	78.00	85.00	80.00	74.00	67.00	61.00	54.00	51.00	46.00	42.00	40.00	40.00
FzCM	15.00	15.00	15.00	15.00	15.00	29.00	43.00	76.00	88.00	91.00	94.00	95.00	95.00	92.00	90.00
FzCM1	14.00	14.00	14.00	19.00	55.00	77.00	84.00	87.00	85.00	80.00	75.00	72.00	69.00	65.00	63.00
FzCM2	16.00	16.00	16.00	16.00	17.00	23.00	40.00	77.00	85.00	90.00	92.00	95.00	92.00	94.00	91.00

Fig. 7. The best validation set classification results from the UMD data set

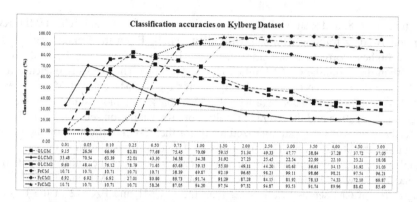

Fig. 8. The best validation set classification results from the Kylberg data set

	0.01	0.05	0.10	0.25	0.50	0.75	1.00	1.50	2.00	2.50	3.00	3.50	4.00	4.50	5.00
GLCM	9.15	26.56	66.96	82.81	77.68	75.45	70.09	59.15	51.34	49.33	47.77	38.84	37.28	37.72	37.05
GLCM1	33.48	70.54	63.39	52.01	43.30	36.38	34.38	31.92	27.23	25.45	22.54	22.99	22.10	23.21	18.06
GLCM2	9.60	48.44	76.12	78.79	71.65	65.63	59.15	55.80	49.11	44.20	40.63	36.61	34.15	31.92	31.03
FzCM	10.71	10.71	10.71	10.71	10.71	38.39	69.87	92.19	96.65	98.21	99.11	98.66	98.21	97.54	96.21
FzCM1	6.92	6.92	6.92	27.01	80.80	89.73	91.74	91.29	87.28	84.15	81.92	78.13	74.33	72.10	69.87
FzCM2	10.71	10.71	10.71	10.71	58.26	87.05	94.20	97.54	97.32	94.87	93.53	91.74	89.96	88.62	85.49

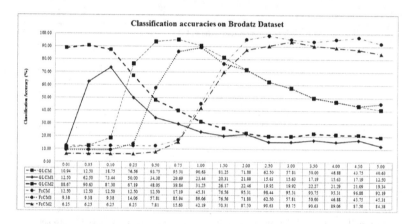

Fig. 9. The best validation set classification results from the Brodatz data set

	0.01	0.05	0.10	0.25	0.50	0.75	1.00	1.50	2.00	2.50	3.00	3.50	4.00	4.50	5.00
GLCM	10.94	12.50	18.75	76.56	93.75	95.31	90.63	81.25	71.88	62.50	57.81	50.00	48.88	43.75	40.63
GLCM1	12.50	62.50	73.44	50.00	34.38	29.69	23.44	20.31	21.88	15.63	15.63	17.19	15.63	17.19	12.50
GLCM2	88.67	90.63	87.30	67.19	48.05	39.84	31.25	26.17	22.46	19.92	19.92	22.27	21.29	21.09	19.34
FzCM	12.50	12.50	12.50	12.50	12.50	17.19	45.31	76.56	95.31	98.44	95.31	93.75	95.31	96.88	92.19
FzCM1	9.38	9.38	9.38	14.06	57.81	85.94	89.06	76.56	71.88	62.50	57.81	50.00	46.88	43.75	45.31
FzCM2	6.25	6.25	6.25	6.25	7.81	15.63	42.19	70.31	87.50	90.63	93.75	90.63	89.06	87.50	84.38

Table 1. The best correct classification and the average result on the UIUC validation set

Method	RBF σ	Best classification result	Average classification result
GLCM	1.0	49.00	44.30
GLCM1	0.3	41.00	37.10
GLCM2	0.8	53.00	46.60
FzCM	4.0	74.00	69.10
FzCM1	5.0, 2.0	72.00	64.50
FzCM2	4.0	77.00	72.30

Table 2. The best correct classification and the average result on the UMD validation set

Method	RBF σ	Best classification result	Average classification result
GLCM	0.5	80.00	75.70
GLCM1	0.1	72.00	67.10
GLCM2	0.25, 0.5, 3.5	85.00	79.00
FzCM	4.0	95.00	92.10
FzCM1	1.5	87.00	82.80
FzCM2	3.5	95.00	93.30

Table 3. The best correct classification and the average result on the Kylberg validation set

Method	RBF σ	Best classification result	Average classification result
GLCM	0.25	82.81	81.74
GLCM1	0.05	70.54	68.64
GLCM2	0.25	78.79	76.47
FzCM	3.0	99.11	97.57
FzCM1	1.0	91.74	90.67
FzCM2	1.5	97.54	96.47

Table 4. The best correct classification and the average result on the Brodatz validation set

Method	RBF σ	Best classification result	Average classification result	Blind test data set classification result
GLCM	0.75	95.31	89.84	23.24
GLCM1	0.1	73.44	65.82	83.40
GLCM2	0.05	90.63	89.77	85.74
FzCM	2.5	98.44	92.58	29.69
FzCM1	1.0	89.06	81.84	84.38
FzCM2	3.0	93.75	86.33	92.19

5   Conclusion

In this work, we incorporate the fuzzy C-means into the gray level co-occurrence matrix. We build eight fuzzy co-occurrence matrices for each direction. We only compute contrast, correlation, energy, and homogeneity features from each fuzzy co-occurrence matrix. We then test our features with the multiclass support vector machine using the one-versus-all strategy on the UIUC, UMD, Kylberg, and Brodatz data sets. We also compare the classification result with the same set of features extracted from the gray level co-occurrence. We found out that the features extracted from our fuzzy co-occurrence matrix yields a better classification result than those from the regular gray level co-occurrence matrix. The best results on the validation set using features computed from the proposed fuzzy co-occurrence are 77%, 95%, 99.11%, and 98.44% on the UIUC, UMD, Kylberg, and Brodatz, whereas those on the same data sets using features from the gray level co-occurrence are 53%, 85%, 82.81% and 95.31%, respectively. The best classification rate on the blind test set of the Brodatz data set using the proposed fuzzy co-occurrence is 92.19%, whereas that from the gray level co-occurrence is 85.74%. Please note that the SVM used in the experiment is not the one with soft margin optimization. The result is still quite good although there are classes that may be overlapped. Since, this blind test data set is a rotated version of the training data set, the results suggest that our proposed fuzzy co-occurrence features provide the property of rotation invariance.

References

1. Julesz, B.: Visual pattern discrimination. IRE Trans. Inform. Theory 8(2), 84–92 (1962)
2. Haralick, R.M.: Statistical and structural approaches to texture. Proceedings of the IEEE 67(5), 786–804 (1979)
3. Haralick, R.M., Shanmugam, K., Dinstein, I.: Textural features for image classification. IEEE Transactions on Systems, Man, and Cybernetics SMC-3, 610–621 (1973)
4. Haralick, R.M., Shanmugam, K., Dinstein, L.: On some quickly computable features for texture. In: Proc. 1972 Symp. on Comput. Image Processing and Recognition, vol. 1, 12-2-1–12-2-10 (1972)
5. Haralick, R.M., Shanmugam, K.: Computer classification of reservoir sandstones. IEEE Trans. Geosci. Electron. GE-11, 171–177 (1973)
6. Haralick, R.M., Shanmugam, K.: Combined spectral and spatial processing of ERTS imagery data. J. of Remote Sensing of the Environment 3, 3–13 (1974)
7. Chien, Y.P., Fu, K.S.: Recognition of X-ray picture patterns. IEEE Trans. Syst., Man, Cybern. SMC-4, 145–156 (1974)
8. Chen, P., Pavlidis, T.: Segmentation by texture using a co-occurrence matrix and a split-and-merge algorithm. Tech. Rep. 237, Princeton University, Princeton, NJ (1978)
9. Leemaitre, G., Rodojevic, M.: Texture segmentation: Co-occurrence matrix and Laws' texture masks methods, http://g.lemaitre58.free.fr/pdf/vibot/scene_segmentation_interpretation/cooccurencelaw.pdf
10. Barrera, M.A., Andrade, M.T.C., Yong Kim, H.: Texture-based fuzzy system for rota-tion-invariant classification of aerial orthoimage regions. In: Proceeding of the 4th GEOBIA. Rio de Janeiro, Brazil (2012)
11. Walker, R.F., Jackway, P.T., Longstaff, I.D.: Recent developments in the use of the co-occurrence matrix for texture recognition. Digital Signal Processing Proceedings, DSP 97(1), 63–65 (1997)
12. Baraldi, A., Parmiggiani, F.: An investigation of the textural characteristics associated with grey level cooccurrence matrix statistical parameters. IEEE Transactions on Geoscience and Remote Sensing 33(2) (1995)
13. Bartels, P., Bahr, G., Weid, G.: Cell recognition from line scan transition probability profiles. Acta Cytol. 13, 210–217 (1969)
14. Wied, G., Bahr, G., Bartels, P.: Automatic analysis of cell images. In: Wied, Vahr (eds.) Automated Cell Identification and Cell Sorting, pp. 195–360. Academic Press, New York (1970)

15. Tou, J.Y., Tay, Y.H., Lau, P.Y.: One-dimensional Grey-level Co-occurrence Matrices for texture classification. In: International Symposium on Information Technology, vol. 3 (2008)

16. Benco, M., Hudec, R., Matuska, S., Zachariasova, M.: One-dimensional color-level co-occurrence matrices. In: Proceedings of ELEKTRO 2010, pp. 18–21 (2010)

17. Chen, W.-S., Huang, R.-H., Hsieh, L.: Iris recognition using 3D co-occurrence matrix. In: Tistarelli, M., Nixon, M.S. (eds.) ICB 2009. LNCS, vol. 5558, pp. 1122–1131. Springer, Heidelberg (2009)

18. Ong, H.-C., Khoo, H.-K.: Improved image texture classification using grey level co-occurrence probabilities with support vector machines post-processing. European Journal of Scientific Research 36(1), 56–64 (2009)

19. Tao, R., Chen, J., Chen, B., Liu, C.: GLCM and fuzzy clustering for ocean features classi-fication. In: International Conference on Machine Vision and Human-machine Interface, pp. 538–540 (2010)

20. Harrabi, R., Braiek, E.B.: Colour image segmentation using the second order statistics and a modified fuzzy C-means technique. Scientific Research and Essays 7(17), 1734–1745 (2012)

21. Mukane, S.M., Bormane, D.S., Gengaje, S.R.: Wavelet and co-occurrence matrix based ro-tation invariant features for texture image retrieval using fuzzy logic. International Journal of Computer Application 24(7) (2011)

22. Cheng, H.D., Chen, C.H., Chiu, H.H.: Image segmentation using fuzzy homogeneity crite-rion. Information Sciences 98(1-4), 237–262 (1997)

23. Jawahar, C.V., Ray, A.K.: Fuzzy statistics of digital images. IEEE Signal Processing Let-ters 3(8) (1996)

24. Sen, D., Pal, S.K.: Image segmentation using global and local fuzzy statistics. In: Annual IEEE India Conference (2006)

25. Dunn, J.: A fuzzy relative of the Isodata process and its use in detecting compact, well-separated clusters. Journal of Cybernetics 3(3), 32–57 (1973)

26. Bezdek, J.: Pattern Recognition with Fuzzy Objective Function Algorithms. Plenum Press, New York (1981)

27. Abe, S.: Support Vector Machines for Pattern Classification. Springer-Varlag London Li-mited (2005)

28. Cristianini, N., Shawe-Taylor, J.: An Introduction to Support Vector Machines and Other Kernel-based Learning Methods. Cambridge University Press (2000)

29. Lazebnik, S., Schmid, C., Ponce, J.: A sparse texture representation using local affine re-gions. IEEE Transactions on Pattern Analysis and Machine Intelligence 27(8), 1265–1278 (2005)

30. Xu, Y., Ji, H., Fermuller, C.: A projective invariant for texture. In: IEEE Conference on Computer Vision and Pattern Recognition, pp. 1932–1939 (2006)

31. Kylberg, G.: The Kylberg Texture Dataset v. 1.0, Centre for Image Analysis, Swedish University of Agricultural Sciences and Uppsala University, External report (Blue series) No. 35, http://www.cu.uu.se/_gustaf/texture/

32. Brodatz, P.: Textures: A Photographic Album for Artists and Designers. Dover, New York (1966)

Improved Flow Shop Schedules
with Total Completion Time Criterion

Dipak Laha and Dhiren Kumar Behera

Mechanical Engineering Department Jadavpur University, Kolkata 700032, India
dipaklaha_jume@yahoo.com, dkb_igit@rediffmail.com

Abstract. The paper addresses the permutation flow shop scheduling problem with the minimization of the total completion time objective. Two heuristics are presented combining two note-worthy constructive heuristics. Exhaustive computational results on standard benchmark problems are presented. It is shown that the solutions generated are better than those obtained by the well - known heuristic in the literature. Statistical significance of better results produced by the proposed heuristics is also reported.

Keywords: Scheduling, flow shop, total completion time criterion, heuristics, complexity, optimization.

1   Introduction

The problem of flow shop scheduling has long received the attention of researchers due to its both complexity and practical importance. In flow shops, n jobs are processed by m machines in the same technological or machine order to optimize a certain performance measure. One important performance measure, which is applied here, is the total completion time, which leads to rapid turn-around of jobs and minimization of work-in-process inventory cost [1].

In the flow shop scheduling problem, the order of the machines is fixed. We assume that a machine processes one job at a time and a job is processed on one machine at a time without preemption. Let $t_p(i,j)$ denote the processing time of job j on machine i, and $t_c(i,j)$ denote the completion time of job j on machine i. Let J_j denote the j-th job and M_i the i-th machine. The completion times of the jobs in a schedule are obtained as follows:

For $i = 1, 2, \ldots , m$ and $j = 1, 2, \ldots, n$

$$t_c(M_1, J_1) = t_p(M_1, J_1)$$

$$t_c(M_i, J_1) = t_c(M_{i-1}, J_1) + t_p(M_i, J_1)$$

$$t_c(M_1, J_j) = t_c(M_1, J_{j-1}) + t_p(M_1, J_j)$$

$$\vdots$$

$$\therefore t_c(M_i, J_j) = max\{t_c(M_{i-1}, J_j), t_c(M_i, J_{j-1})\} + t_p(M_i, J_j) \tag{1}$$

B. Murgante et al. (Eds.): ICCSA 2013, Part III, LNCS 7973, pp. 258–270, 2013.

The total completion time is defined as the sum of completion times of all the jobs in a schedule, $\sum_{j=1}^{n} t_c \left(M_m, J_j \right)$. The goal is to obtain the n-job schedule that minimizes the total completion time of jobs. In this context it should be mentioned that a second major line of research in flow shop scheduling involves designing improved heuristics (e.g., [2-6]) for minimizing the makespan which is defined as the completion time of the last job, $t_c(M_m, J_n)$.

The terms sequence (schedule) and partial sequence (partial schedule) will be used throughout this paper. Suppose, for instance, that the problem involves six jobs -- labeled J1 through J6. Examples of complete sequences (schedules) include {J2, J5, J1, J3, J6, J4} and {J1, J2, J3, J4, J5, J6}. Again, {J2, J3, J5, J1} is a 4-job partial sequence for the same problem. The following are all possible 2-job partial sequences involving J1, J4 and J6: {J4, J6}, {J6, J4}, {J1, J4}, {J4, J1}, {J1, J6}, and {J6, J1}.

For n jobs, the search space (for makespan minimization or total completion time minimization) consists of n! possible job schedules (n! permutations of n distinct objects). The problem of flow shop scheduling is NP-complete [7] and exhaustive enumeration of all n! schedules is computationally prohibitive. Therefore, heuristic approaches are the most suitable ones to solve scheduling problems especially involving a large number of jobs.

Heuristics for solving flow shop scheduling problems can be broadly divided into two categories: constructive heuristics and improvement heuristics. A constructive heuristic generates a schedule of jobs in a series of steps, starting first with just a few jobs, and appending new jobs to a partial schedule at every step, eventually arriving at the complete schedule. An improvement heuristic, on the other hand, starts with a (possibly low-quality) complete schedule and improves upon it, by altering it at successive iterations of the algorithm. A third category – a hybrid approach – is also possible, where a combination of constructive and improvement techniques is applied. This hybridization opens up myriad possibilities -- two (or more) heuristics (constructive and/or improvement) may be combined in many ways, including combinations of global search and local search algorithms.

Examples of important constructive heuristics (with the total completion time minimization criterion) include Gupta [8], Ho [9], Wang et al. [10], Rajendran and Chaudhuri [11], Rajendran [12], Rajendran and Ziegler [13], Woo and Yim [14], Liu and Reeves [15], Framinan and Leisten [16], Laha and Sarin [17], and Laha and Chakravorty [18]. Liu and Reeves [15] also presented and empirically analyzed what they called composite heuristics for total completion time minimization consisting of a bunch of methods where the result produced by a constructive heuristic is further refined by some form of local search. In this paper, we present two hybrid heuristics that employ a combination of two constructive heuristics [2, 17].

The remainder of the paper is organized as follows: Section 2 provides a brief review of related work; Section 3 presents the heuristics developed. Section 4 analyzes empirical results of the proposed schemes and other competing methods. Conclusions are drawn in Section 5.

2   Relevant Work

In the present paper we refer to the following four papers rather frequently: Nawaz et al. [2] (or NEH for short), Framinan and Leisten [16], Laha and Sarin [17] and Liu and Reeves [15] (or LR thereafter in the paper). A brief discussion of the main contributions of these papers is in order.

2.1   NEH Heuristic

The NEH heuristic [2] minimizes makespan, not total completion time. Despite the existence of a plethora of flow shop scheduling heuristics for minimizing makespan, NEH continues to be one of the best constructive heuristics because of its simplicity, solution quality and time complexity. In NEH, an initial schedule of jobs termed the "seed" schedule is developed by arranging jobs in the descending order of total processing time on all machines. The first two jobs are picked from the seed schedule and the best 2-job partial schedule is selected. Inserting the other (unscheduled) jobs from the seed schedule one by one at all possible positions of the current best partial schedule, complete schedules are generated.   Finally, the best n-job schedule among the generated schedules is selected.

2.2   Heuristic of Framinan and Leisten

The method proposed by Framinan and Leisten [16] minimizes total completion time in permutation flow shop scheduling by the principle of optimizing partial schedules. The concept draws inspiration from the NEH heuristic method [2] for the makespan criterion.

2.3   Heuristic of Laha and Sarin

The heuristic (H) of Laha and Sarin [17], builds the n-job schedule incrementally, and is therefore a constructive method. It uses the idea of generating partial schedules based on the principle of job insertion similar to that of NEH. What leads to its improved performance is its use of a group of promising partial solutions at each stage (i.e., as each new job is added to the schedule).

2.4   Composite Heuristics of Liu and Reeves

Liu and Reeves [15] proposed a new constructive heuristic that uses a specially designed index function to choose which job from the list of unscheduled jobs is to be selected to be appended to the already scheduled jobs. The index function consists of the weighted sum of two parts: the total machine idle time and the artificial total completion time. They compared four versions of their new constructive heuristic, namely H(1), H(2), H(n/10) and H(n) (n being the number of jobs) with those of Ho [9], Wang et al. [10], Rajendran and Ziegler [13], and Woo and Yim [14].   Three versions of the constructive heuristic H(1), H(n/10), H(n) were then combined with six different local search methods to build a set of composite heuristics. Three variants each of two neighborhood schemes forward pairwise exchange and backward pairwise exchange were used in the local search. Thus 3 × 6 = 18 types of composite heuristics were obtained. It was empirically shown that the composite heuristics are more effective than the constructive heuristics on Taillard's [19] benchmark problems, but at the cost of additional computation time.

3   Heuristics

We present two heuristics, H-1 and H-2, by combining heuristic H [17], and heuristic [2]. Each of the proposed heuristics H-1 and H-2 starts with an initial schedule produced by the FL heuristic and then improves upon it in several iterations. In each iteration, the NEH method is invoked followed by H. The number of iterations was empirically chosen as five, as a quick-and- dirty compromise between solution quality and the time to find the solution (no attempt was made to find the optimal number of iterations). H-1 and H-2 are similar, differing mainly in how, in each iteration, the output of the NEH heuristic is (or is not) fed as input to H. The pseudocode of H-1 and H-2 is given below. In the pseudocode, a schedule is better than another if the total completion time of the former is less than that of the latter. For H-1, inside the main loop the schedule produced by NEH is unconditionally used as the start schedule of H (the idea is to allow for the possibility of generating a good solution in future from a not-so-promising current solution). Note that in the procedure for H the first two steps were needed to create the initial schedule of jobs when used within H-1 or H-2, the code would be slightly modified: the output of the first two steps in the pseudocode for H is to be replaced with the given schedule, best-so-far-schedule.

3.1   H-1

```
best-so-far-schedule ← FL heuristic's schedule;
do k times
{
  current-best-schedule ← best-so-far-schedule;
  execute NEH using best-so-far-schedule as the initial
  schedule;
  best-so-far-schedule ← NEH's schedule;
  execute H using best-so-far-schedule as the initial
  schedule;
  if
    H's schedule is better than the better of current-
    best-schedule and best-so-far-schedule
  then
    best-so-far-schedule ← H's schedule;
  else
    Best-so-far-schedule ← better of NEH's schedule and
    H's schedule;
}
output best-so-far-schedule;
```

3.2   H-2

```
best-so-far-schedule ← FL heuristic's schedule;
do k times
{
  current-best-schedule ← best-so-far-schedule;
  execute NEH using best-so-far-schedule as the initial
  schedule;
  if
    NEH's schedule is better than best-so-far-schedule
  then
    best-so-far-schedule ← NEH's schedule;
    execute H using best-so-far-schedule as the initial
    schedule;
  if
    H's schedule is better than best-so-far-schedule
  then
    best-so-far-schedule ← H's schedule;
  if
    best-so-far-schedule is the same as current-best-
    schedule (i.e., if neither NEH nor H produced a sche-
    dule better than current-best-schedule)
  then
    best-so-far-schedule ← better of NEH's schedule and
    H's schedule;
}
output best-so-far-schedule;
```

4   Experimental Results

The heuristics H-1 and H-2 were run on nine different problem sizes (n = 20, 50, 100 and m = 5, 10, 20) from Taillard's [19] benchmarks. For each problem size, ten independent instances were created following Taillard's function *unif* ( ) and *time seeds* [19]. Thus there are 90 problem instances each of which corresponds to a new processing time matrix.

The H-1 and H-2 heuristics were coded in C and run on a Pentium 4, 256 MB, 2.8 GHz PC. The proposed methods are compared with the composite heuristics of Liu and Reeves [15]. We did not code the LR heuristics; we used the results published in [15] for our comparative analysis. No single composite heuristic of Liu and Reeves performs best for all the problems in the benchmarks and in that paper more than one composite heuristic have been shown to produce the best completion times in several cases. Therefore, in the present paper, we use the phrase "the composite heuristics of Liu and Reeves" to mean any of the Liu-Reeves composite heuristics that produced the best solution for a particular problem instance. Specifically, the LR completion times used in Tables 1 – 6 of the present paper are obtained from the "Best sol. found" column in Table 6 of [15].

Table 1 compares the three methods using the following performance measures: average relative percentage deviation (ARPD) and the maximum percentage deviation (MPD). The ARPD and MPD are defined as follows (*NP* represents the number of problem instances for a problem size, *tft* stands for total completion time of jobs, and H represents the heuristic):

$$ARPD = \frac{100}{NP} \sum_{i=1}^{N} ((tft_{H,i} - besttft_i)/besttft_i) \qquad (2)$$

$$MPD = max_i ((tft_{H,i} - besttft_i)/besttft_i) \times 100 \qquad (3)$$

Clearly, the best possible performance corresponds to both ARPD and MPD being zero. The "besttft" used in the ARPD and MPD calculations is the best among the three solutions. Results of Table 1 show that the proposed heuristics outperform LR for almost all problem sets. However, for 100 jobs and 5 machines, the performance of H-1 and H-2 is marginally inferior to LR. Note that BES (LR) refers to the best performing composite heuristic among many composite heuristics proposed by Liu and Reeves [15].

Table 1. ARPD and MPD with respect to the best solution (over 10 instances) for each problem size

n	m	BES (LR)		H – 1		H – 2	
		ARPD	MPD	ARPD	MPD	ARPD	MPD
	5	0.8221	2.1664	0.2700	1.6953	0.2792	1.3117
20	10	0.8788	1.4155	0.1711	0.6902	0.1612	0.4475
	20	0.5019	1.3648	0.0613	0.3686	0.2022	0.8994
	5	0.3471	0.7628	0.2412	0.9041	0.1742	0.5766
50	10	1.0401	1.9278	0.3748	1.2211	0.1559	0.6151
	20	0.8644	2.2664	0.1589	0.8929	0.2990	0.7294
	5	0.1501	0.4289	0.1915	0.3924	0.2558	0.7089
100	10	0.2421	0.9840	0.3140	0.8242	0.2193	0.6866
	20	0.8412	2.1420	0.2335	1.052	0.2739	1.052

Table 2. Comparison of H-1 and BES (LR) (over 10 instances) for each problem size

n	m	BES(LR)		H − 1	
		No. of best solutions	Mean completion time	No. of best solutions	Mean completion time
20	5	4	14123	7	14066
	10	2	20290	8	20177
	20	1	33315	9	33137
50	5	3	67443	7	67258
	10	1	88425	9	87880
	20	1	124869	9	123844
100	5	5	243228	5	243343
	10	3	295772	7	295178
	20	3	387303	7	385512

Table 3. Comparison of H-2 and BES (LR) (over 10 instances) for each problem size

n	m	BES(LR)		H − 2	
		No. of best solutions	Mean completion time	No. of best solutions	Mean completion time
20	5	4	14123	6	14067
	10	1	20290	9	20175
	20	4	33315	6	33189
50	5	1	67443	9	67210
	10	1	88425	9	87684
	20	1	124869	9	124386
100	5	6	243228	4	243507
	10	3	295772	7	294880
	20	2	387303	8	385664

Tables 2 and 3 show the success rates of the LR heuristics and the proposed methods H-1 and H-2. For each problem instance, the best of the solutions found by LR, H-1 in Table 2 and LR, H-2 in Table 3 are marked as the "best" for that instance, and then the number of the "bests" found by a given algorithm for each 10-instance set is noted. This number provides a measure of the success rate of a given algorithm on a given problem. Note that it is possible for more than one method to find the same "best" for a given instance. Tables 2 and 3 also show the mean of the 10 completion times (not all of which are the "best") for each problem size. Both the heuristics are superior to LR for all problem sizes except $n = 100$, $m = 5$. Tables 4 - 6 present empirical results of total completion times for each of the ten instances of each problem, obtained by LR, H-1 and H-2. The data in these tables show that the proposed heuristic H-1 performs better than LR in 68 instances out of 90. H-2 is better than LR in 67 instances out of 90. If the combined result of H-1 and H-2 is considered, the performance improves to 77 instances out of 90.

Tables 4 - 6 also show results of statistical tests of significance for two separate cases (LR versus H-1 and LR versus H-2). Note that the phrase "LR versus H-1" really means "the best algorithm from among the bunch of algorithms collectively called LR versus H-1". (The algorithms that we here call LR may even include heuristics not attributed to Liu-Reeves: there are six "Woo" entries in Table 6 in [15].) Each test suite gives us 10 pairs of completion time values and we thus have a paired comparison. For each test suite, the mean and the standard deviation of the ten differences in completion times are obtained. The difference in each instance is obtained by subtracting the completion time of the proposed scheme from that of LR. We now test the hypothesis that the population corresponding to the differences has mean, μ, zero. Specifically, we test the (null) hypothesis $\mu = 0$ against the alternative $\mu > 0$. We assume that the completion time difference is a normal random variable, and choose the significance level $\alpha = 0.05$. If the hypothesis is true, the random variable

$$t = \sqrt{N}((\bar{X} - \mu_0)/S) \tag{4}$$

has a t-distribution with N - 1 degrees of freedom, where N = sample size, $\bar{X}$ = sample mean, S = sample standard deviation, and $\mu 0 = 0$ [20]. The critical value c is obtained from the relation Probability ($t > c$) = $\alpha = 0.05$. From the standard tables of t-distribution, we have for 9 degrees of freedom c = 1.83. From Tables 4 - 6 we see that in most of the cases, H-1 or H-2 or both are statistically significantly better than LR. There is only one case (n = 100, m = 5) where the mean difference is negative for both H-1 and H-2, but even in this case LR is not statistically significantly better than our approach.

To provide an idea of the amount of time involved in the computation, the execution time for a single instance of the problem (obtained as the average of 10 independent instances) of size n = 50, m = 10 is approximately 10 seconds. The times for H-1 and H-2 were approximately the same. For 100 jobs and 20 machines (the last row in Table 6), the main loop was executed only two times, instead of five (hence the relatively shorter time). As can be seen from the pseudocode for H-1 and H-2, the major part of both of these algorithms is spent in the outermost loop, and the NEH algorithm accounts for a substantial part of the loop cost. The experiments reported here used the original NEH algorithm which is known to have a time complexity of $O(mn3)$. A better variant of NEH, such as Taillard's [21] improvement of NEH that runs in $O(mn2)$ time, would result in significant savings in the run time. Moreover, 50% computation time for the NEH heuristic and 33.3% computation time for the FL heuristic in the proposed composite heuristics can be saved by implementing the general flowtime computation method of Li et al. [22] by computing only temporal parameters of changed partial sequences.

Table 4. Total completion time values for H-1, H-2, and BES (LR). The benchmark problems ($n = 20$) are due to Taillard [18].

n	m	LR	H-1	Difference Mean	Difference Std. Dev.	t	H-2	Difference Mean	Difference Std. Dev.	t
20	5	14226	14093				14094			
		15446	15309				15305			
		13676	13415				13313			
		15750	15758				15730			
		13633	13633	56.80	130.68	1.37	13637	55.20	148.01	1.18
		13265	13123				13174			
		13774	13665				13665			
		13968	13948				14047			
		14456	14461				14506			
		13036	13257				13207			
20	10	21207	20958				20911			
		22927	22830				22912			
		20072	20163				20131			
		18857	18821				18786			
		18939	18739	113.00	118.28	3.02	18773	115.10	105.08	3.46
		19608	19439				19526			
		18723	18478				18500			
		20504	20478				20354			
		20561	20353				20396			
		21506	21515				21464			
20	20	34119	33995				33995			
		31918	31996				31996			
		34552	34393				34393			
		32159	31698				31698			
		34990	34840	178.20	165.87	3.40	34840	115.92	218.38	1.68
		32734	32693				32745			
		33449	33053				33141			
		32611	32470				32617			
		34084	33768				33644			
		32537	32465				32829			

Table 5. Total completion time values for H-1, H-2, and BES (LR). The benchmark problems ($n = 50$) are due to Taillard [18].

n	m	LR	H-1	Difference		t	H-2	Difference		t
				Mean	Std. Dev.			Mean	Std. Dev.	
50	5	65663	65365				65589			
		68664	68731				69006			
		64378	63996				64365			
		69795	69755				69347			
		70841	70389	185.10	285.16	2.05	70168	233.00	298.36	2.47
		68084	67816				68036			
		67186	67519				66914			
		65582	65701				65392			
		63968	63505				63505			
		70273	69806				69782			
50	10	88770	89854				89316			
		85600	85035				84702			
		82456	81418				81770			
		89356	87719				87997			
		88482	88217	545.10	739.80	2.33	88389	741.20	622.90	3.76
		89602	88994				88200			
		91422	90925				90613			
		89549	88249				88082			
		88230	88109				87676			
		90787	90283				90097			
50	20	129095	128045				128979			
		122094	121479				121479			
		121379	118182				118979			
		124083	124182				123324			
		122158	120881	1024.70	850.73	3.81	121264	852.20	825.43	3.26
		124061	123493				124297			
		126363	125564				125924			
		126317	125306				124197			
		125318	124493				124859			
		127823	126819				126867			

Table 6. Total completion time values for H-1, H-2, and BES (LR). The benchmark problems ($n = 100$) are due to Taillard [18].

n	m	LR	H-1	Difference Mean	Difference Std. Dev.	t	H-2	Difference Mean	Difference Std. Dev.	t
		256789	256626				256412			
		245609	246608				246135			
		241013	240567				241248			
		231365	231009				230120			
100	5	244016	244618	-15.70	653.38	-0.08	245746	-279.70	971.54	-0.91
		235793	236623				235753			
		243741	244342				245007			
		235171	234723				235223			
		251291	251733				252721			
		247491	246587				246711			
		306375	306205				304845			
		280928	279820				281483			
		296927	295608				293871			
		309607	310789				308248			
100	10	291731	292941	594.20	1778.40	1.06	291876	892.00	1376.00	2.05
		276751	276313				276460			
		288199	286618				287607			
		296130	297780				297504			
		312175	307714				309827			
		298901	297994				297083			
		383865	378169				381219			
		383976	384047				384047			
		383779	380775				382391			
		384854	385519				382915			
100	20	383802	379823	1791.30	2855.00	1.98	378952	1639.20	2375.70	2.18
		387962	385349				385349			
		384839	388913				388913			
		397264	394606				394124			
		387831	387150				386250			
		394861	390769				392481			

5   Conclusion

Two heuristics with the total completion time objective in permutation flow shop scheduling problems have been addressed. Results of computational experiments on standard benchmark problems have been presented. The heuristics have been shown to generate better quality solutions than those produced by the composite heuristics of Liu and Reeves [15]. This study shows that the proposed heuristics can be more attractive alternative to the more traditional simple heuristics for flow shop scheduling problems, especially when the quality of solutions is important. A downside of the hybrid procedures is the larger computational times. However, using computationally efficient versions of the basic algorithms that construct the hybrid method, the computational effort of this procedure can be brought down to acceptable levels. The main advantage of the proposed procedures over the heuristic [15] is that either of the two heuristics presented in this paper performed very well on almost all problems.

References

1. Baker, K.R.: Introduction to sequencing and scheduling. John Wiley, New York (1974)
2. Nawaz, M.E., Enscore, E., Ham, I.: A heuristic algorithm for the $m-$ machine, $n-$job flow-shop sequencing problem. Omega 11, 91–95 (1983)
3. Rajendran, C., Ziegler, H.: Ant-colony algorithms for permutation flowshop scheduling to minimize makespan/total flowtime of jobs. European Journal of Operational Research 155, 426–438 (2004)
4. Laha, D., Chakraborty, U.K.: An efficient stochastic hybrid heuristic for flowshop scheduling. Engineering Applications of Artificial Intelligence 20, 851–856 (2007)
5. Agarwal, A., Colak, S., Eryarsoy, E.: Improvement heuristic for the flow-shop scheduling problem. European Journal of Operational Research 169, 801–815 (2006)
6. Sarin, S., Lefoka, M.: Scheduling heuristic for the n-job m-machine flow shop. Omega 21, 229–234 (1993)
7. Gonzalez, T., Sahni, S.: Flow shop and job shop scheduling: complexity & approximation. Operations Research 26, 36–52 (1978)
8. Gupta, J.N.D.: Heuristic algorithms for multistage flow shop problem. AIIE Transaction 4, 11–18 (1972)
9. Ho, J.C.: Flowshop sequencing with mean flow time objective. European Journal of Operational Research 81, 571–578 (1995)
10. Wang, C., Chu, C., Proth, J.M.: Heuristic approaches for n/m/F/ΣC_i scheduling problems. European Journal of Operational Research 96, 636–644 (1997)
11. Rajendran, C., Chaudhuri, D.: An efficient heuristic approach to the scheduling of jobs in a flowshop. European Journal of Operational Research 61, 318–325 (1991)
12. Rajendran, C.: Heuristic algorithm for scheduling in flowshop to minimize total flow time. International Journal of Production Economics 29, 65–73 (1993)
13. Rajendran, C., Ziegler, H.: An efficient heuristic for scheduling in a flowshop to minimize total weighted flowtime of jobs. European Journal of Operational Research 103, 129–138 (1997)
14. Woo, D.S., Yim, H.S.: A heuristic algorithm for mean flow time objective in flowshop scheduling. Computers and Operations Research 25, 175–182 (1998)
15. Liu, J., Reeves, C.R.: Constructive and composite heuristics solution of the $P \| \sum C_i$ scheduling problem. European Journal of Operational Research 132, 439–452 (2001)
16. Framinan, J.M., Leisten, R.: An efficient constructive heuristic for flowtime minimization in permutation flowshops. Omega 31, 311–317 (2003)
17. Laha, D., Sarin, S.C.: A heuristic to minimize total flow time in permutation flow shop. Omega 37, 734–739 (2009)

18. Laha, D., Chakravorty, A.: A new heuristic for minimizing total completion time objective in permutation flow shop scheduling. International Journal of Advanced Manufacturing Technology 53, 1189–1197 (2010)
19. Taillard, E.: Benchmarks for basic scheduling problems. European Journal of Operational Research 64, 278–285 (1993)
20. Kreyszig, E.: Advanced Engineering Mathematics. John Wiley, New York (1972)
21. Taillard, E.: Some efficient heuristic methods for the flow shop sequencing problem. European Journal of Operational Research 47, 65–74 (1990)
22. Li, X., Wang, Q., Wu, C.: Efficient composite heuristics for total flowtime minimization in permutation flowshops. Omega 37, 155–164 (2009)

Towards a Real Architecture of Wireless Ad-Hoc Router on Open-Source Linux Platform

Quan Le-Trung and Minh-Son Nguyen

Wireless Embedded Internet Group, School of Computer Science & Engineering
International University at Ho Chi Minh City, Vietnam
{ltquan,nmson}@hcmiu.edu.vn

Abstract. Mobile ad-hoc networks (MANETs) have numerous applications in disaster rescue and military due to the infrastructure-less and multi-hop characteristics. Research in MANETs has been in-progress towards real-time implementation of key components in MANETs, such as routing protocols, and device drivers for wireless cards. However, there has still been the lack of a complete real architecture of MANET nodes, i.e., the wireless ad-hoc routers. This paper presents and discusses such a real architecture of wireless ad-hoc routers on open-source Linux platform.

Keywords: MANETs, wireless ad-hoc router, routing protocol, device driver.

1 Introduction

Mobile ad hoc network (MANET) has attracted extensive research interests over the past several years due to its multi-hop characteristic and the introduction of a new application domain. A mobile ad hoc network is a self-organized, multi-hop, wireless network. It is formed automatically by a set of mobile nodes without any infrastructure like wireless LAN or cellular system. Thus, the main applications of MANET are instant applications or fast deployment of networks without any infrastructure, both in the commercial and the military area. Some typical examples are the load balancing of the increasing number of mobile users during a festival, or the formation of a network group between rescue force members in an emergency scenario. The latter is a strong application domain of MANET for information and communications technology (ICT) in disaster management [1]-[2].

However, far less effort has been done on the real-world basis, with extensive evaluations through simulations. The reasons are many-folds. Firstly, the lack of system support and programming abstraction in general purpose operating systems (such as Unix/Linux). Secondly, ad-hoc network protocols often employ new network models or have special requirements that are not directly supported by the current operating systems. Without proper systems support and convenient programming abstractions, implementation is forced to do low-level system programming, and often end up making unplanned changes to the system internals in order to gain the additional functionality required for ad-hoc routing. Finally, not only is this a non-trivial task, but in practice it can also lead to unstable systems, incompatible changes (by different implementations), and un-deployable solutions [3].

B. Murgante et al. (Eds.): ICCSA 2013, Part III, LNCS 7973, pp. 271–286, 2013.

Thus, in this paper, we focus on developing, implementing, and testing **real MANET nodes**, i.e., **wireless ad-hoc routers**, for protocol evaluations and performance measurements of MANET applications in realistic scenarios. These cost-effective wireless ad-hoc routers are developed from: *i)* low-cost devices, e.g., desktop/laptop attached with wlan cards using Atheros chipsets [4]; *ii)* re-use of existing open-source codes, including *aodv-uu* [5], *olsr-unik* [6], and *ath5k* [7]; and *iii)* the open-source Linux platform to allow the free source code access of wireless ad-hoc router through the whole network protocol stack.

This paper is structured as follows. Section 2 will present in details the real architecture and source code analysis of our implemented wireless ad-hoc routers. Section 3 shows the real testing scenarios and discusses on the testing results. Section 4 shows the related work. Finally, Section 5 ends this paper with conclusions and future work.

2   Real Architecture of Wireless Ad-Hoc Router

2.1   The Implemented Ad-Hoc Router Architecture

Fig. 1 shows the architecture of a real wireless ad-hoc router that we have developed, using existing implementation of ad-hoc routing protocols *(aodv-uu* [5], *olsr-unik* [6]) and wlan device driver *(ath5k* [7])*. The main concentration is two-fold. Firstly, it is on the linking of a universal wlan *ath5k* device driver that can work over multiple Atheros chipsets, into the Linux kernel for: *i)* packet transmission and reception; *ii)* setting wireless card parameters from tools/utilities such as *iwconfig, iwlist*; and *iii)* the interaction with the wlan network interface (hardware) via *OpenHAL* [8]. Secondly, it is on the linking of two ad-hoc routing daemons *(aodvd, olsrd)* into the Linux kernel for controlling ad-hoc routing packets and updating kernel routing tables. While the former is briefly described here, the latter is presented and discussed in more details in the *Section 2.2*.

Fig. 1 also shows the functional descriptions in the kernel space of our developed wireless ad-hoc router, organizing into different layers: *i)* the operation system (OS) network *tcp/ip* stack, together with corresponding frameworks *(netfilter, plugins)* to hook ad-hoc routing daemons *(aodvd, olsrd)* into the kernel (more details in the *Section 2.2*); *ii)* the protocol driver *mac80211* [9] with the corresponding wlan configuration API *(cfg80211* [10], or wireless extension-*wext* [11])* for controlling the packet transmission, packet reception, and wireless configuration; *iii)* the chipset driver *(ath5k)*; and *iv)* wifi Atheros chipset. Finally, the */proc/net/wireless* file is used to log the wireless specific statistics on each wireless interface in the system.

Since the protocol driver *mac80211* implements *cfg80211 callbacks* for Soft-MAC devices, *mac80211* depends on cfg80211 for both registration to the networking subsystem and for configuration device. As *ath5k* is a *mac80211-based* driver, the wireless configuration is handled by *cfg80211* (through the netlink interface *nl80211* [12]) and the *wireless extensions-wext*. The different between *wext* and *cfg80211* is that: *wext* uses *ioctl system call* [13] while *cfg80211* replaces it to use *"callbacks"*. What are callbacks? User space enters commands to configure the device driver. The kernel analyzes these commands, takes commands' values and passes values to *mac80211* (through *cfg80211*). So, *callbacks* are data structures in *cfg80211* and *mac80211* that store user space's commands. In Fig. 1, *callbacks* are *cfg80211_ops* and *ieee80211_ops*.

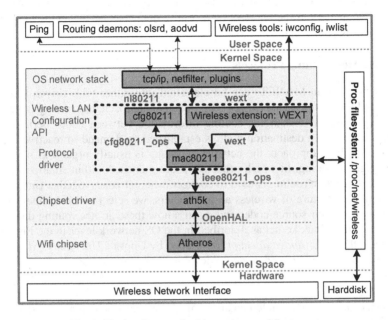

Fig. 1. The implemented ad-hoc router architecture

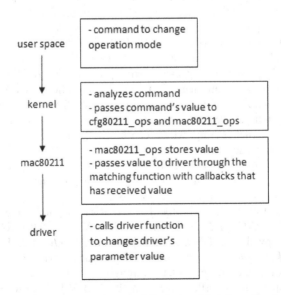

Fig. 2. The configuration of ath5k device driver by callbacks

Each *callback* is pointed by a matching function which is implemented in the device driver. This matching function uses callbacks' values to configure the corresponding parameters in the wireless card. Fig. 2 describes needed steps to show how *ath5k* driver is configured by callbacks.

2.2   Ad-Hoc Routing Protocols

Routing protocol in MANETs can be classified into either table driven (proactive), on-demand (reactive), or hybrid [14]. In proactive approach, the routing table in each node is updated by exchanging routing information between nodes periodically. In contrast, a route to a destination node is established on-demand in reactive approach. Finally, in hybrid approach, the network topology is usually organized into the hierarchy, e.g., clustering, in which proactive routing is used within small zones (intra-cluster) and reactive routing is used to connect different zones together (inter-cluster). In the real architecture of wireless ad-hoc routers, we select two implemented protocols to analyze their source codes and discuss how these ad-hoc routing daemons are hooked into the Linux kernel as described in the OS network layer in the Fig. 1: *i)* the reactive AODV-UU *(aodvd daemon)* developed by Uppsala University [5], and *ii)* the proactive OLSR-UniK *(olsrd daemon)* developed by UniK [6].

AODV-UU
AODV-UU is a Linux implementation of the ad hoc on-demand distance vector (AODV) routing protocol [5]. AODV runs as a user-space daemon *(aodvd)*, maintaining the kernel routing table, and was written using C programming language. For all tasks related to packet processing, AODV uses *netfilter* [15], a Linux kernel framework for mangling packets. For each network protocol, certain hooks are defined. These hooks correspond to well-defined places the protocol stack and allow custom packet mangling code to be inserted in form of kernel modules. Fig. 3 shows the netfilter hooks for IP protocol, of which *NF_IP_PRE_ROUTING*, *NF_IP_LOCAL_OUT*, and *NF_IP_POST_ROUTING* are used by AODV-UU for inserting its packet processing code. In particularly, packets arriving on the *NF_IP_PRE_ROUTING* or *NF_IP_LOCAL_OUT* hook are queued in user-space to allow AODV-UU to process them, while packets arriving on the *NF_IP_POST_ROUTING*, i.e., packets should be sent out by the system, are re-routed to ensure usage of the latest routing information available from the kernel routing, which can be changed as an effect of AODV-UU operations, e.g., route discoveries [16].

In the packet handling, AODV-UU distinguishes between data packets and AODV control messages, and handles data/control packets separately using different software modules. Fig. 4 shows the packet handling of AODV-UU.

When a packet traverses the protocol stack, it is caught by the netfilter hooks *(NF_IP_PRE_ROUTING, NF_IP_LOCAL_OUT, NF_IP_POST_ROUTING)* that have been set up by the AODV-UU kernel module, *kaodv-mod*. The *aodv_hook()* function of the *kaodv-mod* module identifies the packet type, and either tells the netfilter to accept the packet, i.e., to let it through and allow the system to process it on its own, or to queue it for further processing by AODV-UU in user-space [16].

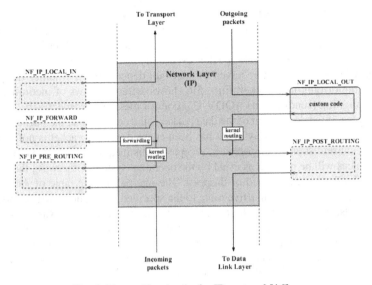

Fig. 3. The netfilter hooks for IP protocol [16]

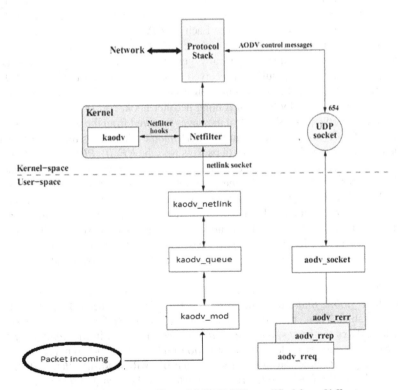

Fig. 4. Packet handling of AODV-UU, modified from [16]

Packet processing is performed by the *kaodv_queue_set_verdict()* function of the *kaodv-queue* module. If the packet is an AODV control message (route request-*RREQ*, or route reply-*RREP*, or route error-*RERR*), an accept verdict is returned to the *kaodv-netlink* module so that the packet eventually will end up on the AODV control message UDP socket, to be received or sent out, depending on whether the packet is an incoming or outgoing packet. Figs. 5-6 shows different flows of incoming (receiving), or outgoing (sending out) AODV-UU control messages, respectively.

Otherwise, if the destination of the data packet is the current host, the packet is a broadcast packet, or Internet gateway mode has been enabled and the packet is not a broadcast within the current subnet, the packet is accepted. This means that the packet under these circumstances will be handled as usual by the operating system. Otherwise, the data packet should be forwarded, queued or dropped. The internal routing table of AODV-UU is used for checking whether an active route to the specified destination exists or not. If such a route exists, the next hop of the packet is set and the packet is forwarded. Otherwise, provided that the data packet was generated locally, the *kaodv-queue* module is used for queuing indirectly the packet until AODV-UU has decided on an action, and a route discovery is initiated. If the data packet was not generated locally, and no route was found, it is instead dropped and a *RERR* message is sent to the source of the packet [16].

AODV control message are received on a UDP socket (on port 654) and processed by the *aodv_socket* module. The type field of the AODV message is checked, the message is converted to the corresponding specialized message type, and the correct handler function is called in the appropriate module. Each AODV control message generated by AODV-UU is sent out on the AODV control message UDP socket. Such a message will be caught by the netfilter hook for locally generated packet, *NF_IP_LOCAL_OUT*, queued by the *kaodv-mod* module and received by the *kaodv-queue* module of AODV-UU. The *kaodv-queue* module will return an accept verdict to *kaodv-netlink*, and the packet will then be caught by the post routing netfilter hook, *NF_IP_POST_ROUTING*. Fig. 5 shows the flows of AODV control messages received from the network drivers.

Firstly, the *kaodv_hook()* function in the *kaodv-mod.c* file catches the control packets from the network driver. Then the control packets pass the *NF_IP_PRE_ROUTING* hook before they are processed by the routing code. The *kaodv_queue_set_verdict()* function in the *kaodv-queue.c* file will verdict these packets. It can decide to receive or send out these packets. The *kaodv_netlink _receive_peer()* function in the *kaodv-netlink.c* file processes these packets. It can add, delete, configure the routes. The *kaodv_netlink_init()* function initializes the socket. After control packets passed the socket, these control packets addressed to the local computer will pass the *NF_IP_LOCAL_IN* hook. Then the *recvAODVUU()* function in the *aodv_socket.c* file will call the corresponding functions to handle packets.

If AODV control message is a *RREQ* message, the *rreq_route_discovery()* function and the *rt_table_find()* function in *aodv_rreq.c* file will check the route entry in the routing table. Then, the *kaodv_update_route_timeout()* function in the *kaodv-mod.c* file updates the routes, and the *rreq_process()* function in the *aodv_rreq.c* file will check if the current node is the destination node. If yes, the *rrep_create()* function and *rrep_send()* function in the *aodv_rreq.c* file will send the *RREP*. If not, the *RREP* is only sent if a fresh enough route is existed, by the *rrep_send()* function in the *aodv_rreq.c* file. If a fresh enough route is not existed, the *RREQ* is forwarded to the neighbors by the *rreq_forward()* function in the *aodv_rreq.c* file.

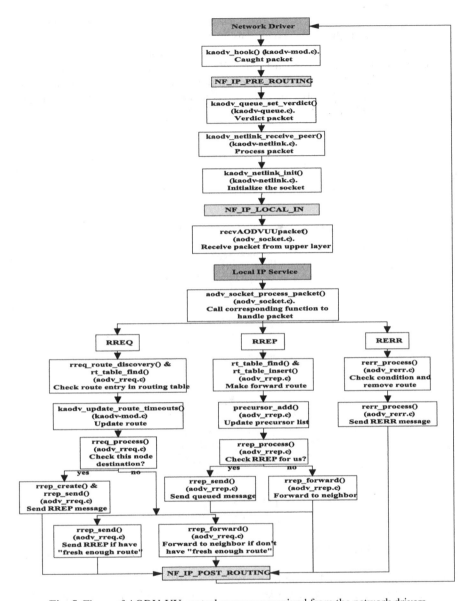

Fig. 5. Flows of AODV-UU control messages received from the network drivers

If AODV control message is a *RREP* message, the *rt_table_find()* function and the *rt_table_insert()* function in the *aodv_rrep.c* file will create the forwarding routes and update the precursor list by the *precursor_add()* function in the *aodv_rrep.c* file. Then, the *rrep_process()* function in the *aodv_rrep.c* file will check the *RREP* message for the current node or not. If yes, the *rrep_send()* function in the *aodv_rrep.c* file will send the queued messages. Otherwise, the *rrep_forward()* function in the *aodv_rrep.c* file will forward the RREP to the neighbors.

If AODV control message is a *RERR* message, the *rrer_process()* function in the *aodv_rerr.c* file will check the conditions to remove routes. If any routes are removed, the *rerr_process()* function in the *aodv_rerr.c* file will send *RERR* message to neighbors in the precursor list. Finally, the control packets pass the *NF_IP_POST_ROUTING* hook. This function represents the last chance to access all outgoing (forwarded or locally created) packets before they leave the wireless ad-hoc router over the wireless card.

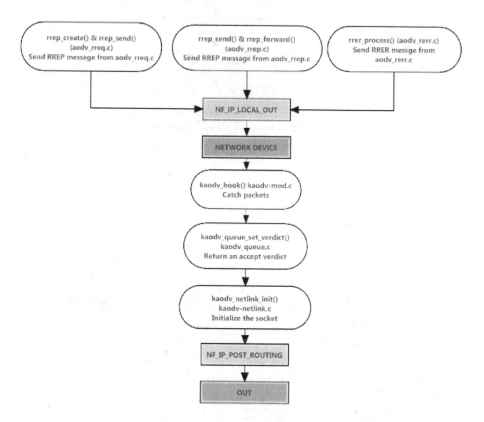

Fig. 6. Flows of outgoing AODV-UU control messages

All outgoing AODV-UU control packets (the *RREP* message specified in the *aodv_rrep.c* file, and the *RRER* message specified in the *aodv_rerr.c* file) created in the local computer (locally generated packet) will pass the *NF_IP_LOCAL_OUT* hook. Then the *kaodv_hook()* function in the *kaodv-mod.c* file will catch these packets, and the *kaodv_queue_set_verdict()* function in the *kaodv-queue.c* file will return an accept verdict for each of message. The *kaodv_netlink_init()* function in the *kaodv-netlink.c* file will then initialize the socket to send control packets out. Finally, the control packets pass the *NF_IP_POST_ROUTING* hook. This function represents the last chance to access all outgoing (forwarded or locally created) packets before they leave the computer over a network device.

OLSR-UniK

The *OLSR-UniK* daemon *(olsrd)* [6] is implemented in c language, and supports IPv4/IPv6, dynamic updates of NICs, and dynamically loadable libraries *(plugins)* at runtime. It provides very flexible configuration options. Fig. 7 shows the implementation overview of *olsrd*.

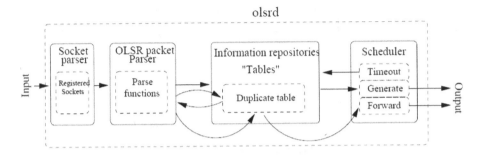

Fig. 7. The implementation overview of *olsrd* [17]

olsrd implementation consists of four components: *i)* the *socket parser*; *ii)* the *packet parser*; *iii)* the *information repositories*; and *iv)* the *scheduler*. The *socket parser* is used to check for incoming traffic, then calls the functions associated with the sockets that has incoming data. Sockets and their corresponding functions are registered at run-time. The *packet parser* receives all incoming traffic from the output of the socket parser, and processes only incoming olsr traffic. It either discards the incoming olsr packet if the packet is found to be invalid, or processes the packet according to given instructions, or forwards the packet according to the default forwarding algorithm. Parse functions for messages are registered with the packet parser at run-time. In *information repositories (routing tables)*, fresh information is kept and all calculations of routes and packets are done based on these repositories. Finally, the event scheduler runs different events at different intervals. To transmit a message at a given interval, one can register a packet generation function with the scheduler. Timing out of tables is also triggered by the scheduler. To maintain an information repository that is timed out on a regular basis, one can register a timeout function with the scheduler [17].

The incoming OLSR data is first handled by the *socket parser*. This entity is responsible for listening for data on a given set of UDP sockets (port 698). Sockets and their corresponding parser functions, are registered with the socket parser at runtime. The socket parser uses the familiar *select(2)* system call to detect when data is available on any socket in a given set of sockets. The socket parser functionality is illustrated in Fig. 8. This modular design allows for multiple entities to listen for multiple types of data without the need for a program flow, often a thread, of their own. So whenever an entity wishes to listen for data on a socket, it calls the *add_olsr_socket()* function in the *socket_parser.h* file. If the connection is lost, the entity removes the socket from the socket parser set using the *remove_olsr_socket()* function.

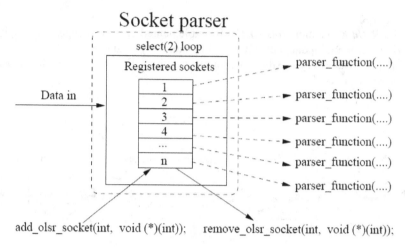

Fig. 8. The socket parser [17]

olsrd registers all control traffic sockets with the socket parser at startup. These sockets are registered with a packet parser function to be called whenever data is available. This way the packet parser receives all broadcasted traffic received on UPD port 698 on either interface. Once the packet parser receives OLSR packets, it checks if the reported size in the OLSR header matches the received amount of data. If so, it parses the packet into messages. These messages are passed on to registered message-parsing functions. If the received size does not match the size read from the OLSR header, the packet is silently discarded [17].

Message-parser functions can be registered and removed for all message types dynamically. One can also register functions with a packet type of *PROMISCUOS* (defined in the *parser.h* file) such functions will receive *all* incoming messages. The *olsr_parser_add_function()* function defined in *parser.h* file, is used to add a message-parser function. If none of these functions are registered to forward the message, then the packet parser forwards the message according to the default forwarding algorithm [17].

olsrd also supports loading of dynamically linked libraries, called *plugins*, for generation and processing of private package-types and any other custom functionality. *Plugins,* which can be written in any language that can be compiled as a dynamic library, allows users or developers to add custom packages or new functionalities into the *olsrd* without changing the original *olsrd* source code. The design of the various entities of *olsrd* allows one to easily add special functionality into most aspects of the program. One can both register and unregister functions with the socket parser, packet parser and scheduler, and one can update many variables, manipulate incoming and outgoing traffic and more. This opens up for possibilities like intercepting current operation and replacing it with custom actions. The plugin can unregister the default functions used by olsrd and replace them with its own. This relationship is illustrated in Fig. 9.

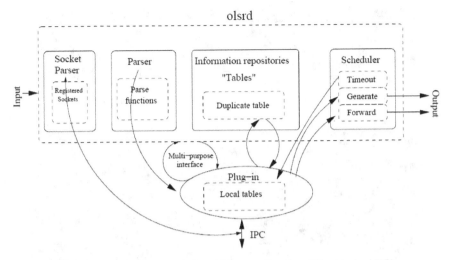

Fig. 9. A plugin framework for adding new functionalities to *olsrd* [17]

For a *plugin* framework like this to work, one needs a well defined and easy-to-expand interface for communication between the OLSR daemon and the *plugin*. The olsrd plugin interface is mainly based upon the *olsr_plugin_io()* function in the *plugin.c* file. To be able to access the *olsr_plugin_io()* function, the *plugin* needs to be initialized from *olsrd*. The file *plugin_loader.c* implements the plugin loader code. For the plugin loader to be able to set up the needed pointers, the plugin must provide the *register_olsr_data()* function. This function is called from the *olsr plugin* loader passing a pointer to a struct *olsr_plugin_data* which contains the pointers to *olsrd* functions that the *plugin* needs to use to be able to set up all needed data pointers. After this, the *plugin* is responsible for fetching all needed pointers from the *olsrd* daemon [17].

3   A Real Testing Scenario

This Section introduces a real scenario to test our implemented wireless ad-hoc router. Three laptops with attached wlan D-Link DWL-G650 cards using *Atheros 5213 chipsets* are used. On each laptop, *ath5k* is loaded as the device driver for wlan card, and AODV-UU is loaded as the ad-hoc routing daemon *(aodvd)* to updating the kernel routing table. For each laptop to take the role of a wireless ad-hoc router, we need to change the wlan interface into the *ad-hoc mode* using the *iwconfig* command. The IP addresses of three wlan interfaces on the corresponding laptop A, B, C is *192.168.30.140/24*, *192.168.30.111/24*, and *192.168.30.56/24*, respectively. Figs. 10-11 show the ad-hoc mode settings and IP address settings of laptop A as an example.

```
wlan0     IEEE 802.11bg  ESSID:"aodv.org"
          Mode:Ad-Hoc  Frequency:2.462 GHz  Cell: 8A:78:04:BB:23:AA
          Tx-Power=20 dBm
          Retry  long limit:7   RTS thr:off   Fragment thr:off
          Encryption key:off
          Power Management:off
```

Fig. 10. Ad-hoc mode setting of wireless ad-hoc router using *iwconfig*

```
root@tramu-laptop:/home/tramu/Desktop# ifconfig
lo        Link encap:Local Loopback
          inet addr:127.0.0.1  Mask:255.0.0.0
          inet6 addr: ::1/128 Scope:Host
          UP LOOPBACK RUNNING  MTU:16436  Metric:1
          RX packets:28 errors:0 dropped:0 overruns:0 frame:0
          TX packets:28 errors:0 dropped:0 overruns:0 carrier:0
          collisions:0 txqueuelen:0
          RX bytes:1904 (1.9 KB)  TX bytes:1904 (1.9 KB)

wlan0     Link encap:Ethernet  HWaddr 00:24:2c:3a:c1:8c
          inet addr:192.168.30.140  Bcast:192.168.30.255  Mask:255.255.255.0
          inet6 addr: fe80::224:2cff:fe3a:c18c/64 Scope:Link
          UP BROADCAST RUNNING MULTICAST  MTU:1500  Metric:1
          RX packets:0 errors:0 dropped:0 overruns:0 frame:0
          TX packets:35 errors:0 dropped:0 overruns:0 carrier:0
          collisions:0 txqueuelen:1000
```

Fig. 11. IP address setting of wireless ad-hoc router using *ifconfig*

Firstly, laptop A and B are turned on, setup, and located in each other transmission range. Once the *aodvd* daemon run on each laptop, it broadcasts AODV control messages to its neighbors and receives/updates its neighbors into its routing table. Fig. 12 shows how laptop A updates and inserts IP address of laptop B into its routing table.

```
root@tramu-laptop:/home/tramu/Desktop# aodvd -l -r 3 -w
19:52:22.454 host_init: Attaching to wlan0, override with -i <if1,if2,...>.
19:52:22.555 aodv_socket_init: RAW send socket buffer size set to 262142
19:52:22.556 aodv_socket_init: Receive buffer size set to 262142
19:52:22.556 main: In wait on reboot for 15000 milliseconds. Disable with "-D".
19:52:22.556 hello_start: Starting to send HELLOs!
19:52:22.556 nl_kaodv_callback: NLMSG_ERROR, error=-16 type=Configuration
19:52:37.387 rt_table_insert: Inserting 192.168.30.111 (bucket 0) next hop 192.1
68.30.111
19:52:37.387 nl_send_add_route_msg: ADD/UPDATE: 192.168.30.111:192.168.30.111 if
index=3
19:52:37.387 rt_table_insert: New timer for 192.168.30.111, life=2100
19:52:37.387 hello_process: 192.168.30.111 new NEIGHBOR!
19:52:37.387 nl_kaodv_callback: NLMSG_ERROR, error=-16 type=Add route
19:52:37.556 wait_on_reboot_timeout: Wait on reboot over!!
```

Fig. 12. Inserting routes into the routing table of wireless ad-hoc router

Two laptops A and B are now reachable, and Fig. 13 shows the ping results.

```
tramu@tramu-laptop:~$ ping 192.168.30.111
PING 192.168.30.111 (192.168.30.111) 56(84) bytes of data.
64 bytes from 192.168.30.111: icmp_seq=1 ttl=64 time=6.59 ms
64 bytes from 192.168.30.111: icmp_seq=2 ttl=64 time=1.85 ms
64 bytes from 192.168.30.111: icmp_seq=3 ttl=64 time=1.35 ms
64 bytes from 192.168.30.111: icmp_seq=4 ttl=64 time=1.24 ms
64 bytes from 192.168.30.111: icmp_seq=5 ttl=64 time=1.12 ms
64 bytes from 192.168.30.111: icmp_seq=6 ttl=64 time=1.18 ms
^Z
[11]+  Stopped                 ping 192.168.30.111
```

Fig. 13. Two testing machines are reachable

We are now testing for the multihop connection. Laptop B is now moved away the transmission range of laptop A, and the connection is disconnected. Each laptop now updates its routing table. Figs. 14-15 show the results.

```
19:53:07.158 hello_timeout: LINK/HELLO FAILURE 192.168.30.111 last HELLO: 2050
19:53:07.158 neighbor_link_break: Link 192.168.30.111 down!
19:53:07.158 nl_send_del_route_msg: Send DEL_ROUTE to kernel: 192.168.30.111
19:53:07.158 rt_table_invalidate: 192.168.30.111 removed in 15000 msecs
19:53:07.158 nl_kaodv_callback: NLMSG ERROR, error=-16 type=Delete route
```

Fig. 14. Laptop A removes stale route out of its routing table

```
tramu@tramu-laptop:~$ ping 192.168.30.111
PING 192.168.30.111 (192.168.30.111) 56(84) bytes of data.
From 192.168.30.140 icmp_seq=16 Destination Host Unreachable
From 192.168.30.140 icmp_seq=17 Destination Host Unreachable
From 192.168.30.140 icmp_seq=18 Destination Host Unreachable
From 192.168.30.140 icmp_seq=19 Destination Host Unreachable
From 192.168.30.140 icmp_seq=20 Destination Host Unreachable
From 192.168.30.140 icmp_seq=21 Destination Host Unreachable
^Z
[14]+  Stopped                 ping 192.168.30.111
```

Fig. 15. Two tested laptops are unreachable

At this moment, we turn on, setup, and located the laptop C between laptop A and B. Fig. 16 shows how laptop C inserts routes to laptop A, B into its routing table.

```
root@htram:/home/htram/Desktop# aodvd -l -r 3 -w
19:54:14.115 host_init: Attaching to wlan0, override with -i <if1,if2,...>.
19:54:14.217 aodv_socket_init: RAW send socket buffer size set to 262142
19:54:14.218 aodv_socket_init: Receive buffer size set to 262142
19:54:14.218 main: In wait on reboot for 15000 milliseconds. Disable with "-D".
19:54:14.218 hello_start: Starting to send HELLOs!
19:54:14.218 nl_kaodv_callback: NLMSG ERROR, error=-16 type=Configuration
19:54:14.763 rt_table_insert: Inserting 192.168.30.140 (bucket 0) next hop 192.1
68.30.140
19:54:14.763 nl_send_add_route_msg: ADD/UPDATE: 192.168.30.140:192.168.30.140 if
index=3
19:54:14.763 rt_table_insert: New timer for 192.168.30.140, life=2100
19:54:14.763 hello_process: 192.168.30.140 new NEIGHBOR!
19:54:14.763 nl_kaodv_callback: NLMSG ERROR, error=-16 type=Add route
19:54:14.763 nl_rt_callback: NLMSG ERROR, error=-17 type=24
19:54:29.218 wait_on_reboot_timeout: Wait on reboot over!!
19:54:48.396 rt_table_insert: Inserting 192.168.30.111 (bucket 0) next hop 192.1
68.30.111
19:54:48.396 nl_send_add_route_msg: ADD/UPDATE: 192.168.30.111:192.168.30.111 if
index=3
19:54:48.396 rt_table_insert: New timer for 192.168.30.111, life=2100
19:54:48.396 hello_process: 192.168.30.111 new NEIGHBOR!
19:54:48.396 nl_kaodv_callback: NLMSG ERROR, error=-16 type=Add route
```

Fig. 16. Inserting routes into the routing table of laptop C

Laptop A, B also insert new route to laptop C into its routing table, and the connection between laptop A and B is now reachable through laptop C.

4   Related Work

In this Section, we show a current review on existing implementation of MANET ad-hoc routing protocols, network testbeds, or semi-testbeds. We do not consider other simulators and/or emulators used for testing and measuring MANET.

mLab [18] allows users to automatically generate arbitrary logical network topologies in order to perform real-time performance measurements of routing protocols and network applications. By changing the logical topology of the network, users can conduct tests in an ad-hoc network without having to physically move the nodes. The mobility generation is built upon the *iptables* [19] rules of every node. To emulate a real MANET, *mLab* uses wired interface to transfer files needed for its operations to and from the node, and manipulate its networking elements in such a way that will create the logical topology. That leaves the wireless interface free of any interference and, most importantly, emulates an actual MANET. *piconet* [20] implements an ad-hoc routing protocol, i.e., dynamic source routing (DSR) [14], for IP on Linux which will allow a dynamic network of mobile devices to be self-organizing and self-configuring. It assumes that the network interface is up before installing the DSR module. *kernel-aodv* [21] is a loadable kernel module for Linux. It implements aodv routing between computers equipped with WLAN devices. Like *piconet*, *kernel-aodv* only implements a single ad-hoc routing protocol. However, in this package, it additionally adds multiple characteristics, e.g., supporting for multiple interfaces and internetworking with the Internet. *awds* [22] implements a layer 2 routing protocol for wireless mesh networks, based on link-state routing, similar to the olsr routing. For communicating with applications, the *awds* daemon creates a virtual Ethernet interface (i.e., awds0). The device emulates an Ethernet interface, and thus all kinds of protocols that work with Ethernet can be used through virtual Ethernet tunnel established over wireless link. While *mLab*, *piconet*, *kernel-aodv*, and *awds* focus only on the network topology and routing issues and makes an assumption on the underlying infrastructures, e.g., device driver for wlan cards, or emulate the wlan cards, our work in this paper considers both, and also specifies in more details how these components are linked into the Linux kernel.

Mesh-toolkit [23] contains Microsoft research's mesh connectivity layer (MCL) and virtual WiFi software in both source and binary form along with documentation. The MCL package includes a loadable Windows XP driver that implements a virtual network adapter, so the ad-hoc network appears as an additional (virtual) network link. Routes are established using Microsoft's link quality source routing (LQSR) protocol, a modified version of DSR protocol. MCL is implemented as an interposition layer between the link layer and network layer. To higher layer software, MCL appears to be just another Ethernet link, albeit a virtual link. To lower layer software, MCL appears to be just another protocol running over the physical link. Developers can modify and build the XP version using Microsoft® driver development kit (DDK) and the CE version using platform builder. However, since Windows are closed-source operating systems, some problems like compatibilities or black-box functions will limit the portability of Mesh-toolkit over multiple platforms and the integration of other network protocols.

APE [24] is a universal ad-hoc testbed for protocol evaluation in realistic scenario, and is the most similar to our work in this paper. It can be installed in an existing Linux or Windows environment (with DOS support), without re-partitioning or complex installation procedures. APE distribution also comes pre-configured and contains tools for data gathering at both IP and Ethernet level. Moreover, APE source code allows building customized APE distribution packages, with new scenarios, protocols, and much more, it also contains analysis tools for analysis of gathered data. There are many popular ad-hoc routing protocols integrated into APE, including aodv, dsr, olsr. Logical topology and mobility scenarios can be created through the usage of Mackill tool, which can force different connectivity configurations in the ad-hoc network without nodes required to be physically separated. Additionally, a modification of the wireless extensions "spy mode" of the IEEE 802.11 WaveLAN drivers, called "superspy" enables logging of signal level and signal noise for all packets picked up by a wireless interface (the interface must be put in the promiscuous mode). This Ethernet level data can be used for several purposes, e.g., estimating the density of the network, or the connectivity. However, APE only works for WLAN cards over either orinoco_cs [25] or wavelan_cs [26] using Hermes/Spectrum24/Prism chipsets, which are not currently supported by most well-known manufacturers. Modifications of APE to work over multiple platforms are still open issues, leading to our contributions in this paper, i.e., porting APE framework over the universal ath5k device driver for Atheros chipsets, which are supported by most wlan device manufacturers nowadays.

5   Conclusions and Future Work

This paper has presented and discussed our implementation and testing of the real wireless ad-hoc routers on the open-source Linux platform. We focus on how to link the existing ad-hoc routing daemons, i.e., aodvd, olsrd, and a universal wlan driver driver, i.e., ath5k, into the Linux kernel to implement our wireless ad-hoc routers. A real multihop testing scenario shows that our wireless ad-hoc routers work well.

Our future work is two-fold. Firstly, we will continue testing our wireless ad-hoc routers for other issues, such as internetworking with the Internet, performance, overhead of signaling. Secondly, we will port the tools/utilities for logging and collecting wireless statistics on APE framework into our wireless ad-hoc routers to work over ath5k device driver.

Acknowledgement. The work in this paper is funded by Vietnam National University-Ho Chi Minh city (VNUHCM), Vietnam, under the contract number: B2012-28-11/HĐ-ĐHQT-QLKH.

References

1. Conti, M., Giordano, S.: Multihop Ad Hoc Networking: The Theory. IEEE Communications Magazine, 78–86 (April 2007)
2. Conti, M., Giordano, S.: Multihop Ad Hoc Networking: The Reality. IEEE Communications Magazine, 88–95 (April 2007)
3. Ha-Son, H., Le-Trung, Q., Nguyen, M.-S.: ManetPRO: A Protocol Evaluation Testbed over m-Platforms. In: ACM IIWAS, Ho Chi Minh city, Vietnam (2011)
4. The Qualcomm Atheros company, http://www.atheros.com/
5. aodv-uu open-source project, http://sourceforge.net/projects/aodvuu/
6. olsrd, an adhoc wireless mesh routing daemon, http://www.olsr.org/
7. The madwifi/ath5k project, http://madwifi-project.org/
8. The OpenHAL, http://madwifi-project.org/wiki/About/OpenHAL
9. Linux wireless mac80211, http://wireless.kernel.org/en/developers/Documentation/mac80211
10. Linux wireless cfg80211, http://wireless.kernel.org/en/developers/Documentation/cfg80211
11. Tourrilhes, J.: Wireless Extensions for Linux, http://www.hpl.hp.com/personal/Jean_Tourrilhes/Linux/Linux.Wireless.Extensions.html
12. Linux wireless nl80211, http://wireless.kernel.org/en/developers/Documentation/nl80211
13. Linux programmer's manual: ioctl – control device, http://www.kernel.org/doc/man-pages/online/pages/man2/ioctl.2.html
14. IETF MANET charter, http://datatracker.ietf.org/wg/manet/charter/
15. The netfilter/iptables project, http://www.netfilter.org/
16. Wiberg, B.: Porting AODV-UU Implementation to ns-2 and Enabling Trace-based Simulation. Master Thesis at Uppsala University, Sweden (2002)
17. Tønnesen, A.: Impementing and Extending the Optimized Link State Routing Protocol. Master Thesis at UniK, University of Oslo, Norway (2004)
18. The mLab project, http://csrc.nist.gov/groups/SNS/manet/projects.html
19. The netfilter iptables project, http://www.netfilter.org/projects/iptables/index.html
20. The piconet project, http://sourceforge.net/projects/piconet/
21. The kernel-aodv project, http://w3.antd.nist.gov/wctg/aodv_kernel/
22. The awds project, http://awds.berlios.de/
23. The mesh-toolkit project, http://research.microsoft.com/en-us/projects/mesh/
24. The APE project, http://apetestbed.sourceforge.net/
25. The Linux ORiNOCO driver, http://www.nongnu.org/orinoco/
26. The Linux pcmcia-cs package, http://pcmcia-cs.sourceforge.net/

A Study of Robot-Based Context-Aware Fire Escape Service Model

NamJin Bae, KyungHun Kwak, Sivamani Saraswathi,
JangWoo Park, ChangSun Shin, and YongYun Cho[*]

Information and Communication Engineering, Sunchon National University,
413 Jungangno, Suncheon, Jeonnam 540-742, Korea
{bakkepo,supersdar,saraswathi,csshin,jwpark,yycho}@sunchon.ac.kr

Abstract. Most of buildings in modern society become bigger and higher, the fire can cause serious damage in such as a large and complex building. Because of this, escape environment is very important when the fire occurs in the building. This paper suggests Robot-based Context-Aware Fire Escape Service Model that can provide optimal escape routes considering fire information and context to user through user's smart phone with real time. The service model provides safe escape routes to user who did not find an escape route from disaster situation through smart phone and robot. In addition, the service model can expect rapid fire suppression and minimal expected casualty by teaching the present information in building to the rescue party.

Keywords: Ubiquitous Agriculture, Context-Aware, Optimal Route, Smart Service.

1 Introduction

During the last ten-year period, you can see to increase expected casualty and property loss through incidence of fire accidents each year. In case of fire in high-rise building, the sober and quick judgment is required for safe escape. But, most people are difficult to maintain a calm and quick judgment when the fire occurs because people get into a panic from the insecurities and the fear. But, it is common for people to cause fear, anxiety, disorientation, confusion and panic due to stress. Smoke and blackout from the fire incident causes visual obscuration which leads to the biggest cause of clouded judgment. Also, if the people were misguided during the evacuation, it may increase the fear and stress [2].

The use of the aerial ladder truck is difficult in the building when fire occurs. According to the structural characteristics of high-rise complexes, the fast spreading flame can increase the damage enormously. Thus, the initial suppression and the securing of the safe escape path are important to minimize the damage for humans in the high-rise building, large building and underground. [3].

---

[*] Corresponding author.

B. Murgante et al. (Eds.): ICCSA 2013, Part III, LNCS 7973, pp. 287–293, 2013.

This paper suggests the optimal escape path based on the reliable data of the initial fire and context-aware service model for fire escape support to reduce fear, anxiety and panic.

The rescue worker is difficult to enter into the building after five minutes from the fire occurs because the combustion of the fire spread speed and damage area is rapidly increasing [4]. Thus, this system can minimize the damage that not the people of inside building but the rescuer as providing and guiding the optimal escape path.

2   Related Work

2.1   Fire Disaster Robot [5]

The field robot for firefighting is a robot that supports firefighting and search of life in the scene of the fire and the risk scene needed of long-term. The robot consists of firefighting monitors, body part with camera and driving part with caterpillar. The robot is able to remote control using high performance camera equipped with wireless remote control devices. In addition, it is able to search of life and fire spot using thermo-graphic camera in smoke. A firefighting assist is a kind of search robot to know the initial fire situation before firefighter start extinguishing the fire. This robot is throwaway robot with strong structure, various function and devices such as real-time transport of scene screen, escape inducement, various sensors.

2.2   Study on the Fire Escape Simulation Software [6]

The refuge is the act of escape to a safe place more in case of emergency. Escape path is often blocked by the rapid spread of smoke when people escape from fire. Therefore, a simple idea such as quickly evacuate to the open air can make a dangerous situation. Escape behavior should be simple and clear in the light of various situations because the escape is in a difficult state such as physical and psychological. Mostly people go towards the direction of the light or follow the crowd blindly because they can't recognize the escape path. Therefore, the escape paths have to compose considering psychological characteristics of the people.

3   Service Model Design

3.1   LEGO MindStorm NXT

Using robot as prototype of the service model proposed in this paper is Mindstorms NXT of LEGO. Mindstorms NXT can communicate with Bluetooth, Wi-Fi. It consists of various sensors, interactive servomotor of 8bit and a brick with possible microprocessor to the use of high-level programming language such as C and Java. It has been developed so that user can run a test of robot system such as autonomous mobile robot and walking robot. Because of this, intelligent brick can easily implement a variety of complex robot [8].

3.2   The Optimal Path Search Algorithm [9]

A* is a computer algorithm that is widely used. It is one of the shortest distance searching algorithms between source node and destination node. This algorithm investigates neighboring nodes from the source node to the destination node. At the same time, using evaluation function $f(n)$ is defined about each node find path of the lowest cost.

The evaluation function $f(n)$ expressed by the equation (1) [7].

$$f(n) = g(n) + h(n) \qquad (1)$$

Here,

$g(n)$: The cost from the source until n

$h(n)$: The cost from n until destination

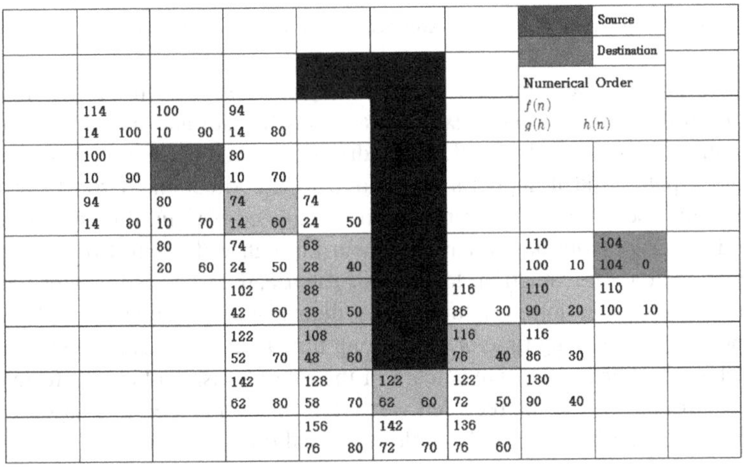

Fig. 1. The Example of A* Algorithm

3.3   Service Architecture

Figure 2 sis a structure of the fire escape support service model. The user installs the fire escape support service application using the smart device by scan QR-Code in the building. The user installs the application. At the same time, the information of user send to the server and the server stores this information in a database.

The server receives data such as temperature, smoke, and population density through the sensor in the building. Later, in case of emergency such as the fire, the robot or the user using A* algorithm inform a safe escape path suited to the individual. The evacuated user's location is automatically deleted in the database. And the users who do not move for a long time or take a long time to evacuate are shown separately on the web control page.

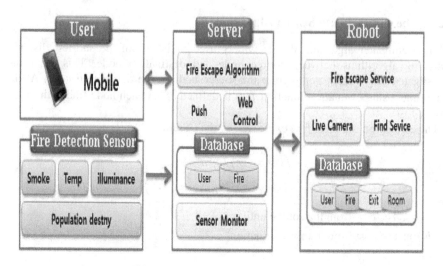

Fig. 2. Service Model Architecture for Fire Escape Support

Sensor information and the fire point in the server send to the robot. Later, the robot using information received explores an optimal path and the robot secures an optimal path. The robot finds people using driving method of path recognition Based on secured a path. And it supports evacuation in fire situation. If robot cut off the communication then robot operate based on recognized path until communication is restored. The server from hours that fire occur through web control page support an information such as fire situation, location of robot and user to the user or the robot. The escaped user's location is automatically deleted in the database. And the users who do not move for a long time or take a long time to evacuate are shown separately on the web control page. The live camera of the robot transmits visual information to the web control page. And the received visual information is used to control the robot or to show the inside situation of the building in real time.

3.4   Service Flow

The exception of the unexpected can happen when the robot is driving. The exception processing are largely path reset, transmission of emergency rescue local coordinate, self-optimal path search. Figure 3 shows the exception processing about drive of the robot and exceptions. The exceptions of path reset include the new ignition point realization in the escape point, the high density of the personnel in the escape path, path breakaway of the robot and blocked path. Self-path search is exception routine that searching path based on the situation of the last received building when the robot is incommunicado from situation of the building. "Follow me" Service using a voice use when the user does not follow the robot or injured person arise. In addition, the coordinate of the emergency relief local transmit to the rescue team for emergency when the rescue team arrived. Additionally, there are many exceptions but the exceptions are divided into three divisions (path reset, self-path search, coordinate transmission for emergency rescue) in this paper.

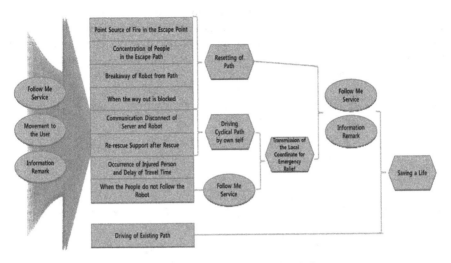

Fig. 3. Exception Process of Driving Robot

3.5   Simulation

Figure 4 shows the implementation of the simulator to test the proposed service that looked like the real situation. The simulator can design the user's location, the fire's area and so on. And provide optimal evacuation path according to the setting.

If the server is running, then the escape path is immediately displayed on the screen.

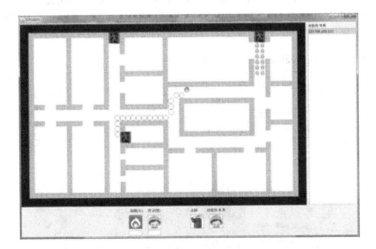

Fig. 4. Simulator for Fire Escape Support Service

The server received all the data needed for the calculation of the coordinate from the simulator. And then, the optimal escape path sends to the users currently registered in the database. At this point, the service is consistently path search in case of changing the user's location or the situation. The proposed service has been designed to synchronize between the simulator and the client. Even if there is a change in the user's location or the navigation path, it is designed to ensure the synchronization between the simulator and the client. This way, at any rate the user is kept updated of the optimal path.

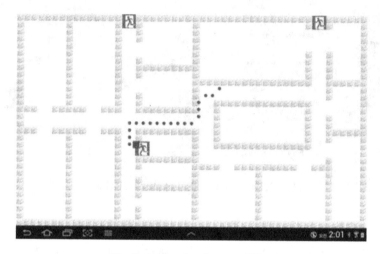

Fig. 5. Application Screen of User

Figure 5 is an application that is provided to the user.

The application provided to the users, receives the coordinate data from the server and displays it on the screen which will run as the background service. The application automatically displays the escape path on the screen without special event of user and then transmits the warning screen to the user in case of the first situation occurs.

Figure 6 shows the real test.

Fig. 6. The Actual Simulation

4   Conclusion

In the modern society, the robots are utilized in many fields such as medical, industrial, space exploration, and artificial intelligence. In addition, the robot are used instead of human, in order to explore extremely dangerous areas

In this paper, the fusion of technology, communication and control of the robot and the user's smart devices to transfer the optimal evacuation routes in the event of a fire evacuation supports a safer evacuation. If the users uses the proposed service model then it reduces the loss of life in a dangerous situation.

In future research, the proposed service model in this paper will be verified by making a prototype.

Acknowledgments. This work was supported by the National Research Foundation of Korea (NRF) grant funded by the Korea government. (MEST) (No. 2012-0003026).

References

1. National Emergency Management Agency Information System (2011), e-fire statistics, http://www.nfds.go.kr/index.jsf
2. Sim, C.B.: A Study on the design for fire safety via analysis of fire hazard in high rise buildings.a master's thesis, Graduate School of Urban Sciences University of Seoul (2004)
3. Kim, Y.M., Kim, C.K.: A Study on the Alternative Evaluation of the High-rise Building Structural System. Computational Structural Engineering Institute of Korea 23(4) (2010)
4. Lee, D.K., Do, C.W., Park, Y.S.: Study on rapid arrival at the scene of fire to protect life and property of the Daegu'citizen when fire occurs 21 (2009), http://naver.nanet.go.kr/SearchDetailView.do?cn=KINX2010048411&sysid=nhn
5. Kim, K.R., Kim, J.T.: A research of the development plan for a highly adaptable FSR(Fire Safety Robot) in the scene of the fire. Korean Institute of Fire Science & Engineering 23(3) (2010)
6. Kim, Y.H.: A study on the development of the simulation software for the fire evacuation. a master's thesis, Hanyang University (December 2002)
7. Jeong, H.S., Joo, M.J., Yoon, S.H., Jeong, J.Y., Cho, Y.Y.: A Design of the fire escape service model based on contexts. Korea Information Processing Society 18(2) (2011)
8. Park, J.J., Chun, C.H.: Development of Location Estimation and Navigation System of Mobile Robots Using USN and LEGO Mindstorms NXT. Jornal of Institute of Control, Robotics and Systems 16(3) (2010)

Co-creating Urban Development:
A Living Lab for Community Regeneration
in the Second District of Palermo

Jesse Marsh[1], Francesco Molinari[1], and Ferdinando Trapani[2]

[1] Network for Social and Territorial Innovation (NeSTI)
Via M. Bonello n. 31, 90134 Palermo, Italy
jesse@atelier.it, mail@francescomolinari.it
[2] Department of Architecture – University of Palermo
Viale delle Scienze Edificio n. 14, 90128 Palermo, Italy
ferdinando.trapani@unipa.it

Abstract. The characterisation of urban 'smartness' emerges as a product of social mobilisation, which marks the pathway towards collective technology adoption and policy innovation. This paper highlights the didactic and critical aspects that relate to the use of participatory solutions – namely the electronic Town Meeting and others, such as weblogs and the "Planning for Real" scheme – which start within the dimension of social animation and serious gaming and are only later oriented to urban planning. The Palermo pilot of the PARTERRE ICT-PSP project, based on the Territorial Living Lab approach, documents one possible transition from the stage of a free relationship with scenarios and visions, to the definition of a social demand for planning, specifically within the framework of a real experience of the citizenship life. From a policy making perspective, Participation in planning does not come to an end but continuously tends to a gradual improvement, both in the quality of the projects and in those cohesion factors, which lead to the constitution of spontaneous partnerships.

Keywords: Urban Planning, eParticipation, Living Labs.

1   Introduction

The EU funded project PARTERRE [1] recently demonstrated the potential for spatial policy design and territorial development of two e-Participatory solutions:

1. The Electronic Town Meeting - eTM, a deliberative democracy methodology and platform combining the advantages of small group discussion with electronic voting in public assemblies [2]; and
2. The DEMOS-Plan solution for the management of formal and informal consultations of citizens and stakeholders in the context of spatial planning.

Key features of the eTM are that:

- The participants are briefed in detail, several days before the event, on the topics to be dealt with – which makes the discussion informed and politically correct;

B. Murgante et al. (Eds.): ICCSA 2013, Part III, LNCS 7973, pp. 294–308, 2013.
© Springer-Verlag Berlin Heidelberg 2013

During the day, they can see their opinions reflected in the summaries of contributions that are continuously displayed on a maxi screen – which makes the discussion inclusive and improves the climate of collaboration;

- At the end of the day, the participants receive an "instant report" summarising what was discussed during the assembly and including the results of the voting sessions – which enhances their confidence in the utility of the whole exercise;
- The observed satisfaction rate is always about 90% in any survey – which contributes to restoring the reputation of the public agency that organized the event.

Key features of DEMOS-Plan are that:

- All public authorities and agencies involved in the process take benefit from documented savings in the printing and shipping of maps and accompanying documents to the other parties being consulted;
- The solution enables workflow management and can be easily integrated into the existing IT infrastructure of the agency;
- Both formal (i.e. mandatory by law) and informal (optional, e.g. pre-emptive) consultations of citizens and stakeholders can be handled by the system;
- Every participant can receive a formal response to their application or contribution by the public body in charge of the process.

The main aim of the PARTERRE project was to refine the two technical solutions introduced above within a number of pilot environments (Hamburg, Germany; Larnaca and Voroklini, Cyprus; Sicily and Tuscany, Italy; the Turku Archipelago, Finland; Ulster, UK) that have involved real citizens and businesses in discussions on real planning and programming issues, in compliance with the Territorial Living Lab approach [3].

The latter is a variant of the more popular Living Lab approach, which was born in the ICT research domain and officially endorsed in 2006 by the Finnish EU Presidency as a new, truly European model of "co-creation of innovation in public, private and civic partnership". Briefly, the term Living Lab broadly refers to a set of quantitative and qualitative methodologies and tools for the ideation, design, development and validation of innovation together with (and by) the end users within real-world environments. In these open settings, people are taken across the different roles played during a normal day and allowed to contribute on a peer basis with the developers in prototyping, evaluating and testing novel ICT solutions; thereby, innovation becomes human-driven, in contrast to technology-driven. Further to that, trial activities go on round the clock: this means that solution developers get the opportunity to gain understanding of a new product or service in its 24/7 usage dimension. Differently from User Centred Design, end users are integrated within all stages of product/service development, from ideation to design, from development to validation, from testing to evaluation.

Scientific ancestors of the Living Lab concept possibly include: the late William J. Mitchell from MIT Medialab, who is said to be the inventor of the term itself, born in an urban planning and architecture design context; Eric Von Hippel, again from MIT, with his elaboration of the 'Lead Users' paradigm; Henry Chesbrough from Berkeley, known by many as the 'father' of Open Innovation theory; and the Canadian business strategist Don Tapscott, who first introduced the notion of 'Prosumer' (i.e. Consumer + Producer) as key actor of the globalised markets in the Web 2.0 era. From an initial, purely technological (and especially ICT pervasive) field of experimentation, Living Lab applications are now increasingly migrating towards broader socio-economic, environmental and even governance related contexts, including the co-creation of open data, open government and innovation policies. In this sense, Living Labs may represent a valid instrument to support territorial development policies, by assuming three possible configurations [4]:

1.   As vertical tools for promoting user driven research, development and innovation in a given application domain (such as eHealth, eInclusion, or eParticipation);

2.   As intermediaries between citizens, governments and other stakeholders of the PPPP (Public Private People Partnership) that supervises the whole experiment;

3.   As behavioral and improvement guidelines for public administration officials, wanting to exploit the 'mixture' of technological, social and organizational innovation to valorize local intellectual capital and to increase available knowledge for development.

In this paper, we report about the way the PARTERRE project implementation has adopted and adapted the Territorial Living Lab approach to establish and manage eParticipatory trials in the respective thematic domains. We will particularly focus on one of the pilots, which took place in 2011-2012 in the City of Palermo, Sicily, culminating with the celebration of an Electronic Town Meeting (eTM) on 18th February 2012, together with a representative sample of the population of the Second City District of Brancaccio. We will analyse the role played by the Territorial Living Lab Sicily in promoting social mobilisation and by the chosen PARTERRE ICT solution in supporting and structuring people's participation to a bottom-up instantiation of the process planning. We will draw lessons that are possibly worth considering in the reflection on both the future of urban planning and of eParticipation in Europe.

2   A Big Question

The recent evolution of urban planning practice has certainly added new facets to the historically well-enforced interest of private citizens and local businesses in being individually included in the spatial planning process as far as its implications directly have impact on their property or building or development rights. Namely, it has created or strengthened a sort of "extended interest" of the local community and its organized representatives (which we sometimes call "civil society") in taking active part to major planning decisions for the defense of collective rights, such as the quality of air, soil and water, or the preservation of cultural and architectural heritage, or the possibility to run a successful business in a service rich environment.

Conventionally, the former interest has been declined by the public authorities in charge mainly as a right to participate in the *concluding* stage of planning – after some preliminary decisions have been taken on zoning and destination of land and formal consultations must be run with the affected people as required by extant legislation – while the latter has emerged mainly as a right to be informed and informally consulted during the *initial* stage of planning, when strategic visions of the future and alternative development scenarios are laid down and discussed by the elected officials in charge of taking decisions about them.

Protecting both of these interests requires the adoption of different participatory tools. In fact, the "individual right" to formal participation basically relies on transparency and open access to the information contained in the drafted plan: such as the publication of accurate, complete and comprehensible maps, datasets, and documents, which enables any citizen to gain the necessary knowledge of the planning proposals, in order to interact with the competent public authority more efficiently and effectively. Conversely, the "collective right" to informal participation in planning, which obviously needs to be exercised with the same transparency and accessibility as the other, requires a further layer of "enabling tools" to be enforced and become effective: tools that actually stimulate and organize the public debate in order to build consensus on a common perspective towards the future.

This shifts the focus from information sharing to shared visioning, by far a more complex issue, which requires new methodologies and tools, able to blur the information boundaries that normally separate urban designers/experts from ordinary citizens – the ones who "know how" from the ones that "do not" – including the latter in all stages of the planning process: from the discussion on alternative options, to the construction of alternatives and the making of selected choices.

In support of this view comes the growing perception and awareness of "vision sharing" as a precondition for acceptance and ultimately legitimization of the decisions taken according to that same vision. Following Innes and Booher [5], a list of motivations for public participation in decision-making may include:

- Making fairness and justice advance in the community;
- Ensuring legitimacy of public decisions;
- Collecting public preferences to take them into account in decisions;
- Incorporating citizens' "localised knowledge" in the decisions affecting them;
- Fulfilling the requirements for public participation that are in scope of extant legislation on planning.

Apart from law requirements, we can split the remaining purposes in two groups:

a. Reputational gains, including compliance with broader legitimization, accountability or simply visibility goals of the government agency involved (and its elected or employed officials);

b. Content related gains: in fact, it is well known that public participation regularly increases the quantity and quality of results. It helps take richer decisions that go beyond the narrow stakeholder interests and improve over the logic of "short term" and "short sightedness".

A third group of motivations have started to emerge, however, from the practical experience we report about in this paper, and which has to do with the implications of a wider community acceptance for the chances of good implementation of any given plan or programme. This should be facilitated, rather than not, by the fast and pervasive diffusion of ICT (Information and Communication Technologies) in the practice of urban planning, as much as in other contexts of government and daily life.

The big question surrounding this paper, as others in the same research stream (e.g. [6, 7]), is why the advantages of enhanced participation are not yet extensively and systematically gathered within the EU planning processes. This even despite the considerable push received from the Community policy in that regard. For instance, the URBAN I programme – based on the UK Neighborhood Initiative of the 1980's – has been for countries like Italy and France the first occasion to structurally include participatory urban design principles in their respective legal frameworks. Or the diffusion and establishment of the "acquis communautaire", particularly in, but not limited to, the New Member States joining the Union from Eastern Europe at the beginning of 2000's, has promoted a more uniform vision as it is now included in the so-called Territorial Agenda of the EU, as well as in the ESDP (European Spatial Development Perspective) and in the SEA (Strategic Environmental Assessment) concept and procedure.

In fact, the original inspiration that motivated PARTERRE has been that "spatial planning and environmental assessment are in the best position to achieve a paradigm shift in the way electronic participation and social capacity building are practised in Europe. This for at least three good reasons:

- Their legal framework is completely defined at EU level, based on a reasonable distribution of competences across Member States, Regional and local institutions, and on a sustainable combination of mandatory and optional participation procedures;
- The migration from 'offline' to 'online' participation can be supported by a sound business model, showing up the efficiency and quality advantages usually advocated by supporters of electronic democracy for other key processes of public administration;
- A multitude of successful trials exist in this domain – some of which funded by the EC under the ICT Framework Programmes, ERDF/INTERREG and/or the Preparatory Actions on eParticipation – which have demonstrated the above advantages, not only in a political sense, but also in a financial perspective". (see [1], p. 5)

3   Exploring the Potential of Living Lab Partnerships

There are three possible working hypotheses built into the PARTERRE project regarding the potential of Living Labs for spatial planning (henceforth: SP) and strategic environmental assessment (henceforth: SEA):

1.   SP and SEA are compatible with the kind of institutional partnership-based innovation that Living Labs represent and the evolution of the PARTERRE local partnerships should be evidence of this;
2.   The Living Lab approach for user-driven innovation is as an integral part of the PARTERRE driven planning model, and the dynamics of co-design processes in the pilot projects should be evidence of this;
3.   The Living Lab approach can provide significant added value to the conventional "eParticipation in planning" model and the work carried out in the pilots should be evidence of this.

These three assumptions can be supported on the basis of the PARTERRE results, as a baseline upon which a broader evaluation of impact can be performed. In this Section, we specifically look at the likely impact of Living Lab Public-Private-People Partnership (henceforth: PPPP) formation processes on SP and SEA.

In our perception, SP ans SEA are both moving in a direction that increases the complexity of the thematic issues addressed as well as towards both vertical and horizontal topic integration. Ever broader partnerships are formed both to set the agenda and to identify priorities, although this exercise is often carried out as a distinct moment from citizen focused participatory processes.

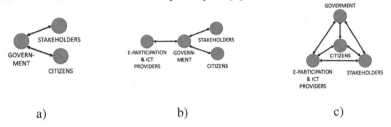

a)   b)   c)

Fig. 1. a) "Traditional" Partnership Model in SP and SEA; b) Traditional" e-Participation Partnership Model; c) Living Lab PPPP Model For SP And SEA

In turn, eParticipation generally works by introducing ICT tools and the associated methodologies for participation into this trend, generally maintaining the partnership structure, aiming to increase the ability of governments to "listen" to stakeholders in participatory processes.

As stated previously, the Living Lab approach introduces ICT into the equation but also different forms of partnership. Traditional Living Labs have ICT and the development of new products and services as their starting point, and generally follow a "triple helix" model (University, research centre, public authority), while Territorial Living Labs introduce ICT into more territorially driven partnership models with a greater role for citizens, in part as a means of sparking off innovative dynamics among the actors.

This second model is clearly more amenable to integration into SP and SEA processes, as it aligns with the trend towards an ever-greater institutional role for the citizen at the centre of the model.

The key evaluation question that arises here is thus the degree to which Living Lab partnerships emerged or evolved throughout the PARTERRE project and the link between these dynamics and developments in the SP and SEA processes. It is interesting to note that although all of the pilot sites have ultimately become part of the European Network of Living Labs (ENoLL), and two in fact did so during the PARTERRE project's lifetime, not all of the formally constituted Living Labs did follow the Territorial approach.

The partnerships for the pilot projects depended to a good degree on the baseline technology being piloted. Namely, DEMOS-Plan provides a more functional added value to the formal planning process and therefore implies the engagement of all those directly involved in its management. The eTM addresses more sectorial and strategic aspects, as well as involving a significant preparatory and outreach phase, so it is natural that it implies the engagement of a broader variety of stakeholders.

Table 1. ENoLL Living Labs in the PARTERRE Pilots

Pilot setting	PARTERRE Partner	Living Lab name	ENoLL year of entry	Partnership model
Ulster	TRAIL	TRAIL	2007	Triple helix
Turku	TUAS	Turku Archipelago	2006	Territorial
		TUAS	2009	Triple helix
Palermo	UNIPA	TLL-Sicily	2007	Territorial
Tuscany	RT	eToscana	2009	Region-driven
Hamburg	Hamburg	TuTech Innovation	2011	Business-driven
Cyprus	VCC	TLL-Kypros	2010	Territorial

Table 2. Impact of Living Lab Partnerships on SP and SEA

Pilot setting	Nature and role of partnership	Potential impact in follow-up
Ulster	Multiple inter-sectorial partnerships for strategic sectorial planning.	Stakeholders' "whetted appetites" for collaboration and participation.
Turku	Broad citizen participation as part of consolidated tradition	Openness of municipalities undergoing change to new approaches.
Palermo	Pilot aligned with bottom-up aggregation of community partners.	Emergence of a permanent partnership for the pilot area, also based on potential for realization of ideas and effective incorporation into the new City plan.
Tuscany	Institutional partnership for multi-level planning.	Institution of regional "Cabina di regia" (coordination panel).
Hamburg	Focused on extended implementation of BOP.	Incorporation into ordinary planning procedures.
Cyprus	Defined by need for new strategic plan.	Replication of eTM through permanent partnership.

A final aspect is the impact of the PPPP developed in the PARTERRE Living Labs on the SP and SEA processes and their potential for the future. While it is premature to fully assess the institutionalization processes after more or less a year after pilots' execution, it is possible to interpret the first signals coming from the partnerships.

We can thus conclude that Living Lab partnerships do have emerged from the PARTERRE project, sometimes but not necessarily based on pre-existing Living

Labs, primarily as a means for capitalisation of outcomes and governance of the future planning and implementation stages.

4   Living Lab Co-design Processes in PARTERRE

In this section we look at the likely impact of Living Lab user-driven innovation on SP and SEA through the co-design processes occurred in the PARTERRE pilots.

Generally, Living Lab co-design processes are considered in relation to an ICT based product or service being designed or developed. This aspect is addressed in PARTERRE WP3 "Platform Customisation and Integration", where for example the Living Lab co-design process has strongly influenced the development of DEMOS-Plan as a multi-lingual eParticipation platform. Here instead the issue of concern is the degree to which Living Lab co-design processes introduce innovation into the planning model itself.

Intuitively, this innovation is not necessarily the same thing as increasing citizen interaction with government, an innovation already widespread in spatial planning, but rather the Living Lab-type interaction between all of the stakeholders involved, driven by citizens. The way this unfolds depends much on the ICT solution being tested in the different PARTERRE pilots, since eTM and DEMOS-Plan have very different processes with different degrees of margin for Living Lab co-design processes.

Of course, the conventional planning model does not need to be entirely re-designed; there are many aspects that require innovation, such as a need for ownership in order to speed up implementation, while there remain many aspects where it is useful to reinforce existing methods, especially related to citizen participation. The Living Lab co-design processes are thus of interest primarily where their innovations bring added value to planning, which we can summarize as follows:

Table 3. Added Value of Living Lab for Innovation in Planning

Pilot setting	Innovation nature and role of co-design process	Living Lab added value for innovation in planning
Ulster	Co-design of agenda setting and thematic development of multiple sectorial eTM pilots.	Increased ownership, cross-sectorial coherence.
Turku	Co-design of village planning practices and procedures in municipal re-organization.	Recognition of value of village planning and tenant democracy.
Palermo	Pilot aligned with bottom-up aggregation of community partners.	Consolidation of shared value and relevance to City planning process. University designs based on eTM outcomes.
Tuscany	Institutional partnership for multi-level planning.	Coherence and consensus in planning process.
Hamburg	Focused on extended implementation of BOP rather than innovation in planning.	Relevance to stakeholder needs.
Cyprus	Defined by need for new strategic plan.	Discovery of positive value of participatory process.

From the above table, it can be seen that the specific added value of Living Lab co-design processes was quite different in the different pilot settings, much as a function of the purpose of the pilot and the "innovation need" of the planning process in question.

We can conclude that the Living Lab co-design processes in PARTERRE appear to adapt to the specific contexts and settings of eParticipation in planning, shifting the innovation potential in the direction of those aspects of the SP and SEA models where innovation is mostly required.

Now in order to assess the added value of the Living Lab approach for conventional eParticipation, we present in more detail and then discuss the results of the Palermo pilot of PARTERRE.

5   The Palermo Pilot of PARTERRE

The starting point of the experience was created by local cultural associations with a request for help to the University of Palermo, Faculty of Architecture, in relation to the problem of gaining access to the Castle of Maredolce [8, 9, 10, 11], located in the Second District of Brancaccio, which had been restored by the competent regional authorities along a period of more than twenty years, but was somehow "guarded" by the local affiliates of criminal organizations. These prevented the population from using the Castle and its surrounding park, and more importantly from giving a different external image of the Brancaccio District than that of delinquency and Mafia (right there, about twenty years ago, the Catholic Priest Don Puglisi had been killed by some anonymous murderers).

Initially, the University took some action in response to the request received, by looking upon the Castle. However, it soon became clear that the proposed issue had to be placed into a wider framework, and specifically within the broader topic of urban regeneration, to be meant as an instantiation (and maybe a turning point) of all accessibility-related problems that affect Palermo, the inland and Eastern Sicily. Here, the term "accessibility" has to be understood both in a physical sense (e.g. as lack of transportation, social services, services for enterprises, etc.) and as immaterial aspect (e.g. lack of quality education, evidence of denied identity rights, social marginality, etc.). Likewise, the actions undertaken by public institutions (such as the Regional Office for Cultural Heritage, the Department of Fine Arts, Local Police and School) to increase "accessibility" to the Second Municipal District - although sometimes very productive – had showed a lack of cohesion, both between the institutions involved and with the neighbourhood associations, parishes, and stakeholders in general.

The listening phase carried out by the eTM organising committee led to identify the following five priorities for urban and community regeneration: a) Restore the cohesion of groups; b) Rationalize the economic activities of Brancaccio; c) Create a network of ecosystem components; d) Reuse the Maredolce Castle for a new identity of the neighbourhood; e) Reconnect the Brancaccio District to overcome the current urban and rural marginal conditions.

These propositions were collected and organized so as to become the knowledge base of the eTM (Electronic Town Meeting) held in Palermo on 18th February 2012. The outcome of the eTM was the production of a programmatic document, voted by all the participants, which was formally handed over to the Municipality of Palermo in August, to accompany the General Directives of the City Council in preparation for the new urban plan, and to contribute to the participatory activities that are embedded in the SEA process, which will accompany the formation of the new General Master Plan.

In addition to cultural heritage and tourism stakeholders, the people who attended the eTM were invited from all the main economic, social and religious expressions of the District of Brancaccio: small industry, handicrafts and trade, but also representatives of the Church and the self-help communities. A further contribution to the definition of a complete and shared model of intervention was provided by a "Planning for Real" [12] experiment, which involved University students, professors and relevant members of public administration in Palermo, to produce a structured list of real projects "from the City, to the City". These have mapped the following key issues: Accessibility, Housing; Policy-Making; Public Safety; Architectural Design; Social Solidarity; Environment and Agriculture; Water and its responsible use.

The thematic focus of the PARTERRE pilot was to help societal actors in the territory of the Second Municipal District of Palermo – an area characterized by the dominant presence of Mafia – to initiate the qualification and urban regeneration process of this and other similar neighbourhoods located in the South-Eastern part of the City. Where a Living Lab already exists, like in this case, or could be created from scratch, the endowment of social capital would be reinforced by the emergence of the PPPP and through carrying out a number of concrete activities for awareness creation and elicitation of individual and collective requirements within the population involved.

Evidence collected from the Sicilian pilot largely confirms this assumption, leading to a bottom-up approach to urban re-design, supported by participatory ICT solutions (namely the eTM, but also an Internet weblog that accompanied the entire process). This gave life to a new kind of co-created planning process in a weak and peripheral urban context [13, 14, 15], involving a number of key steps:

1. Social mobilisation of local community actors;
2. Engagement of citizens and stakeholders in territorial development issues;
3. Building of partnership relations in relation to planning priorities and agendas;
4. Participatory activities to capture new potential for the spatial context;
5. New institutional arrangements to capture those change processes;
6. Associated spatial arrangements to the strategic ones;

In this framework, technology has played a triple role:

a. Promoting functional efficiency (i.e. what we expect of technology);
b. Stimulating collective creativity (by mapping local actors' expectations, enabling open and transparent debates, empowering the participants themselves)
c. Transforming human and institutional relations (by the legitimisation of new rules and processes, challenging the status of information with respect to power, and reshaping current network structures).

The problem of efficacy after participatory processes also emerges during strategic planning [16]. The coordination that was carried out in Palermo between institutional actors on the one hand and the participant NGOs, private businesses and general public on the other hand, can constitute a new and innovative framework of urban and territorial transformation [17, 18, 19, 20]. To reach this kind of multi-actor coordination it may be necessary to pursue the (very difficult) integration of policies and programmes [21] not only in economic sense, but also in spatial planning [22]. The implementation of ICT based on social innovation demands, can help processes of deliberative democracy [23, 24] grow, especially in weak and peripheral urban contexts.

6   Smartness and eParticipation

Quoting a Kevin Mattson's book on the Progressive Era in America - which ran from the end of 19th century through the 1920's - Amin and Thrift [25] propose seven principles of direct democracy in the context of planning, namely: have courage of experimenting in response to real needs by a combination of idealism and pragmatism; set out clear objectives for the purpose of regulation; enhance civic culture as well as social autonomy; implement projects of social transformation that work in a public dimension to change behaviours through training; give to participants the power of control and involve their creative energies; mix sociality with political activity; institutionalize the emergent processes; recognize local actions in the framework of institutional policies.

The principles of Amin and Thrift show some pre-conditions for the achievement of political objectives, in terms of direct democracy, which can also be seen as particularly appropriate to inform the "Smartness" of a City. In fact, a City can also be viewed as a rapid movement of thoughts and practices that technology can only partially capture and determine. In order to function properly, a Smart City does not only require the adherence, either manifest or hidden, of its key institutional actors: social bodies, public agencies and their representatives, who are normally supported in the formulation of their wills within "participatory urban governance platforms". It also needs to get the political consensus and active engagement of (a wide majority of) the population as a whole. This seems especially to be the case for spatial transformation effects, which are largely produced because of a myriad of micro-decisions that socio-economic actors take "from the inside out" of their fields of influence and expertise, using the knowledge base available here and now, and interacting with each other in many ways [26, 27]. The aim to frame and control the actions of these "spatial transformation agents" by the followers of the traditional, "decision-centred" vision of planning [28], not only collides against the evidence, but even prevents from taking benefit of the convergent actions of this multitude, who can be seen as an additional key component (besides ICT infrastructure and policy) of an extended concept of City Smartness [29, 30].

In the case of Palermo eTM, the power of social capital [31, 32] and its mobilization have proven crucial to determine the process and influence the outcomes of a successful planning experiment, which was indeed seen like a revolution in public decision making for this marginalised part of the City.

Therefore, a new sense of Smartness is appreciable within the Living Lab experience of Palermo, which also brings some implications for the knowledge exchange occurring in specific co-creative ecosystems [3, 33] including the pilots supported by the ICT-PSP programme [34]. The final outcome of this integration of the territorial Living Lab approach into spatial and strategic planning at City level will be the so-called PARTERRE service [5], candidating to take advantage of sustainable eParticipation methods and tools to strengthen and consolidate service outcomes in an evolved Smart City environment. As a comprehensive framework, the concept remains open to the future incorporation of additional eParticipation solutions, using

Living Lab co-design for their optimal take-up and to maximize innovation of local government policies and services.

The typical case of regulatory frame within the rational action in respect to the aim is constituted by the freely agreed decision; the association is the institution that is based on the decided order or, where a binding apparatus permanently penalizes the original agreement, the institution [22].

The set-up of a citizenship highlights the central role of the achievements of participation, together with its constitutive limits: positive aspects are given by the co-creativity that is implied in the relations, both advanced and emerging, between individuals and groups. Negative aspects are represented by the necessity of sectorial authorisation for the achievement or even only implementation of project paths aiming at shared objectives. The difference between institutional planning and other approaches (place based programs, strategic/integrated plans, etc.) was discussed in several occasions and by many disciplines [36, 37, 38] but the hypothesis of an open, plural and shared social interaction oriented to inform institutional planning, is preferred to strategic planning as a better way of planning.

The relationship between participation and negotiation [39] and its central role within the practices and policies for integration, particularly in local development, are much wider issues however. Participation and negotiation, actually used in the world of public decision-making [40], represent diverse universes of discourse. Mazza [41] explained that in defined ideas of society, citizenship, equality and democracy, the different tools for negotiation and participation are eminently and inherently political, whatever the intents, objectives and outcomes as well the level of awareness of involved actors.

7   Conclusions

We started this paper by asking ourselves the somehow rhetoric question of why the advantages of public participation (and particularly eParticipation) are not yet extensively and systematically gathered in the spatial and strategic planning processes at EU level. An instrumental answer can be a certain lack of awareness from the public authorities in charge. In that respect, Community policy can play an extraordinary role for the creation and enlargement of "mandatory spaces" for civic engagement in a revised governance system at Member State level in Europe. Just to make a trivial example, the "right to information" principle introduced in Article 2 of the Aarhus Directive (2003/35/EC), as well as in most national planning systems, could be easily reinforced in such a way that the use of electronic media becomes a "standard means of publicity". Eventually, this might pave the way to introducing the new principle that "electronic participation tools, where available" should integrate the technical means available to the purpose of engaging citizens or stakeholders in thematic consultations and deliberative sessions on territorial development.

An additional contribution to awareness rising has come from the PARTERRE project through its pilot experiments, which demonstrated the potential of the Living Lab approach for spatial planning and environmental assessment in three key directions:

- Living Lab partnerships allowing to improve citizen participation and stakeholder relationships, which leads leading to greater ownership of planning outcomes and thus smoother implementation.

- Living Lab co-design processes allowing to address innovation needs in planning processes where they emerge, particularly in the integration and prioritisation of stakeholder interests.
- Living Lab tools and methods being complementary to eParticipation solutions for planning, leading to an integrated PARTERRE service concept [7] covering the entire process and its iterative cycles.

With the Sicilian pilot in particular, it has been possible to use the European funding for the methodological and technical support of a spontaneous partnership between self-organising local actors and an intermediary organisation such as the University of Palermo, with the primary goal of building concrete and shared visions of the future from within the Brancaccio District population and contributing to shape the new Master Plan of the City. Associations and movements of one of the most degraded neighbourhoods of the City of Palermo gained the University support in promoting local energies to contrast the trend of decline and to fight against the Mafia by a symbolic appropriation of the main landmark of the District – the Maredolce Castle. To this purpose, a gradual construction of a public-private-people partnership (PPPP) was achieved, which leveraged the existing experience of the Territorial Living Lab (TLL) Sicily, borrowing its model of governance and plans for the future.

The Palermo pilot of the PARTERRE ICT-PSP project, based on the Territorial Living Lab approach, documents one possible transition from the stage of a free relationship with scenarios and visions, to the definition of a social demand for planning, specifically within the framework of a real experience of the citizenship life. From a policy making perspective, participation in planning does not come to an end but continuously tends to a gradual improvement, both in the quality of the projects and in those cohesion factors, which lead to the constitution of spontaneous partnerships. In this sense, the Living Lab PPPP gains a political, not only socio-technical, dimension, which has to be assessed for the sake of transferability and replication within the different legal frameworks of EU27 Member States, in its capacity of promoting and protecting the various interests and rights - of businesses, citizens and other stakeholders - on a peer basis with government officials and policy makers. These different aspects shaped the expectations of the PARTERRE partners concerning the role of Living Labs in the co-creation of urban development as well as the construction of local partnerships for the Palermo pilot.

Acknowledgements. This paper is grounded on the results of the EU-funded (CIP ICT-PSP) project PARTERRE. However, the opinions expressed here are solely of the authors and do not engage the European Commission. We would specially thank the following project partners: B. Galbraith, S. Martin, M. Mulvenna & J. Wallace (University of Ulster - UK), T. Ferm & O. Ojala (Turun Ammattikorkeakoulu - FI), J. Heaven (TuTech Innovation GmbH - DE), S. Besteher (Freie und Hansestadt Hamburg - DE), I. Romano & E. Galetto (Avventura Urbana s.r.l. - IT), M. Andreou, E. Balamou & M. Zanos (Anetel – CY), A. Theodosiou (Voroklini Community Council) and especially the project coordinator A. Marcotulli (Regione Toscana – IT).

References

1. PARTERRE DoW (Description of Work), amended version of October 2011, http://www.parterre-project.eu
2. Garramone, V., Aicardi, M. (eds.): Democrazia partecipata ed Electronic Town Meeting. Incontri ravvicinati del terzo tipo, Franco Angeli Milano (2011)
3. Marsh, J.: Living Labs and Territorial Innovation. In: Cunningham, P., Cunningham, M. (eds.) Collaboration and the Knowledge Economy: Issues, Applications, Case Studies. IOS Press, Amsterdam (2008)
4. Molinari F.: Living Labs and Pre-Commercial Public Procurement: A Marriage of Interest? In: Proceedings of the 1st EIBURS-TAIPS Conference, University of Urbino, Italy (2012)
5. Innes, J.E., Booher, D.E.: Reframing Public Participation: Strategies for the 21st Century. Planning Theory & Practice 5(4), 419–436 (2004)
6. Concilio, G., Molinari, F.: Citizen Participation in Urban Planning. Looking for the "E" Dimension in the EU National Systems and Policies. In: Proceedings of ECEG11 Conference,, Ljubljana, Slovenia, June 16-17, pp. 177–186 (2011)
7. Molinari, F.: eParticipation that works. Evidence from the Old Europe. JeDEM 4(2), 245–264 (2012)
8. Braida, S.: Il castello di Favara. Studi di restauro. Architetti di Sicilia, n. 5-6. Palermo, 21–34 (1965)
9. Montagna, C. (ed.): Maredolce. Studiare il territorio di Maredolce/Brancaccio e valorizzarlo come distretto culturale e turistico, Unicoop Firenze (2011)
10. Prescia, R., Trapani, F.: Il posto di Maredolce. Un paradiso a Brancaccio. Strategie per la riqualificazione dell'area industriale di Palermo. In: Niglio, O. (ed.) Paisaje Cultural Urbano e Identitad Territorial, Aracne Roma, pp. 377–393 (2011)
11. Scognamiglio, M., Corselli D'Ondes, G.: Il castello di Maredolce. In: I Georgofili, Atti dell'Accademia dei Georgofili, serie VIII – vol.I°, tomo II°, Firenze, 609-616 (2005)
12. Gibson, T.: Planning for Real: The Approach of the Neighbourhood Initiative Foundation in the UK, RRA Notes 11, 29–30 (1991), http://pubs.iied.org/pdfs/G01376.pdf
13. De Spuches, G.: Brancaccio come terreno d'azione. Sguardi geografici su un quartiere delle periferie di Palermo. In: Archivio di Studi Urbani e Regionali, n. 90, Franco Angeli Milano, 183–189 (2007)
14. Picone, M.: Inquadramento geografico e urbanistico. In: Le città nella città. Politiche urbane, disagio e devianza minorile alla periferia di Palermo. Rapporto di Ricerca Programma Operativo Nazionale Sicurezza per lo sviluppo del Mezzogiorno d'Italia, Palermo, 10–21 (2008)
15. Notari, G. (ed.): Marginalità narrate. Palermo (2007)
16. Armondi, S., Fedeli, V., Pasqui, G.: La valutazione dei piani strategici delle città italiane: contesti, intenzioni, esiti. Rapporto preliminare. Milano (2009)
17. Fera, G.: Comunità, urbanistica, partecipazione. Materiali per una pianificazione strategica comunitaria. Franco Angeli Milano (2008)
18. Ferraresi, G.: La costruzione sociale del piano. Urbanistica 103, 105–112 (1994)
19. Forester, J.: Planning in the Face of Power. University of California Press, Berkeley & Los Angeles (1989)
20. Forester, J.: The Deliberative Practitioner: Encouraging Participatory Planning Processes. MIT Press, Cambridge MA & London (1999)

21. Bianchi, T., Casavola, P.: I progetti integrati territoriali del QCS obiettivo 1 2000-2006, teorie, fatti e riflessioni sulla policy per lo sviluppo locale. Materiali Uval 17, Ministero dello Sviluppo economico, Dipartimento per le Politiche di Sviluppo, Roma (2008)
22. Trapani, F.: Verso la pianificazione territoriale integrata. Il governo del territorio a confronto delle politiche di sviluppo locale. Franco Angeli Milano (2009)
23. Elster, J.: Deliberative Democracy. Cambridge University Press, Cambridge (1998)
24. Paba, G., Perrone, C.: Cittadinanza attiva. Il coinvolgimento degli abitanti nella costruzione della città. Alinea Firenze (2006)
25. Amin, A., Thrift, N.: Cities. Remaigining the Urban. Polity Press, Cambridge (2001)
26. Crosta, P.: Politiche: quale conoscenza per l'azione territoriale. Franco Angeli Milano (1998)
27. Lindblom, C.E., Cohen, D.K.: Usable Knowledge: Social Science and Social Problem Solving. CT Yale University Press, New Haven (1979)
28. Faludi, A.: A Decision–centred View of Environmental Planning. Pergamon Press, Oxford (1987)
29. Concilio, G., De Bonis, L., Marsh, J., Trapani, F.: Towards a deep integration of socio-economic action and spatial planning. In: IFKAD-KCWS 2012: Knowledge, Innovation and Sustainability: Integrating Micro and Macro Perspectives, Matera (Italy), June 13-15 (2012)
30. Concilio, G., De Bonis, L., Marsh, J., Trapani, F.: Urban Smartness: Perspectives Arising in the Periphèria Project. Journal of the Knowledge Economy 4(2), 205–216 (2013)
31. Coleman, J.C.: Social capital in the creation of human capital. American Journal of Sociology 94, 95–120 (1988)
32. Putnam, R.: Bowling Alone: America's Declining Social Capital. The Journal of Democracy 6(1), 65–78 (1995)
33. Marsh, J.: The Territorial Dimension of Innovation and the MedLab Project. In: MEDLAB in Sicily. An opportunity for social and territorial innovation, pp. 39–57. Gulotta Palermo (2011)
34. European Commission: Living Labs for user-driven open innovation. An Overview of the Living Labs methodology, activities and achievements, Luxembourg (2008).
35. Habermas, J.: Theorie des kommunikativen Handelns, Suhrkamp, Frankfurt am Main (1981), Eng. Vers.: The theory of communicative action. Beacon Press Boston (1984)
36. Friedmann, J.: Hong Kong, Vancouver and Beyond: Strategic Spatial Planning and the Longer Range. In: Friedmann, J., Bryson, J., Hyslop, J., Balducci, A., Wiewel, W., Albrechts, L., Healey, P. (eds.) Strategic Spatial Planning and the Longer Range, Planning Theory & Practice, vol. 5(1), pp. 50–56. Routledge, London (2004)
37. Healey, P.: Urban Complexity and Spatial Strategies. Routledge, London and New York (2007)
38. Hillier, J.: Plan(e) Speaking: a Multiplanar Theory of Spatial Planning. Planning Theory 7(1), 24–50 (2008)
39. Lo Piccolo, F.: Consultazione, concertazione, partecipazione: i gradini mancanti. In: F. Trapani (a cura di) Urbacost. Un Progetto Pilota Per la Sicilia Centrale. Urbanizzazione Costiera, Centri Storici e Arene Decisionali: Ipotesi a Confronto. Franco Angeli Milano, 247–256 (2006)
40. OECD: Engaged Citizens in Policy-making: Information, consultation and Public Participation; (2001), http://www.oecd.org/dataoecd/24/34/2384040.pdf
41. Mazza, L.: Distribuzione e giustificazione nei processi di pianificazione. In: Moroni, S. (ed.) Territorio e giustizia Distributiva, Franco Angeli Milano, pp. 47–54 (1994)

Semantic Interoperability
of German and European Land-Use Information

Hartmut Müller and Falk Würriehausen

FH Mainz, University of Applied Sciences,
Lucy-Hillebrand-Str. 2, 55128 Mainz, Germany
{mueller,wuerriehausen}@geoinform.fh-mainz.de,
www.i3mainz.fh-mainz.de

Abstract. An operational spatial information infrastructure needs feasible spatial information exchange between different stakeholders. When building spatial information infrastructures at the international level interoperability between the national level and supra-national level data becomes particularly important. The paper presents a study analyzing spatial information interoperability in the field of land-use information. In particular, the case study addresses information exchange between German and European land-use planning information. Spatial Planning in Germany is regulated by the country-specific Federal Building Code.

A spatial application schema XPlanGML was developed to serve as a standard for information exchange of spatial planning documents in the national e-government processes. At the European level the ongoing European INSPIRE initiative seeks to establish a framework to enable interoperable information exchange in many themes, one of which is the theme of land-use. The paper studies interoperability aspects between the recently released German standard as defined in XPlanGML and the current status of corresponding INSPIRE data specifications in detail.

Keywords: Spatial Information Management, Interoperability, Land-use Planning, Standardization, E-government, Semantics.

1   Introduction

Interoperability is a central concept in dealing with spatial data. Existing spatial data are collected, analyzed and managed in various distributed systems. It obviously makes much more sense to share existing records rather than to acquire new data for each application separately. The need to use heterogeneous and distributed data becomes evident particularly in current developments of Spatial Data Infrastructures (SDI) and already existing geo-portals on the Internet. From the sharing of heterogeneous data in the context of cross-border SDI, new value creating insights are gained. Concepts such as metadata description of spatial data for geo-services according to ISO 19115 and ISO 19119 [7] offer users a feasible tool to describe the contents of administrative records both in technically and administratively correct way.

B. Murgante et al. (Eds.): ICCSA 2013, Part III, LNCS 7973, pp. 309–323, 2013.
© Springer-Verlag Berlin Heidelberg 2013

The provinces and municipalities in Germany are required to implement the legal framework and rules of the European INSPIRE directive which seeks to establish an infrastructure for spatial information in Europe [1]. A considerable share of land-use information is maintained at the level of local government. Annex III of the INSPIRE Directive addresses the theme Land-use. Local government, therefore, is responsible at large to provide local level spatial data sets and services as well as metadata to meet the requirements of Annex III of the INSPIRE Directive.

At present, however, there is hardly any realization of a comprehensive management and deployment system available to provide capabilities of interoperable spatial data handling within a local administrative environment. Small-scale administrative structures resulted in many specific features and, consequently, in heterogeneous databases. This heterogeneity substantially hinders inter-municipal cooperation at the operational level as well as information exchange between the different levels of public administration both at the national German and at the European scale. Particularly needed are agreed rules for the provision of Web Map Services (WMS), as well as for download services which are able to process comprehensive spatial data based on different, even municipal, spatial data themes. Integrated data concepts and data models can help to guarantee for the smooth cooperation even in a heterogeneous environment of organization units. In depth analysis of data structures, therefore, is imperative.

2   E-Government in Germany

2.1   E-Government Framework

The economic success of a country intertwines with the efficiency of its public administration. E-Government has the potential to considerably improve the efficiency of administration processes. At present time, horizontal and vertical integration of administration processes where different administration agencies are involved is difficult in Germany because many heterogeneous IT systems are driven at the German Federation level, at the German Federation States level comprising 16 states, at the County level comprising more than 300 counties and, finally, at the Municipality level comprising more than 13.000 municipalities. Actually, in Germany several initiatives are in progress to improve the integrated electronic support of workflows to support the German E-Government requirements. [8]

Figure 1 shows the components of the E-Government framework as defined by the German national government. Horizontal and vertical integration of administration processes where different administration agencies are involved is difficult at the moment because many heterogeneous IT systems are driven at the local and federal state level. E-Government 2.0 has the potential to considerably improve the efficiency of administration processes. In its program E-Government 2.0 the German Federal Government formulates the goal 'to ensure that public administrations are fully and comprehensively accessible by electronic means and that businesses and administrations can cooperate seamlessly using electronic means' [3]. In the context of this paper it is of particular interest that the National E-Government Strategy as decided by IT Planning Council [8] explicitly includes the commitment to adapt international standards by stating.

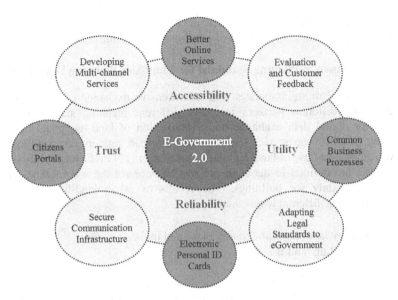

Fig. 1. Components of German E-Government 2.0, adapted from [3]

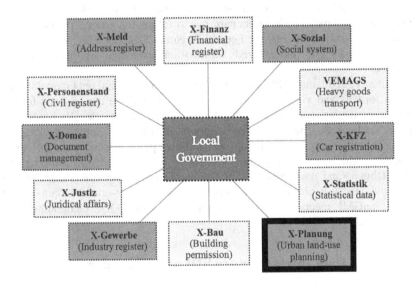

Fig. 2. Governmental standards of X-Type available in XML format in Germany, adapted [12]

Goal 19: International standards, especially for interoperability, are applied, and Germany plays an active role in the EU and internationally in defining these standards.

Up to now 20 standards of X-Type, which means that they realize the internationally adopted XML format were developed. One of the standards shown in Figure 2 is X-Planung. X-Planung specifically addresses the theme of Regional and Urban Land-use planning and, therefore, is of particular importance for the discussion of interoperability aspects in the context of Land-use information exchange.

2.2   The German X-Planung Standard for Land-Use Planning Documents

Urban land-use planning associates with an extensive exchange of plans and maps between many different partners in many different planning steps. Missing IT standards for digital data exchange and visualization of land-use plans hinder the installation of electronic services, which otherwise can support very efficiently the approval, change and use of land-use plans via Internet.[11] Procedures of urban land-use planning are defined in detail to prepare and control the use of land within a municipality, mainly for buildings. The results of urban land-use planning are documented at two different levels, namely in

- the preparatory land-use plan (zoning plan) and in
- the legally binding land-use plan.

Both kinds of German land-use plans are often prepared and provided by the local government in forms, which are not suited for further automatic processing of the contents of textual and cartographic documents. Local planning activities in Germany, however, are closely linked to the higher-level planning frameworks consisting of the regional planning level which, on its part, has to follow the general planning rules as given at the Federation level by the Federal Regional Planning Act. [4] This situation urgently demands for standardized information exchange of plans and other documents.

The initiative X-Planung [20] was taken to develop a national data model including exchange formats and visualization standards which shall be the IT basis for future services to particularly enable access to the preparatory and the legally binding land-use plans via Internet. The work to be done with regard to semantic and cartographic modeling of the features occurring in land-use plans bases upon the existing regulations, which essentially are recorded in the Federal Building Code (Baugesetzbuch) and the regulation for the cartographic symbols (PlanzV90) to be used in the land-use plans [2]. The corresponding object-oriented data exchange format XPlanGML meets the Open Geospatial Consortium OGC Geography Markup Language GML Standard [14].

3   Spatial Data Infrastructure SDI and Interoperability Levels

3.1   Hierarchy Levels of Spatial Data Infrastructure

Seeing Land-use planning from a European perspective the following sections highlight the present situation concerning SDI implementation in Germany by describing a specific part of one local SDI as a subset of the overall German National SDI. Over time the following classification of SDI hierarchy levels achieved acceptance, see Rajabifard et al [15] for instance:

- Global Spatial Data Infrastructure (GSDI)
- Regional Spatial Data Infrastructure (RSDI)
- National Spatial Data Infrastructure (NSDI)
- State or Provincial Spatial Data Infrastructure (SSDI)
- Local Spatial Data Infrastructure (LSDI) and
- Corporate Spatial Data Infrastructure (CSDI).

The lead project to construct an operational Regional Spatial Data Infrastructure in the European region (RSDI) is the EU INSPIRE initiative which progresses continuously. At the national level many efforts of the EU Member States take place to build up their own NSDI within the INSPIRE framework, according to the specific and specified needs of their respective countries. The SDI stakeholders in the Federal Republic of Germany operate under the umbrella name GDI-DE, which covers the SDI of the federation as well as the SDIs of all 16 Federal States (Fig. 3). Reflecting the structure of the public administration in Germany many SDIs already are or will have to be implemented at the lower administration levels. [17]

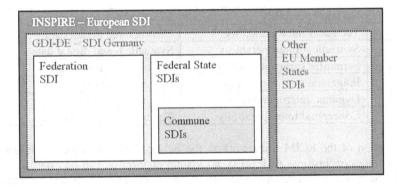

Fig. 3. Spatial Data Infrastructure SDI Germany within the European framework. Source: adapted from [16].

Particularly at the lowest hierarchy level of local SDIs in Germany a very heterogeneous structure has to be stated (see the discussion in section 2). At the local SDI level a wide range of well-functioning operational SDIs are available in some large cities but, at the same time, a huge number of thousands of very small municipalities with few if any resources to build up an SDI can be found.

3.2   Levels of Conceptual Interoperability

Interoperability can be described in accordance with Kubicek et al [10] from three main perspectives: the technical, the semantic and the organizational interoperability. Referencing these tree aspects of interoperability, Wang et al [19] present a more detailed model called Level of Conceptual Interoperability Model (LCIM) which they originally design as a framework for conceptual modeling.

LCIM can be used as a descriptive as well as a prescriptive model. In that way the model can be used to describe the interoperability of systems not only in the originally intended modeling and simulation context, but also as a general purpose framework to describe the interoperability of machine systems in general or, more specifically, of computer systems. Machines gain understanding, if they have a consistent system description at their disposal, which includes meta-data of data, processes and constraints. Entities and attributes, the behavior of entities and change of attributes, constraints concerning the values of the attributes and the behavior of the system have to be addressed in the context of spatial information processing alike. That is why it seems appropriate to apply the specified seven LCIM levels from 'no interoperability' to 'conceptual interoperability' to the tasks of spatial information exchange between different systems (see Table 1).

Table 1. Levels of Conceptual Interoperability (LCIM) [19] and Level based Spatial Data and Implementation Requirements

Level	Type of Interoperability	Spatial Data Requirements
0	No interoperability	Unstructured analogue data
1	Technical Interoperability	Unstructured digital data
2	Syntactical Interoperability	Syntactic based digital data
3	**Semantic Interoperability**	**Semantic model based**
4	Pragmatic Interoperability	Implementing rules
5	Dynamic Interoperability	Monitoring and Reporting
6	Conceptual Interoperability	Generic Conceptual model

Adaptation of the LCIM framework to the field of land-use information leads to the following conclusions. At level 0, no system connection can be established. At level 1, the technical level, physical connectivity is established allowing digital but unstructured data to be exchanged. At level two, the syntactical level, data can be exchanged in standardized formats, i.e., the same file formats, like pdf, xls, shp, dxf, are supported. At level 3, the semantic level, not only data but also its context, i.e. semantic information, can be exchanged. The unambiguous meaning of data is defined by a common semantic model. At level 4 and 5, the pragmatic/dynamical level, information and its use and applicability can be exchanged by different members. Applicability of information in the context of land-use information means binding implementation rules and monitoring and reporting over time. At level 6, the generic conceptual level, a common view of the world is established. [18] This level comprises the implemented knowledge and the interrelations between different elements, themes and stakeholders.

The highest interoperability level could be reached if all member states in Europe would support a Generic Conceptual Model. This may be seen as a long-term goal of interoperability. In the following sections we will analyze the current situation for the theme land-use information within the context of INSPIRE and X-PLANUNG. Without any binding concepts spatial data infrastructures generally cannot be realized. Activities to establish the corresponding rules encompass experiments that demonstrate searchability, usability, reusability, composability and interoperability of digital content within the context of the existing legal framework. [13] Three defined terms are to be considered in this context:

- **Integratability** addresses the realm of physical/ technical connections between systems, which include hardware and firmware, protocols, etc.
- **Interoperability** addresses the software- and implementation details of interoperations, including exchange of data elements based on a common data interpretation, etc.
- **Composability** addresses the alignment of issues at the modeling level. The underlying models consist of feasible abstractions of reality which are used for the conceptualization to be implemented by the resulting systems.

In the INSPIRE context several components of a generic conceptual model of interoperability and composability can be found in the INSPIRE directive itself and in the INSPIRE data specifications (Fig.4, see also [9]). In every day practice, however, the highest level of interoperability which can be achieved regularly is the semantic level resulting in a model-based database. For that reason we will concentrate in the following sections on the discussion of semantic interoperability to be realized between the local administration level and the European administration level. Our considerations specifically will address the realm of land-use information.

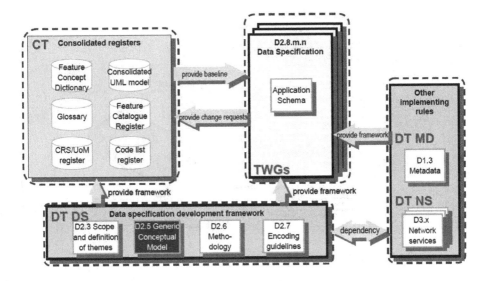

Fig. 4. The Generic Conceptual Model (D2.5) as part of the interoperability framework in the European SDI [6]

As a result of a criteria-based investigations in 23 local administration offices in the German State Rhineland-Palatinate we identified a number of spatial data requirements in the light of the INSPIRE requirements. Table 1 shows the different levels of identified spatial data requirements in connection with the seven levels of interoperability as defined in LCIM. Many spatial data in the administration offices were found to be available only at level 1 or even at level 0, some at level 2. Progress at levels 4 to 6 of conceptual interoperability was identified to be subject of further conceptual and practical research. Level 3, the semantic interoperability, was estimated to be achievable in the foreseeable future. The following section will give further details on this issue.

4   The Case Study of Interoperability

The Open Geospatial Consortium OGC continuously has developed a comprehensive framework of spatial standards which builds part of the European INSPIRE specification base. In Germany the e-government project X-PLANUNG [20] was initiated which resulted in the OGC compliant application schema XPlanGML of the Geography Markup Language (GML) standard [14].

In a pilot study we created a local development plan 'Brühl' for a small municipality in Germany called 'Gau-Algesheim' in compliance with the XPlanGML standard (Fig.5). Based on the planning standard 'XPlanGML' it will be shown, that by using GML-based models which are integrated into a spatial data infrastructure, the required semantic interoperability from local SDI to INSPIRE can be achieved.

Fig. 5. Local development plan 'Brühl' in German XPlanGML standard

The ultimate goal of the pilot study is to show if and how the XPlanGML schema is compatible with the current INSPIRE Data specifications. At the semantic level the correspondence between the two semantic data models XPlanGML and INSPIRE PLU has to be identified as precisely as possible for all elements to be included in the transformation rules.

4.1   Conceptual Transformation Process for Planned Land-Use Values

Planned Land-use is regulated by spatial planning documents elaborated at various levels of administration. The documentation of Land-use regulation for a geographical area is composed of an overall strategic orientation, a textual regulation and a cartographic representation. Finally adopted spatial planning documents result from a well-defined spatial planning process, with a number of stakeholders being involved. The INSPIRE Land Use Data Specification is an example on how the exact spatial dimension of all elements a spatial plan may consist of can be defined.

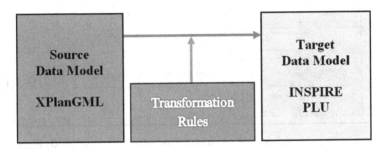

Fig. 6. Conceptual Transformation Process between a defined Source Data Model (XPlanGML) and a Target Data Model (INSPIRE PLU Planned Land Use)

To achieve interoperability between the INSPIRE Land Use Data Specification on the one hand and the implementation of existing German Land-Use Data Models on the other hand adequate transformation rules have to be formulated (Fig. 6). The goal is to show if and how the XPlanGML schema is compatible with the current INSPIRE Data specifications.

4.2   Mapping and Transformation Rules XPlanGML to INSPIRE Land Use Classification

A data model mapping step is assumed as a pre-requisite towards integration and is dealt with as a separate problem. The needed mapping and transformation rules are those between any local source data model and the common target data model, like INSPIRE. Analyzing of all feature types of the local level source data model and the target data model is necessary to fulfill the requirements of INSPIRE.

In Figure 7 a simplified presentation of X-Planung and INSPIRE data types is given. From the Figure it becomes evident that the data models are different at 'featureType' level. A simple example shall illustrate the consequences. The German XPlanGML schema allows for a spatial plan (<<featureType>>XP_Plan) comprising several areas (<<featureType>>XP_Bereich) whereas in the European INSPIRE specification a spatial plan is a single feature (<<featureType>>SpatialPlan).

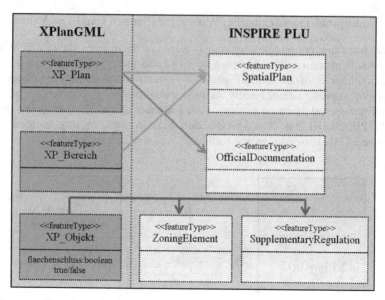

Fig. 7. One-way Transformation Concept from XPlanGML (left) to INSPIRE Planned Land-use Feature Types (right)

To transfer data from the German standard to the European standard the relation in the German data set must be dissolved. The capability of the German standard to manage spatial plans by updating only parts of it while maintaining the relation to the complete 'XP_Plan' is often used. As a consequence data transfer of one German plan will generate a more or less huge number of European Spatial Plan features, whilst the link to the original plan even might get lost. Another issue of concern in this context is the geometric property of spatial features. In the German standard features of <<featureType>>XP_Objekt obtain their geometric feature type from the attribute value flaechenschluss true/false. If the attribute value of a feature is 'true' this feature is a closed polygon. If the attribute value is 'false', the considered feature may be a point or line feature, instead. The transformation rules have to handle such cases. There are other differences between both models which are not critical. To give an example for that, in INSPIRE all official documents of a spatial plan will be recorded in a feature (<<featureType>>OfficialDocumentation) rather than in X-Planung XPlanGML where such information is recorded as an attribute of a plan (<<featureType>>XP_Plan). With formulating the corresponding transformation rules, like shown in Table 2, can easily solve this problem.

4.3   Interoperability Problems and Solutions

Our first attempts to create a set of transformation rules in order to create a framework of one-way transformation from the German X-PLANUNG XPlanGML to the European INSPIRE were based on versions XPlanGML 3.0 and INSPIRE data specifications 2.0. In this initial phase several problems occurred. Certain types of areas being part of plans in compliance with the German standard could not be transferred into the European model; transfer of attribute values was not possible in a number of cases due to semantic inconsistencies which could not be managed within the given standards, and so on. As a consequence of such shortcomings the German Working group which links the German and the European level for the theme Planned Land-use insisted on an update of the German and the European standards alike. As a result of these activities updated versions XPlanGML 4.0 and INSPIRE data specifications 3.0 were generated.

Again processing the already mentioned legally binding land-use plan 'Brühl' of Gau-Algesheim by using the updated standards versions we succeeded in establishing the one-way transformation from German land-use information to the European level. The whole framework of transformation rules results in a large table consisting of all XPlanGML objects with attribute values which correspond to INSPIRE PLU elements.

Table 2. Extract of Transformation Rules for a 'Regulated Water Management Area' and an 'Area of Agricultural Use'; Overlay Object in XPlanGML and INSPIRE PLU [6]

XPlanGML		INSPIRE PLU
BP_Wasserwirtschafts Flaeche	**flächen schluss**	**Supplementary Regulation Value**
Hochwasser Rueckhaltebecken	false	2_RiskExposure
Ueberschwemmungsgebiet	**false**	**2_1_FloodRisk**
Versickerungsflaeche	false	2_RiskExposure
Entwaesserungsgraben	false	2_RiskExposure
Sonstiges	false	2_RiskExposure
BP_Landwirtschaft		**Zoning Element Value**
LandwirtschaftAllgemein	**true**	**1_1_agriculture**
Ackerbau	true	1_1_1_CommercialAgricultureProduction
WiesenWeidewirtschaft	true	1_1_1_CommercialAgricultureProduction
GartenbaulicheErzeugung	true	1_1_1_CommercialAgricultureProduction
Obstbau	true	1_1_1_CommercialAgricultureProduction
Weinbau	true	1_1_1_CommercialAgricultureProduction
Imkerei	true	1_1_1_CommercialAgricultureProduction
Binnenfischerei	true	1_4_2_ProfessionalFishing

Table 2 shows an extract of transformation rules for a 'Regulated Water Management Area' and an 'Area of Agricultural Use', with attributes of the land-use plan 'Brühl' in bold. As can be seen from the list, the German Land-use classification schema shows a more detailed resolution of planned land-use than does the INSPIRE schema. Four different German land-use classes will be transferred to the one Supplementary Regulation Value '2_RiskExposure'. In future, INSPIRE will hold so-called national code lists which will give the opportunity to the national level administrations to map their national land-use classifications schemas to INSPIRE in a more sophisticated way than what they can do today. The code lists will be part of <<featureType>>SupplementaryRegulation. In that way mandatory and optional INSPIRE Planned Land-use elements will be kept separately in a clear way.

Another issue of concern is the cartographic visualization of spatial plans. For INSPIRE visualization purposes, simple rules for default portrayal are given by specifying a single color value attached to each class of the 'Hierarchical INSPIRE Land Use Classification System' HILUCS. The Spatial plan extent is drawn by a black line of 2 pixel size. In contrary to the simple INSPIRE cartographic presentation model, the German planning law sets a detailed regulation framework called 'PlanzV90' which defines symbolisms, signatures and surface representations of legal plans in a very sophisticated way. This issue will have to be subject of future work.

In this study we restrict our considerations to the INSPIRE theme Planned Land-use solely. In practice single national spatial data sets often provide information not only for one but also for other INSPIRE themes. Reversely requirements at the European level for one theme may request national data not only from one but from several national data sets. More work will have to be done to develop feasible solutions for these problems.

4.4   Creating INSPIRE-Compliant Metadata Sets

Metadata defined according to ISO 19115 and ISO 19119 descriptions of spatial data for geo-services [7] give users technical and administrative information on the contents of administrative and other records. Metadata can be reported for individual spatial objects (spatial object-level metadata), for complete datasets or for dataset series (dataset-level metadata). In a standards based environment spatial object-level metadata are fully described in the application schema.

If data quality elements are used at spatial object level, the documentation shall refer to the appropriate definition. For some dataset-level metadata elements, in particular on data quality and maintenance, a more specific scope can be specified. The purpose of the INSPIRE Metadata Validator is to test the compliancy of INSPIRE metadata with the Metadata Regulation.

The validator accepts metadata which follow the Metadata Technical Guidance encoded in EN ISO 19139 Schema and generates a report on its compliancy. As part of the validation process the validator will check whether the document is well-formed. In our case study which addresses spatial planning in the German state Rhineland-Palatinate a concept was developed to test Web Map Service WMS metadata of spatial development plans against the requirements of INSPIRE. Figure 8 shows an example in which way the INSPIRE Metadata Validator [5] checks metadata for compliancy with the INSPIRE requirements. In the shown case the test was passed successfully.

Fig. 8. INSPIRE GeoPortal – Example Plan Metadata validation response [5]

5   Conclusions

Spatial planning has taken place in European countries for a long time. The national spatial planning systems are reflected in corresponding regulations and, consequently, databases. At the European level the INSPIRE directive sets the framework to establish a European wide Planned Land-use schema. Unavoidably such a situation will generate problems of interoperability of the different systems. Common standards are needed to ensure smooth data exchange within a spatial data infrastructure at all levels of public administration. Standards guarantee for sustainable data exchange between different administration bodies. Technical standards are requested to implement the semantic data requirements of all administration levels. Large scale land-use planning essentially takes place at the level of local administration.

The case study focuses on the development of a strategy for national local authorities in Germany how to build up local spatial data models to meet not only national requirements but also the specific requirements as given by INSPIRE. As a result of the efforts many administration units now create spatial datasets of their own public information by using the German standard 'XPlanGML'. In that way they are able to meet the future INSPIRE requirements without having to spend additional resources.

The municipalities in Germany takes practical benefit from that capability by avoiding to re-enter the plan for INSPIRE reporting tasks rather than to transfer the output data set in XPlanGML automatically to the INSPIRE classification. It can be stated that the technical and semantic provision of digital development plans with XPlanGML also supports fulfilling the legal requirements of INSPIRE in Germany.

Land-use covers only a small piece of European spatial information as defined by INSPIRE. To give some examples, in the context of Spatial Information Management the fields of spatial planning, of land registry and of real estate cadastre play a major role and have to be interlinked to INSPIRE in practice. Currently in Germany several initiatives seek to support the integration of spatial information management including spatial data processing in the named fields. Many local IT standards were developed and meanwhile were adopted by many institutions. The adoption of existing local IT standards will continue to be a driving force for the need to establish semantic interoperability at and between different levels of public and private administration.

References

1. European Union: Directive 2007/2/EC of the European Parliament and of the Council of 14 March 2007 establishing an Infrastructure for Spatial Information in the European Community (INSPIRE) (2007), http://inspire.jrc.ec.europa.eu/directive/1_10820070425en00010014.pdf
2. Federal Building Code (Baugesetzbuch, BauGB), Online version (2012), http://www.iuscomp.org/gla/statutes/BauGB.htm (accessed November 8, 2012)
3. Federal Ministry of Interior, eGovernment 2.0, The Programme of the Federal Government (2007), http://www.verwaltung-innovativ.de/cln_349/nn_684674/SharedDocs/Pressemitteilungen/1125281__english__version__egovernment__2__0,templateId=raw,property=publicationFile.pdf/1125281_english_version_egovernment_2_0.pdf
4. Federal Ministry of Transport, Building and Urban Affairs, http://www.bmvbs.de/en
5. INSPIRE Metadata Validator: http://inspire-geoportal.ec.europa.eu/validator/
6. INSPIRE Thematic Working Group Land Use: D2.8.III.4 Data Specification on Land Use – Draft Guidelines, (July 4, 2012), http://inspire.jrc.ec.europa.eu/documents/Data_Specifications/INSPIRE_DataSpecification_LU_v3.0rc2.pdf
7. International Organization for Standardization: ISO 19115:2003 Geographic information – Metadata (2007), http://www.iso.org/iso/catalogue_detail.htm?csnumber=26020
8. IT Planning Council: National E-Government Strategy, IT Planning Council decision of 24 September (2010), http://www.it-planungsrat.de/SharedDocs/Downloads/DE/Strategie/National_E-Government_Strategy.pdf?__blob=publicationFile

9. Joint Research Centre–European Commission, IES, JRC Reference Reports, Tóth, K, et al: A Conceptual Model for Developing Interoperability Specifications in Spatial Data Infrastructures (2012), http://inspire.jrc.ec.europa.eu/documents/Data_Specifications/IES_Spatial_Data_Infrastructures_(online).pdf

10. Kubicek, H., Cimander, R., Scholl, H.J.: Organizational Interoperability in E-Government, Lessons from 77 European Good-Practice Cases. Springer, Heidelberg (2011)

11. Müller, H., Siebold, M.: Land use control and property registration in Germany–procedures, interrelationships, IT systems. In: Proceedings of the XXX FIG General Assembly and Working Week, Hong Kong SAR, May 13-17 (2007), http://www.fig.net/pub/fig2007/papers/ts_7d/ts07d_01_mueller_siebold_1248.pdf

12. Network eVerwaltung.net (2010), http://www.everwaltung.net/

13. Novakouski, M., Lewis, G.: Interoperability in the e-Government Context. Technical note CMU/SEI-2011-TN-014, Carnegie Mellon University, 35pp. (2012), http://www.sei.cmu.edu/reports/11tn014.pdf

14. OGC – Open Geospatial Consortium, Inc.: GML - the Geography Markup Language (2012), http://www.opengeospatial.org/standards/gml

15. Rajabifard, A., Chan, T.O., Williamson, I.P.: The Natureof Regional Spatial Data Infrastructures. In: Proceedings of AURISA 1999, Blue Mountains, Australia, November 22–26 (1999) AURISA 99:CD-ROM

16. Schilcher, M., Fichtinger, A., Jaenicke, K., Kraut, V., Stahl, J., Straub, F.: INSPIRE, Fundamentals, Examples, Test Results. (2009), http://www.cagi.cz/files/INSPIRE_Broschuere_V4_en_final_web_231109162434.pdf

17. Seifert, M.: AAA-The Contribution of the AdV in an Increasing European Spatial Data Infrastructure - the German Way. In: Proceedings of the XXIII FIG Congress, Munich, Germany, October 8-13 (2006), http://www.fig.net/pub/fig2006/papers/ts59/ts59_03_seifert_0449.pdf

18. Tolk, A., Muguira, J.: The Levels of Conceptual Interoperability Model. In: Proceedings of the 2003 Fall Simulation Interoperability Workshop, September 14-19. Simulation Interoperability Standards Organization, Orlando (2003)

19. Wang, W.G., Tolk, A., Wang, W.P.: The Levels of Conceptual Interoperability Model: Applying Systems Engineering Principles to M&S. Spring Simulation Multiconference, SpringSim 2009, San Diego, CA, USA. (2009), http://arxiv.org/ftp/arxiv/papers/0908/0908.0191.pdf

20. X-Planung: Forschungszentrum Karlsruhe Institute for Applied Computer Science (2012), http://www.iai.fzk.de/www-extern/index.php?id=679

The Representation for All Model:
An Agent-Based Collaborative Method
for More Meaningful Citizen Participation
in Urban Planning

Maria-Lluïsa Marsal-Llacuna[1] and Josep-Lluís de la Rosa-Esteva[2]

[1] University of Girona, Department of Architecture and Urban Planning
Campus Montilivi, Politècnic Building III, Room 121, 17003, Girona
luisa.marsal@udg.edu
[2] University of Girona, TECNIO Centre EASY,
Campus Montilivi, P-IV, Room 018 17003, Girona
joseplluis.delarosa@udg.edu

Abstract. Our Model is designed to greatly increase public participation in urban planning and make it more citizen-friendly. We use an agent technology consisting of a pair of opinion-miner recommender agents which, through mining of the opinions of citizens, make recommendations to planners on the design of the master plan. The advantages of using recommender agent technology in our DSS Model are that it accelerates acceptance of planning proposals and creates more participatory urban planning. A particularly innovative feature of our Model is that public participation occurs both before and during the development of the master plan, and in a citizen-friendly way. With our Model, planners come up with citizen-sensitive proposals and are able to more accurately predict the reaction of citizens to them. The case of the redesign of the Diagonal Avenue in Barcelona is provided as a concluding example.

Keywords: decision support system, urban planning, public participation, opinion-miner recommender agents.

1   The Problem: Public Participation in Urban Planning

Public participation in the processes involved in the design and updating of urban master plans need to be improved. As things stand, it is not easy for citizens to become involved in these processes. This is due in large measure to the fact that participatory channels open too late in the timeline of master plan development. This results in a very low level of participation, which can raise doubts about the validity and legitimacy of the plan, and even lead to its rejection.

The processes for ensuring participation during the design and approval of master plans have a similar structure in all countries: citizens are requested to give their opinions once the design process is over and the stage is set for the approval process

B. Murgante et al. (Eds.): ICCSA 2013, Part III, LNCS 7973, pp. 324–339, 2013.
© Springer-Verlag Berlin Heidelberg 2013

to begin. The approval process of master plans typically has three chronological steps, each with an increasingly binding status. In the first step, immediately after the design phase is over, the authors of the plan present it to the municipal council for its initial approval. Council members can ask for changes in the proposals of the plan before they officially approve it. Once this first approval is given – the so-called initial approval - public participation starts. The council organizes a public presentation conducted by the authors of the plan. Presentation is followed by a one to two month period for public participation and input. During this period, citizens can only make suggestions about the proposals of the plan, but not give new ideas for the city. In a very real sense, the role of citizens is reduced to mere validators of the proposals already included in the plan. This limited role of citizens is a constant in all participatory steps of the master plan. Once this period for suggestions ends, suggestions are assessed by the authors of the plan. Some suggestions will be favorably considered and included in a new version of the plan prepared by the same team of urban planners.

In countries where urban planning is within the remit of the municipality, it is the responsibility of the municipal council to approve this second version of the plan. In countries where urban planning is within the remit of the regional authority or the national government, this second version will be considered by the responsible body in each case, which will either approve it or ask the planners to make further amendments. The approval of this second version of the plan is known as provisional approval. At this point, another two-month and more binding process of public participation starts. During this participatory period, the plan is available to citizens on the municipal web site and at the town hall urban planning offices. However, once again, citizens can only express their objections to it. They simply serve as validators of the proposals and, in this phase, their role is even reduced. Unlike suggestions, objections are not opinions on the proposals of the plan. Instead, they simply reject the proposals of the plan and/or point out illegalities, irregularities, or other anomalies. Once the two-month period of public consultation is over, the authors of the plan collect all the objections, assess them, and consider their inclusion in what will be the third and final version of the plan.

The third version of the plan is then presented to the competent authority. As in the second phase, the municipal, regional, or national controlling authority, depending on the legal system of the country concerned, either approves or requests amendments to the authors of the plan. This final step is usually described as the definitive approval. Once the third version of the plan is official, citizens that still object to any part of it can only lodge an appeal, thus initiating a legal process that will finish in court.

The described public participation in urban planning does not fairly respect or incorporate the opinion of citizens. It is not fair because it happens once the design phase of the master plan is over, limiting the role of the citizens to mere validators of the proposals of the plan. Moreover, some citizens that could participate do not do so because the institutional structure and procedures utilized are discouraging, demotivating, and very unpleasant if they finally result in citizens having to confront their local authorities before the court system. Participatory processes in urban planning should be more citizen-friendly, and must provide equal opportunity of expression of proposals for the city, instead of adhering to a bureaucratic process geared only towards validation. Public consultation should serve citizens so that they can identify with and take ownership of the design of the master plan which, after all, creates the city they will use and live in.

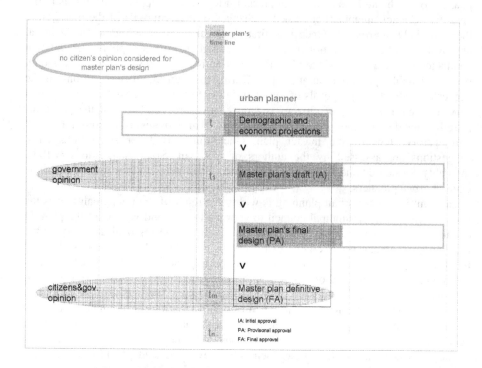

Fig. 1. Current elaboration process of a master plan

A master plan is a tool with which to effect change in a city: new residential developments provide new housing resources, new industrial and tertiary sites generate new job opportunities, new public spaces offer more leisure opportunities, new facilities provide more services, etc. It does not make sense that citizens should suffer rather than enjoy playing a part in the creation and design of all these resources for their city. They should be involved in the design phase, not only in the approval process. Moreover, to draft three versions of the master plan (with their corresponding approval processes) is inefficient. Reaching final approval can easily take a year or more, even if it is not interrupted or frozen before completion due to unforeseen circumstances such as political changes in one of the authorities involved, lack of resources, etc. For all described reasons, urban planning participatory processes need to be reformulated in order to allow the public to be involved in the early design phase of master plans. In addition, this should be done by employing technological citizen-friendly participatory methods.

Computer technology has helped to develop new mechanisms which incentive public participation in urban planning. More recently, the so called 'immersive planning' [1] focuses on the depth and breadth of user experience. Commonly web-based, the immersive planning techniques can be framed in three categories of immersion: challenge-based, sensory and imaginative. Geographic information systems (GIS), planning support systems, virtual environments, and 3D digital games are all methods of obtaining user immersion in one or a combination of these categories. The latest developments on these techniques comprise the following: Bugs [2] et al. analyze the impact of collaborative Web 2.0 tools applied to public participation GIS applications in urban planning actions. Brabham [3] uses a web-based crowd sourcing model typically used in distributed problem solving and production model for business as a public participation support system for urban planning. Dai et al. [4] use advanced virtual reality not only to incentive public participation in urban planning purposes but in the design of public buildings. Bogdahn et al. [5] enable users to explore in a playful way urban planning proposals in 3D models. Howard [6] ads interaction mechanisms between users inside the modeled environment.

Our novel contribution to the existing ecosystem of technologies helping to incentive public participation in the urban planning activity is an innovative two-step agent-based decision-support system (DSS) which involves the public both before and during master plan development. Under our method the public participation consists of citizens expressing as many ideas and opinions about their city as they choose, and before the design of the master plan starts. The proposed DSS uses agent technology to mine the opinions and ideas of citizens. We propose a pair of opinion-miner agents. These agents belong to a class of recommender agents whose aim is to formulate recommendations out of a mining process, and submit these recommendations to the urban planners in charge of the master plan. We will describe our method after reviewing existing agent-based models for urban planning purposes and outlining recommender agent technology.

The paper is structured as follows: in section two, existing implementation of agent technology used in urban planning are reviewed. In section three, after providing an overall idea of recommender agent techniques, we explain our Model. In the fourth and last section we use an unsuccessful experience in urban planning public consultation, the renewal of the Diagonal Avenue in Barcelona, as a concluding example of how our Model reinvents participatory processes in urban planning.

2   The Technique: Agent Technology in the Urban Context

In terms of functionalities, what characterizes urban planning is its predictive capacity. By definition, prediction means obtaining a model for the future by modeling the present. To obtain the future urban model, the urban planning practice uses demographic and economic projections, normally with a twenty-year time horizon. Thus, the main urban planning function is to predict what a city will need in twenty years, based on demographic and economic projections. And a master plan is the graphical representation of such predictions, (through the zoning) depicting how the city will look like in two decades.

Agent technology it is useful for urban assessment and decision support as it relies on simulation. Agent technology for urban planning purposes must incorporate predictive functionalities.

The following definition will help to explain how simulation works for urban assessment and decision support purposes using agent technology. "An agent is a computer system that is capable of independent action on behalf of its user or owner in order to satisfy design objectives. An intelligent software agent has to be autonomous, reactive, proactive, and social (capable of interacting with other agents to communicate and negotiate). Intelligent agents can learn and adapt to new situations. (…). A multi-agent system consists of a number of agents that interact with one another, cooperate in order to accomplish different tasks and are able to negotiate and solve conflicts" [7].

In an agent-based urban simulation, different scenarios are created to find a solution to a specific problem. In such scenarios, agents act on behalf of people, so simulation is a fairly realistic virtual approach to better assessing a real problem. Simulation allows different scenarios to be generated by changing the macro-characteristics of the model which is being assessed. The functionality that enables the creation of different testing scenarios makes it an appropriate technique for decision support systems (DSS) assessing physical realities. As examples in the context of physical urban assessment, Ligtenberg, A. et al. [8] have developed a model to decide on the location and spatial distribution of new development areas. The model combines a multi-agent simulation (MAS) approach with cellular automata (CA). CA is used as a support to provide the knowledge an agent needs. It includes individual actor behavior with different levels of participation. Spatial intentions and related decision-making actions are defined by agents. The modeling concept is intentional and based on the Schulz DSS model where agents interact with the environment rather than between themselves. The same author [9] has developed a similar model where interactions between actors are stronger. Here the MAS also simulates a multi-actor interactive spatial DSS process for allocation purposes. It pursues knowledge sharing between participating actors in a learning approach designed to create a common view that minimizes decision conflicts when selecting areas eligible for development.

Besides assessing physical realities, agent-based DSS models are also used to assess the strategy for growth in a city, testing different approaches such as balanced growth as a whole, or uneven growth favoring some parts of the city. Kou et al. [10] adopt an MAS based on Swarm simulation to assess the process of urban growth in an artificial city. The effects of organizing the city into one or more city-centers are tested. Still in the research area of city-center simulation, Wagner [11] has developed a topological and metrical categorization of different city core areas to assess their level of attraction. The research combines an MAS of random walking agents and axial graphs. In other cases, cellular automata and agent concepts have been used to model and simulate different scenarios of growth and dissemination of slum areas as a basis for assessing control measures. Chen et al. [12] have developed a tool for urban growth control. Li and Liu [13] have combined agent technology with CA and GIS (geographical information system) techniques in a single exploratory tool to simulate different urban development patterns in order to assess their level of sustainability.

Agent –based DSS models for urban assessment have also been used at the urban building level to simulate how citizens use a city by changing the built environment, either in terms of building functions and other urban elements [8] or performance indicators [9]. Other authors have been more concerned about assessing social sustainability in a changing built environment [10].

A final point to mention is that agent-based DSS models have been developed not only for urban assessment but also for larger-scale spatial decisions. Arentze et al. [11] have developed a spatial DSS using agent-based model (ABM) techniques for land-suitability and facility-needs analysis. The authors highlight the innovation of their analysis as it combines the two related approaches of land-use-allocation and land-use-needs modeling. The resulting integrated model is able to generate different scenarios of allocated land-use alternatives by changing the features of facilities.

Agent-based models (ABMs) are useful for simulating urban, spatial, and all other kinds of complex systems [12]. Applications can vary from the simulation of socio-economic systems through the elaboration of scenarios for logistics optimization and biological systems to the urban and spatial ones already seen. Interestingly, ABMs provide the means to include real decision making without losing the strength of the concept of self-organization [13] inherent in complex systems.

Despite the powerful simulation functionalities of ABMs that allow them to explore all kinds of complex systems, the addition of predictive capabilities makes them useful for the urban planning activity. In this terrain, to highlight the works of Batty [14]. He conceptualizes a bottom-up urban planning process in which the outcomes are always uncertain. This is combined with new forms of geometry associated with fractal patterns. Batty's model begins with the use of cellular automata (CA) to simulate urban dynamics through the local actions of automata. Next, he introduces agent-based models (ABM) in which agents move between locations. These models relate to many scales, from the scale of the street to the urban region. Finally, Batty attempts to understand and predict different urban phenomena with a focus in spatial modeling changing to dynamic simulations of the individual and collective behaviors of involved actors at such scales

Finally, a conceptual clarification is necessary. Prediction and probability are completely different techniques [15], and demographic projections are much more than just an extrapolation of population to a certain time-horizon [16]. We will revisit these concepts when we present our method in the next section.

3   The Representation for All Method: A Pair of Opinion-Miner Agents That Recommend the Ideas of Citizens to Planners

Recommender agents have been used extensively in very different fields, wherever an intelligent mechanism is needed to filter out large quantities of available information [17], and to provide customers or users with the means to effortlessly find the items they are probably looking for based on their history of prior actions [18].

Monitoring agents are used for location and allocation purposes. Their recommendations help tourists to find sites to visit [19] or simply provide information using location awareness technology [20]. The agents in ABM models for urban assessment described in the previous section can be considered monitoring agents.

Miner agents are now being used for e-learning, a new and important area of application. Miner agents recommend educational materials [20], [21], [22] or personalized resources [23], [24], [25], to groups of students based on their profiles.

This technology is also being used in the growing area of social and other networks on the Internet to provide users with useful internet and website suggestions [26], [27] (see Montaner et al. [28] for a complete taxonomy of internet agents) and to help find potential network members [29].

Hybrid and customized agents have more specific applications, varying from agents for personal diet recommendations [30] to recommender agents that help software developers to select task-relevant software libraries [31].

The most appropriate profile of agent for our Model, whose purpose is to enhance and enrich public participation in urban planning by making the opinions of citizens more prominent, is the miner-recommender agent:

A pair of opinion-miner agents extract and classify data contained in the opinions of citizens to make recommendations to planners during the design of the master plan. The first agent mines web-requested opinions from citizens before the design of the master plan begins. The first agent assists the planner, helping prepare a more citizen-centered first draft of the master plan. Next, during the elaboration of the master plan, the second agent mines the Internet seeking opinions of citizens from other cities where similar plan proposals have been made. The second agent informs the planners regarding the success or failure of similar proposals found, helping to address the actual proposals being elaborated in the master plan.

In both cases the agents mine the web, since both opinion-mining processes will be web-based. Opinion-mining is a process similar to data-mining, which mines text seeking keywords instead of mining data. The techniques, algorithms, and methodologies used in web-based opinion-mining encompass those specific to data-mining, mainly because there is a great deal of unstructured data on the web, and data changes are frequent and rapid [32], [33].

Our first agent mines the opinion of citizens from a website specifically designed for that purpose. The second agent uses hyperlinks to discover websites containing opinions of citizens about planning proposals similar to the ones contained in the master plan under elaboration. In the second agent tasks, a complete web-mining process applies, from web-structure-mining to web-content-mining and web-usage-mining [7].

Our pair of opinion-mining recommender-agents is involved in a three-phase operational process under which the recommendation tasks are executed, the retrieval (1), the sorting (2), and the reporting (3) steps. These steps are explained in the following paragraphs.

In the first phase, the retrieval step, a mining algorithm is used. Because it is about text-mining, the unsupervised Apriori algorithm is selected. The Apriori algorithm successively finds large item sets which satisfy the desired minimal support. Association rules are then constructed, based on the elements which satisfy the desired minimal confidence [34]. Text-mining techniques can deal with noise and data uniformity by implementing correlation rules between the selected keywords and their descriptions [35]. Nowadays, text-mining techniques are extremely advanced, and can even be used to mine scientific texts [36]. Still in this first phase, the next step is to create a structured representation of the information that has been mined using an ontology. An ontology must be defined, so that every piece of information can be represented in structured formats. Each information piece is mapped into the ontology as an instance.

The second phase, the sorting step, is to implement an algorithm to classify and map the information into the ontology. Unstructured data can be automatically classified and mapped [37], again using text-mining techniques, but based on decision rules [38]. There are several algorithms that can be used for text classification purposes. Bayesian classifiers, neural networks, and decision trees are among the most common [37]. The ratings of what is being assessed are computed using the information already mapped into the ontology.

The third and final phase of the process, the reporting step, is to make recommendations in response to the assessment request. Recommendations are made using efficient prediction algorithms. In the context of agent technology, a prediction can be defined as a value that expresses the predicted likelihood that a user will have an interest in an item [39]. Accordingly, a recommendation is defined as the list of n items in the top n predictions from the information set available [39]. To ensure quality recommendations, once the recommender agent has been implemented, it must be trained using a test data set (unseen data or training data) which is different from the data that is being assessed [40].

Still in the reporting phase and in order to select the appropriate predictive method for recommendation, we have to briefly describe two different techniques: the content-based recommendation (CBF) method [31] and the collaborative-filtering (CF) recommendation method [41]. Hybrid techniques have also been proposed by some authors [31]. CBF methods are mostly used when documents, web pages, publications, news, or similar pieces of information are to be recommended [39]. The agent provides the recommendation according to the preferences of the user and/or interests which have been either explored by the agent or preset by the user.

CF methods aim to identify users that have compatible interests and preferences by calculating similarities and dissimilarities between user profiles [41]. This technique is useful when users need refined information, and it is to their benefit to consult the opinions of other users who share similar interests and whose opinions can therefore be trusted [39]. Within CF methods there are two major categories: memory-based and model-based [42]. Memory-based CF uses nearest-neighbor algorithms that determine a set of neighboring users who have rated items similarly, and combine the preferences of neighbors to obtain a prediction for the active user [43]. Model-based collaborative recommenders do not use the user-item matrix to make recommendations directly, but generalize a model of user ratings using a machine learning approach, and then use this model to make predictions [44]. In this category the subgroup of demographic recommenders is of interest to our Model. Demographic recommenders aim to categorize users based on their personal attributes in terms of belonging to stereotypical classes. Instead of applying learning techniques to acquire user models, these agents are based on stereotype reasoning [44]. In the case of demographic recommenders, a user model is a list of demographic features that represent a class of users. The representation of demographic information in a user model can vary greatly. For example, Pazzani [43] extracts features from home pages to predict the preferences of users for certain restaurants, while Krulwich [45] uses demographic groups for market research to suggest a range of products and services.

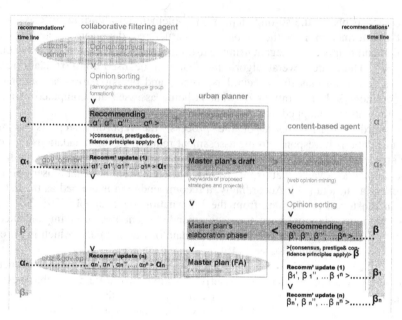

Fig. 2. Recommendations time line of the collaborative filtering and content-based agents. The picture above represents how recommendations operate over time in parallel with the elaboration of the master plan.

According to the descriptions of the content-based recommendation and the CF methods, in our pair of agents, the second opinion miner provides CBF recommendations based on the opinions of citizens of other cities. Clustering techniques are implemented in order to group the collected opinions [46]. The recommendation task of the agent is to alert regarding the positive or negative opinion of citizens of other cities regarding proposals which are similar to those made in the master plan under preparation. The first agent of the pair, although it also mines opinions, uses collaborative-filtering techniques because of the demographic recommender sub-technology that helps when working with groups. This first demographic recommender agent mines the opinions of citizens concerning their city, focusing on citizens in the demographical group. A survey asking them about their degree of satisfaction with public spaces, facilities and services, the housing market, job opportunities etc., and requesting their ideas and proposals for the city is conducted on-line. To achieve maximum coverage, additional tools besides the website are employed: phone surveys, hard copy questionnaires, etc. All non-digital survey answers are converted into digital text to allow the agent to execute its opinion-mining function. Using the demographic collaborative-filtering functions of the agent, different stereotype groups based on the survey answers are created. Finally, this first agent reports an organized selection of ideas for the plan to its authors. Here, the recommendation task is to select from all the comments and ideas those that apply to planning purposes, and to dismiss those which are not related.

In our pair of agents, both content-based and CF agents will provide their recommendations to the planner based on the principles of consensus, prestige, and confidence. A consensus opinion is generated amongst the agents conforming the pair, and synthesized recommendations are sent to the planner. Consensual will be weighted according to the status of the opinion source (e.g. the reputation of the webpage) and ranked according to the source's credibility (reliability of information).

We are aware that the opinion of citizens changes over time, and have taken this into account in our Model. The first agent, the CF agent, will report recommendations to the planners before design of the master plan starts, and then also throughout the elaboration of the master plan. Similarly, the second, content-based agent will provide recommendations as long as the master plan is under development. Recommendations will be provided until the final approval of the plan, in order to anticipate public reaction to the eventual changes and updates in the plan.

The architecture and protocols of the miner agents in the search of opinions that are relevant to every feature of the Master Plans will be accessed by social search through a web of trust and aggregated with simple weighted averages taking into account again the trust that every agent has declared in its contact list that happens to be a same contact list of citizens participating in the discussion of the Master Plan.

The agents emulate the social features of the human social networks as a sort of social machines, dealing with their own contact lists, handling with requests and interactions in human like language, and supporting their users in their daily operations in and among the open social networks. This is a fully bottom-up approach that contains a design of the types of social machines that will interact similarly to people, under their roles and rules of interaction, sharing their contact lists. The first approaches of social search are 'Sixearch' [47] as a search peer to peer, 'AskNext' [48], search based on agents 'topsy.com' looking for tweets (on twitter), 'factoetum.com' for the automation of contributions to blogs, and heystack.com as an example of collaborative search [49]. In the case of multi-agent referral system (MARS). Yu et al. [50] proposed a P2P application in which agents assist users in obtaining and following referrals to find experts who might answer their questions.

Regarding the contact lists, further works examine the benefits of using social networks like the FOAF (Friend-of-a-Friend) resource description framework (RDF) ontology that has recently gained popularity [51]. FOAF predicates consist of a person's properties such as name, email address, group memberships, employer, gender, birthday, interests, projects, and acquaintances. By 'spidering' the Semantic Web and collecting the information contained in FOAF files, one can build a large collection of data about people and their interests. This information can be used to email people with a given interest, people who know people who know a particular person, and so on.

Regarding issues in designing the architecture, we use the bottom-up methodology which starts with a rigorously predefined set of rules for the individual behaviors and local information and allows the agents to decide with whom to associate, what roles they can take, what types of questions and answers they want to exchange and address, in what languages, with what particular or general ontology. Coordination-related issues to other agents using protocol-based conversations expressed in a coordination language (called the Agent Communication Language) implemented over SOAP. The growth of automated services on the Web, like those here implementing the collaborative-filtering and content based filtering agents as well as the miner agents, has facilitated this new software development paradigm.

Finally, the Agent Oriented Methodology that will be used for the design of the agents and the full multi-agent system is INGENIAS-MESSAGE. The MESSAGE modelling language is a meta-modelling technique based on MOF (meta-object facility). It also extends UML concepts of class and association. The most important concepts in message are agent, organization, role, goal, task, protocol and interaction. INGENIAS is an extension of MESSAGE, which improves its meta-models with five meta-models: agent, organization, interaction, environment, and goals. The free design tool is available http://grasia.fdi.ucm.es/main/?q=es/node/61.

4   A Concluding Example of the Usage of the Representation for All Model: The Redesign of Diagonal Avenue in Barcelona

In March 2010, the Barcelona City Council launched a public consultation process to select one of two designs for the Diagonal Avenue renovation master plan (see pictures below showing designs A and B). Both of these options included new tram lanes to connect with existing lines. The survey made little impression on the citizens, since no opinion and only validation of the options was requested. Participation was only 10%. Interestingly, the option which received most support was a third one, C, which was a rejection of the other two. The victory of option C destroyed any chances that the project would go ahead.

The Diagonal Avenue consultation followed the traditional and unsuccessful public participation process in urban planning, detailed in the problem description section of this paper, under which the public gives its opinion of the proposals presented by a master plan at the very end of the design of the plan (when there is no scope for change). This undemocratic and elitist approach can sometimes lead to extremely long approval processes, or even the rejection of master plans when public reaction is negative, as in the case of Diagonal Avenue in Barcelona. It is quite clear that introducing wider participation during a plan's preparatory and/or design phases would be a step in the right direction for obtaining the opinion of concerned citizens. In non-official blogs and opinion websites created during the Diagonal Avenue case, some people expressed the view that only a bus lane was needed, and that there was not enough financing available to justify a huge investment in new tram lines. Others preferred more investment in basic services and public spaces in the city, rather than spending significant funds on complete renovation of the Avenue.

Following our Model, a public participation survey would be conducted before design of the urban project begins. This extensive survey would involve asking for specific ideas concerning the objectives of the project as well as more general opinions on related urban aspects affecting the project. In the case of the Diagonal Avenue, these related aspects would revolve around the quantity and quality of public spaces and facilities in Barcelona. Such general opinions would help the planners produce more focused and useful proposals that besides helping design the specific project, would end up satisfying the more general urban needs of citizens.

The first type of agent of our pair, the demographic, collaborative-filtering agent would retrieve the relevant opinions of individual citizens through text-mining from a website created for that purpose doing social search so that the deep web composed of opinions of all kinds would be accessible. Recommendations would be reported to the planner in the form of lists of opinions, positive and negative, ordered by relevance and importance while taking into account the trust and influence of those who are giving their opinion away in the form of comments in open social networks. The planner would develop one or more project solutions based on these recommendations and economic and demographic projections.

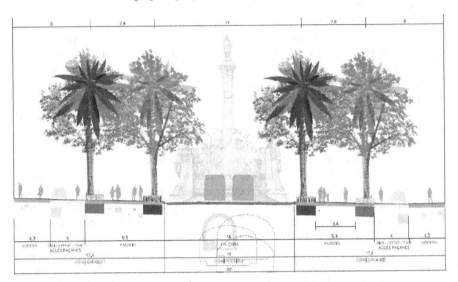

Fig. 3. Redesign of the Diagonal Avenue in Barcelona. Option A Barcelona Municipality [59].

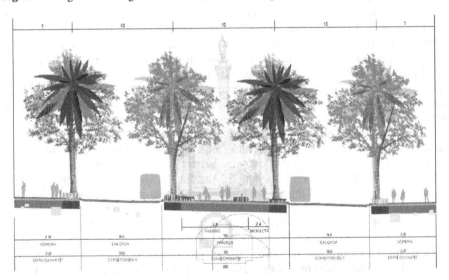

Fig. 4. Redesign of the Diagonal Avenue in Barcelona. Option B Barcelona Municipality [59].

Before presenting the project solutions to the public, the second type of agent, the content-based agent, would come into play and mine the Internet for reactions to similar project solutions in other cities following a similar scheme of social search but instantiated with other similar projects. The similarity of projects is assessed by the features of those projects (through the content based filtering process) and will proceed to aggregate the opinions of those projects with the forehead mentioned social search process. Recommendations on the opinions that similar proposals raised in other cities would be reported to the planner. Armed with this information, the planner can carefully consider and compare actual previous reactions, responses and potential future responses. Finally, the planner can adjust the project solutions being designed before they are public with proper estimates of their opinions and relevance.

Returning to our case study in Barcelona, through a key word search using "new tram line, public consultation", the content-based agent would have found that several public consultations had already been conducted on whether or not a new tram line was needed in a given city, and which other design options should be considered. Indeed, there is a great deal of experience in other cities. For example, In Edinburgh, the Scottish Parliament conducted a public consultation in 2003 regarding the new tramway network. The people indicated that only three new tram lines were needed, in combination with feeder bus services or shuttles, to reduce the cost [52]. In Dublin, in December 2005, the Railway Procurement Agency [53] started a public consultation exercise on a new route to connect with the existing Red and Green lines. Citizens were asked about their preferred route for such a connection. Similarly, in London in 2007, the local authority carried out a survey on the route options for the new Cross River Tram (CRT) line. The CRT followed a long route covering four boroughs. It was divided into five sections to facilitate ease of comment during the public consultation [54]. Finally, there is the case of Manchester, where Transport for Greater Manchester worked closely with Manchester City Council to ensure a new tram line fitted in with popular opinion and also fulfilled regeneration aspirations for the area around the new tram line [55]. Interestingly, in all these new tram line consultations, public participation began with consideration of whether a new tram line was necessary in the first place. This differs from the approach in Barcelona.

The agent-based implementation of such model must scale properly to the open nature of comments of citizens about public plans in the social networks. In the end, with the implementation of our agent-based Model for public participation in the Diagonal Avenue case, a more citizen-centered solution would have been possible.

Our Model respects the classical three-step participatory process in urban planning, but considerably must reduce the social efforts and the economic costs. Our model has more social value because it allows real participation and involves citizens in the planning process from the very beginning. And finally, its citizen-friendly approach must be cost-effective and must save resources because it manages opposition and reduces the chances that citizens will object or resort to litigation in response to the final design of a master plan. Its implementation as a multi-agent system that do social search for the mining of comments for getting the relevant opinions regarding an on-going design of a master plan is useful for scaling up properly at the internet scale and a modern way of taking advantage of not only the existing discussing happening along the exposition of the current plan but including preceding discussions of similarly plans all over the world that usually are of relevance for that specific plan, with high chances of recommending the proper features of the plans in a way that people might understand and valuate its benefits or drawbacks according to the recommendations of former existing experiences found in the web in an automatic way. The novelty of the approach deserves further work to validate with a proof of concept that is going to be developed as future work.

Acknowledgements. Thanks to Mark Evan Segal for the editorial assistance, valuable comments and suggestions. This research is partly funded by the EU ITEA-2 for CARCODE n°11037, ITEA2-Call 6, Platform for Smart Car to Car Content Delivery.

References

1. Gordon, E., Schirra, S., Hollander, J.: Immersive planning: a conceptual model for designing public participation with new technologies. Environment and Planning B-Planning & Design 38, 505–519 (2011)
2. Bugs, G., Granell, C., Fonts, O., Huerta, J., Painho, M.: An assessment of Public Participation GIS and Web 2.0 technologies in urban planning practice in Canela, Brazil. Cities 27(3), 172–181 (2010)
3. Brabham, D.: Crowdsourcing the Public Participation Process for Planning Projects. Planning Theory 8(3), 242–262 (2009)
4. Dai, C., Li, T.: Research on the Application of Virtual Reality Technology in the Large-scale Public Construction Planning Dynamic Analysis, Editor(s): China Civil Eng. Soc., Chinese Acad. of Eng./DCHAE; Tongji Univ., China State Construction Eng. Co. Ltd. Proceedings of Shanghai Intl. Conf. on Tech. of Architecture and Structure, 580–586 (2009)
5. Bogdahn, J., Coors, V., Sachdeva, V.: A 3D tool for public participation in urban planning. In: Coors, V., Rumor, M., Fendel, E., Zlatanova, S. (eds.) Urban and Regional Data Management. Procs. and Monographs in Eng., Water and Earth Sciences, pp. 231–236 (2008)
6. Howard, T.L.J., Gaborit, N.: Using virtual environment tech. to improve public participation in urban planning process. J. Urban Planning and Dev., ASCE 133(4), 233–241 (2007)
7. Dzitac, I., Moisil, I.: Advanced AI Techniques for Web Mining. In: Mastorakis, N.E., Poulos, M., Mladenov, V., Bojkovic, Z., Simian, D., Kartalopoulos, S. (eds.) Mathematical Methods, Computational Techniques, Non-linear Systems, Intelligent Systems. Mathematics and Computers in Science and Engineering, pp. 343–346 (2008)
8. Ligtenberg, A., Bregt, A., Van Lammeren, R.: Multi-actor-based land use modelling: spatial planning using agents. Landscape and Urban Planning 56, 21–33 (2001)
9. Ligtenberg, A., Beulens, A., Kettenis, D., Bregt, A., Wachowicz, M.: Simulating knowledge sharing in spatial planning: An agent-based approach. Environment and Planning B-Planning & Design 36, 644–663 (2009)
10. Kou, X., Yang, L., Cai, L.: Artificial Urban Planning: Application of MAS in Urban Planning Education. In: 2008 Intl. Symp. on Computational Intelligence and Design, vol. 2, pp. 349–353 (2008)
11. Wagner, R.: On the metric, topological and functional structures of urban networks. Physical-statistical Mechanics and its Applications 387(2008), 2120–2132 (2008)
12. Chen, J., Jiang, J., Yeh, A.: Designing a GIS-based CSCW system for dev. control with an event-driven approach. Photogrammetric Eng. and Remote Sensing 70, 225–233 (2004)
13. Li, X., Liu, X.P.: Embedding sustainable development strategies in agent-based models for use as a planning tool. Intl. Journal of Geographical Information Science 22, 21–45 (2008)
14. Batty, M.: Cities and Complexity: Understanding Cities with Cellular Automata, Agent-Based Models, and Fractals. The MIT Press (2007)
15. Aschwanden, G., Haegler, S., Bosche, F., Van Gool, L., Schmitt, G.: Empiric design evaluation in urban planning. Automation in Construction 20, 299–310 (2011)
16. Dijkstra, J., Timmermans, H.: Towards a multi-agent model for visualizing simulated user behavior to support the assessment of design performance. Automation in Construction 11, 135–145 (2002)

17. Steinhöfel, J., Anders, F., Köhler, H., Kalisch, D., König, R.: Computer-Based Methods for a Socially Sustainable Urban and Regional Planning – CoMStaR. In: Taniar, D., Gervasi, O., Murgante, B., Pardede, E., Apduhan, B.O. (eds.) ICCSA 2010, Part I. LNCS, vol. 6016, pp. 152–165. Springer, Heidelberg (2010)

18. Arentze, T.A., Borgers, A., Ma, L.D., Timmermans, H.: An agent-based heuristic method for generating land-use plans in urban planning. Environment and Planning B-Planning & Design 37, 463–482 (2010)

19. Casali, A., Godo, L., Sierra, C.: Validation and Experimentation of a Tourism Recommender Agent based on a Graded BDI Model. In: Alsinet, T., PuyolGruart, J., Torras, C. (eds.) AI Research and Dev. Frontiers in Artificial Intelligence and Applications, vol. 3, pp. 41–50 (2008)

20. Yang, H.-L., Yang, H.-F.: Recommender agent based on social network. In: Okuno, H.G., Ali, M. (eds.) IEA/AIE 2007. LNCS (LNAI), vol. 4570, pp. 943–952. Springer, Heidelberg (2007)

21. Haubl, G., Murray, K.: Double agents. MIT SLOAN Management Review 47 (2006)

22. Rosaci, D., Sarne, G.: Dynamically Computing Reputation of Recommender Agents with Learning Capabilities. In: Badica, C., Mangioni, G., Carchiolo, V., Burdescu, D.D. (eds.) Intel. Distributed Comput., Systems & Appl. SCI, vol. 162, pp. 299–304. Springer, Heidelberg (2008)

23. Zaiane, O.: Building a recommender agent for e-learning systems. In: Kinshuk, X., Lewis, R., Akahori, K., Kemp, R., Okamoto, T., Henderson, L., Lee, C.H. (eds.) Proceedings of International Conference on Computers in Education (I –II), pp. 55–59 (2002)

24. Cristian Mihăescu, M., Burdescu, D.D., Ionascu, C.M., Logofatu, B.: Intelligent Educational System Based on Personal Agents. In: Damiani, E., Jeong, J., Howlett, R.J., Jain, L.C. (eds.) New Directions in Intelligent Interactive Multimedia Systems and Services - 2. SCI, vol. 226, pp. 545–554. Springer, Heidelberg (2009)

25. Karoui, H., Kanawati, R., Petrucci, L.: COBRAS: Cooperative CBR system for bibliographical reference recommendation. In: Roth-Berghofer, T.R., Göker, M.H., Güvenir, H.A. (eds.) ECCBR 2006. LNCS (LNAI), vol. 4106, pp. 76–90. Springer, Heidelberg (2006)

26. Garruzzo, S., Rosaci, D., Sarné, G.M.L.: MARS: An agent-based recommender system for the semantic web. In: Indulska, J., Raymond, K. (eds.) DAIS 2007. LNCS, vol. 4531, pp. 181–194. Springer, Heidelberg (2007)

27. Takenouchi, T., Kawamura, T., Ohsuga, A.: Development of knowledge-filtering agent along with user context in ubiquitous environment. In: Enokido, T., Yan, L., Xiao, B., Kim, D.Y., Dai, Y.-S., Yang, L.T. (eds.) EUC-WS 2005. LNCS, vol. 3823, pp. 71–80. Springer, Heidelberg (2005)

28. Montaner, M., Lopez, B., de la Rosa, J.: A taxonomy of recommender agents on the Internet. Artificial Intelligence Review 19, 285–330 (2003)

29. Kim, J.H., Kwon, H.J., Hong, K.S.: Location awareness-based intelligent multi-agent technology. Multimedia Systems 16, 275–292 (2010)

30. Lee, C.S., Wang, M.H., Hagras, H.A.: Type-2 Fuzzy Ontology and its Application to Personal Diabetic-Diet Recommendation. IEICE Trans. on Info. and Systems 18, 374–395 (2010)

31. McCarey, F., O'Cinneide, M., Kushmerick, N.: A recommender agent for software libraries: An evaluation of memory-based and model-based collaborative filtering. In: Nishida, T., et al. (eds.) 2006 IEEE/WIC/ACM Intl Conf. on Intelligent Agent Technology (2006), pp. 154–162 (2006)

32. Kantardzic, M.: Data Mining: Concepts, Models, Methods, and Algorithms. John Wiley & Sons, New Jersey (2003)

33. Liu, B.: Web Data Mining. Exploring Hyperlinks, Contents, and Usage Data. Springer, Heidelberg (2007)

34. Lau, R.: Belief revision for adaptive recommender agents in E-commerce. In: Liu, J., Cheung, Y.-M., Yin, H. (eds.) IDEAL 2003. LNCS, vol. 2690, pp. 99–103. Springer, Heidelberg (2003)

35. Kusumura, Y., Hijikata, Y., Nishida, S.: NTM-Agent: Text mining agent for net auction. IEICE Transactions on Information and Systems E87D, 1386–1396 (2004)
36. Blanke, T., Hedges, M., Palmer, R.: Restful services for the e-Humanities - web services that work for the e-Humanities ecosystem. In: 3rd IEEE International Conference on Digital Ecosystems and Technologies, pp. 290–295 (2009)
37. Chiang, J.H., Chen, Y.C.: An intelligent news recommender agent for filtering and categorizing large volumes of text corpus. Intl J. of Intelligent Systems 19, 201–216 (2004)
38. Aciar, S., Zhang, D., Simoff, S., Debenham, J.: Informed recommender agent: Utilizing consumer product reviews through text mining. In: Butz, C.J., et al. (eds.) IEEE/WIC/ACM Intl Conf. on Web Intelligence and Agent Technology, pp. 37–40 (2006)
39. Papagelis, M., Plexousakis, D.: Qualitative analysis of user-based and item-based prediction algorithms for recommendation agents. Eng. Applications of AI 18, 781–789 (2005)
40. Smeureanu, I., Diosteanu, A., Delcea, C., Cotfas, L.: Business Ontology for Evaluating Corporate Social Responsibility. Amfiteatru Economic 13, 28–42 (2011)
41. Herlocker, J., Konstan, J.A., Terveen, L., Riedl, J.: Evaluating collaborative filtering recommender systems. ACM Transactions on Information Systems (TOIS) 22 (2004)
42. Sarwar, B., Karypis, G., Konstan, J., Reidl, J.: Item-based coll. filtering recommendation algorithms. In: 10th Intl. Conf. on World Wide Web, pp. 285–295. ACM Press, NY (2001)
43. Pazzani, M.: A framework for collaborative, content-based and demographic filtering. Artificial Intelligence Review 13, 393–408 (1999)
44. Kobsa, A., Koenemann, J., Pohl, W.: Personalized hypermedia presentation techniques for improving online customer relationships. The Knowledge Eng. Review 16, 111–155 (2001)
45. Krulwich, B.: Lifestyle finder: intelligent user profiling using large-scale demographic data. Artificial Intelligence Magazine 18, 37–45 (1997)
46. Wang, Z., Liu, A.: Toward Intelligent Web Collaborative Learning System. In: Hu, Z.B., Liu, Q.T. (eds.) Proceedings of the First International Workshop on Education Technology and Computer Science III, pp. 467–471 (2009)
47. Lele, N., Wu, L., Akavipat, R., Menczer, F.: Sixearch.org 2.0 peer application for collaborative web search. In: Conference on Hypertext and Hypermedia archive, Proceedings of the 20th ACM Conference on Hypertext and Hypermedia Table of Contents, pp. 333–334 (2009)
48. Trias, A., Mansilla, J., de la Rosa, J.L.L.: Asknext: An Agent Protocol for Social Search. Information Sciences 190, 144–161 (2012)
49. Smyth, B., Briggs, P., Coyle, M., O'Mahony, M.: Google Shared. A Case-Study in Social Search. In: Houben, G.-J., McCalla, G., Pianesi, F., Zancanaro, M. (eds.) UMAP 2009. LNCS, vol. 5535, pp. 283–294. Springer, Heidelberg (2009)
50. Yu, B., Singh, M.P., Sycara, K.: Developing Trust in Large-Scale Peer-to-Peer Systems. In: First IEEE Symposium on Multi-Agent Security and Survivability, pp. 1–10 (2004)
51. Kassoff, M., Petrie, C., Zen, L.M., Genesereth, M.: Semantic Email Addressing, The Semantic Web Killer App? IEEE Internet Computing, 30–37 (2009)
52. Scottish Transport Organization, http://www.scottishtransport.org/new_trams_for_edinburgh/edinburgh_tram_i
53. Dublin Municipality. Railway Procurement Agency, http://www.rpa.ie/en/Pages/default.aspx
54. London transport authority, http://www.tfl.gov.uk/assets/downloads/final-report.pdf
55. Transport for Greater Manchester, http://www.tfgm.com/2009_news.cfm?news_id=9006950?submenuheader=3

Smart Cities as "EnvironMental" Cities

Luciano De Bonis

University of Molise, Department of Biosciences and Territory
Via Duca degli Abruzzi, 86039 Termoli (CB), Italy
luciano.debonis@unimol.it

Abstract. Starting from the references to Smart Cities and Living Labs included in the Strategic Research Framework of the European Joint Programming Initiative "Urban Europe", the paper tries to outline the basic notions of a more limited reference framework, specifically aimed at fostering an approach to Smart City through an Urban/Territorial Living Lab concept, able to bring out the latent potential of innovation eco-socio-technological (i.e. "EnvironMental") linked to the rhetoric of "urban smartness". To this end, the paper also describes briefly some issues emerging from a paradigmatic experience of Living Lab for Smart Cities and Territories (the Perphèria Project), highlighting their interest for an innovative spatial planning research perspective.

Keywords: Smart Cities, Urban/Territorial Living Labs, Urban Europe.

1   Introduction

"Smart Cities" (SCs) is an emerging issue at the moment, both at political-administrative level and in the field of science and technology including, but not long ago, the urban and regional research.

The Strategic Research Framework (SRF) of the European Joint Programming Initiative (JPI) "Urban Europe", for example, refers to Smart Cities in terms of possible «operational solutions to be envisaged to realize ecological sustainability in cities», and it seems also (indirectly) in terms of the contribution of new technologies «to a reinforcement of the urban growth potential» (Research C.1. Sustainability of Urban Systems, pp. 13 and 59). Even stronger are in the above cited Research Framework the references to the theme of Living Labs, both in terms of «conceptual and operational models and tools (...) needed for shaping the transformation of city towards desirable urban vision ...» - together with «...envisioning, experiments, complex urban system models» - and in terms of testbeds for «new concepts and research instruments toolkits» (Research Theme A.2. Urban Indicators, Strategic Information Systems and Models of Urban Development, pp. 11 and 56). Furthermore, the references of the Framework to the Living Labs are not confined under the "Issues and Operational Research Themes for Urban Europe" but are also found in one of four interconnected urban images from which the research challenges arise. This is precisely the image defined as "Entrepreneurial City 2050", which assumes that «in the current and future local and global competition, Europe can only

B. Murgante et al. (Eds.): ICCSA 2013, Part III, LNCS 7973, pp. 340–350, 2013.
© Springer-Verlag Berlin Heidelberg 2013

survive, if it is able to maximize its innovative and creative potential in order to gain access to emerging markets outside Europe ...». (p. 8). In other words, while in the Strategic Research Framework of "Urban Europe" the theme of Smart Cities is treated only at the level of the research issue "Sustainability of Urban Systems" - even if indirectly connected to the "urban growth potential" - the references to the Living Labs seem more transversal, and above all they appear directly connected to the image of the so called "Entrepreneurial City".

Taking into account the above cited references, although they are not connected to each other in the Framework Research of "Urban Europe", the next sections of this paper try to outline the basic notions of a more limited reference framework, specifically aimed at fostering a planning approach to Smart Cities via an Urban/Territorial Living Lab concept. In order to both overcome what look like deficits of the Joint Programming Initiative "Urban Europe", and to enhance the potential of real "eco-socio-technological" innovation, linked either to the new rhetoric of "urban smartness" [1] [2] or to some critical polemic against the rhetoric itself [3].

2   Cities and Territories as "EnvironMental" Systems

In my opinion, a viable reference framework for a spatial planning research approach to the theme of Smart Cities through an Urban/Territorial Living Lab concept has to try first of all to treat in a unified way - although obviously metaphorical - ecological systems, territorial, social and technological (especially digital).

The concept of Socio-Ecological System (SES) [4] goes already in this direction, but is in the field in which Smart Cities and Living Labs are more rooted, i.e. the research on ICT, that is possible to retrieve the elements for extending the metaphor of the ecosystem also to the technological aspects, following the not dominant but fruitful legacy of studies dating back to the so-called "second-order cybernetics" and to the complexity theory [5].

Basing on the recognition gained under those studies, of a common (not unique) cognitive and communicative mode of the environmental and socio-technical spheres, it can be considered permissible to use the environmental ecosystem metaphor, territorial and digital [2], as partly already done for the so-called Digital Business Ecosystems (DBEs). To be more precise it must be said that in the case of DBEs the metaphor deliberately does not involve environmental issues, but refers instead to the balancing effect that a higher level of integration between socio-cultural context and economic activities should have on the economic vitality long-term of a region [6].

The here proposed extension of the metaphor of ecosystems to a wider sphere, at the same time ecological, territorial and techno-digital, is however based on the same conceptual cornerstones, as indicated below [6].

- The fallacy of proceeding in a direction in which the designers establish what should or should not be possible in a system, in favour of systems with great intensity of knowledge, able to build complexity from a set of elementary components [7].

- The belonging of the territorial research, despite its specificity, to a broad research domain, or better to a domain of interaction between research fields, inaugurated by the Palo Alto school [8] [9], into which can converge philosophy of science, epistemology, second-order cybernetics, information theory, linguistics, communication theory, biology, etc.
- As to the second-order cybernetics in particular, the applicability to environmental ecosystems, spatial and digital of the following two key principles (derived from Piaget):
  - o knowledge is not passively received through the senses or other means of communication, but is actively constructed by the knower;
  - o the function of knowledge is adaptive (in respect to the environment) and it is aimed at organizing the experiential flow by the knower/perceiver, not at discovering any objective ontological reality [10] [11].
- The validity for the above ecosystems of the autopoiesis principle [12], according to which a system can be described simply in terms of self-productive machine, able to self-reproduce recursively, creating, modifying or destroying itself in response to inputs and to external perturbations.

What emerges is essentially a kind of paradigm of interaction that tends to obscure the exogenous control paradigm, still lasting in the field of physical planning. In this emerging paradigm both the organism and the social systems are recognized as real agents, interacting each other and with another agent, the observer [13] and/or planner.

This is pretty close to the image of the "self-guiding society" of *Inquiry and Change* by C.E. Lindblom [14]: you can try to solve a problem (or improve a situation), by means of reflection and analysis (i.e. basing on some kind of knowledge), or you can take action to promote social interaction. In the latter case the result is not dependent on any analysis of the problem or elaboration of solutions, but it is with the action in relation to that problem that you move towards the solution, i.e. towards the kind of preferred situation. Furthermore the interaction - in which can be used some complementary form of knowledge - does not mean that a decision is taken, aimed at solving a problem explicitly recognized as such, by some "authority" [14] [15].

3   Smart Cities and Urban/Territorial Living Labs

To this kind of interactionist paradigm can be assimilated also the Living Labs (LLs). It is therefore the case of focusing on the spatial declension of LLs - the so-called Territorial Living Labs (TLLs) - and on its relations with the emerging wave of the Smart Cities. And finally on the relationships that may exist with innovative planning practices.

The core of the Living Labs is the idea of co-design, which exceeds the concept of user-centred design in the direction of a participation of the users in the design process from its very early stages [16]. Even in the field of greater application in Europe of the Living Labs' method, i.e. that of ICT R&D, this should mean a deep rooting of the labs in the places of the community of reference [17]; and therefore a close relationship with the everyday problems of cities and territories. The territorial Living Labs can therefore be conceived and defined not in terms of mere extroversion of research from the laboratory to the real world, that is quite usual in the field of territorial research, but more precisely in terms of the "transverse" role, rather than only sectorial in the ICT field, that they can play in relation to the aforementioned problems and then to a more general "local innovation", intended as simultaneously technological, social, cultural, economic and institutional [17].

It is evident the similarity of the concept of Territorial Living Lab described above with that of Urban Living Lab (ULL) of the 1st Call "Urban Europe", whose Annex C provides the following definition: «...a forum for innovation, applied to the development of new products, systems, services, and processes, employing working methods to integrate people into the entire development process as users and co-creators, to explore, examine, experiment, test and evaluate new ideas, scenarios, processes, systems, concepts and creative solutions in complex and real context». In addition, the call specifies that:

- «An Urban Living Lab is based on users and other stakeholders being co-creator in a systematic way in the innovation process...»;
- «Users in an Urban Living Lab may refer to both end users of o product or service, or those involved in service provision...»;
- «An Urban Living Lab is located right where the process being addressed take place, in real time and in the real context of the process...»;
- «An Urban Living Lab brings in expertise from several academic disciplines, and integrates them deeply together...».

In turn it is consonant with this concept of Urban/Territorial Living Lab the vision of the Smart City as a kind of «complex urban attitude that considers several areas of urban development and management (...) asking smart technology solutions be aligned with urban issues and interests, driven in a targeted manner based on existing expertise and not adopted from the outside without further reflection and effectively developing existing assets, (competencies, know-how, competitive focus activities) in cooperation with social, economic, and political forces» [18]. In other words, according to this line of interpretation, Smart Cities are deeply rooted on their problems and challenges so to overcome administrative borders and including the surrounding regions as far as the problems require, as well as relying on the specificities of involved users and citizens more and more recognized as key actors of the innovation processes [19]. Consequently it is recognized that altogether, Future Internet, Smart City and Living Lab form an intelligent innovation ecosystem comprising users/citizens, ICT companies, research scientists and policy makers, in which the Future Internet represents the technology push, Smart Cities represent the application pull and Living Labs (here we would say better Urban/Territorial Living Labs) are the meeting ground between push and pull. I.e. they can in turn be seen as an ecosystem partnership, working not as simple test-beds of new technologies - as stated in the Strategic Resarch Framework of "Urban Europe" - but rather providing opportunities to users/citizens to co-create innovative scenarios based on equally innovative technology platforms, in this way contributing also to the co-creation of new content and/or applications [19]. That is to say that «Living Labs are standing at the crossroads of different paradigms and technological streams such as Future Internet, Open Innovation, User co-Creation, User Content Creation and Social Interaction (Web 2.0), Mass Collaboration (i.e. Wikipedia), Internet Networking and Network Computing» [20] [21]. Allowing them to perform a function of true interface between new technologies and the city (and territories).

I think, in summary, that the still largely unrealized potential of innovation in planning, deriving in my view substantially from Lindblom's vision of a "self-guiding society", would certainly be married with the urban-territorial declension of Living Labs, considered as part of a broader ecosystem, environmental, territorial and socio-techno-digital, and in particular with their common feature of interactive sets, rather than command and control.

4   The Image of the "Smart" City

4.1   Entrepreneurial or "EnvironMental"?

The image of the "Entrepreneurial City 2050: economic vitality and innovation", directly linked to the Living Labs by the Strategic Research Framework of "Urban Europe", is interconnected to three other images provided in Framework itself, namely:

- Connected city 2050: smart logistic & sustainable mobility
- Pioneer City 2050: social participation & social capital
- Liveable City 2050: ecological sustainability

Despite this declared interconnection, and although it is not at all approached to the idea of Smart City, the image of the Entrepreneurial City is strongly exposed to the neo-Marxist critiques to the rhetoric of Smart Cities. The subtitle itself of an article about Smart Cities [3]- "intelligent, progressive or entrepreneurial" – clearly puts in contrast the intelligent and progressive features of the Smart City and the entrepreneurial ones. In particular Hollands explores to what extent labelled smart cities can be understood as a high-tech variation of the 'entrepreneurial city' [22], evidently to be intended as a city ruled by those tenets of urban entrepreneurialism [23] which include «...seeking out the most effective means to maximize profit, creating jobs in both existing and emerging markets through speculative projects, engaging in coordinative partnerships with private businesses in order to increase consumerism, and generating exposure for investment by extolling specific place-based features and associated imageries» [24]. The result of the exploration of Hollands is that «...it might be argued that beneath the emphases on human capital, social learning and the creation of smart communities, lay a more limited political agenda of 'high-tech urban entrepreneurialism'». Although «this assertion requires further study and in-depth analyses of specific urban cases» [3].

Beyond the likelihood of Hollands' findings, it seems to me that the image of an "EnvironMental City", i.e. of an Eco-Socio-Technological Urban and Regional System could be the most appropriate for both the construction and the evaluation of any experiment of Smart City.

4.2   No Social Innovation without Economic Innovation

The appropriateness of the image of an "EnvironMental City" for an approach through Urban/territorial Living Labs to Smart Cities is even more clear if the term eco-system - which could also mean eco-logical and eco-nomic together - is taken (as in the Digital Business Ecosystems) as a biological metaphor aimed at highlighting the interdependence between all the actors in the economic environment, which co-evolve with their capacities and roles, generating a changing environment, self-organizing and self-optimizing, where cooperation and competition are balanced, although within the free market dynamics [25] [6].

In this regard I noted elsewhere [26] that it is now time to recognize that self-organizing socio-economic actors can and must act directly on the pursuit of territorial development and cohesion. Or, in other words, that the broadest possible number of socio-economic actors must be involved, to this end, in co-creating products, services and innovative content.

It is exactly what could be done through an Urban/Territorial Living Labs approach to Smart Cities. And it is precisely also the contribution of innovation, I believe, that this approach could provide to spatial planning.

4.3   No Smart Cities through Planning without Smart Planning

In the traditional view, town and country planning is intended as a separate instance of top-down control of the territorial transformation processes, aimed at achieving a state of order and formal balance, or social, or environmental, etc. This top-down view of planning has been by long time subjected to heavy criticism in and out the discipline. Nevertheless planning theory (and practice) – participatory planning included - is still strongly influenced by a vision of the planning as a decision support technique, which relentlessly forces it into a scheme in which one (government) or more (governance) enabled subjects have to take decisions "with the participation" or the "involvement" of other "not enabled" subjects. In other words, it is always a sharing "from outside". But spatial transformation effects, as some scholars have rightly pointed out [27] [15] [28], are largely produced because of a myriad of micro-decisions that actors take "from inside" their fields of action and expertise, using the knowledge available *hic et nunc*, and variously interacting each other. The claim to control the actions of others of the classical planning, but also the involvement "from outside" of the participatory and strategic forms of planning, not only collide against this evidence, but they even prevent to implement the policies pursued by institutional decision makers themselves, profiting from the actions of the multitude of "spatial transformation agents" [26].

This is a very important issue to me, in a sense much more disciplinary than economic-political (if detachable): the Urban/Territorial Living Lab approach to the Smart Cities is able to overcome at least some of the epistemological obsolescence of planning, provided to shun from any neo-interventionist and neo-statist temptation; where for "State" of course has to be intended any level of government, including the local one [28]. And no criticism of the 'entrepreneurial city', though plausible and acute [24] [22] [3] [23], could justify in my opinion a such evident methodological regression.

But does it exist a "smart planning" research and/or operational topic in the field of spatial planning? Indeed, there are three main meanings of "smart planning" to which the literature and some current practice refer [29].

In the first meaning "smart planning" is essentially identified with those planning practices that are less or more successfully endorsing the principles of "smart growth" [30] [31], as intended particularly in the North American context.

In the second meaning, with particular reference to some practices emerging in the European context rather than to literature, "smart planning" consists simply in planning (no matter which way) for Smart Cities, intended as cities "technologically enhanced".

In the third and last current meaning, again referring mainly to the practices rather than to literature, the stress on technological aspects is shifted from cities to "new" technologies (generally GIS-based) used for city planning (that we could then call "technoloGIS"...).

It seems to me that within the spatial planning research field are not retrievable explicit proposals of "smart planning" able to go beyond the three current meanings of the term mentioned above.

In my view, however, the term "smart" remains closely associated with an emphasis on technological aspects. But those aspects have to be intended as bearers of intrinsic socio-cultural values (or dis-values) and as such they have not to be confined in only one polarity of the pair planning/city. In the sense that you cannot pursue any socio-technical innovation in the city through planning, unless you simultaneously pursue some form of socio-technical innovation within city planning [32] [21].

4.4   No Technological Innovation without Social Innovation and Vice Versa

To clarify why "urban smartness" remains in my opinion closely associated with technological innovation, or better with socio-technological innovation, I have to say that I totally agree with the McLuhan's [33] view of technology as including any human artefact and as expressive of the relationships, allowed by the artefact itself, between human senses, between man and man and between man and environment, according to the well-known slogan "the medium is the message". On this basis I consider misleading, and once more methodologically and epistemologically regressive - as well as operationally ineffective if not counterproductive - any opposition between technology and uses of technology, included the so-called "social uses" which, as McLuhan has made clear, are nothing but technologies in turn. Similarly, I cannot join to any conceptualization which makes the one (technology) subordinated to the others (uses), and vice versa of course.

The fact is that any change can occur only in the form of technological change (in the McLuhan sense) and that emerging new technologies, especially web technology, are inherently bearers of intra- and inter-subjective relational schemes, and between Man and his environment, of great innovative potential, and progressive, even in the social and economic field.

It doesn't seem to me a novelty, in this sense, the attempt to compress if not to cancel this great potential, and to turn it in favour of dominant forces and socio-economic interests, even with adaptive forms of transmutation of these latter. But I consider even less innovative and culturally appropriate to deny the above potential, just because it is or it could be subservient to other purposes, postulating a similar subservience to some other "higher use", the nature of which in turn strictly technological, although perhaps social, remains hidden in the eyes of those selves postulating it.

5   Smart Cities (and Planning) for Real

Towards the end of the above cited article Hollands says that «...the 'real' smart city might use IT to enhance democratic debates about the kind of city it wants to be and what kind of city people want to live in a type of virtual 'public culture', to redefine a term from Sharon Zukin (1995)» [3, p. 315].

This paper is not aimed to identify and describe any interesting experiences - although existing [29] - of "smart planning" through an Urban/Territorial Living Labs concept, which also seem to correspond to the criteria provided by Hollands to discriminate the "real" Smart Cities. Rather, I will refer here to the case I know best (as member of its "Observatory"), i.e. the "Perrphèria Project - Networked Smart Peripheral Cities for Sustainable Lifestyles", to extrapolate from it the most interesting aspects in the perspective of an innovative spatial planning research [1].

Perrphèria is a 30-month action funded by the European Commission under the Information Communication Technology Policy Support Programme (ICT PSP), part of the Competitiveness and Innovation framework Programme (CIP). It aims at deploying convergent Future Internet (FI) platforms and services for the promotion of sustainable lifestyles in and across emergent networks of "smart" peripheral (in respect to the "centre" of Europe) cities in Europe. The core-network of Perrphèria consists of five pilot cities (Athens, Bremen, Genoa, Malmö, Palmela) and of seven sponsoring cities (Lisbon, Helsinki, Rio de Janeiro, Budapest, La-Ferté-sous-Jouarre, Larnaca, Malaga, Malta, Palermo).

In my opinion is worth to emphasize here in particular two salient aspects of the initiative, to be considered in the framework of Living Lab environments, i.e in the framework of environments within which different subjects of the involved communities interact each other in co-creation processes of social and territorial innovation.

First, in Perrphèria the desired behavioural changes are let emerge through replicable, scalable and transferable patterns of individual and collective innovation, called "Behavlets"; supported and/or induced by corresponding elementary units of technological innovation, called "Urblets", consisting of physical (urban) devices "powered" by FI applications. For example, in the case of Malmö, the Urblet is intended as a set of devices embedded in the neighbourhoods, able to detect and publicly display the energy consumption of common areas, so as to give rise to more virtuous energy behaviours (Behavlets) not confined, as usual, in the purely private sphere. With regard to this first aspect it is important to emphasize the inseparability of the pair Urblets/Behavlets, that in my interpretation could mean the full integration of social and technical aspects in a higher socio-technical entity, as well as the overcoming of the dichotomy technology/use of technology [33].

The second salient aspect to be highlighted, even more relevant for the territorial research, is that the co-design of the pairs Urblets/Behavlets - or rather of the Urblets intrinsically and potentially bearers of Behavlets - should occur in Periphèria in the context of so-called "structured scenarios" [34]. To interpret, in my opinion, as "visions" emerging from the interactions but also able to reflect and refract in the individual projects and strategies, so that to polarize the molecular processes towards collective dynamics [35].

These important aspects of Periphèria are in my opinion directly related to the above view of the Smart City as an EnvironMental City. In the sense that the abandonment of the paradigm of exogenous control, in favour of the acceptance of the immanence of planning in the EnvironMental System with which it interacts, should result in particular in the participation (of the planner) to the local generation and to the trans-local interconnection of the above cited visions. This latter intended as dynamic configurations, able to support and contextualize the co-design of technological devices embedded in cities and territories, inherently expressive of new relationships between people and environment.

6   Conclusions

It is possible, and useful, to approach the emergent theme of Smart Cities through an Urban/Territorial Living Lab concept. To do that it is necessary to outline the basic notions of a research reference framework aimed first of all at treating in a unified way environmental systems, territorial, socio-economic and technological (especially digital).

The deriving image of the Smart City as an "EnvironMental City" is based on the following pillars:

- the self-organized nature of the system, namely the inability to proceed in a direction in which the designers decide what should or should not be possible in the system itself;
- the consequent abandonment of the paradigm of exogenous control in favour of that one of mutual interaction between all elements of the system, planner included, i.e. the substantial acceptance of the image of the "self-guiding society";
- the inseparability, and especially the need not to subordinate each other, of the elements of the pairs social innovation/economic innovation, social innovation/technological innovation, technology/ use of technology, spatial innovation/planning innovation, etc.

From a point of view of spatial planning research, the main interest of the experience of Living Lab for Smart Cities and Territories, considered here as a paradigmatic case, lies in the possibility, in the Periphèria Project, to completely immerse the planning activities in an interactive flow of eco-socio-technological innovation, through the participation of planners in the generation and continuous regeneration of visions supporting the co-design of technological devices, inherently expressive of new human relations, social, environmental and political, embedded in cities and territories.

To freely paraphrase R. Buckminster Fuller [36], the function of what he called – but I don't - "design science" is to introduce into the environment new artefacts, the availability of which will induce their spontaneous employment by humans and thus, coincidentally, cause humans to abandon their previous problem producing new behaviours and devices.

Maybe that the function (or at least one function) of planning is precisely to support this kind of co-creative design.

References

1. Concilio, G., De Bonis, L., Marsh, J., Trapani, F.: Urban Smartness: Perspectives Arising in the Perihèria Project. J. Knowledge Economy (2012) (published online)
2. De Bonis, L.: EnvironMental, Planum 25 (2012) (published online)
3. Hollands, R.G.: Will the real smart city please stand up? In: Hogan-Brun, G., Jung, U.O.H. (eds.) City, vol. 12(3), pp. 303–320. Routledge, London (2008)
4. Gunderson, L.H., Holling, C.S. (eds.): Panarchy: Understanding Transformations in Human and Natural Systems. Island Press, Washington (2002)
5. De Bonis, L.: Dal trasferimento alla retroazione tecnologica. In: 13th National SIU Conference. Città e crisi globale. Clima, Sviluppo e convivenza, Rome (2010)
6. Nachira, F., Dini, P., Nicolai, A.: A Network of Digital Business Ecosystems for Europe: Roots, Processes and Perspectives. In: Nachira, F., et al. (eds.) Digital Business Ecosystems, pp. 1–20. Office for Official Publications of the European Communities, Luxembourg (2007)
7. Kumar, S., Bentley, P.J.: Biologically Inspired Evolutionary Development. In: Tyrrell, A.M., Haddow, P.C., Torresen, J. (eds.) ICES 2003. LNCS, vol. 2606, pp. 57–68. Springer, Heidelberg (2003)
8. Watzlawick, P., Beavin, J.H., Jackson, D.: Pragmatics of Human Communication: A Study of Interactional Patterns, Pathologies, and Paradoxes. Norton, New York (1967)
9. Bateson, G.: Steps to an Ecology of Mind. Chandler Publishing Company, San Francisco (1972)
10. Von Glasersfeld, E.: The Reluctance to Change a Way of Thinking. The Irish Journal of Psychology 9(1), 83–90 (1988)
11. Von Glasersfeld, E.: Radical Constructivism. A Way of Knowing and Learning. Routledge, London (1996)
12. Maturana, H.R., Varela, F.J.: The Tree of Knowledge. Shambhala, Boston (1987)
13. Heylighen, F.: Second Order Cybernetics. In: Heylighen, F., Joslyn, C., Turchin, V. (eds.) Principia Cybernetica Web
14. Lindblom, C.E.: Inquiry and Change. Yale University Press, New Haven (1990)
15. Crosta, P.L.: Politiche. Quale Conoscenza Per l'azione Territoriale, Franco Angeli, Milano (1998)
16. Sanders, E., Stappers, P.J.: Co-creation and the new landscapes of design. CoDesign 4(1), 5–18 (2008)
17. Marsh, J.: Living Labs and Territorial Innovation. In: Cunningham, P., Cunningham, M. (eds.) Collaboration and the Knowledge Economy: Issues, Applications, Case Studies. IOS Press, Amsterdam (2008)
18. Acatech: Smart Cities. German High Technology for the Cities of the Future. Tasks and Opportunities", Acatech bezieht Position, vol. 10. Springer

19. Komninos, N., Schaffers, H., Pallot, M.: Developing a Policy Roadmap for Smart Cities and the Future Internet. In: Cunningham, P., Cunningham, M. (eds.) Proceedings on eChallenges e-2011 Conference (2011)

20. Pallot, M., Trousse, B., Senach, B., Schaffer, S.H., Komninos, N.: Future Internet and Living Lab Research Domain Landscapes: Filling the Gap between Technology Push and Application Pull in the Context of Smart Cities. In: Cunningham, P., Cunningham, M. (eds.) Proceedings on eChallenges e-2011 Conference (2011)

21. De Bonis, L.: Is Planning 2.0 a Mashup? In: Rabino, G., Caglioni, M. (eds.) Planning, Complexity and New ICT, Alinea, Firenze, pp. 205–214 (2009)

22. Harvey, D.: From Managerialism to Entrepreneurialism: The Transformation in Urban Governance in Late Capitalism. Geografiska Annale 71B(1), 3–17 (1989)

23. Jessop, B.: The Entrepreneurial City: Re-imagining Localities, Redesigning Economic Governance or Restructuring capital. In: Jewson, N., McGregor, S. (eds.) Transforming Cities, pp. 28–41. Routledge, London (1997)

24. Gillen, J.: The Co-Production of Narrative in an Entrepreneurial City: an analysis of Cincinnati, Ohio. In 'turmoil', Geografiska Annaler, Series B, Human Geography 91(2), 107–122 (2009)

25. Moore, J.F.: The Death of Competition. HarperBusiness, New York (1996)

26. De Bonis, L., Concilio, G., Marsh, J., Trapani, F.: Towards a deep integration of socio-economic action and spatial planning. In: Schiuma, G., Spender, J.C., Yigitcanlar, T. (eds.) Knowledge, Innovation and Sustainability: Integrating Micro and Macro Perspectives, Proceedings E-Book of IFKAD-KCWS 2012, 7th International Forum on Asset Dynamics and 5th Knowledge Cities World Summit, pp. 1323–1328 (2012)

27. Lindblom, C.E., Cohen, D.K.: Usable Knowledge: social science and social problem solving. Yale University Press, New Haven (1979)

28. Crosta, P.L.: Di cosa parliamo quando parliamo di urbanistica? In: Tosi, M.C. (ed.) Di Cosa Parliamo Quando Parliamo di Urbanistica?, Meltemi, Roma (2006)

29. Concilio, G., De Bonis, L.: Smart Cities and Planning in a Living Lab Perspective. In: Campagna, M., De Montis, A., Isola, F., Lai, S., Pira, C., Zoppi, C. (eds.) Planning Suppport Tools: Policy Analysis, Implementation and Evaluation, FrancoAngeli, Milano (2012)

30. Edwards, M.M., Haines, A.: Evaluating Smart Growth: Implications for Small Communities. J. Planning Education and Research 27(1), 49–64 (2007)

31. Grant, J.: Theory and Practice in Planning the Suburbs: Challenges to Implementing New Urbanism, Smart Growth, and Sustainability Principles. Planning Theory & Practice 10(1), 11–33 (2009)

32. De Bonis, L.: Planning as Medium vs. Planning as Means. In: Besussi, E., Rizzi, P. (eds.) On the Edge of the Millenium, Franco Angeli, Milano (1999)

33. McLuhan, M.: Understanding Media, M. McLuhan, Toronto (1964)

34. Celino, A., Concilio, G.: Participation in Environmental Spatial Planning: Structuring-scenario to Manage Knowledge in Action. Futures 42(7), 733–742 (2010)

35. Lévy, P.: L'intelligence Collective. In: Pour une Anthropologie du Cyberespace, La Découverte, Paris (1994)

36. Buckminster Fuller, R., Kuromiya, K.: Cosmography. In: A Posthumous Scenario for the Future of Humanity. Macmillan Publishing Company, New York (1992)

Impact of Urban Development and Vegetation on Land Surface Temperature of Dhaka City

Debasish Roy Raja[1] and Meher Nigar Neema[2]

[1] Department of Urban and Regional Planning,
Chittagong University of Engineering and Technology (CUET),
Chittagong-4349, Bangladesh
[2] Department of Urban and Regional Planning,
Bangladesh University of Engineering and Technology (BUET),
Dhaka 1000, Bangladesh

Abstract. This paper addresses both quantitative and qualitative assessment of the relationship between the land surface temperatures (LST) and land cover (LC) changes in Dhaka Metro Area (DMA) using Landsat TM/ETM+ data over the period from 1989 to 2010. The LC map was prepared using supervised classification methods. On the other hand, using the calibration of spectral radiance and emissivity correction LST has been derived from the thermal band of Landsat TM/ETM+. To establish the relationship between LST and LC, GIS based spatial simulation has been conducted. The changing of LST is found to be directly correlated with LC transition. LST has shown to increase in areas with growing urban developments. In contrary, the amount of Vegetation (NDVI) is negatively correlated with LST. The trend of LST and LC transitions indicates that LST will be abruptly increased in near future. The urban LST maps, the analyses of thermal-land cover relationships and the spatial simulated results thus obtained could be used as strategies for quality improvement of urban environment and a smart solution to the reduction of Urban Heat Island (UHI) effect.

Keywords: GIS, Remote sensing, Land cover change (LCC), Land surface temperature (LST).

1 Introduction

Dhaka city is confronted with a significantly high rate of physical and population growth since 1981 [1], [2] which has created tremendous pressure on urban land, utility services, and other amenities of urban life. A substantial growth of built-up areas (*urban development*) is transforming increasingly the landscape from vegetated cover types to Impervious Surface (IS). It is building up Urban Heat Island (UHI), which has adverse effect on the *urban climate change* such as abrupt temperature rise, erratic rainfall, degrading air quality [3], [4], [5], [6], [7], [8], [9]. Therefore, Dhaka city is affected by erratic rainfall and heat stress, resulting calamities like flood, water logging, health outbreak, and water scarcity including greenhouse climate changes [10]. The 0.5^0 C global warming realized over the last century is due mainly to the increase of greenhouse gases, urbanization, and other plausible climatic factors such as desertification [8], [11], [12], [13]. However, only few *qualitative* studies [14], [15] were found to assess the

B. Murgante et al. (Eds.): ICCSA 2013, Part III, LNCS 7973, pp. 351–367, 2013.
© Springer-Verlag Berlin Heidelberg 2013

urban climate change in Dhaka city but *no study* has been performed yet to correlate the changes of both Land Cover Area (LCA) and Land Surface Temperature (LST) with urban development and Vegetation. Therefore, it is undoubtedly important to examine the impacts of urban development and Vegetation on Land surface temperature (LST) in Dhaka city and to find out its consequence on foreseeable future.

The main goal of the study is to examine the impacts of urban development and vegetation on land surface temperature (LST) in Dhaka Metropolitan Area (DMA). The change of LST and LCA has been determined over different time-periods using remote sensing and spatial techniques of GIS and the impact of urban development, NDBI and vegetation (NDVI) on LST also have been analyzed. The study area is confined to Dhaka Metropolitan Area (DMA) because of its high level growth in the last two decades [16] shown in Fig. 1. DMA also cover the core Dhaka city as well as surroundings area of its. As changing of LST with transition of LCA change is derived in the research, the surrounding area of the DMA also considered explaining the facts.

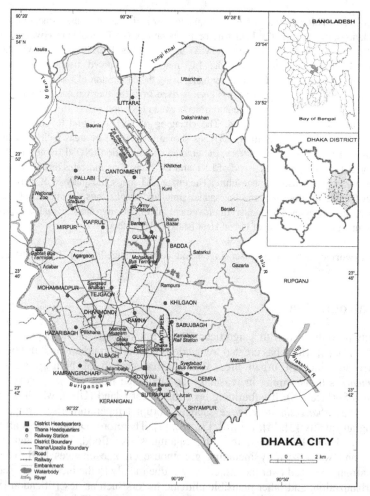

Fig. 1. Location map of the study area [18]

2   Methodology

At first, Landsat Satellite images of DMA area over different time period (1989-2010) were compiled from USGS website as zip format. Landsat satellites images were taken into consideration due to its thermal band from which LST can be determined as well as different period of images are available in free public-domain. However, this kind of image has mid spatial resolution which thermal band resolution is 30 m and others bands are 30m. To obtain better results, images of years 1989, 2000 and 2010 are selected with same season but with limited seasonal variation. The Landsat data used for analysis is shown in Table 1 [17]. The Landsat TM/ETM+ image was collected as Universal Transverse Mercator (UTM) within 46N-Datum World Geodetic System (WGS) 1984.

Table 1. Landsat TM/ETM+ images used in the study [16]

Representative year	Path and Row	Date & Scan time	sensor	Resolution	Cloud %	Remarks
1989	137 & 44	Jan 12, 1989 03:57:14	Landsat4 TM		0	
2000	137 & 44	Feb 28, 2000 04:17:27	Landsat7 ETM+	Thermal band 60m & other bands 30m	0	Low seasonal variation
2010	137 & 44	Jan 30, 2010 04:15:40	Landsat5 TM		0	

Next, to prepare LC map image with required enhancement for the identification and selection of the interest area, contrast enhancement, intensity/ hue/ saturation transformations,   density slicing were adopted. Landsat image has also been enhanced by generating composite band combination. The generation of composite band combination was employed particularly for false color Composite (FCC), true color composite, false natural color composite.

For composite band combination, the following procedure was adopted. Landsat TM/ETM+ image has several bands. Any three bands of the same sensor form an image that is called false color composite (FCC) [19]. Fig. 2 shows several composites of Landsat 5 TM images (DMA) that we have shown with different band combinations. RGB means basic three color red, green and blue and it has been used for band 4, 3 and 2 to make FCC. This FCC represents different color codings: blue for the urban area, red for vegetation, dark blue to black for water bodies, white to brown for soil without vegetation. Using similar strategy, true color composite shows the different LCA as its real color. Fig. 2(e) shows trafficability composite RGB= Band 6, 4 & 3) which emphasizes the traffic lane and built-up area as dark purple color. Therefore, different LCA features can be separated by choosing color of the cell. More color intensity increases the probability of any LCA types of the composite image. These band composite images were used to identify and select the interest area for building the signature of image classification.

Image classification is generally a process of sorting pixels into a finite number of individual classes, or categories, of data based on their values [20]. We employed supervised classification method as follows:

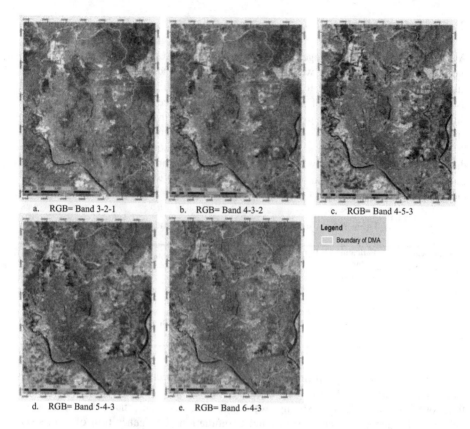

a. RGB= Band 3-2-1   b. RGB= Band 4-3-2   c. RGB= Band 4-5-3

Legend
☐ Boundary of DMA

d. RGB= Band 5-4-3   e. RGB= Band 6-4-3

Fig. 2. Composite Band Combination: (a) true color composite, (b) false color composite, c) false color composite, (d) false natural color composite, and (e) traficability composite

a) Ground data, secondary data and composite band images are used to digitize the known LCA so that the building signature value can be collected. Signature values contain the Digital Number (DN) value of each band of this image such as the reflectance value of the cell. It is called signature development. For this research, five types of land cover have been selected on the basis of their similar character shown in Table 2. A lot of signature values were determined for each land cover types and a statistical mean value was also derived for each land cover from this value known as combined signature or merged signature.

b) As collected signature values from the composite band image may contain deviation or error, it was evaluated from signature mean graph. The signature value range of each land cover types is separated from others. From the combined signature mean Fig. 3, it is found that each land cover types mean signature value is different from others. Though the difference of land cover type signature mean value is subtle for one band but in other band the difference was found to be high.

Table 2. Selected criteria for land cover classification

LC type	Criteria
Water and wetland	River, Permanent open water, pond, canal, lake, reservoir, permanent and seasonal wetland
Vegetation Type	Trees , agricultural land, grassy land, park and playfield etc.
Built-up Area	All type of infrastructure such as residential, commercial , industrial, road, village settlement etc.
Earth Fill or Sand	Constructions site, development land, earth filling or sand
Bare soil	low land, marshy land, vacant land

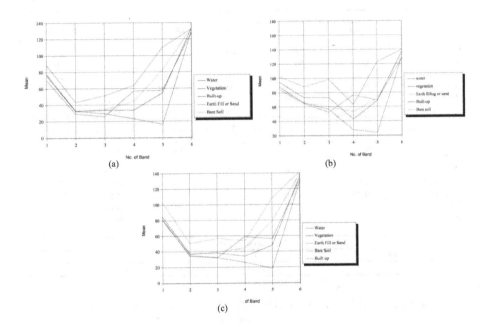

Fig. 3. Combined signature mean: (a) 1989, (b) 2000, and (c) 2010

c) The map accuracy or error estimation is derived from the signature error matrix on the base of reference data. The map accuracy is given in the Table 3. From the table it is found that the selected signatures are not overlapped with other signature. At least 95% of the each class data are accurate for all year. The overall accuracy is about 99.60% for 1989, 98.28% for 2000 and 99.32% for 2010.

d) Landsat TM/ETM+ image pixels were converted to units of absolute radiance using the following equation [19-20].

$$\text{Radiance, L}\lambda = (\text{QCAL}/255) *((\text{LMAX-LMIN}) + \text{LMIN} \quad \text{(For Landsat TM)} \quad (1)$$

Table 3. Error estimation

| Classified Data | Reference Data | | | | | |
	Water (%)	Vegetation (%)	Earth fill (%)	Built-up (%)	Bare soil (%)	Row Total
Year 1989						
Water	99.91	0.03	0.00	0.00	0.00	28618
Vegetation	0.06	98.64	0.00	0.00	2.68	5981
Built-up	0.02	0.00	99.74	0.00	0.85	3411
Earth Fill	0.00	1.33	0.00	98.71	0.73	470
Bare soil	0.01	0.00	0.26	1.29	95.73	801
Column Total	28641	6023	3408	389	820	39281
Year 2000						
Water	97.13	0.00	0.00	0.01	0.00	6829
Vegetation	0.16	98.77	0.00	0.00	0.38	10927
Earth fill	0.00	0.31	100.00	0.00	0.00	1941
Built-up	0.46	0.00	0.00	97.70	0.44	10924
Bare soil	2.26	0.93	0.00	2.29	99.17	6996
Column Total	7030	11027	1907	11119	6534	37617
Year 2010						
Water	99.66	0.00	0.00	0.00	0.29	11593
Vegetation	0.07	98.76	0.00	0.18	0.40	3716
Earth fill	0.00	0.00	99.56	0.00	0.18	1584
Built-up	0.00	1.24	0.44	99.82	0.16	1181
Bare soil	0.28	0.00	0.00	0.00	98.97	7249
Column Total	11612	3723	1578	1118	7292	25323
Map accuracy (%)						
1989		2000		2010		
99.60%		98.28%		99.32%		

$$\text{Radiance, } L\lambda = ((LMAX-LMIN)/(QCALMAX-QCALMIN))*(QCAL-QCALMIN)+LMIN \text{(For Landsat ETM+)} \quad (2)$$

Where,

QCALMIN	= Minimum Digital Number(DN) Value
QCALMAX	= Maximum Image Digital Number(DN)
QCAL	= Digital Number(DN) of the Band 6
LMAX	= Maximum spectral radiances
LMIN	= Minimum spectral radiances

For conversion of Image digital number (DN) to spectral radiance following data is used shown in Table 4.

e) Satellite Brightness temperature or black body temperature is derived from spectral radiance by the following formulas [16, 18]. Constant values are given in the table 5.

$$T_B = \frac{K_2}{ln\left(\frac{K_1}{L_\lambda}+1\right)} \tag{3}$$

Where

T_B = At- satellite Brightness temperature
L_λ = Spectral Radiance

Table 4. Conversion of the image digital number (DN) values to spectral radiance [17]

Value	Year		
	1989(TM)	2000 (ETM+)	2010 (TM)
QCALMIN	1	0	1
QCALMAX	255	255	255
QCAL	Image Digital Number	Image Digital Number	Image Digital Number
LMAX	15.303	17.040	15.303
LMIN	1.238	0.000	1.238

Table 5. ETM+ and TM Thermal Band Calibration Constants [19]

	K_1 (Wm^{-2} sr^{-1}μm^{-1})	K_2 (K)
TM	607.76	1260.56
ETM+	666.09	1282.71

The Land Surface Temperature (LST) was derived from the following formula for emissivity correction. Each land cover types have different emissivity. For this study, a simple grouping, that is, 0.95 for vegetative areas and 0.92 for non-vegetative areas [21], [22], [23] were used to calculate LST.

$$S_t = \frac{T_B}{1+(\lambda.T_B/\rho).ln\,\varepsilon} \tag{4}$$

Where

S_t = Land Surface Temperature
λ = 11.457μm
ρ= 1.438× 10^{-2} mK
ε= Emissivity

Derived LST map of different years are shown in Fig.s. 5-7. Table 6 shows the statistical information of LST.

f) Remote sensing Normalized Difference of Vegetation Index (NDVI) was used to measure the density of vegetation. To measure the NDVI, following formula was used to calculate this index value [24]

$$NDVI = (band\ 4 - band\ 3)/\ (band\ 4 + band\ 3) \tag{5}$$

Built-up area as well as impervious surface can be measured by Normalized Difference Built-up Index (NDBI).

NDBI of different years are derived using the following formula [25]:

$$NDBI= (band\ 5 - band\ 4)/ (band\ 5 + band\ 4) \tag{6}$$

Obtained statistical information of NDVI and NDBI is presented in Table 6.

Table 6. Statistical Information of LST, NDVI & NDBI

Year	LST				NDVI			NDBI		
	Minimum	Maximum	Average	Standard deviation	Range	Mean	Standard deviation	Range	Mean	Standard deviation
1989	20.42	33.97	24.74	1.67	-0.63 to 0.65	0.155	0.139	-0.698 to 0.554	0.069	0.143
2000	22.02	39.47	28.26	2.44	-0.41 to 0.36	-0.087	0.131	-0.459 to 0.502	0.112	0.122
2010	21.82	36.13	26.81	2.17	-0.27 to 0.50	0.057	0.092	-0.488 to 0.588	0.087	0.137

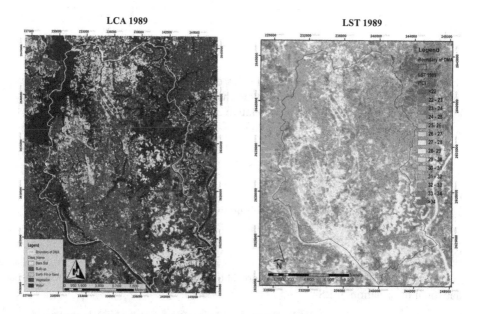

Fig. 4. LCA and LST MAP, 1989

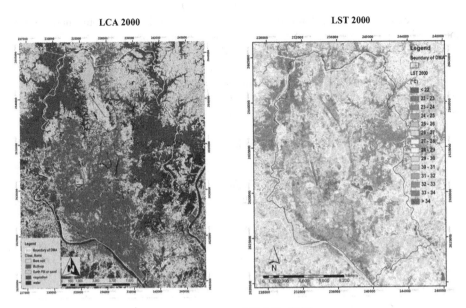

Fig. 5. LCA and LST MAP, 2000

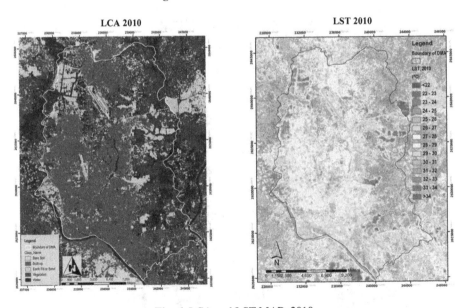

Fig. 6. LCA and LST MAP, 2010

3   Results and Discussion

Table 7 summarizes how the land covers change during the period of 1989 to 2010. In 1989, the LC areas were vegetation (105.22 sq. km) and water body (76.00 sq. km) and built-up area was found to be 59.39 sq. km. In 2000, both vegetation and water bodies were decreased to 70.14 sq. km and 19.45 sq. km respectively and the built-up area was increased to 95.42 sq. km. But the scenario was significantly changed in 2010. The vegetation and water bodies were increased slightly such as 5.82 % and 5.17 % respectively. This may be attributed to seasonal variation. In contrary, the growth of built-up area remains the same. The total scenario of changing land cover has been shown in the Fig. 7 during the period of 1989 to 2010. During this period the water body is decreased in 13.42% of total area where as built-up area is increased in 23.18%.

Table 7. Land Cover (LC) change in DMA (1989 to 2010)

Land cover types	Year			Loss & Gain of LC (1989 to 2010)		
	1989 (km²)	2000 (km²)	2010 (km²)	1989-2000 (%)	2000-2010 (%)	1989-2000 (%)
Water	76.00	19.45	35.17	-18.59	5.17	-13.42
Vegetation	105.22	70.14	87.85	-11.53	5.82	-5.71
Built-up	59.39	95.42	129.89	11.85	11.33	23.18
Earth fill or sand	30.53	32.36	15.09	0.60	-5.68	-5.08
Bare soil	33.01	86.79	36.15	17.68	-16.65	1.03
Total Area	304.15	304.15	304.15	-	-	-

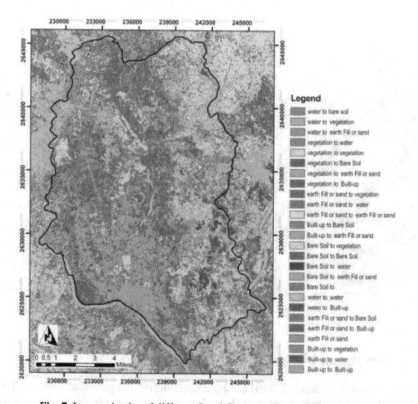

Fig. 7. Loss and gain of different Land Cover types (1989 to 2010)

Fig. 7 also shows how built-up area has been growing up during the period of 1989 to 2010. 30.17 km² and 11.00 km² of vegetation and bare soil lands respectively have been turned into built-up area during the year 1989 to 2000 is illustrated in the Table 8 and Fig. 8. In this decade highest amount of vegetation land has been converted into built-up area in the south and east portion of the city shown in Fig. 8 (1989 to 2000) as green color.

Table 8. Conversion of Land Cover Types into Built-up area in DMA (1989 to 2010)

Land cover types	1989 (Km²)	1989 to 2000 (Km²)	2000 to 2010 (Km²)
Water to Built-up	-	4.9383	2.3643
Vegetation to Built-up	-	30.1716	8.2989
Built-up to Built-up	-	43.0254	85.6332
Earth fill or sand to Built-up	-	6.2793	11.7054
Bare soil to Built-up	-	11.0007	21.8862
Total Built-up Area	59.3874	95.4153	129.888

Fig. 9 represent the attention to the trend of land covers type changing. All types of land cover's growth rate are changeable in either negative or positive rate except built-up area. Only built-up area is increasing in a constant positive rate. In the Fig. 9, a linear trend line is added and an equation is also derived for the future estimation of built-up area. The co-relational value, r^2 of this equation is 0.999. Value of r^2 indicates that the variable of this linear equation is highly co-related.

The LST can be changed for pattern and density of land cover type. The spectral resolution of the Lansat TM/ETM+ images are 30 m. So, 900 sq. m areas do not contain the same LC property. Greater percentage of land type determines what will be the LC type for the pixel of Landsat image. For this reason, same type of LC's LST can be varied for same year. Again the LST value of year 2000 is higher than the others because of seasonal variation. The derived mean LST of different LC type has been plotted in Fig. 10 where the LST line of the year, 1989 shows almost a liner line among built-up, Earth fill or sand and bare soil. But the vegetation has lower LST value than other categories. In 2000, the line of LST line is abrupt and the built-up LC retains hotter than other land cover classes. The mean LST of built-up is 30.65°C which is the highest LST in the Fig.10. Whereas the second highest heat contains the Earth fill or sand type land cover and its mean LST is 30.44°C. Bare soil and water are the other land covers which are 28.62 and 27.19°C respectively. In the year 2010, the LST character was found to be different than year 2000. Earth fill or sand land cover type is the highest temperature which is 29.72°C. But vegetation land has the lowest LST value. It can be inferred from the above discussion that LST of vegetation value is always lower than other categories and built-up or Earth fill or sand have the higher LST than the others. So it can concluded form the above discussion that the LST is increasing along changing of LCA within this period which can be the effects of urban heat island.

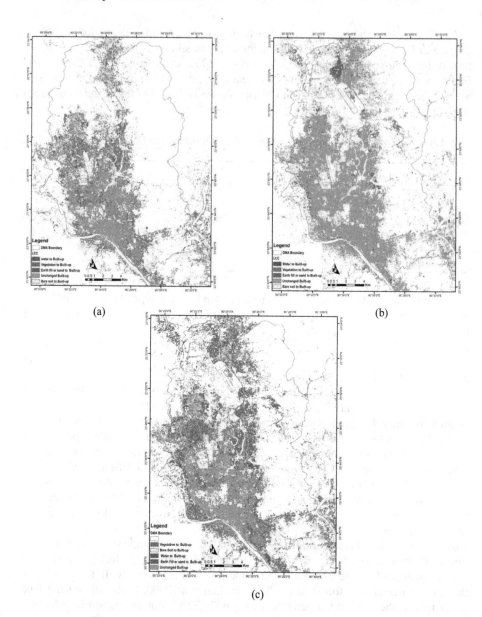

(a)

(b)

(c)

Fig. 8. Conversion of other land cover types into built-up area for the period of: (a) 1989 to 2000, (b) 2000 to 2010, and (c) 1989-2000

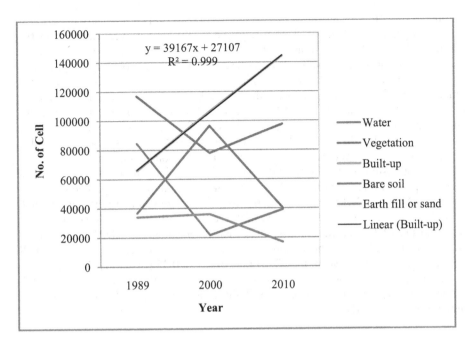

Fig. 9. Trend of land cover change over 1989 to 2010

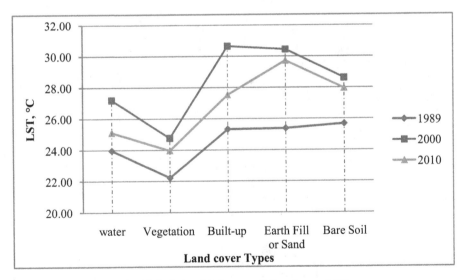

Fig. 10. Relationship between LC and mean LST

The relationship of LST and NDVI of different period is shown in the simulated Fig.11. Fig.11(a) represents the relationship between the NDVI and LST of year 1989. In this year the highest positive NDVI value is about 0.65 and most of the Pixel value is plotted in the positive site. The density of positive NDVI value indicates that the more probability of available of vegetation, so the vegetation density of this year is found high. The trend line of LST is negative Co-related with vegetation. From the simulated Fig. 11(b), it is found that the density of the cell value is more in the Negative portion of NDVI value. So it indicates that the less probability of available of vegetation and negative NDVI value represents highest LST value in the simulated graph. From Fig. 11 (c), it is found that the value of NDVI is about 0.5 and the density of the cell value is found between the NDVI ranges of -0.1 to 0.15 in the simulated graph which indicate that the density of vegetation is low. From those graph it is clear that the LST of different year is decreasing with the increasing value of NDVI as well as vegetation. So, categories of vegetation land impact on LST.

The relationship of the NDBI and LST is shown in the simulated Fig. 12. Form those graph it is found that the LST is increasing with the increasing value of NDBI. Positive value of NDBI represents the Built-up area, Bare soil and Earth fill or sand. From Fig. 12 (a), it is found that desity of the cell vule is between the rage of -0.15 to 0.23 and the LST trend line is less slopy than other simulated NDBI graph. In the simulated Fig.s 11(b)-(c) show more sloppy trend line than year 1989 and the cell density is increased between the NDBI rage of 0.1 to 0.35. Generally the rage of 0.1 to 0.35 is sensitive for built-up area. so it can be conculed that LST is incresing with the growing built-up areas. So growing built-up area is one of the major factor of urban heat island(UHI) effect.

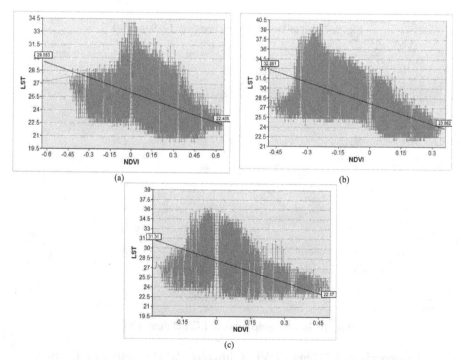

(a)

(b)

(c)

Fig. 11. Relationship of NDVI and LST: (a) 1989, (b) 2000, and (c) 2010

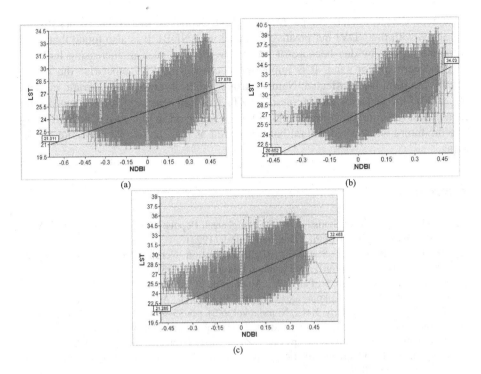

Fig. 12. Relationship of NDBI and LST: (a) 1989, (b) 2000, and (c) 2010

4   Conclusions

In this paper, a successful correlation has been established between the LST changing of DMA area with the transition of LCA during the period of 1989 to 2010 using remote sensing image-based analysis. Land covers areas (LCA) of DMA have seen to change rapidly during the period of 1989 to 2010. The highest amount of vegetation land (33.17 sq. km) is converted to built-up area in center city during the period of 1989 and 2000. But the character of converting built is changed during the period of 2000 to 2010. It happens with more spread out within DMA. Firstly, the categories of water land are converted into Earth fill or sand / Bare soil then it is converted into built-up area. Category of Water land was found to decrease (-13.42%) in the highest rate in this study period. On the other hand, category of built-up is grown up (23.18%) in constant growth rate during this period. From the analysis of NDBI and LST, it is found that the LST value is increasing with the growing built-up areas. So, it creates the urban heat island (UHI) effect. If the built up area is increasing in this rapid growth rate, UHI effect will be in critical situation which adversely affect on microclimate. Average LST is also correlated with the LCA changing using NDVI and NDBI. NDVI such as vegetation has also an effect on LST. In 2000, the highest average LST of built-up area is found because of low vegetation density. The LST of built-up is comparatively low in other years than the other LC types because of better vegetation density. Analysis of LC and LST indicates that LST is negatively correlated with the NDVI but positively related with NDBI. It is also found that LST is increasing day by day because of urban development such as increasing built-up area and earth fill or sand. It is important to control the LST otherwise the micro climate has been badly affected. Natural hazard like heat stress, flood, erratic rainfall etc. will be the cause of this increasing LST. This can also negatively affect of our economy of the country. If the changing LST is going on in this way, it is difficult to face the future unexpected natural hazard as well as loss the livable environment for living.

References

1. Bangladesh Bureau of Statistics (BBS): Bangladesh Population Census 1991 Urban Area Report Dhaka: Ministry of Planning, Bangladesh (1997)
2. Bangladesh Bureau of Statistics (BBS): Population Census 2001, National Report (Provisional) Dhaka: Ministry of Planning, Bangladesh (2003)
3. Ifatimehin, O.O., Ishaya, S., Fanan, U.: An analysis of temperature variations using remote sensing approach in Lokoja area. Nigeria Production Agricultural and Technology, Nasarawa State University, Keffi, Nigeria, vol. 6(2), pp. 35–44 (2010) ISSN: 0794-5213
4. Mayer, H., Matzarakis, A., Iziomon, M.G.: Spatio-temporal variability of moisture conditions within the urban canopy layer. Theor. Appl. Climatol. 76, 165–179 (2003)
5. Hossain, S.: Rapid urban growth and poverty in Dhaka city. Bangladesh e-Journal of Sociology 5(1) (2008)
6. Atkinson, B.W.: Numerical modeling of urban heat-island intensity, Department of Geography, Queen Mary, University of London, London E1 4NS, U.K (2002)
7. Dewan, A.M., Yamaguchi, Y.: Land use and land cover change in greater Dhaka, Bangladesh: using remote sensing to promote sustainable urbanization. Applied Geography 29(3), 390–401 (2009)
8. Alam, M., Rabbani, M.D.G.: Vulnerabilities and responses to climate change for Dhaka. Environment & Urbanization 19(1), 81–97 (2007)
9. Dixon, P.G., Mote, T.L.: Patterns and causes of Atlanta's urban heat island initiated precipitation. Journal of Applied Meteorology 42(9), 1273–1284 (2003)
10. Monsur, S.K.: Dhaka's vulnerability to climate change. The Daily Star, Karwan bazaar, Dhaka (July 15, 2011), http://www.thedailystar.net/newDesign/news-details.php?nid=194224
11. Nasrallah, H.A., Balling, R.C.: Spatial and temporal analysis of middle eastern temperature changes. Clim. Change 25, 153–161 (1993)
12. Kukla, G., Gavin, J., Karl, T.R.: Urban Warming, vol. 25, pp. 1265–1270 (1986)
13. Wood, F.B.: Comment: On the Need for Validation of the Jones et al. Temperature Trends with Respect to Urban Warming. Clim. Change 12, 292–312 (1988)
14. Basak, P.: Spatio- temporal trend and dimensions of urban form in central banglasesh: A GIS and remote sensing analysis, unpublished MURP thesis, department of urban and regional planning, BUET. Dhaka, Bangladesh (2006)
15. Rahman, A., Mallick, D.L.: Climate change impacts on cities developing countries: A case study of dhaka. C40 Tokyo Conference on Climate Change Adaptation Measures for Sustainable Low Carbone Cities, Japan (n.d.), http://www.kankyo.metro.tokyo.jp/en/attachement/dl_mallick.pdf

16. Rabbani, M.D.G.: Climate Change Vulnerabilities For Urban Areas In Bangladesh: Dhaka As A Case, ICLEI, Bone, Germany (2010), `http://resilient-cities.iclei.org/fileadmin/sites/resilient-cities/files/docs/B4-Bonn2010-Rabbani.pdf`
17. US Geological Survey (2012)
18. Banglapedia: Dhaka (Geology) (2012), `http://www.banglapedia.org/HT/D_0147.HTM`
19. NASA: Landsat 7 Science Data Users Handbook (2010), `http://landsathandbook.gsfc.nasa.gov/pdfs/Landsat7_Handbook.pdf`
20. Erdas Imagine Manual: Leica Geosystems Geospatial Imaging, LLC, USA (2006)
21. Zhang, Z., Ji, M., Shu, J., Deng, Z., Wu, Y.: Surface Urban Heat Island In Shanghai, China: Examining The Relationship Between Land Surface Temperature And Impervious Surface Fractions Derived From Landsat ETM+ Imagery. The International Archives of the Photogrammetry, Remote Sensing and Spatial Information Sciences XXXVII(Pt. B8) (2008)
22. Weng, Q.: A Remote Sensing-GIS Evaluation of Urban Expansion And Its Impact On Surface Temperature in the Zhujiang Delta. International Journal of Remote Sensing 22(10), 1999–2014 (2001)
23. Nichol, J., Wong, M., Sing, F., Christopher, L., Kenneth, K.M.: Assessment Of Urban Environmental Quality In A Subtropical City Using Multispectral Satellite Images. Environment and Planning B: Planning and Design 33, 39–58 (2006)
24. Chen, X., Zhao, H.M., Li, P.X., Yin, Z.Y.: Remote sensing image-based analysis of the relationship between urban heat island and land use/cover changes. Remote Sensing of Environment 104, 133–146 (2006)
25. Zha, Y., Gao, J., Ni, S.: Use of normalized difference built-up index in automatically mapping urban areas from TM imagery. International Journal of Remote Sensing 24(3), 583–594 (2003)

Design of a Team-Based Relocation Scheme in Electric Vehicle Sharing Systems*

Junghoon Lee and Gyung-Leen Park**

Dept. of Computer Science and Statistics,
Jeju National University, Republic of Korea
{jhlee,glpark}@jejunu.ac.kr

Abstract. This paper designs a team-based relocation scheme for electric vehicle sharing systems in which stock imbalance can lead to serious service quality degradation. For an efficient operation plan for a relocation team consisting of multiple service staffs, stations are geographically grouped first and the planner merges as many equivalent relocation pairs as possible, considering the number of simultaneously movable vehicles. To obtain a reasonable quality plan within a limited time bound using genetic algorithms even for a large-scale sharing system, each relocation plan is encoded to an integer-valued vector. A vector index points to a overflow station while the vector element points to an underflow station, according to predefined association maps built by the number of surplus and lacking vehicles. The experiment result obtained by a prototype implementation shows that each addition of a service staff cut down the relocation distance by 38.1 %, 17.2 %, and 12.7 %, respectively.

Keywords: Electric vehicle sharing system, relocation team schedule, genetic algorithms, equivalent pairs, relocation distance.

1   Introduction

Carsharing is capable of reducing the number of vehicles and thus the amount of greenhouse gas emissions [1]. Moreover, considering that electric vehicles, or EVs in short, will penetrate into our daily lives in the near future, EV sharing is a very economic way to overcome their high cost and long charging time. In EV sharing services, EVs are parked and maintained in sharing stations [2]. The service designer selects the location of sharing stations according to many factors including demand density, space availability, and easiness-to-access. Customers rent out and returns EVs from and to a station on necessary basis. As shown in Figure 1, where stations are installed in airport, shopping mall, residential area, tour place, and business area, renters can either drive just between two stations or drop by some other places during their trips. After all, carsharing lies between public and private forms of transport.

---

* This research was financially supported by the Ministry of Knowledge Economy (MKE), Korea Institute for Advancement of Technology (KIAT) through the Inter-ER Cooperation Projects.
** Corresponding author.

B. Murgante et al. (Eds.): ICCSA 2013, Part III, LNCS 7973, pp. 368–377, 2013.

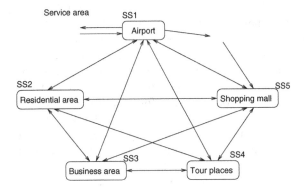

Fig. 1. EV sharing model

Even though the one-way rental is most convenient to customers, a stock imbalance is easily foreseen due to temporally and spatially asymmetric travel patterns [3]. EVs can get stuck in stations having low demand density, while some hot-spot stations runs out of EVs, worsening the serviceability. Accordingly, explicit vehicle relocation is indispensable to solve this problem. EVs need to be moved from overflow stations to underflow stations, either by human staffs or by vehicle transporters. The relocation by human staffs is more appropriate when EVs are scattered over the multiple stations. Moreover, an efficient operation schedule, generated by an intelligent computer algorithm [4], can reduce the relocation distance and time for the given number of service staffs. The relocation is the process of matching two items in each of both parties, one for EVs in overflow stations and the other for underflow stations.

If there are multiple relocation teams, it is necessary to determine a set of routes for each team, and this route planning is a kind of multiple traveling salesman problem (mTSP) [5]. For the case of relocation planning, we can practically consider the *cluster-first and route-second* approach, as the driving distance can be quite long for a relocation vehicle to cover the entire service area and the sharing service area can be geographically decomposed in practice. A series of decomposition criteria can be defined, including spatial affinity and temporal dependency. Moreover, the stations can be decomposed according to the specific system goal such as access priority, commute support, and the like. Then, relocation planning is reduced down to operation planning for a single relocation team. For EV relocation, each station set allocated to a single team is not necessarily disjoint. The overlapped stations can be used for intercluster route adaptation.

Essentially, the time complexity is important as the number of EVs in the sharing system can grow too much, while computational intelligence can cope with this problem. In this regard, this paper designs an EV relocation scheme based on genetic algorithms, which can obtain reasonable (suboptimal) solutions within a controllable time bound even for problems having large search space [6]. The proposed scheme begins with the assumption that the sharing stations are grouped and each group is assigned to a relocation team which consists of 2-5 service staffs and that a single relocation vehicle. Each relocation plan, namely, the set of relocation pairs of an EV in an overflow station and an underflow station itself, is represented by an integer-valued vector to run the genetic operators such as crossover, selection, reproduction, and mutation. In addition, the fitness function evaluates the quality of a relocation plan, taking into account the number of EVs simultaneously movable between two stations.

The rest of this paper is organized as follows: Section 2 reviews some related works. Section 3 describes the proposed scheme in detail, focusing on the encoding scheme and the cost function for our genetic algorithm design. After Section 4 shows the performance measurement results, Section 5 concludes this paper with a brief introduction of future work.

2 Related Works

As a recent example of a vehicle sharing system, Green Move creates a flexible EV sharing service, taking advantage of social networking to design a service satisfying the user-side expectations and figure out the dynamics in the user-side demand via real-time feedback and rating mechanisms [7]. Here, Green e-Box provides easy access to users and integrative coordination to management centers. Its key features lie in intermodality, multi-ownership, multi-business, and mobility credits. This ongoing project is funded by Regione Lombardia and a system test is on the way in the Milan area, especially focusing on accessibility and usability. Its promising ideas include different fares to help car relocation, real-time car pooling within the Green Move community, and education through social networks. The relocation scheme gives incentives to those trips bound for lacking stations by fare discount.

[8] presents how to solve a station distribution and route planning problem for general delivery services, based on 3 separate phases. The first phase divides the stations into a predefined number of groups according to proper decomposition techniques. The decomposition process reduces the problem size and the capacity clustering algorithm is developed based on the well-known k-mean scheme. Here, in addition to the standard station types such as normal stations and centroids, the centroid of the centroids is further defined for upper-level distribution center clustering. The second phase finds optimal or suboptimal routes within each group by means of existing heuristic algorithms. Finally, the third phase improves the routes between groups. This step modifies routing sequences among interrelated groups to get closer to the global optima, exploiting the simulated annealing technique combined with a 3-opt heuristic.

For EV relocation in carsharing systems, our previous work has designed a single-vehicle relocation strategy which reduces the relocation distance and time. The primitive relocation operation is assumed just for two staffs in a single vehicle as follows: Two staffs move to an overflow station driving a single service vehicle. One drives an EV to the assigned underflow station while the other follows by the service vehicle. Then, two head for the next overflow station together in the service vehicle again [9]. For the current and target distributions, the relocation scheme determines overflow and underflow stations, calculating how many EVs will be moved or replenished for each station. Then, the EVs in overflow stations are assigned to underflow stations based on the stable marriage problem model. Here, the preference is specified by the distance from each EV to an underflow station. The matching result can efficiently reduce the relocation distance, but this scheme cannot work in the case of more than two service staffs in a relocation team.

3 Relocation Scheduler

3.1 Problem Definition

The system operator triggers the relocation procedure usually according to a given rule, for example, during the nonoperation hours, when the service ratio is expected to drop shortly, and so on [10]. In addition, the target distribution, which is the EV distribution after relocation, can be determined by an appropriate strategy, possibly combined with the future demand forecast. With m relocation teams, the stations are partitioned into several groups, mainly by geographical factors, and each group is assigned to a relocation team. This assumption is practical and reduces the overall relocation distance, as most EVs are relocated within a group. Here, sometimes, it is necessary to move EVs between different groups. This inter-group relocation can be accomplished through intermediary stations which belong to multiple groups. Namely, the relocation team in an overflow group moves EVs to an intermediary station, as if it is an underflow station. Next, the intermediary station will be an overflow station in the underflow group.

Basically, the problem formulation is the same as our previous work [11]. For a station set, $S = \{S_1, S_2, ..., S_n\}$, let $C = \{C_1, C_2, ..., C_n\}$ be the current distribution and $T = \{T_1, T_2, ..., T_n\}$ be the target distribution, where n is the number of stations. Then, the relocation vector, $\{V_i\}$, can be calculated by $C_i - T_i$ for all S_i. If V_i is positive, S_i is an overflow station having surplus EVs which should be moved to other stations. As contrast, if V_i is negative, S_i is an underflow station and wants to receive EVs. If an EV, say E_k, currently belonging to a station, say S_i, will be relocated to another station, S_j, it is denoted by (S_i, S_j), and it is equivalent to (E_k, S_j). If there are u EVs to be relocated, its relocation plan consists of u pairs, namely, $|\{(S_i, S_j)\}| = u$. Here, EVs are all different, but an underflow station appears as many times as the number of EVs to replenish. The relocation distance is the sum of distances for all station pairs in a relocation plan. To calculate the relocation distance, it is necessary to be aware of distances between all individual station pairs.

3.2 Encoding and Fitness Function

To find an efficient relocation team plan taking advantage of genetic algorithms, it is necessary to encode a relocation plan to the corresponding integer-valued

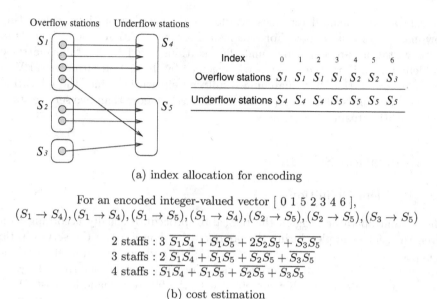

(a) index allocation for encoding

For an encoded integer-valued vector $[\,0\;1\;5\;2\;3\;4\;6\,]$,

$(S_1 \rightarrow S_4), (S_1 \rightarrow S_4), (S_1 \rightarrow S_5), (S_1 \rightarrow S_4), (S_2 \rightarrow S_5), (S_2 \rightarrow S_5), (S_3 \rightarrow S_5)$

$$2 \text{ staffs} : 3\,\overline{S_1 S_4} + \overline{S_1 S_5} + 2\overline{S_2 S_5} + \overline{S_3 S_5}$$
$$3 \text{ staffs} : 2\,\overline{S_1 S_4} + \overline{S_1 S_5} + \overline{S_2 S_5} + \overline{S_3 S_5}$$
$$4 \text{ staffs} : \overline{S_1 S_4} + \overline{S_1 S_5} + \overline{S_2 S_5} + \overline{S_3 S_5}$$

(b) cost estimation

Fig. 2. Encoding of relocation plans

vector called a chromosome. This encoding scheme can be explained by an example shown in Figure 2. There are 5 stations from S_1 to S_5, while S_1, S_2, and S_3 are overflow stations. Each relocation plan includes 7 relocation pairs, as there are 7 EVs to relocate. To represent a relocation plan by an integer-valued vector, overflow stations are indexed sequentially with duplication as shown in Figure 2(a). Each station appears as many times as the number of surplus EVs. Now, each index is bound to a specific overflow station. Next, each underflow station is assigned a set of unique sequence numbers according to the number of lacking EVs. In the example of Figure 2(a), 0, 1, and 2 are given to S_4, as S_4 needs to replenish 3 EVs.

A relocation pair can be represented by an index and the number located at the index, while a relocation plan is represented by a sequence of distinct numbers within the index range. For example, in $[0\;1\;5\;2\;3\;4\;6]$, number 0 appears at location index 0. It denotes the relocation pair which moves an EV from the overflow station bound to index 0, namely, S_1, to the underflow location associated with the number 0, namely, S_4. In addition, number 4 appears at location index 5. Index 5 is bound to S_2, and number 4 is bound to S_5. Now, the relocation problem is to find the best sequence having the minimal cost, just like the well-known traveling salesman problem. This encoding scheme makes it possible to run standard genetic operators such as crossover, selection, reproduction, and mutation. Even if the number of relocation pairs increases, genetic algorithms can adjust the iteration loop length to create the relocation plan within a reasonable time bound.

Genetic loops iteratively evaluate and improve the fitness of each solution. For EV relocation, relocation distance, defined by the distance the relocation vehicle moves, is the most important criteria as it can indicate energy saving and service staff efficiency. The shorter the distance, a better the relocation plan. It can be calculated by the sum of the distance between two stations for each relocation pair. However, if a relocation team consists of m service staffs, $(m-1)$ EVs can be moved simultaneously between the same overflow and underflow stations. Namely, m staffs move to an overflow stations while one staff drives the relocation vehicle. $(m-1)$ staffs drives $(m-1)$ EVs to an underflow station, respectively, while the driver of the relocation vehicle follows them. Then, they will go to the next overflow station until the relocation procedure completes. Actually, m is restricted by the maximum capacity of a relocation vehicle.

As can be inferred from the above example, we can expect relocation distance reduction with more than 2 staffs, only when a relocation plan includes equivalent relocation pairs. If they are all different, the relocation distance cannot be reduced regardless of the number of relocation staffs, as all EVs must be moved one at a time. With sharing station grouping, the number of stations in a group gets smaller. It enhances the probability of creating the same relocation pairs in a relocation plan. Moreover, the inter-group relocation is performed through the intermediary nodes which belong to part of groups or entire groups. This relocation creates many equivalent relocation pairs, as destinations within an overflow group will be an intermediary station even if their final destinations are different in the underflow group. The genetic iteration gives precedence to those relocation plans to make them survive.

4   Experiment Result

This section measures the performance of the proposed scheme via a prototype implementation using Microsoft Visual Studio 2012. The prototype version exploits the Roulette wheel selection method and initial population of chromosomes is selected randomly. For better population diversity, it does not permit duplicated members, and they will be replaced by new random ones during mutation operations. The fitness value, or interchangeably the relocation distance of a relocation plan, is calculated when the plan is first generated for better efficiency. The genetic parameters such as the number of iterations and the population size can be freely selected by experiments. For a relocation plan, the fitness function merges the pairs having the identical overflow and underflow stations first, and then each of m-1 pairs will be packed to one. In addition, the experiment fixes the number of stations to 10 and inter-station distance exponentially distributes with the average of 3 km.

The performance metric is the relocation distance, while performance parameters consist of population size, the number of EVs, the number of moves, and the number of underflow stations, respectively. Underflow and overflow stations are selected randomly out of 10 stations according to the given number of underflow stations. The experiment also randomly sets the number of surplus EVs in overflow stations as well as the number of lacking EVs in underflow stations, respectively. The random numbers are seeded by the system clock for each experiment run. For each parameter setting, 30 sets are generated and the results are averaged. In each experiment, the relocation distance will be plotted for the cases of 2, 3, 4, and 5 staffs. As mentioned previously, with m service staffs, m-1 EVs can be moved simultaneously. Here, we do not consider the intergroup route adaptation.

The first experiment measures the effect of population size to the relocation distance. Large population allows better diversity, but not necessarily leads to a better solution. Even with large population, different initial population can result in a little bit poor quality relocation plans. Here, the number of EVs is set to 100, 5 out of 10 stations are underflow ones, each relocation plan consists of 30 relocation pairs. As shown in Figure 3, when the population size is 60, the relocation distance is smallest. Moreover, this figure indicates that the population size hardly affects the relocation distance, especially for the case of 5 staffs. However, for the case of 2 staffs, the distance gap reaches 11.3 km. The addition of a service staff cut down the relocation distance from 64.6 km to 37. 7 km, when the number of staffs is changed from 2 to 3 and the population size is 60. It corresponds to 31.3 % reduction. However, when the number of staffs increases from 4 to 5, the performance improvement remains at 9.8 %.

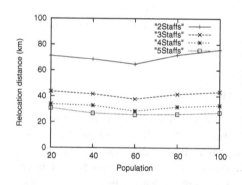
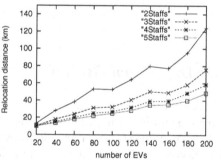

Fig. 3. Population size effect   **Fig. 4.** Effect of the number of EVs

The second experiment measures the effect of the number of EVs, while the number of underflow stations is set to 5. Here, 30 % of EVs need to be relocated. From now on, the population size is fixed to 50 and the genetic loop iterates 1,000 times. The results are plotted in Figure 4. The proposed scheme more improves the relocation distance when there are more EVs for all cases. If we change the number of staffs from 2 to 3, the performance improvement is 22 % when there are 20 EVs, but it will jump up to 39.1 % for the case of 200 EVs. In addition, for the change from 4 staffs to 5 staffs, the relocation distance is improved by 0.1 % on 20 EVs and by 7.9 % on 200 EVs. Here again, each addition of a service staff reduces the relocation distance by 38 %, 17.2 %, and 10.5 % on average. For large m, the planner is less likely to find m identical pairs.

Next, Figure 5 plots the effect of the number of moves, that is, the number of relocation pairs, to the relocation distance. If EVs neither enter nor get out of a relocation group, the number of relocation pairs is equal to the number of surplus EVs as well as the number of lacking EVs in the group. Moreover, the relocation plan embracing the intermediary stations makes all the numbers equal. Here, the number of EVs is set to 100, while 5 out of 10 stations are underflow stations. As contrast to the case of Figure 4, the performance gain does not change too much according to the number of moves for all cases of 2, 3, 4, and 5 staffs. The curve looks more linear, compared with Figure 4, as the total number of EVs is constant. Except for the case of 10 moves, the improvement remains around 38.1 %, 17.2 %, and 12.7 %, respectively. The performance enhancement is more influenced by the number of EVs.

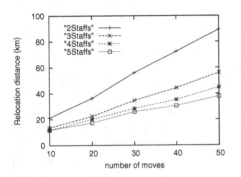

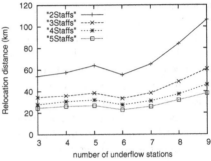

Fig. 5. Effect of the number of moves   **Fig. 6.** Underflow station effect

Finally, Figure 6 shows the effect of the number of underflow stations to which EVs will be moved. Here, if the number of underflow stations is 9, there is just 1 overflow station, from which the surplus EVs are relocated to other 9 stations. The number of total EVs is set to 100 and the number of moves is to 30. The number of underflow stations ranges from 3 to 9. When the number of underflow stations is close to the number of overflow stations, the relocation distance gets smaller. However, the performance gap between the adjacent two curves increases, as the skewed distribution creates more identical relocation pairs which can be merged for a simultaneous relocation. The relocation distance is cut down by up to 42.9 % when the number of service staffs increases from 2 to 3 and by up to 17.1 % when the number of service staffs increases from 4 to 5.

5 Conclusions

With the advent of smart grids, EVs are expected to replace gasoline-powered vehicles for better energy efficiency and eco-friendliness in transportation. EV sharing is a prospective business model to cope with not only high price of EVs for personal ownership but also difficult management stemmed from long charging time. Skewed rent-out and return patterns result in stock imbalance, which deteriorates service ratio seriously. Targeting at a sharing system with many stations scattered over a wide area, this paper has designed an intelligent relocation scheme taking advantage of genetic algorithms. Here, each plan is encoded to an integer-valued vector by means of two maps, one from vector indexes to overflow stations and the other from vector values to underflow stations. For a relocation team consisting of multiple service staffs, the planner merges as many equivalent relocation pairs as the number of simultaneously movable vehicles. The performance of the proposed scheme has been measured in terms of the distance ratio according to the number of relocation pairs and the like. According to the experiment, each addition of a service staff cut down the relocation distance by 38.1 %, 17.2 %, and 12.7 %, respectively.

As future work, we are planning to design a staff allocation scheme for multiple sharing station groups, as the relocation load is highly likely to be different group by group. For the sake of balancing the load in each group and reducing the relocation overhead, it is necessary to assign service staffs according to the estimation on the relocation distance and time. Moreover, the relocation sequence is also important when intergroup relocation is needed. That is, the relocation to the intermediary station in an overflow station must precede that in underflow stations. Actually, each relocation procedure within a group can start from any relocation pair for a given sequence. So, we think that it is possible to efficiently synchronize multiple relocation actions in respective groups. In the mean time, the matching scheme can integratively take into account the driving distance between each relocation pair, not just the relocation distance specified the relocation pairs [12].

References

1. Cepolina, E., Farina, A.: A New Shared Vehicle System for Urban Areas. Transportation Research Part C, 230–243 (2012)
2. Correia, G., Antunes, A.: Optimization Approach to Depot Location and Trip Selection in One-Way Carsharing Systems. Transportation Research Part E, 233–247 (2012)
3. Weikl, S., Bogenberger, K.: Relocation Strategies and Algorithms for Free-Floating Car Sharing Systems. In: IEEE Conference on Intelligent Transportation Systems, pp. 355–360 (2012)
4. Ipakchi, A., Albuyeh, F.: Grid of the Future. IEEE Power & Energy Magazine, 52–62 (2009)
5. Bektas, T.: The Multiple Traveling Salesman Problem: An Overview of Formulations and Solution Procedures. International Journal of Management Science 34, 209–219 (2006)
6. Giardini, G., Kalmar-Nagy, T.: Genetic Algorithm for Combinational Path Planning: The Subtour Problem. Mathematical Problems in Engineering (February 2011)
7. Lue, A., Colorni, A., Nocerino, R., Paruscio, V.: Greem Move: An Innovative Electric Vehicle-Sharing System. Procedia-Social and Behavioral Sciences 48, 2978–2987 (2012)

8. Lian, L., Castelain, E.: A Decomposition Approach to Solve a General Delivery Problem. Engineering Letters 18(1) (2010)

9. Lee, J., Kim, H., Park, G.: Relocation Action Planning in Electric Vehicle Sharing Systems. In: Sombattheera, C., Loi, N.K., Wankar, R., Quan, T. (eds.) MIWAI 2012. LNCS, vol. 7694, pp. 47–56. Springer, Heidelberg (2012)

10. Kek, A., Cheu, R., Meng, Q., Fung, C.: A Decision Support System for Vehicle Relocation Operations in Carsharing Systems. Transportation Research Part E, 149–158 (2009)

11. Lee, J., Park, G.: Genetic Algorithm-based Relocation Scheme in Electric Vehicle Sharing Systems. In: Submitted to International Conference on Information Technology Convergence and Services (2013)

12. Lee, J., Park, G.: Planning of Relocation Staff Operations in Electric Vehicle Sharing Systems. In: Selamat, A., Nguyen, N.T., Haron, H. (eds.) ACIIDS 2013, Part II. LNCS, vol. 7803, pp. 256–265. Springer, Heidelberg (2013)

Qualitative Analysis of Volunteered Geographic Information in a Spatially Enabled Society Project

Jarbas Nunes Vidal-Filho[1], Jugurta Lisboa-Filho[1],
Wagner Dias de Souza[1], and Gerson Rodrigues dos Santos[2]

[1] Departamento de Informática,
[2] Departamento de Estatística,
Universidade Federal de Viçosa
Viçosa, Minas Gerais, Brazil, 36570-000
{jarbas.filho,jugurta}@ufv.br,
{wagner.supremo,prof.gersonrodrigues}@gmail.com

Abstract. The increase of data sources on the Internet and Web 2.0 advances have contributed to significant changes in the way we produce spatial information. The citizen is using collaborative environments to produce their own data, whether in the area of public security, infrastructure or for simple fun. The voluntary contribution is essential to make a spatially enabled society and help in decision-making process at all levels of a society, whether governmental, private or by the citizen. Analyzing the information gain that is recorded in collaborative Web systems is essential to know what is not being recorded in official data sources. The aim of this study is to evaluate the information gain generated by the user on a project to transform a conventional computerized municipal management for a spatially enabled society.

Keywords: Volunteered Geographic Information, Web Systems, Geographic Information System, Public Security.

1 Introduction

Most Brazilian cities have gone through many problems in the public security area. Several cases of violence are constantly reported in newspapers, magazines and other media. Given this, the public security issue is being treated as a priority of governments and has become a research subject in several areas. The academy and the public security organs have advanced the development of new tools that can support decision making in the public security sector, whose aim is to fight crime.

The use of Geotechnologies applied in the area of public security is utilized by various organizations to combat violence. Therefore these geoprocessing tools have become essential in the process of decision making. Public security agencies have obtained good results with the use of tools of Geographic Information System (GIS). These results have been useful in improving the activities in police departments. However, some departments have problems that prevent the access to spatial information or the dissemination of spatial data.

B. Murgante et al. (Eds.): ICCSA 2013, Part III, LNCS 7973, pp. 378–393, 2013.

Mossoró is a Brazilian city located in the western Rio Grande do Norte state, with about 260.000 inhabitants [1]. In recent years, the Technical Institute of Police (ITEP) has recorded alarming numbers of homicides (e.g. murder), Drug traffic, among others [2]. From this, it was realized the need to create a computing platform that would help in making decision process of the police in the city of Mossoró. This platform would approximate people to the police and would make users the producers and reviewers of data about public Security in their neighborhood or street. Each user can make denunciations, record occurrences that he/she witnessed or heard by others and share any other useful information to security departments.

The idea is to make Mossoró a spatially enabled society in the context of public security, where the citizen and the police authorities may have access to user-generated content in real time and independent of computer platform that they are using. In this society, the citizen acts as a "*voluntary human sensor*" [3], producing information that may help to minimize violence and reviewing other contributions. Based on this, the project Mossoró Spatially Enabled emerged containing as computing platform, the collaborative Web system called MossoróCrimes, [4], connecting people to data.

In a spatially enabled society, location, services and spatial information are available and accessible to citizens, private companies and governments as a means of organizing information and activities in the decision making process [5]. According to [6], this kind of society provides a coordinated effort in the production, storage, dissemination and use of spatial data at all levels of that society. This type of environment should allow users access regardless of the computational platform and can be improved with the voluntary participation of users in the production of geospatial data.

For the development of this society, it is necessary the use of spatial data. They need to be sufficiently reliable, easily accessible and available in real time [7]. The developed system allows people with Internet access to query the recorded data and voluntarily collaborate with geospatial data about the public security of the city of Mossoro. The aim is to generate data that are not often recorded in the databases of the agencies responsible for public security. According to [8], this possibility that allows users to search and contribute to spatial information is crucial for a vision of a spatially enabled society.

The production of spatial data by any citizen of a society happened due to the advances of Web 1.0 to Web 2.0. At Web 1.0, it was able to view more information consumers than producers. The producers were largely people specialized in computer science or geosciences. With advances in Web 2.0, there was a greater number of data producers. Anyone with access to the Internet can produce information [9]. Some typical examples of the use of Web 2.0 are: Wikipedia, blogs and social networks.

The collaborative systems are crucial for the production of information. In this kind of system there is greater interaction among a user or groups of users to produce information [10]. This type of collaborative environment has been essential to the production of spatial data in a spatially enabled society. Increasingly users are generating content based on their geographic coordinates to aid in the process of decision making by a particular organization.

With advances in Web, users without technical knowledge began producing spatial data. Thereby, the term Volunteered Geographic Information (VGI) emerged to this new data production process. It was defined by Goodchild [3] as a new phenomenon to describe information generated by the user, combining elements of Neogeography, Collective Intelligence and Web 2.0. The concept of VGI can be described as a set of spatial data that are not produced by individuals and organizations specialized in the production of spatial data, but by people who use Web 2.0 tools to disseminate geospatial information [11]. In MossoróCrimes system, the user himself spends some of his/her time to produce his/her VGI data, reporting or recording police incidents that occur in his/her region. It is expected ethics from users of MossoróCrimes in the production of VGI.

Geography took other paths with the use of Web 2.0 techniques. It allowed that people without technical knowledge could create and overlay their own data. Thus, the term Neogeography was determined by [12] to describe this new way of working the geography along with Web 2.0 resources. In collaborative systems, data can be produced from voluntary contributions among people. These, together, exchange knowledge for data production, such as those that are found in the Wikis [13] systems. So, this exchange of knowledge among users in collaborative environments is called Collective Intelligence [14].

The aim of this study is to collect VGI data in order to statistically analyze the benefits of VGI for the public security departments in the city of Mossoró. Contributions received were considered in isolation, in order to know the information gain in the system. Furthermore, VGI data were compared with the official data from the Integrated Center of Public Security Operations (CIOSP) of Mossoró, which is the agency responsible for recording the police incidents reported by the citizen via telephone. The rest of the paper is organized as follows. Section 2 describes the project Mossoró Spatially Enabled and presents the system MossoróCrimes. Section 3 presents the mechanisms to ensure the quality of VGI. Section 4 shows the results of statistical analyzes on the data of MossoróCrimes and the comparison with data from CIOSP. Section 5 presents some conclusions and future studies.

2   Project of a Spatially Enabled Society

Mossoró Spatially Enabled is a project that consists of creating a spatially enabled society in the public security area using Web 2.0 tools to provide services of voluntary contribution and access to VGI data. The launch of the project took place in the city of Mossoro on October 10, 2012 and recorded the presence of local and regional press, authorities of police departments, professors, researchers and students interested in supporting research on public security in Mossoró. The platform developed in this project is the system MossoróCrimes (*www.ide.ufv.br/mossorocrimes*). It remains open for contributions record, but for this study, it was defined a period of two months (October 10, 2012 to December 10, 2012) to conduct statistical studies.

2.1   The System MossoróCrimes

The collaborative Web system known as MossoróCrimes is a computational platform connecting people to the data on the project Mossoró Spatially Enabled. This system uses the Google Maps platform to provide the digital map of the city of Mossoró and anyone with Internet access can fulfill a voluntary collaboration. Using programming languages HTML and CSS, it was developed a simple interface to facilitate access by the users with devices with limited resources, and it was also used the JavaScript language to perform the customization of the Google Maps API v3. Figure 1 shows the system interface.

Fig. 1. Interface of the system MossoróCrimes

The system has a module of collaboration that allows users to register the data of VGI (Figure 2). This module was developed using programming languages PHP, JavaScript, HTML and XML. The collaboration module receives textual information that represents the general data (date, time, subject, description, etc.) about the occurrence, multimedia information (video and photo) and extra files that may contribute to the collaboration. All information is stored and managed by database management system Mysql Server.

The system MossoróCrimes also has some services to help in the decision making process of police departments such as the filtering service, the service of areas of risk analysis, the statistical services and the *news* service. The filtering service filters contributions by type or category. The service of risk analysis is to identify the critical points of the city. This service can be used in conjunction with filtering service, so it is possible to analyze the areas of risk of each type of collaboration. The service of risk analysis is implemented based on the algorithm of Kernel Map [15]. The system implements a service that displays statistics of registered contributions and of the users that collaborate more on the system. The idea is to evaluate what has been informed and who is collaborating. Then, there may be a study on the user that collaborates more, checking if his/her information follow some standard and seeking to ban users with false information. The *"news"* service is a table containing the latest contributions that were recorded. This table is updated in real time when a collaboration is registered. The update of this table allows the police authorities to accompany the record of contributions in real time.

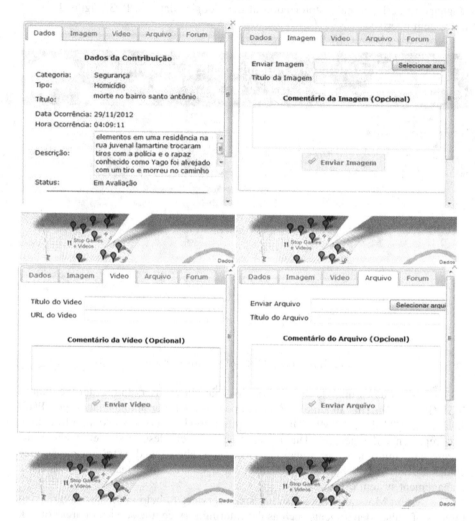

Fig. 2. Module of Collaboration

3   Quality of Volunteered Geographic Information

The use of VGI is still questioned nowadays because it is a data type which not much is known about its production. It is neither known who produced it nor the method or technique used in production for example. But according to the researcher [16], the risk for end users when using a VGI data is the same when using the data of poor quality of commercial or governmental producers. For example, in the database of CIOSP - Mossoró virtually all data are supplied via telephone calls, and most of them are anonymous phone calls. Therefore, there is also a risk when using official data. Thus, one should not dismiss the use of VGI data, as these can be useful in the decision making process, in the improvements in police activities and in the violence prevention.

Several projects that use data generated by users are able to gain credibility for their end users such as Wikipedia, Wikimapia and OpenStreetMap. The latter two perform the world mapping collaborative. They went through a long process of acceptance. Now they use mechanisms to monitor the quality of VGI and require user awareness in collaboration to achieve a level of overall satisfaction.

According to Goodchild [17], the geographic aspects of VGI have some degree of quality not found in other types of user generated content. Geographic data are widely used to perform analysis on a certain location. Sometimes there are no spatial data available for the manager to perform the procedures of decision making. For [11], when there is no data, VGI gains greater importance, since it makes possible to perform some kind of analysis, albeit with questionable quality. Goodchild [18] draws attention to the need to develop mechanisms to ensure the quality of VGI data using Wikis concepts, Collective Intelligence or Crowdsourcing. The latter can be understood as a crowd of people producing data to use in a particular application [19], [20].

Police in the city of Mossoró has an official database about crimes, but it is known that not all data are recorded. The idea of approach the citizen with the police is to make him/her provide data that in most cases the police doesn't receive such as denunciations about drug traffic , murderers, etc. In collaborative systems there is the problem of being provided false information, which may disturb police work. Thus, the system MossoróCrimes implements two mechanisms to monitor the quality of VGI. These mechanisms are described in the following subsections.

3.1   Mechanism Based on Assigning Grades

The first mechanism is the service of grades. It consists in the user assigning grades from zero to five for each collaboration. The system through a simple algorithm computes and displays the average of the recorded grades. This service of grades is a mechanism for providing quality to the VGI. Each collaboration has a interface as shown in figure 3.

3.2   Mechanism Based on Concepts of Collective Intelligence

The second mechanism introduced in the system MossoróCrimes is a forum for each collaboration. The idea is to use the concepts of collective intelligence to monitor data quality. It is possible to improve collaboration recorded through the exchange of knowledge between different users, but it can also destroy the credibility of it.

Fig. 3. Service of grade for evaluation the collaboration

Each collaboration has a forum to receive extra information about it from other users (Figure 4). In this forum users exchange information among themselves, providing complementary data to the collaboration recorded or decrease the credibility of a VGI registered. This interaction combining human and machine intelligence creates a more reliable information to the end user, characterizing the collective intelligence.

Fig. 4. Forum of the collaboration

4   Statistical Analysis of VGI Data

The data analyzed in this article correspond to VGI data collected by the system MossoróCrimes in a period of 60 days, starting on the project's launch. In collaboration module the system MossoróCrimes the user initially faces two broad categories (*Security & Others*) regarding the inclusion of contributions. Each category (e.g. *security*) is also classified by type of occurrence (e.g. *theft*). The categories and types are chosen from a window in the shape of *combobox*, i.e., the user simply select the category and type want he/she wants to collaborate with.

The *Security* category represents the types of occurrences directly linked to the issue of public security, such as *theft, robbery, homicide, Drug traffic*, etc. And the *Others* category represents data that have an indirect link with the area of public security, such as *lack of lighting, Public place in poor condition, lack of policing*, etc.

4.1   VGI Data Analysis of the System MossoróCrimes

This section provides a general discussion of the data that the system collected and analyzes the information gain for the system itself with the contributions that were recorded. Furthermore, it presents analysis representing the user's interaction with the contributions already registered. For this, it is analyzed the gain of images, videos, extra files, the average views of contributions, the number of users who used the service of grades and the forum in contributions.

4.1.1 Analysis of Contributions by Category
During the study period (60 days) 74 contributions were collected. 66 contributions in the *Security* category and 8 contributions in the category *Others*. In addition, the system had the registry of 97 users. Table 1 shows the percentage of registration for each category.

Table 1. Percentage of record for each category

Category of system	Percentage (%)
Security	89.2
Others	10.8

The *Security* category received the highest number of contributions, because the types of occurrences associated with this category are the types directly related to public Security. Thus, it was the category most in demand by users to register their VGI data. Furthermore, it can be stated that *42.42%* of the information recorded in the *Security* category were not in the CIOSP databases.

4.1.2 Analysis of Contributions by Period
In the first month the system received 44 contributions related to *Security* and *Others* categories. This represents *59.45%* of the total recorded contributions. Of the 44 contributions, 41 were recorded in the *Security* category and 3 in the *Others* category. Meanwhile, in the second month 30 contributions were recorded, representing *40.55%* of the total registered. Of the 30 contributions, 25 were recorded in the *Security* category and 5 in the *Others* category. Table 2 shows the percentage of registry in each category during the first and second month.

Table 2. Percentage of record of each category during the first and second month

Category of system	October 10, 2012 to November 10, 2012 (%)	November 11, 2012 to December 10, 2012 (%)
Security	93.20	83.35
Others	6.80	16.65

During the first month there was a greater amount of collaborations. This occurred because of the launch of the project in the city of Mossoró, which mobilized much of the press for publication. After launch, the authors used the Blog O Camera (*www.ocamera.com.br*), a police blog, to continue to publicize the system and invite users to collaborate on the platform MossoróCrimes. It was used the free tool Google Analytics to analyze the number of accesses to the system. It was found that in the first month there was more access. The next month the system recorded peaks of access when the system was broadcasted by commenting on the news in the police Blog "O Câmera".

One can also see that in the second month there was an increase in the number of records in the *Others* category. It is believed that, over time, people begin to learn more about the tool and find that they can collaborate with other information, such as *lack of policing*, *Public place in poor condition* and other contributions that are indirectly linked to the area of public security. Therefore, it can be stated that *62.5%* of the information recorded in the *Others* category were not in the CIOSP databases.

4.1.3 Analysis of Contributions by the Type of Incident

Each category registered in the system contains a kind of incident associated. In the *Security* category types were defined with the aid of a policeman of CIOSP - Mossoró.

In the *Others* category, it was defined types of occurrences based on requests from some users that sent suggestions via e-mail. Table 3 shows the amount of registration by type of occurrence in the *Security* category. Furthermore, it also displays the percentage of record related with the data registered in the *Security* category and with the total data recorded in the system.

Table 4 uses the same methodology of Table 3. It only shows the setting of analysis used in the *Security* category for the category *Others*.

In the MossoróCrimes system, in order to perform a collaboration it is required to provide only textual information, such as category, type, time, subject and description. But each contribution can receive extra information, such as photos, videos, extra files and extra comments that are entered in the forum. In the subsections below, it is possible to see the gain of contributions recorded in the system by collecting extra information. Furthermore, it is analyzed the user interaction with the contributions, i.e., the percentage of users who cooperated with the system, who assessed contributions and other analyzes.

Table 3. Statistical returns by type of collaboration in the category Security

Types of occurrences in the system MossoróCrimes	Count of the number of collaborations	Type of collaboration/Category Security (%)	Type of collaboration/Total contributions (%)
Point of consumption of drugs	3	4.54	4.05
Murder	25	37.87	33.78
Drug traffic (Illegal Drug Trade)	15	22.72	20.27
Theft	1	1.51	1.35
Robbery (others)	2	3.03	2.70
Attempted murder	8	12.12	10.81
Point of sale of drugs	5	7.57	6.75
Robbery of cargo transported	1	1.51	1.35
firearms shooting	1	1.51	1.35
Disturbance of the peace	2	3.03	2.70
Theft or Robbery of vehicles	1	1.51	1.35
Denunciation	2	3.03	2.70

4.1.4 The Gain with Photo and Video Information

It is observed that of the 74 contributions received during the period of 60 days, 27 collaborations possessed extra information in image format. Each contribution accepts one photo, so *36.48%* of the contributions of the system received images. These represent extra or additional information on collaborations. This represents more information to police agencies try to solve a case based on the VGI data and provides more credibility to VGI.

Nowadays it is common to press submit news about violence, using videos made by the citizen with phone or other digital equipment. Based on this fact, the system MossoróCrimes provides a module capable of receiving video provided by users. However, the system has not received any video as extra information. It is believed that this is linked to the fear of users of producing videos related to an act of violence, because usually there is some risk to the citizen.

4.1.5 The Gain with Information of Extra Files

A registered contribution may contain a photo or video. If another user has one more picture, video or any other file to insert into the collaboration, he/she should insert on the tab files of the contribution. The collaboration supports much information upon registration. So it can happen that user is not able to enter all data collaboration. Analyzing the 74 recorded contributions, only 1 received extra files, then, merely *1.35%* of the collaborations received extra files. It may seem few, but for a sector where information is practically scarce, it can be considered that this contribution as a reasonable gain depending on the quality of information.

Table 4. Statistical returns by type of collaboration in the category Others

Types of occurrences in the system MossoróCrimes	Count of the number of collaborations	Type of collaboration/Category Others (%)	Type of collaboration/Total contributions (%)
Traffic accident	3	37.5	4.05
Public place in poor condition	1	12.5	1.35
Disappeared person	1	12.5	1.35
Poor illumination	1	12.5	1.35
Lack of policing	1	12.5	1.35
Environmental crime	1	12.5	1.35

4.1.6 The Gain with Information from Comments of the Forum

The forum in each contribution functions as a mechanism of quality of the VGI. This forum aims to receive additional information about the occurrences, similar to the idea of collaboration on collaboration, as in blogs, where the collective intelligence is used to discuss the data and improve the quality of information. Analyzing the registered contributions, *24.32%* of them received extra comments.

4.1.7 Number of Users That Collaborated

During the experiment, the system received the registry of 97 users, which provided a *name*, *email* and *password*. Among them, only 20 users inserted a contribution to the system, which represent *20.60%* of the registered users. Analyzing the records of each user, it was found that each region of the Mossoró map had one or two users to record information about their neighborhood. By the user identifier in the database, it was possible to assess how behaved the records of a particular user in the city. And by the Google Analytics tool, it was noticeable that most users just visualized the home of the website, perhaps for fear of register and have their data stored, even with the possibility of creating a fictitious name.

4.1.8 Evaluation of Collaborations

As seen in a previous section, each contribution can be evaluated only once per user. The evaluation is another mechanism to guarantee the quality of VGI and gives users a way to review or validate recorded contributions. Only 33 collaborations were assessed, representing *44.59%* of them. And of the 97 registered users, it was noted that only *7.20%* of users rated the collaborations. Analyzing the database, among collaborations evaluated, *75.75%* were evaluated only one time, while *24.25%* were evaluated more than once. The low number of users evaluating contributions may have occurred for two reasons. They didn't know any information about the occurrences or they didn't have the necessary knowledge of this tool.

4.1.9 Viewed Collaborations

The system records the number of viewing of each occurrence as a way to track whether users are querying them and to create a ranking of most viewed collaborations. Analyzing the database, it was found that *100%* of recorded contributions were viewed. The number of views for most recent collaborations was around 15 *views* per contribution. And for older ones, it exceeded the number of 50 *views* per each.

The project was well publicized initially and received a higher number of users in the first month. This factor explains why older collaborations were more accessed, because they were the collaborations that users were initially. Analyzing Google Analytics, over time the number of users accessing the system was much lower, which made the number of views of collaborations decrease. The system needs a validation of the police trying to solve some registered occurrences, so users tend to believe more in the tool. Therefore there may be a continuous participation by users.

4.2   Comparing Data of the System MossoróCrimes with Official Data of CIOSP

The Integrated Center of Public Security Operations (CIOSP) of Mossoró is responsible for receiving via telephone and record police occurrence reported by the citizen. After receiving an occurrence report, the CIOPS place the order for the agency responsible for answering and resolving the occurrence. The police of Mossoró works with this database and ITEP statistical data to assist in the activities. Therefore, it is intended to show in this paper that by using the platform MossoróCrimes, police has another source of data that can aid in fighting crime. The idea is to show the gain with the data of the system MossoróCrimes to police of Mossoró, showing, by statistical analyzes, the amount of data that were not recorded in the CIOSP data.

During the study period, 74 contributions of VGI on the system MossoróCrimes were recorded and, in the same period, 5650 police occurrences were reported at the base of CIOSP. However, comparing the two databases, it is observed that 33 contributions of the system were not recorded in the CIOSP data, representing that *44.59%* of VGI data recorded in the system MossoróCrimes were not in the CIOSP official data.

Analyzing these 33 contributions that have not been registered at the CIOSP base and were at the base of MossoróCrimes, there was 28 unregistered collaborations on *Security* category, and 5 collaborations in *Others* Category. This occurred because this category is responsible for receiving the types of occurrences that are directly related to public security, so it was much more used. The following subsections make an analysis of the registration number of each type of occurrence on *Security* category of the system with CIOSP official data.

4.2.1 Analyzing the Security Category

The *Security* category collected all types of occurrences that were previously shown in Table 3. Each type in this category was compared with the corresponding type in the CIOSP database to know the number of records for each type of occurrence in the database. Figures 5 and 6 show the number of records for each type of occurrence in MossoróCrimes and CIOSP databases.

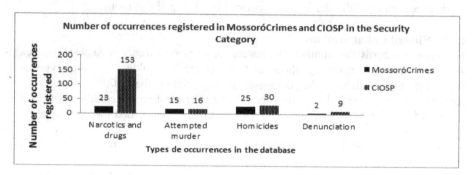

Fig. 5. Comparing the number of record in databases of MossoróCrimes and CIOSP

The CIOSP has a kind of occurrence called *Narcotics and drugs* which includes 3 types of occurrences of *Security* category on MossoróCrimes which are: *Point of sale of drugs*, *Point of consumption of drugs* and *drug traffic*. To perform this comparison, we adopted a join operation in the 3 types of occurrences of MossoróCrimes turning it into a single type of base like CIOSP. Figure 5 shows a well-balanced record number between the two bases. It is because they are the most common types and alarming of occurrences in the city of Mossoró. Drug traffic is directly reflected in the number of homicides and attempted murder that occur in the city of Mossoró and the denunciations made to the system were related to the subject drug or homicide. Therefore, there has been a major concern for users to record these types of occurrences.

In figure 6 it can be seen that there is a greater discrepancy between the data recorded in the MossoróCrimes system and the CIOSP database. It is believed that the fact that the system has received fewer records regarding the types of occurrences of Figure 6 is explained by these occurrences requiring more urgency to be solved. These are incidents recorded by the police phone and the citizen usually requires the presence of the police in time to solve the case.

Theft and *robbery* of figures 6 and 7, it was adopted a join operation used in *narcotic and drugs* because according to Table 3 there are more than one record of *thefts* and *robbery* in the MossoróCrimes system. Then, *thefts* and *robberies* represent the union of their records in the system related to each type.

It is noticed that there was a greater gain in the *narcotic and drug* type as this is the main problem of violence in the city of Mossoró. Constantly, police arrest people linked to Drug traffic in the city. Therefore, there has been a major concern for users to report such occurrence.

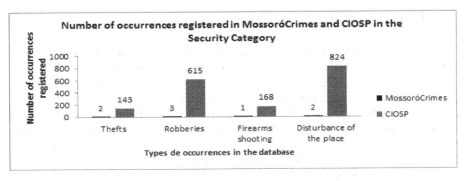

Fig. 6. Comparing the number of record in databases MossoróCrimes and CIOSP

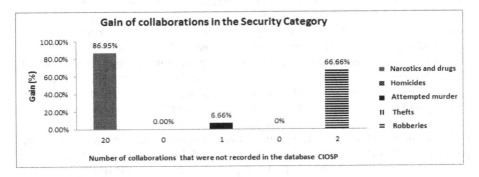

Fig. 7. Gain of Information in Security Category for CIOSP

Mossoró has excelled with alarming numbers of homicides. There is a greater concern in mapping this type of occurrence. Thus, all homicide records were in the CIOSP database. This kind of data was recorded with the help of the police searching to make decisions to reduce the numbers of homicides from previous years. In 2012 there was a drop in homicides. The *attempted murder* type relates to the *homicide* type, so it was expected that all data of *attempted murder* type were recorded because they are incidents that the police is normally aware of just like the *homicide* type. But only one record of the *attempted murder* type was not on the CIOSP base, which already represents a gain for the security agencies of the city.

The *thefts* type got no gain, since the only recorded data were occurrences of June. Then, when comparing with the database in the analyzed period they were not found. In the category *robberies*, there was a gain of 2 records, representing that *66.66%* of information of *robberies* registered on MossoróCrimes were not recorded in the CIOSP base. The only data that was in the CIOSP database was the *robbery of a cargo transported of mail* and this record was first inserted in the system and later in the police Blog O Camera. This blog has information in real time of CIOSP, and when reading the news on the blog, it was found that the time of registration in the system was sooner than the report of the occurrences on the blog.

The types of occurrences such as *denunciation, firearms shooting* and *Disturbance of the peace* have received few records and all records were not in the CIOSP database.

5   Conclusions and Future Work

The collection of information provides from citizens is important for the process of decision making by all levels of society. The rapprochement between citizens and the police is helpful to solve problems of public security, because in most situations the citizen has important information and is afraid to inform via phone as they may have their phone number registered. Web collaborative systems give an idea of obscurity for the producer of the data. Nobody knows him/her, so this can make it easier for users to contribute with more information. But it is necessary to give a positive feedback to the citizen, i.e., show that the tool is useful.

According to Goodchild, when talking about VGI data it is difficult to talk about its credibility, because it is not known anything about the producer. The proposed work shows interesting results in gaining information that could be used by police departments of Mossoró. Although they may be questioned about their quality, it is pertinent to state that VGI data in the system follow a pattern quite acceptable when compared with other academic studies conducted using CIOSP official data.

Analyzing the recorded data, there is much important information. It is informed names of alleged murderers or people involved in Drug traffic, denunciations of places where drugs are sold and used. Much information show that citizens know their region and daily live with these situations. So, overall, there was a considerable and important gain for security departments. They could adopt the platform as another tool to combat violence by encouraging citizens to contribute, but without ignoring phone use, as this is still more efficient to receive and dispatch the police cars.

It is intended in future studies to develop a mechanism of VGI quality using the concept of Wiki, which would make the citizen the reviewer of the data, as it is done in Wikipedia. Furthermore, it is intended to compare the VGI data with the data of Police Reports (BO's) of the city of Mossoró. The BO's are responsible to register an incident in an official way. For example, if the citizen is stolen, he/she can make a BO about the act occurred. If his/her object is retrieved, the citizen has the BO to prove that the object belongs to him/her. Finally, it is intended to make a comparison between the Kernel Maps of VGI, CIOSP and BO's, seeking to know the standard of data recording of VGI in relation to official databases.

Acknowledgments. Project partially financed by CNPq, CAPES and Fapemig. Authors also thank the CIOSP-Mossoró for providing the data for this research and the company Sydle.

References

1. Instituto Brasileiro de Geografia e Estatística (2013), http://www.ibge.gov.br (in Portuguese)
2. Instituto Técnico-Científico de Polícia (2013), http://www.itep.rn.gov.br/ (in Portuguese)
3. Goodchild, M.F.: Citizens as voluntary sensors: spatial data infrastructure in the world of Web 2.0. International Journal of Spatial Data Infrastructures Research 2, 24–32 (2007)

4. Vidal Filho, J.N., Lisboa-Filho, J., Souza, W.D., Oliveira, D.F.: Starting a Spatially Enabled Society: a Web system for collecting Volunteered Geographic Information in the area of public security. In: The Fifth International Conference on Advanced Geographic Information Systems, Applications, and Services, Nice, France (2013)
5. Williamson, I., Rajabifard, A., Wallace, J.: Spatially Enabling Government – An International Challenge. In: International Workshop on Spatial Enablement of Government and NSDI – Policy Implications, Seoul, Korea, pp. 1–12 (2007)
6. Khoo, V., Yee, L.S.: Spatially Enabled Singapore through Singapore Geospatial Collaborative Environment (SG-SPACE). In: GSDI 12, Singapore, Ásia, pp. 1–7 (2010)
7. Rajabifard, A.: A Spatial Data Infrastructure for a Spatially Enabled Government and Society. In: Crompvoets, J., Rajabifard, A., Loenen, B.V., Fernández, T.D. (eds.) A multi-View Framework to Assess Spatial Data Infrastructures, vol. 1, pp. 11–22. Digital Print Centre, Australia (2008)
8. Coleman, D.J.: Volunteered Geographic Information in Spatial Data Infrastructure: An Early Look At Opportnities and Constraints. In: GSDI 12, Singapore, Ásia, pp. 1–18 (2010)
9. Bugs, G.: Assessment of Online PPGIS Study Cases in Urban Planning. In: Murgante, B., Gervasi, O., Misra, S., Nedjah, N., Rocha, A.M.A.C., Taniar, D., Apduhan, B.O. (eds.) ICCSA 2012, Part I. LNCS, vol. 7333, pp. 477–490. Springer, Heidelberg (2012)
10. Furtado, V., Ayres, L., Oliveira, M., Vasconcelos, E., Caminha, C., D'Orleans, J., Belchior, M.: Collective intelligence in law enforcement the WikiCrimes system. Information Sciences 180, 4–17 (2010)
11. Elwood, S., Goodchild, M.F., Sui, D.Z.: Volunteered geographic information: future research directions motivated by critical, participatory, and feminist GIS. GeoJournal 72, 173–183 (2008)
12. Turner, A.D.: Introduction to Neogeography. O'Reilly Media, EUA (2006)
13. Chaves, A.P., Steinmacher, I., Vieira, V.: Social networks and collective intelligence applied to public transportation systems: A survey. In: Simpósio Brasileiro de Sistemas Colaborativos, Rio de Janeiro, Brazil, pp. 1–8 (2011)
14. Hudson-Smith, A., Crooks, A., Gibin, M.: NeoGeography and Web 2.0: concepts, tools and applications. Journal of Location Based Services 3, 118–145 (2009)
15. Kernel Map. An Introduction to Kernel Methods, http://intellisysdev.enm.bris.ac.uk/cig/pubs/2000/svmintro.pdf/
16. Cooper, A.K., Coetzee, S., Kaczmarek, I., Kourie, D.G., Iwaniak, A., Kubik, T.: Challenges for quality in volunteered geographical information. In: AfricaGEO 10, Town, South Africa, pp. 1–13 (2011)
17. Goodchild, M.F.: The quality of geospatial context. In: Rothermel, K., Fritsch, D., Blochinger, W., Dürr, F. (eds.) QuaCon 2009. LNCS, vol. 5786, pp. 15–24. Springer, Heidelberg (2009)
18. Goodchild, M.: Assuring the quality of volunteered geographic information. Spatial Statistics 1, 110–120 (2012)
19. Howe, J.: Crowdsourcing: Why the Power of the Crowd Is Driving the Future of Business. Crow Business, Washington (2008)
20. Connors, J.P., Lei, S., Kelly, M.: Citizen Science in the Age of Neogeography: Utilizing Volunteered Geographic Information for Environmental Monitoring. Annals of the Association of American Geographers 102, 1267–1289 (2012)

An Innovative Approach to Assess the Quality of Major Parks in Environmentally Degraded Mega-City Dhaka

Antora Mohsena Haque, Md. Rifat Hossain,
Md. Hasan Murshed Farhan, and Meher Nigar Neema

Dept. of Urban and Regional Planning,
Bangladesh University of Engineering and Technology, Dhaka-1000, Bangladesh
mehernigar@urp.buet.ac.bd

Abstract. This study addresses both qualitative and quantitative assessments of the quality of major parks of Dhaka. Four incommensurate factors namely environment, safety and security, landscape and aesthetic factors have been smartly chosen to measure the quality of the parks by formulating new index values. Index value of each factor has been calculated for all the parks. It is shown that the quality of all the parks considered are dispersed. In addition, investigation of universal accessibility of the parks and direct park user's opinion has been accumulated to concretize the results. Based on the major findings of this study a number of recommendations have been provided for the improvement of the quality of parks in Dhaka city. Involvement of local community and establishment of office for the park authority inside the park can be helpful in the quality maintenance of the parks. The findings of this paper will enhance the existing knowledge of city planners a step forward with a-priori knowledge to ensure quality of parks in further city planning.

Keywords: Urban Parks, Quality Assessment, Environment, Landscape, Accessibility, Safety and Security, City planning.

1 Introduction

Urban Parks and open space refers to land that has been reserved for the purpose of formal and informal sport and recreation, preservation of natural environments, for provisions of green space and/or urban storm water management [1, 2]. In any populated and polluted city life, parks work as lungs of the city which not only provide outdoor recreation but also provide a sense of spaciousness and scale [3, 4]. People may use the parks for visual amenity, environmental, educational, health, cultural and recreational purposes. Therefore, it is utmost important to ensure quality of parks. According to [5], the term "Quality" is defined as the "gestalt" attitude towards a service which has been acquired over a period of time after multiple experiences with it [6]. Manning (1986) as cited in [7] suggested that to ensure high quality in outdoor recreation the needs of the visitors must be met [7]. To enhance the quality of parks more natural features should be included, opportunities for social interaction should be increased and level of annoyance should be reduced. This will increase the amount of outdoor activity especially among older generation. Presence of good quality, well maintained public spaces help

B. Murgante et al. (Eds.): ICCSA 2013, Part III, LNCS 7973, pp. 394–407, 2013.

to improve the physical and mental well being of human. These places are a powerful weapon that helps in decreasing obesity and improving ill health [8]. Lam et al. [9] reports that lower values of pollutants were found inside urban parks and open spaces in comparison to the roadside stations in Hong Kong [10]. Urban trees also assist in reducing the "heat island" effect. The USDA Forest Service estimates that every 1% increase in canopy cover results in maximum mid-day air temperature reductions of 0.07°F to 0.36°F (0.04°C to 0.2°C).

Dhaka was a city of 2.8 million in 1981, which rapidly increased alarmingly to 5.3 million in 2001 while expansion of city area was negligible i.e. area increased to 276 sq. km from 208 sq. km in the same period. It can be shown through Landsat Satellite images of 1989, 1999 and 2009 that the land use of Dhaka is changing rapidly. It was reported that from 1989 to 1999 built up areas have increased 20.54% and vegetation has decreased 3.4% [11]. From 1999 to 2009 there have been 16.86% increase in built up area but 3.24% decrease in vegetation. There are about 54 registered parks under Dhaka City Corporation (DCC). But these parks make up only an average of 14.5% of the total land area (17% in north and central part and 12% in old town) whereas any city requires 25% for fresh environment and to maintain a sustainable land ecosystem.

At present the local planning experts recommend that there should be at least 1 acre of parks or open spaces per 1000 population for cities of Bangladesh. If, this standard is to be applied in Dhaka, then the city needs approximately 6 sq. miles of area for recreation purpose [2]. Most of the areas of Dhaka city are so unplanned that there is very little scope for creating a new park or open space to meet the needs of the growing population. In this case it is inevitable that the existing parks need to be improved or developed.

But unfortunately till now no initiatives have been taken to improve the parks of Dhaka city [12]. DCC has failed to continue its responsibility to maintain the greenery of the city and have converted the parks or open spaces into garages, shopping malls or mosques. There are many new unauthorized housing projects that are being developed in Dhaka at present. These will shrink the greenery and wetlands to create extra and unbearable pressure on the over burdened public utility. If conditions remain unchanged then Dhaka will definitely collapse [13].

Allocate more areas for new parks are very difficult as Dhaka is scarce in land. Improvement of the condition of the existing parks seem to be the only the viable solution to meet the needs of the citizens. But to improve the quality of existing parks it is necessary to identify which park is lacking behind in what factors. So that resources can be efficiently allocated to develop the quality of parks of Dhaka.

Few studies have been made earlier on parks or open spaces [2, 14–16]. But, no systematic study has been performed yet to assess the quality of parks in Dhaka city and to provide recommendations for better planning of urban parks. To examine visual quality of the park a study has been carried out in Alanya County, Turkey using photographs [17]. Ter performed a study on Alaaddin Hill, a big tumulus place in the city of Konya which serves as an urban park, to determine what quality criteria are effective in assessment of Quality of urban parks [18]. A study was carried out in parks of two cities of Massachusetts to identify which attributes influence park characteristics more. The total use of parks have been made dependent on four variables namely activity index, amenity index, park size and aesthetic rating [19]. Therefore, in this research an attempt has been taken to qualitatively and quantitatively assess the quality of some major parks of Dhaka city with respect to environmental factors, safety and security factors, landscape factors and aesthetic factor.

The paper is organized in the following ways. First, we provide the details of our innovative approach to assess the quality of parks in section 2. In the next section 3, we first discuss about demographic statistics of users, comparison between adjacent and distant Users, and attractiveness of the Parks. In section 4, we analyze Safety and Security Factor, Environmental factors, Aesthetic Factor, Landscape Factor. Finally, in section 5, we draw some concluding remarks.

2   Methodology

This study has focused only on the 6 major urban parks of Dhaka which are more than 5 acres in size namely Ramna Park, Sohrawardy Uddyan, Osmani Uddayan, Gulshan Lake Park, Gulshan Tank Park and Fazle Rabbi Park. For data collection, at first a reconnaissance survey and a preliminary questionnaire survey have been conducted. From this the final questionnaire and checklist have been prepared by excluding the unnecessary options and variables. Then data have been collected both from primary and secondary sources in accordance with the objective and the study area. To identify the existing features of the park a field survey has been performed by using check-list. Necessary photographs of the park features and amenities have also been taken. The quality of the parks has been identified from user's perspective. A Questionnaire survey has been conducted to get the user's opinion regarding landscape factors, environmental factors and safety and security factors.

Dhaka city's population is specific but the number of users of these parks is unknown. More over these parks are major parks of Dhaka and the area they serve or the numbers of households they cover are not identified. Again due to time and resource constraints huge amount of sample population could not be surveyed. For unknown population 384 samples are surveyed. But to maintain the authenticity 402 samples surveys have been conducted to complete the study (67 surveys per park). Park users have been the target group of the study. The Q-Sort Method is a psychometric technique. It produces reliable and valid interval measurements of people's perceptions about the visual quality of landscape as depicted in photographs. An explicit and valid assumption is inherent in the use of the Q-Sort Method. It is that the visual response to landscape photographs is consistent with visual response to actual landscapes. A panel of 5 senior architecture students has been selected to rate aesthetic features of the park using photographs of the parks. Five park users have been chosen as well to rate the photographs. Their points have been used to determine the aesthetic factor of each park using Q-sort method. Collected data from field and questionnaires have been accumulated for analysis using SPSS software. All data collected from surveys have been checked and reviewed to escape unexpected error. All data compiled from questionnaire and field survey have been analyzed by Microsoft Excel, SPSS and ArcGIS software. Parameters of the objectives have been converted into quantitative value from qualitative data. To evaluate the quality of the parks four factors have been observed. These are: Environmental factor (E),

Safety and Security factor (S), Landscaping factor (L) and Aesthetic factor (A). The following formula has been used to calculate the quality of the parks:

$$Q = E + S + L + A \tag{1}$$

where E= Environmental factor, S= Safety and Security factor, L= Landscaping factor and A= Aesthetic factor.

Landscape Factor: Landscape Design combines nature with culture. It focuses on planning of a property with various landscape elements and plants. Selection of the elements depend on climate, topography and orientation, site drainage and ground water recharge, soil and irrigation, human and vehicular access and circulation, recreational amenities, furnishing and lighting, native plants, property safety and other measurable conditions. For this research, at first field survey has been conducted to identify the landscape design elements that are found in almost all the 6 parks under study. After the survey 12 landscape components have been selected for quality analysis. These components are greenery, water body, seat/bench, lighting, playing instrument, kiosk/shade, paved walkway, dustbin, toilet/washroom, water tap/basin, tea/coffee shop and bridge.

Environmental Factor: Parks are the places which are created for providing a sense of nature in the monotonous city life. Designers always try to bring the touch of natural environment in it. A very good nature is the one where the air is pure, everything is clean and the atmosphere is calm and quiet. In a very good nature usually there is no mosquito and temperature is soothing. Most importantly natural environment is usually free from the crowdedness of the city. In this study 6 components have been considered under Environmental factor namely air quality, noise level, cleanliness, temperature, crowdedness and mosquito. The environment of a park largely depends on its maintenance and its users' behavior and sometimes even on its surrounding areas.

Safety and Security Factor: Safety and security of the parks is very important factor for the visitors' satisfaction and participation. The parks with security guards and comparatively smaller in size have more security than larger parks without security guards or care takers. Sometimes larger parks fail to attract visitors because of the unsafe condition. Parents do not feel safe to send their children to those parks. Even adults especially women do not feel secure to visit such kind of places. 6 components namely mugging, drug dealing, anti-social activity, begging, eve teasing, hawking have been observed to understand the safety and security condition.

Aesthetic Factor: This factor has been used to rate the beauty and appearance of the park from user's perspective. A psychometric technique called the Q-Sort Method is a. It produces reliable and valid interval measurements of people's perceptions about the visual quality of landscape as depicted in photographs. An explicit and valid assumption is inherent in the use of the Q-Sort Method. It is that the visual response to landscape photographs is consistent with visual response to actual landscapes. A panel of 5 senior architecture students has been selected to rate aesthetic features of the park using photographs of the parks. Five park users have been chosen as well to rate the photographs. Their points have been used to determine the aesthetic factor of each park using Q-sort method.

2.1   Index Calculation for Environmental, Landscape, Safety and Security Factors

To study environmental condition of the parks, six variable are considered i.e. air quality/odor, noise level, cleanliness, temperature, crowdedness and invasion of mosquito. For safety and security factor, mugging, drug dealing, anti-social activity, begging, eve teasing and hawker are considered as variables. For evaluating the landscape factor, twelve Landscape elements such as Greenery, water body, seat, lighting, playing instrument, shade, paved walkway, dustbin, toilet, water tap, tea/coffee shop, and bridge are considered. These variables are evaluated on a 5 point scale (1 = Very Bad, 2 = Bad, 3 = Moderate, 4 = Good, 5 = Very Good) according to user's opinion. Variables are prioritized according to the user's opinion. Average value of each variable of each factor will be calculated to find out the overall index. Following formula are used to calculate index for each factor:

- Total point of a variable= $\sum$ (Frequency of users × Weight of scale level)
- Average point of a variable = Total point of a variable/Total Frequency of users
- Weightage of each variable = (Total point of a variable)/$\sum$ (total point of all variables)
- Index of a factor = $\sum$ (Average point of a variable × Weightage of each variable)

Weightage of the variables of each factor are determined based on user's opinion.

2.2   Index Calculation for Aesthetic Factor

Visual appearance of the parks are evaluated by the expert's opinion. A Panel of five senior architecture students judged a series of five color photographs of the parks. The photos represented the typical features of the park. The photographs are unintentionally taken to make them either attractive or unattractive. The photographs are taken without any intension to make them either attractive or unattractive. The students rated each photograph on a 5 point scale from 1(least attractive) to 5 (most attractive). The scores from each photo of a particular park are summed and averaged to obtain a mean aesthetic rating for that park. Five park users are randomly selected to rate the photographs in the same procedure.

1) Calculation of index with ratings given by senior architecture students: The formula used to calculate the average rate of the five selected photographs is

Average Rate of each Photograph = ($\sum$ Ratings given by architecture students)/5
Then aesthetic index of each park are calculated by the following formula:
Aesthetic Index = ($\sum$ Average Rate of each photograph)/5
2) Calculation of Index with ratings given by Park Users: The same procedure are applied to calculate the Aesthetic Index using the ratings given by park users.
3) Calculation of Average Index: An average index for each of the parks is calculated following the formula given below

Average Index = (Aesthetic Index by senior architect students + Aesthetic Index given by park users)/ 2

There are 54 registered parks under DCC area. These parks are located in different wards of DCC and may vary in size and facilities. But for this research only 6 parks, which are more than 5 acres in size, have been selected for quality assessment. These are Ramna park, Sohrawardy Uddyan, Osmani Uddyan, Gulshan Lake park, Gulshan Tank park and Fazle Rabbi park Dhaka North City Corporation (DNCC) area and Dhaka South City Corporation (DSCC) area Fig. 1. The parks that fall into Dhaka North City Corporation have been listed as North Region parks and the parks falling into Dhaka South City Corporation have been listed as South Region parks.

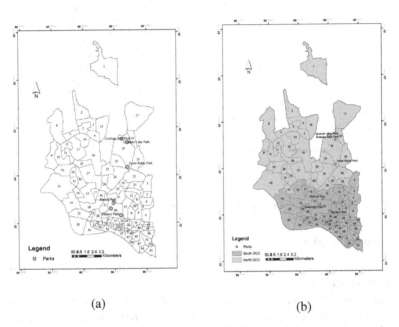

(a)   (b)

Fig. 1. (a) Locations of major parks in Dhaka city and (b) Major parks in the northern and southern part of Dhaka city

3   Results and Discussion

In this study, an attempt has been taken to assess the quality of the parks on the basis of user's perspective. However, for better understanding of the analysis the park users have been categorized into 2 groups on basis of the distance between their houses and the parks they visit. These are adjacent users and distant users. Adjacent users are those who live within 2 km distance from the park under study. Distant users are those who live beyond 2 km distance from the park under study. The buffer zones of adjacent and distance user's residence from parks are shown in Fig. 2(a) and Fig. 2(b) Among the surveyed population 70.4% users live near the parks that they use. Only 29.6% users come from distant places to use the park. It gives an indication that location of the park is an important factor. The tendency to visit the park declines with its increasing distance from the park user's residence.

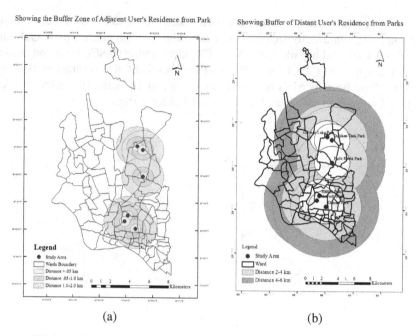

Fig. 2. Buffer zone of: (a) adjacent and (b) distant user's residence from parks

3.1 Demographic Statistics of Users

Out of 402 surveyed population 297 are male (74%) and 105 are female (26%). It indicates that on an average male are dominant users of parks. It can be seen that majority of the park users fall between 21-30 years age which is 33.5% of the total surveyed population. This group consists of the young energetic people who are health conscious and tries to stay fit. But most people of this group go to park for recreation and spending leisure time. The age group of 41-50 and 51-60 are actual health conscious people who visit park for exercise and walking mostly. People above 60 years of age mainly go for walking. The frequency of park users in different age groups is shown in Fig. 3(a). Almost half of the park users are graduates. 18% and 16% users have passed H.S.C and done post graduate respectively. It means that most of the users are educated. Only 3% illiterate users come to parks usually to spend their leisure time and to meet someone. The distribution of education level of park users is shown in Fig. 3(b). Majority of the park users are businessman (27%) followed by the private service holders (21%) who come for walking and physical exercise in the morning or evening. Housewives also come for walking or recreation but a significant portion of them come to parks to spend their leisure time while waiting for their children's school break. Compared to other parks high income users are dominant in Gulshan Lake park, Gulshan Tank park and Fazle Rabbi park. Middle to lower income group are dominant users of the Ramna park, Sohrawardy Uddayan and Osmani Uddayan parks. Fig. 3(c) and Fig. 3(d) respectively shows the occupation of park users and park users of different income level.

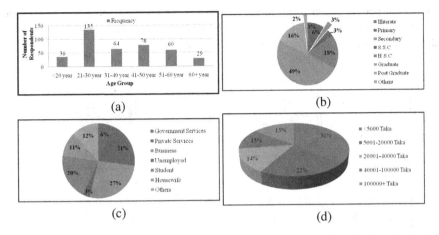

Fig. 3. (a) Frequency of park users in different age groups, (b) distribution of education level of park users, (c) occupation of park users, and (d) park users of different income level

3.2 Comparison between Adjacent and Distant Users

Comparison between adjacent and distant users may help to understand the reason why users prefer to come to distant parks. Among 402 surveyed populations 283 respondents go to the parks that are within 2 km distance from their residence (Fig. 4(a) and Fig. 4(b)). It means that most of the users feel the urge to go to the park if it is located near the house. Users seldom visit the parks that are distant from their residence. In this study it has been found that 119 users visit parks that are at least 2 km far from their house. But the only significant difference between adjacent and distant users is that distant users who stay >2 hours in the parks are 10% more in number than the adjacent users (Fig. 4(c) and Fig. 4(d)). It implies that the distant users stay in the parks for longer time. There is a significant difference in the visiting hour between the distant user and adjacent users. This implies that, in the morning period parks have more distant users and in the evening period parks have more adjacent users (Fig. 4(e) and Fig. 4(f)). In comparison between the two types of user the adjacent users are in a privileged position. Adjacent users have to travel very short distance and for which they do not have to depend on motorized transport that often. So, 40% adjacent users travel on foot while only 1% distant user travels on foot. 44% distant users travel by bus whereas only 1% adjacent user uses bus (Fig. 4(g) and Fig. 4(h)). It indicates that distant users travel more on bus and adjacent users travel more on foot.

Travel cost also varies between adjacent and distant users. Fig. 4(i) and Fig. 4(j) shows that 61% users have no cost of travel as they come by car or on foot. Only 4% users spend 5-10 taka who come by rickshaw and live within 0.5 km distance. 17% users' travel cost is above 20 taka majority of who come by rickshaw from 1-2 km distance. It is observed that most of the distant users spend <20 taka to visit the parks. These users come by bus or rickshaw. Only 5.9% user spend >100 taka to come to distant parks. They usually come by CNG or car. From the comparison it has been realized that people prefer to go to adjacent parks for walking and physical exercise rather than other purposes. But distant users choose to go to distant parks for recreation and spending leisure time rather than other purposes.

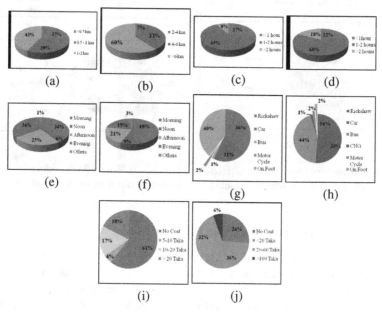

Fig. 4. Percentage of users based on: (a) distance between adjacent user's house and park, (b) distance between distant user's house and park, (c) duration of visit of adjacent user, (d) duration of visit of distant user, (e) visiting hour of adjacent user, (f) visiting hour of distant user, (g) mode of travel of adjacent user, (h) mode of travel of distant user, (i) travel cost of adjacent user, and (j) travel cost of distant user

3.3 Attractiveness of the Parks

Depending on its targeted population size, purpose and location different parks might have different attractive qualities. Its qualities also depend on its maintenance and user behavior. People usually prefer to go to their nearest park if it is well maintained. Otherwise they go to the parks that have better landscape and more open area. User might even prefer to go to distant parks if the park is accessible by better transport facilities. Openness is the dominant characteristics which attract people to visit the parks as 129 out of 402 people visit parks for openness Fig. 5. The parks under study are not attractive for the fact of being accessible by better transport facility. Around 10% people are attracted to come to the parks as these are the only place where they can meet someone or wait before going for some other works.

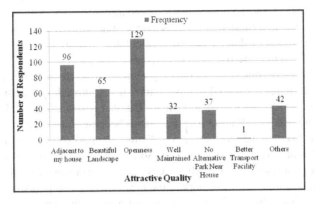

Fig. 5. User's opinion regarding attractive qualities of the park

On an average majority of the users (31%) asked for more secured and safer environment. It has been followed by requirement of proper maintenance as 23% users feel that existing parks are sufficient but due to improper maintenance they are failing to attract more users. 18% user mentioned that lack of playing instruments and physical exercise equipments fail to meet the demand of the children and other users who wants to work out in the park Fig. 6.

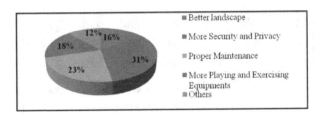

Fig. 6. User's opinion regarding the required qualities to enhance the parks attractiveness

4   Evaluation of Overall Quality of Major Parks

Index value of each of the factors are calculated using formulae of subsection 2.1 and subsection 2.2. The index values have been summerized to get an overall quality of the parks. Fig. 7 shows the overall quality of major urban parks in Dhaka city. It can be understood from the figure that, among the parks under study Environmental condition of Gulshan Lake park is the best as it gets an index value of 4.166 and it is in a Good condition. Environmental condition of Osmani Uddyan is the worst among the study parks as it gets the lowest index value 2.297. It means Osmani Uddyan's overall environment is bad. In case of safety and security factor Sohrawardy Uddyan is in the worst condition as it gets an index value of 1.797. This value is less than the index number of other parks. Sohrawardy and Osmani have no security guards and mugging, drug dealing, anti-social activity, eve teasing and intrusion of hawker is severe in these parks.

The parks of Gulshan and Fazle Rabbi are comparatively safer as their index values are above 4. It is because these parks have security guards and workers who ensure the safety condition. Aesthetic Factor has been calculated according to both senior architecture students who are referred to as experts and park users. It can be seen from the figure that, experts have rated Ramna park to be more attractive than the other parks as it gets the highest index compared to other parks. It is followed by Gukshan Lake and Gulshan Tank park. Experts think that Osmani Uddyan is less attractive than the other parks. According to users, Ramna park has again become more attractive that others and Osmani has become the least attractive. Among the parks Ramna park is the most attractive and Osmani is the least attractive. So measures should be taken to increase Osmani Uddyan's aesthetic beauty. After analyzing all the components, it is found that landscape design of Gulshan Lake park is the best among the 6 parks but it is moderately satisfactory according to users. It means that none of these parks are satisfactory to the users in case of landscape design. The Gulshan Lake park is followed by the Gulshan Tank park and Ramna park which means these parks are also moderate in landscape design. Osmani Uddyan's landscaping is the worst among the parks followed by Sohrawardy and Fazle Rabbi park.

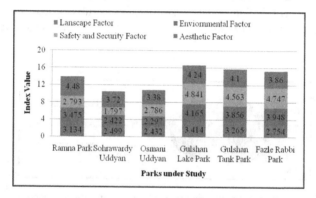

Fig. 7. Overall quality assessment of the parks under study based on their index value

Index value of each of the factors has been summed to get an overall quality of the park as well as ranking of the parks Fig. 8. The overall value of Gulshan Lake park is the highest. Overall index value of Sohrawardy Uddyan and Osmani Uddyan are very much closer to one another but Sohrawardy is the worst among these six parks. It is followed by Osmani Uddyan and Ramna Park. Although Osmani Uddyan's environmental, landscape and aesthetic factor index are lower than those of Osmani but the Sohrawardy Uddyan's safety and security factor is much lower than Osmani Uddyan's. For this reason Sohrawardy has become worse than Osmani in overall quality.

In gist, it has been found that users who come to visit the parks from 2 km distance are 70.4% of the total park users. Dominant users of parks are male and most of the users age between 21-30 years. Overall safety and security condition of all parks are not satisfactory.

Map Showing the Rank of Parks based on Quality

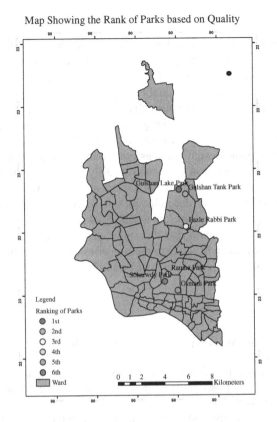

Fig. 8. Ranking of major parks based on the overall quality

Most of the adjacent users come daily to the park and most of them stay for 1-2 hour. Greenery and paved walkway condition of all parks are almost satisfactory as per user's opinion. Playing instruments and toilet of the most of the parks are in bad condition. Gulshan Lake Park has the highest index value in Landscape factor and environment factor and index value of Osmani Uddyan in these two factors are the lowest. In case of Safety and Security Gulshan Lake Park are more Secured than other parks and Sohrawardy Uddyan has the worst condition in comparison to other parks. Aesthetic factor of all the parks are moderate to good but landscape factor is moderate to bad in the parks.

This study has identified the factors in which each park is weak. So to ensure better quality of park the that are lacking behind need to be improved. None of the parks under study are good in landscape design according to the user's opinion. Thus, this study recommends planning for better landscape factor should be given high priority for major parks of Dhaka. Playing instrument are absent in most of all the parks. Some parks having playing instrument but most of them are out of order. So steps should taken to repair and maintain them. Also, proper and regular maintenance of each park will ensure the environmental quality.

5 Conclusions

Dhaka is significantly losing its appeal to ensure good quality city life because it compromises negatively with the quality of parks. Several factors namely environment, safety and security, landscape and aesthetic factors have been considered to qualitatively and quantitatively measure the quality of the parks. Among 54 registered parks in Dhaka city, most of them have not been quality maintained properly.

The obtained overall scenario from our *qualitative* analysis showed that, some parks namely Gulshan Lake park, Tank park and Fazle Rabbi park are superior in safety and security and landscape design whereas other parks such as Ramna, Osmani and Sohrawardi Uddyan are inferior in safety and security and environmental factor. So to ensure better quality of park the factors that are lacking behind need to be improved.

Using our proposed new index formulation, this study has *quantitatively* identified the factors in which each park is weak and a ranking of parks first time is obtained from our results. The results thus obtained in this study will provide very useful metrics for the planning and management of parks to the concerned authority who are responsible for the planning and management of parks of Dhaka city.

References

1. Dhaka Structure Plan, Volume I & II, Dhaka Metropolitan Development Plan, Dhaka, Bangladesh (1995)
2. Chowdhury, A.: Parks in the urban environment an analytical study with reference to urban parks of dhaka, Master's thesis, Department of Urban and Regional Planning, Bangladesh University of Engineering and Technology (2004)
3. Nabi, A.: Study of Open Space in Dhaka City, Dissertation presented to the Development Planning Unit, University College, London (1978)
4. Nabi, A.: Urban Planning Principles Class Lectures, Provati Library, Dhaka (2012)
5. N. Malek, M. Mariapan, M. Shariff, A. Aziz, Assessing the Quality of Green Open Spaces: A review (2010), http://www.hphpcentral.com/wp-content/uploads/2010/09/5000-paper-by-Abdul-Malek.pdf
6. Parasuraman, A., Zeithaml, V.A., Berry, L.: A conceptual model of service quality and its implications for future research. Journal of Marketing 49, 41–50 (1985)
7. MacKay, K.J., Crompton, J.L.: Measuring the quality of recreation services. Journal of Park and Recreation Administration 8, 47–56 (1990)
8. Space, C.: The value of public space: How high quality parks and public spaces create economic, Social and Environmental Value, London (2004)
9. Lam, K., NG, S., Hui, P., Chan, W.C.: Environmental quality of urban parks and open spaces in hong kong. Environmental Monitoring and Assessment 111, 55–73 (2005)
10. Cohen, P., Potchter, O.: Daily and seasonal air quality characteristics of urban parks in the mediterranean city of tel aviv. In: CLIMAQS Workshop 'Local Air Quality and its Interactions with Vegetation', Antwerp, Belgium, January 21–22 (2010)
11. Ahmed, B., Ahmed, R.: Modeling urban land cover growth dynamics using multi-temporal satellite images: A case study of Dhaka, Bangladesh. ISPRS Int. J. Geo-Inf. 1, 3–31 (2012)
12. Alam, S.: Vanishing open spaces, parks and play grounds, The Financial Express (2012). http://www.thefinancialexpress-bd.com/more.phpnews_id=126773date=2012-04-16

13. Hasan, S.: The failing city, New Age (April 23, 2012),
    http://www.newagebd.com/special.phpspid=2&id=8
14. Islam, M., Kawsar, M., Ahmed, R.: Open space in dhaka city: A study on use of parks in dhaka city corporation area, BURP thesis, Department of Urban and Regional Planning, Bangladesh University of Engineering and Technology (2002)
15. Siddiqui, M.: MURP thesis, Recreational Facilities in Dhaka City: a study of existing parks and open spaces, BUET, Dhaka (1990)
16. Nehrin, K., Quamruzzaman, J., Khan, M.: Status of Parks and Garden in old Dhaka, BURP thesis, Department of Urban and Regional Planning, Bangladesh University of Engineering and Technology, Dhaka, Bangladesh (2004)
17. Ter, U.: Evaluation of urban park in alanya county with visual quality assessment method antalya/turkey. International Journal of Natural and Engineering Sciences 6, 71–78 (2012)
18. Ter, U.: Quality criteria of urban parks: The case of alaaddin hill (konya-turkey). African Journal of Agricultural Research 6, 5367–5376 (2011)
19. More, T.: Factors Affecting the Productivity of Urban Parks, Research Paper NE-630 (1990)

Analysis of Potential Factors Bringing Disparity in House Rent of Dhaka City

Taslima Akter, Md. Mehedi Hasan, Akter Uz Zaman,
Md. Rifat Hossain, and Meher Nigar Neema

Dept. of Urban and Regional Planning, Bangladesh University of Engineering and Technology,
Dhaka-1000, Bangladesh
mehernigar@urp.buet.ac.bd

Abstract. Housing problem is one of the most acute problems in the mega-city Dhaka. A recent study of Consumers Association of Bangladesh (CAB) showed that house rent in the city has alarmingly increased to about 350 % during the last 22 years (1990–2012) while the increase was 15.83 % higher in 2011 than in 2010. As a result fixed income city dwellers comprising both middle and lower middle class households are in great trouble to tackle the real-world problem of house rent in the city.

This research thus conducted an extensive study to find the potential factors affecting the house rent by investigating relationship of increasing house rent with a number of important factors namely zonal variations of external appearance of the buildings, surrounding land use type, road distance from house, availability of open space, presence of utility facilities, type of structure, total number of flats, and average area of each unit flat are considered. A survey of 360 different areas on ten different zones of Dhaka City Corporation (DCC) area is conducted.

Using the proposed prioritizing factors through weighted index method, it has been found that owners consider size of the unit as the main factor of determining the house rent. Social status plays the least role before determining the house rent. Increase of house rent in different interval of time has no relation with the distance from the main roads. Rather structures being distant from main road increase higher rent. Increase in price of utility services and daily necessary commodities are the prime reasons behind the house rent increase. Owners claim that house rent increase become a must when there is an increase in price of water supply, gas etc. Owners of the apartment mainly increase the house rent for their own accountability. Presence of house owner organization may reduce the abrupt and irregular increase of house rent.

Building and structural condition is found to be the most influential factor for hiring residence in Dhaka city for the tenants. Accessibility and mobility along with social and community facilities are the next two criteria for choosing the residence. Accessibility is one of the most important factors for incensement of house rent. Safety and security, Proximity of educational institutions and social status also plays important role in this context. Though traditionally road distance and structure types are considered as the major factors determining the house rent, in contrary other factors namely size of the unit and presence of utility services are found to be the most dominant factors.

Keywords: Housing problem, City planning, Land use, Infrastructure services, House rent disparity.

B. Murgante et al. (Eds.): ICCSA 2013, Part III, LNCS 7973, pp. 408–421, 2013.

1   Introduction

Dhaka, the seventh largest populous mega-city in the world, is confronted with a big challenge to cope up with a population of 14 million. The area under the administration of Dhaka City Corporation[1] (both South & North part) is depicted in Fig. 1(a). Dhaka City Corporation (DCC) is further divided into ten broad zones[2] as shown in Fig. 1(b).

The population of Dhaka DCC stands at approximately 7.0 million. The rate of increase of population in the DCC area was about 6 percent in 2001. The total population in Dhaka City grew from only 0.104 million in 1906 to 5.4 million in 2001. The population growth rate was 4.15 percent in 1991 and 7 percent in 2001. The rapid rise in population of Dhaka City has been caused mainly by high immigration of people from rural areas. It has also happened due to territorial expansion and natural growth of the native city population. Density of population is about 5831 persons per sq. km. The city is growing at a faster growing rate of 3.84% per year [1]. Far Eastern Economic Review stated that Dhaka would become a home of 25 million people by the year 2025.

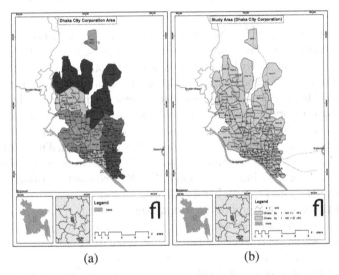

| (a) | (b) |

Fig. 1. (a) North and south part of Dhaka city and (b) Ten zones of Dhaka city

'Urban attractions' and 'rural distractions' has gradually persuaded people to migrate throughout the last decade. Sixty percent of the population of Dhaka city is migrants [2]. The Consumers Association of Bangladesh (CAB), in its recent study showed that house rent in the city has increased about by 350 % during the last 22 years (1990-2012) while the increase was 15.83 % higher in 2011 than in 2010 [3]. An increase over 23 percent in house rent was witnessed in 1991 and the trend continued later on. It was increased by 17.4 percent in 2001, 13.49 percent in 2002, 8.4 per cent in 2003, 9.96 per cent in 2004, 7.89 percent in 2005 and 14.14 per cent in 2006. Further it has increased average

---

[1] Chosen as the study area for this research.
[2] Residential house rent of these ten zones is surveyed randomly to fulfill the objectives of the study.

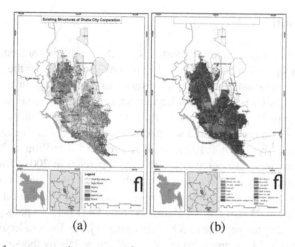

(a)   (b)

Fig. 2. (a) Pucca[3], semi-pucca[4], and kutcha[5] structure in Dhaka city and (b) Land use map of Dhaka city

by 21.48 percent only in 2007. As a result fixed income city dwellers comprising both middle and lower middle class households are in real trouble to tackle the rising house rent in Dhaka city. Lin shows how the housing consumption and housing investment change in Taiwan [4]. Ooi and Lee, focuses on the relationship of residential land with housing rent and house prices [5]. It investigates whether high land prices in urban areas caused high housing prices or whether high housing prices leads to high land prices. A study was conducted by Malpezzi to survey recent research on housing markets and policies in the second and third world countries [6]. DeSalvo presented in an article that a simple model of rent control in which the costs and benefits can be analyzed in terms of Hicksian consumer and producer surpluses [7]. Frankena conducted a whole study which was about monitoring the life cycle of a dwelling unit is a succession of tenants, generally proceeding to lower-income groups [8]. Arnott explained many types of procedures and methods are now using throughout the world in order to maintain and implement control rent policy within the tenancies but the tenancies are free to vary within themselves in the research [9]. According to Iwata, if any landlord is sufficiently risk-averse under asymmetric information on tenure length, the authority reduces the equilibrium quantity of rental housing for that particular house which results increase inefficiency in Japanese rental-housing market [10]. From the research of Gilderbloom, it was found that region, race and climate do not play a significant role for explaining rent differentials in 2000 model as it was done in 1980 model and 1970 model research [11]. Mitra demonstrated that neighborhoods attribute are the most significant factors in determining the house rent in Rajshahi city [12]. According to Rent Control Act 1991, house rent should be assessed at 15 per cent of the total cost of land and construction for a premise, which is in fact ignored in most of the cases. Although DCC has made a list of monthly house rent of 10 different zones all over the city but this is not implemented yet in real life.

---

[3] Structure of temporary building materials.

[4] Structure of permanent building materials.

[5] Structure of semi-permanent building materials.

However, till now no systematic study has been conducted to analyze potential factors which bring disparity in house rent of Dhaka City. Accordingly, this research identifies in details the potential factors and reasons behind increasing the house rent in Dhaka city.

The remainder of this paper is organized as follows. The next section (section 2) describes the methodology of the study. The results and discussion are presented in section 3. Finally, we provide conclusions in section 4.

2   Methodology

The research has three consecutive stages. Initial stage contains study area selection, second stage contains survey work and final stage contains data processing and analysis.

1. Secondary data collection: As the study is based on house rent analysis, residential structure is selected to be best suited with it. A complete house rent data from DCC is collected for the study from website of DCC (2012, May 24). Different map for physical feature survey, land use survey, topographical survey is collected from concerned agency. RAJUK, DCC and different NGOs help to supply this necessary maps and data.

2. Questionnaire and checklist preparation: A questionnaire has been prepared to assess the present condition of the study area and to get a preliminary idea on responsible factors which provide a guideline for necessary surveys required for this study. Also a checklist has been prepared to find out the overall condition of a structure.

3. Sample size determination: To conduct the actual survey a sample size has been required. The ten zones of Dhaka city are divided into 729 sub-zones according to the accessibility to main road by DCC. Fig. 3(a) shows existing road network in Dhaka.. Rent is determined on the basis of structural use as residential, commercial and industrial for all of the sub-zones. Rent of the sub-zones for all structural use is also divided into three categories: structures adjacent to the main road, structures within 300 feet from main road and structures beyond 300 feet from main road. Selected structures are presented in Fig. 3(b). Each of the structures of the categories is also divided into three sub categories: Pucca structure, Semi-pucca and Kutcha structure. About 5680 house rent data (Tk/sq.feet) consisting of nine categories are identified for 729 areas. Among these 5680 house rent data, about 360 house rent data are selected as sample through systematic random sampling by using a 95 % confidence level and 5% confidence interval. 360 owners and 360 tenants have been surveyed for this research. After that about 120 rent data are considered for each of the three broad categories (structures adjacent to the main road, structures within 300 feet from main road and structures beyond 300 feet from main road). The sample size is calculated for each zone using the percentage value of their structure type. Hence, the calculated k value for systematic random sampling is 16. After ascending the house rent data for each sub-category among all areas, the corresponding location of every 16th data (rent/sq.feet) will be considered as systematic random sample. By following this procedure 360 house rent data are selected for this study.

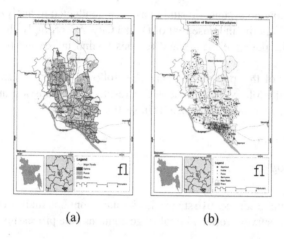

(a)   (b)

Fig. 3. (a) Existing road network in Dhaka city and (b) Selected structures for house rent analysis

3   Results and Analysis

The first objective of the study is to identify the responsible factors and reasons behind increasing the house rent in Dhaka city. Total 720 respondents including both house owners and tenants are interviewed to know the real condition of residential house rent. After completing the questionnaire survey, the information are analyzed in a systematic way to identify the responsible factors and reasons behind increasing the house rent in Dhaka city.

3.1   Analysis of Information Obtained Form Questionnaire Survey

The following analysis of the house rent is performed on the basis of information obtained form questionnaire survey:

1. General information of structure: About 50 percent kutcha buildings are situated in zone-one and it occupies the lion share compare to other zones (Fig. 4(a)). Zone-six occupies the large percentage of pucca and semi-pucca buildings. Zone-ten has also a significant amount of percentage in case of pucca and semi-pucca structures. Apartment structures are surveyed only zone-three, zone-four and zone-five. Among this percentage of zone-five are remarkable.

2. Relationship between rent of each flat/unit and total number of units: A moderate positive relationship between total number of flats and rent of each unit is noticeable. The Pearson correlation value is .367. The relationship shows that if the total number of flats increase in a building, the house rent also increases.

3. Relationship between rent of each unit and availability of open space: There is a weak negative relationship between house rent and availability of open space which is depicted in the analysis. The Pearson correlation value is -.238. It can be inferred that house does not have influence for availability of open space. There may be available open space but it affect negligibly for increase of house rent. The reason behind this factor may be the mental setup of the dweller of the Dhaka city that the lacking of open space is very common and it does not affect in increase in house rent.

4. Relationship between rent of each unit and presence of utility facilities: The presence of utility facilities is common for almost all of the structures (Fig. 4(b)). In pucca and apartment structures, it occupies almost 99 percent utility facilities. In most of the semi-pucca structures (about 90 percent) the utility facilities are present. Almost 58 percent kutcha structures have their utility facilities but rest of the 42 percent structures does not have utility facilities present at their house.

5. Relationship between house rent(per sq.feet) and type of structures: Only semi-pucca and kutcha buildings have house rent of 4-6 tk in per square feet (Fig. 4(c)). Most of the percentage of semi-pucca buildings have house rent within 7-10 tk in per square feet. The highest percentage (about 80 percent) of the pucca buildings are in the range of 11-18 tk in per square feet. Only house rent of apartment and pucca structures lies in above 18 tk in per square feet.

6. Relationship between flat Size(sq.feet) and type of structures: Only kutcha and semi-pucca structure occupies the flat size between 140-300 sq. feet. Most of the percentages of semi-pucca structures have flat size between 301-800 sq. feet. About 92 percent pucca structures have the flat size between 801-1400 sq feet. All of the flat size having above 1400 sq feet belongs to pucca structure. It is found that in Dhaka city smaller unit flat are higher than larger unit flat. Generally in Dhaka city most of the families are single small family who prefer smaller unit.

7. Relationship between house rent (per sq.feet) and road distance: The maximum percentage(about 60 percent) of buildings situated to road distance less than 100 feet from main road has house rent above 18 tk and the second highest percent belongs to 11-18tk (in per square feet) (Fig. 4(d)). Due to increase of road distance from main road, the percentage of buildings having house rent of above 18 tk (in per square feet) also decrease. With the increase of road distance, the percentage of buildings having house rent 7-10 tk and 4-6tk (in per square feet) also increases. So, it can be inferred that the house rent of medium range (in per sq.feet) almost remain same from road distance but the very high range (in per sq.feet) of house rent decreases from the road distance increase.

3.2   Analysis of Information Obtained Form Owner's Perspective

The following analysis of the house rent is performed on the basis of information obtained from owner's perspective:

1. Correlations between size of unit and rent (Tk. /sq. feet) of each unit: Generally it is assumed that big unit possesses big amount of house rent. The correlation analysis

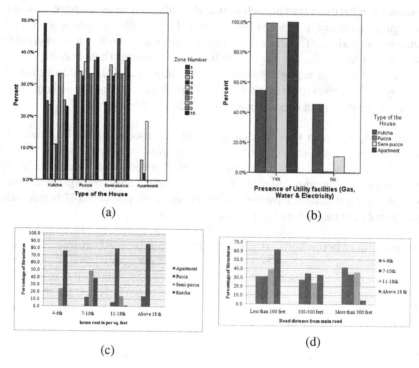

Fig. 4. (a) Structure type varies with zone to zone, (b) Presence of utility facilities according to type of structure, (c) House rent and type of structure, and (d) House rent and road distance

also shows that there is strong correlation between size of unit and the rent in per square feet in Dhaka. Rather smaller unit costs larger rent in per sq. feet. Pearson correlation value here is 0.537 which means it is strongly correlated with each other.

2. Factors considered by the owners to determine house rent: Presence of community facilities increase the house rent. Internal decoration of flat, the floor number and external appearance of the structure are also important factors to determine house rent. Existence of open space and accessibility increase the value of structures. Sometimes owners consider none of these factors. Besides, owners determine house rent considering neighborhood house rent. It is seen from Fig. 5(a) that size of the unit is the first criteria to determine house rent by the owners. Presence of utility services, community facilities and open space are the next prioritized factors. Structures being near to school, college, market have higher house rent than those structures which are distant from the community facilities. This factor is also known as locational advantage. Existence of open space, park, play field etc. increase the house rent of a locality. Social status is the least prioritized factor in the determination of house rent.

3. Relationship between DCC rent code followers and type of houses: It is seen from Fig. 5(b) that maximum house owners do not follow the DCC rent code to collect rent from the tenants. For kutcha structure, the gap between "yes" and "no" respondents is large. This gap is least for the pucca structures. Number of "yes" respondents is higher than the number of "no" respondents for apartment. In zone1, zone2 and zone3 there are very few house owners who follow the DCC rent code. Most of the house owners of pucca structures in zone4 follow the DCC rent code. Most of the house owners of all types of houses in zone5 and zone8 follow DCC rent code. The number of DCC rent code followers and not followers are almost equal in rest of the zones.

4. Temporal analysis of house rent increase: Now a days house rent increases within a short interval of time. It shows that majority of the house owners make conversation before increasing the house rent in all types of houses. The amount of house rent increase is less than tk. 1000 for most of the houses in both kutcha and semi-pucca structures. It is also seen that in both kutcha and semi-pucca houses some of the house owners increase the house rent for tk. 1000-5000 without conversation with the tenants. In pucca structures house owners usually increase house rent more than tk.1000. Kutcha structures being located within 300 feet from main road is seen to increase higher rent with lower interval. The reason behind is that dwellers of kutcha structures prefer to live in this location which is near to their work, like garments industry. For semi-pucca structures increase in house rent is higher in the areas which are near to main road than the distant areas. In some cases, house rent increase is not a problem for structures located beyond 300 feet from main road. There is a quite good interval of house rent increase with a smaller amount of money. The reason is found that the pucca structures and semi-pucca structures of those areas posses very close rent per sq feet. As a result people prefer pucca structures to semi-pucca when the house owners increase large amount of rent in semi-pucca structures. For pucca structures the highest percentage of house rent increase is seen for those structures which are located beyond 300 feet from main road. As the house rent of structures located near main road is already very high, those structures increase rent little comparatively. Since housing become a problem for Dhaka city all types of house rent is started to increase at an alarming rate without following any rule. In a consequence, the pucca structures being located beyond 300 feet from main road increase house rent with a 6 month time interval.

5. Relationship between presence of house owner organization and interval of house rent increase: It is clear from Fig. 5(c) and Fig. 5(d) that house owner organization is hardly seen in kutcha structures. As a result the structures face many problems within the locality and increase house rent abruptly. For pucca structures the presence of house owner organizations reduces irregular increase of house rent. House owner organization also seen for semi-pucca structures. It is seen that they increase rent as a result of increasing price of utility services.

3.3 Analysis of Information Obtained Form Tenant's Perspective

The following analysis of the house rent is performed on the basis of information obtained form tenant's perspective:

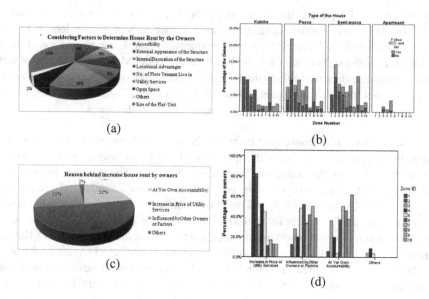

Fig. 5. (a) Factors considered for determining house rent by owners , (b) DCC code rent followers and house type, (c) Reason behind increase of house rent by owners, and (d) Reason behind increase of house rent in different zones by owners

1. Amount of money is paid by the tenant for hiring a residence: In Dhaka city almost every dwellings are hired on monthly basis. Tenants have to pay certain amount of money to the owners or landlords after a month interval. Amount of rent is depends on various factors such as structural condition, external and internal appearances, area or size of the unit, surrounding land uses, locational factors, social status, community and utility facilities, open space, parking lot etc. In general the most of the dwellers of pucca houses have to pay 10,000 to 25000 Tk. per month, where as peak of the residents of kutcha and semi-pucca dwellings have to pay 1000 to 4000 Tk. and 2000 to 6000 Tk. respectively. According to CAB, monthly house rent should not exceed 20% of total household income [13]. But at present scenario in Dhaka is quite different. Situation of pucca house hold dwellers are worst. About 70% of pucca household tenants have to pay 35%- 50% of their income per month as rent. For apartments situation even more intensive. Almost 80% tenants of apartments have to pay around 40% or more as rent of their monthly house hold income. In case of kutcha and semi-pucca structures the situation is not too much different from other categories. Most of the dwellers of these structures have to pay 20% to 40% of their total income as rent. In one word, almost 45% of the tenants of all categories have to pay around 40% of income as rent and 33% tenants are paying more than 40% per month. It causes huge financial burden on the tenants and dwellers.

2. Prioritization of preferences for living in the specific areas: Some certain factors work as key aspect for selecting a house for hiring or living in a specific area. Most common factors for the tenants for making this choice are ranked below through weighted index method. It has been found that building and structural condition and accessibility and move ability are the two top most factors. People used to make decision to hire a residence from the owners on the basis of structural condition. Transportation and communication facilities, distance from the main accessible road, good external and internal decorations, better structural condition, social and community facilities and working place etc. force the people's decision making process tremendously. Usually people prefer to live on those areas having good proximity of community facilities to cut down the social cost for living. Safety and security and proximity of educational institution are ranked as fourth and fifth key factors respectively. In some cases social status become a strong issue for this preferences. High income group specially wants to have their residence on those areas having a good fame of social status with an aim to represent their social status. This type of tendency is also identical among low income people mostly living in kutcha and semi-pucca structures. They like to be grouped with the people of same social status to have a more preferable communal environment.

3. Satisfaction level of the dwellers with regards to rent and other facilities: Most of the dwellers in Dhaka city are not satisfied with the existing rent system. Only 9% of tenants are satisfied with the amount of rent they have to pay for hiring residence. On the other hand, almost half of the residents are extremely unsatisfied with rent. In case of utility facilities about 74% of the dwellers are satisfied with the availability of gas, water and electricity. Satisfaction level with house rent doesn't indicate the satisfaction level of other facilities clearly. Sometimes there appears a irregular presence of satisfaction index of utility and community facilities along with rent. From this study it has been found that, among satisfied tenants with rent, about 77% people are satisfied with regards to utility facilities. In case of community facilities all dwellers are satisfied with existing provisions. About 26% unsatisfied for utility facilities where as 25% are unsatisfied for community facilities among the unsatisfied tenants with rents. But importantly more than 70% of tenants are satisfied with both utility and community facilities. Among the tenants of moderately satisfied with rent, 77% are satisfied with utility facilities and almost 89% are satisfied with community facilities. About 22% tenants are unsatisfied with utility facilities whose are mainly low income group living in kutcha structures.

4. Frequency or duration of last rent increase: At present in Dhaka house rent are increasing very randomly with regarding to its causes and temporal variation. About 84% of total kutcha structures monthly rent has increased just under one year. Most of the time, this increment rate varies in the range of 10%-20% or over 20%. For semi pucca structures, condition of increasing rent is even awful than pucca structures. Rent of 70% households are increased within a year. Most of the cases it increases 5% - 20% of its previous amount of rent. Also a handsome amount (almost 14%) of semi pucca household's rent increase over 30% during last year. In case of increasing house rent most of the house owners and landlords prefer to increase 10%-20% of the previous rent. Most of the dwellers (almost 69%) of pucca structures have experienced house rent increase within a time period of less than one year. In most of the cases (almost 37% of total households), rent of this houses is increased by 10%-30% of the previous one. Only 17.69% and 13.85% of pucca structures have experienced rent increase by an interval of 1-5 years and 5-8 years respectively.In case of apartments overall scenario

of rent increase is comparatively better than pucca structures. More than half (55.56%) of the structure's rent has been increased to 10%-20% (almost 45%) for more than 1-5 years ago, which indicates a slight relief for the tenants.

5. Reasons behind increase in rent from tenants perspective: According to the tenants, the factors and causes for increasing rent for the residence not varies a lot. Conceptions of most of the tenants are similar to the owners and landlords. Almost 35% of the tenants have blamed increasing price of daily necessities along with land value, acting as the major cause for rent increment. On the other hand second major group tenants have no comment on this. Actually they believe there is no real cause or factors for such type of frequent rent increase. They blame the house owners and landlords for taking negative advantages of living space shortage in Dhaka city. As there is no strict enforcement of laws and regulations, some of the tenants think that the house owners are started to consider themselves as all in all. On the other hand, some tenants conceived the locational advantages like near to school, colleges, working place, market, community places, recreational spots etc are also responsible for rent increase.

6. Reasons behind House Rent Increase in Different Zones: Analyzing the survey of the house owners it is identified that increase in price of utility services and a daily necessary commodity is the main reason behind the irregular increase of house rent. Some house owners increase the house rent being influenced by the neighboring house owners. Some also increase the rent at their own accountability. A few house owners stated other reasons i.e. they increase the house rent as a maintenance cost of the structures. The study finds that many kutcha structures' owners increase house rent considering other reasons. The other reasons include the maintenance cost, providing utility services etc.

3.4   Analysis of the Comparison between House Rent Fixed in DCC Rent List and House Rent Obtained from Questionnaire Survey

The second objective of the study is to compare the actual house rent of Dhaka city with Dhaka City Corporation (DCC) fixed rent list. To fulfill the objective a comparision is made between house rent fixed in DCC rent list and house rent obtained from questionnaire survey.

1. Zonal variation of existing house rent with DCC rent list: After analyzing the questionnaire surveys with DCC rent list, it is found that the existing house rent is higher than the DCC fixed rent list in ten zones. It is observed that most of the structures have house rent 50%-100% greater than DCC fixed rent list. Very few houses have house rent greater than 300% of DCC fixed rent list. Less than 50% increase in existing house rent is seen in all zones. The highest share of this increase is seen in zone-5. 50%-100% increase of house rent than DCC rent list is also found in all zones. The highest share of this increase is seen in zone-8. The big share of 100%-200% higher existing house rent is observed in zone-2. Zone-2 also bears the lion share of 200%-300% higher existing house rent. Existing house rent is found above 300% higher than DCC rent list in zone-2 and zone-4, where zone-4 bears the large share of the increase. The spatial differences of average house rent (for different type of structures) between DCC rent list and real condition are shown in Figs. 6(a)– 6(c).

2. Relationship between increase of house Rent with context to previous rent and DCC rent list (per Sq. Feet per Year): Increase of house rent per sq. feet per year has been found through analyzing the present and previous house rent with the time period. The lowest increase in per sq. feet rent is less than tk. 0.5 and it is prominent in semi-pucca structures. On the other hand the highest increase is seen in pucca structures. Even some pucca structures increase house rent above tk. 20 in per sq. feet per year. House rent increase with Tk. 1-2 is mainly found in kutcha structures. DCC rent list has been updated last in 2007. On the other hand the questionnaire survey has been conducted in 2012. These five years play a vast impact on the difference of existing house rent with DCC rent list. It is seen that the lowest increase of house rent is less than Tk. 0.5 (per sq. feet per year) and it is prominent in semi-pucca structures.

3. Relationship between structural variation of existing house rent with DCC rent list and distance of structures from main road: DCC have considered the distance of main road as a major criteria for determining house rent. But the figure shows that distances of structures from main road have a little impact on house rent. The analysis shows that the greater difference is seen in the structures which are distant from main roads. It means that DCC fixed lower house rent in those areas but in reality those areas also have higher house rent because of other factors play bigger role in determining house rent.

4. Relationship between structural variation of existing house rent with DCC rent list and external appearance of structures: It is found that external appearance plays a role to determine higher house rent in pucca and apartment structures only. For kutcha and semi-pucca structures higher house rent is seen rather in houses having poor and moderate external appearance. Relatively low percentage of house rent increase is seen in poor conditioned pucca structures.

5. Relationship between difference of existing house rent from DCC rent list and presence of house owner organization: The presence of house owner organization have little impact on house rent increase. Structures which faces highest increase of house rent have no house owner organization in their locality. It indicates that house owner organization control the abnormal increase of house rent within the locality.

6. Relationship between structural variation of existing house rent with DCC rent list and owners follow DCC rent list: It is clear that owners who do not follow the DCC rent code collect higher house rent from the tenants in all type of structures. A small percentage of house owners have claimed that they follow the DCC rent list, though they collect higher house rent from the tenants. The house owners define the reasons behind that as the outdated DCC rent list. They have claimed that they started following the DCC rent code in 2007. With the time being because of increasing price the house owners become bound to increase the house rent.

7. Relationship between structural variation of existing house rent with DCC rent list and concerned tenant about DCC rent list: The owners cannot collect higher house rent from the tenants who are concerned about the DCC rent list. Rather the owners collect irregular and higher house rent from the tenants who are not concerned about the DCC rent list. In some pucca and apartment structures some tenants pay

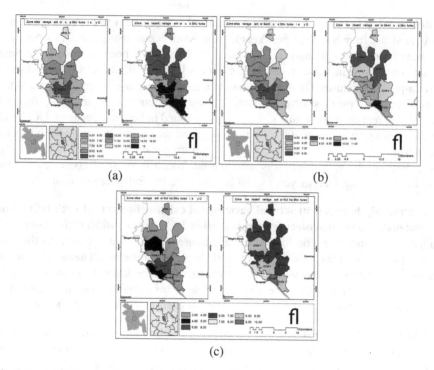

(a)   (b)

(c)

Fig. 6. Spatial difference of average house rent for: (a) pucca structure between DCC rent list and real condition, (b) semi-pucca structure between DCC rent list and real condition, and (c) kutcha structure between DCC rent list and real condition

200%-300% higher house rent though they are concerned about the DCC rent list. Outdated DCC rent list is the main reason behind this case.

8.Relationship between Expected House Rent of Tenant and Existing House Rent and Concerned Tenant about DCC Rent List: The expected house rent of tenants is also higher than the DCC fixed rent. Almost 90% tenants are willing to pay higher house rent than DCC fixed rent. In some cases tenants are willing to pay 200%-300% higher rent, even being concerned about the DCC rent. From the analysis of the expected rent and existing rent, it is seen that tenants are not satisfied with the existing house rent. There are also some tenants who want to pay lower rent than DCC rent.

4   Conclusions

In this study, zonal variations of external appearance of the buildings, surrounding land use type, residential house rent are extensively analyzed. Correlation was sought between house rent (per sq.feet), road distance, total number of flats, availability of open space, presence of utility facilities, type of structure, average area of each unit flat etc. Prioritizing the factors through our proposed weighted index method, it has been found that owners consider size of the unit as the main factor of determining the house rent.

Social status plays the least role before determining the house rent. Increase of house rent in different interval of time has shown to have no significant relation with the distance from the main roads. In contrary, structures being distant from main road increase higher rent. Increase in price of utility services and daily necessary commodities are found to be the prime factors behind the house rent increase. Low income people, living in kutcha and semi pucca structures usually pay 25%-40% of their total income as house rent. From this research it has been found that very few house owners consider number of floor as a factor for determining house rent. Building and structural condition are determined to be the most influential factors for hiring residence in Dhaka city for the tenants. Accessibility and mobility along with social and community facilities have found to be other top two criteria for choosing the residence. Safety and security, proximity of educational institutions and social status are also shown to play important role in this context. In some cases tenants pay 200%-300% higher rent, even being concerned about the DCC rent. From this analysis of the expected rent and existing rent, it is seen that tenants are not satisfied with the existing house rent. From this research, it is obtained that there are many influencing factors which bring the house rent disparity. This research may help the authority to know about the actual factors which affect the house rent in Dhaka city. Therefore, these factors could be given the most priority to find out the revised rent list in sustainable city planning.

References

1. BBS: Statistical Yearbook of Bangladesh, Bangladesh Bureau of Statistics, Planning Division, Ministry of Planning, Government of the People's Republic of Bangladesh (2008)
2. Mahbub, A.Q.M., Islam, N.: Extent and Causes of Migration into Dhaka Metropolis, and the Impact in Urban Environment, UNDP and UNFPA, Dhaka (1990)
3. The News Today: New law sought to rein in rising house rent (2012),
   http://www.newstoday.com.bd/index.php?option=details&news_id=2316315&date=2012-06-25
4. Lin, C.: The relationship between rents and prices of owner-occupied housing in taiwan. Journal of Real Estate Finance and Economics 6, 25–54 (1993)
5. Ooi, L.: Price Discovery Between Residential Land and Housing Markets. Journal of Department of Real Estate, National University of Singapore (2006)
6. Malpezzi: Economic analysis of housing markets in developing and transition economies, Research Report, University of Wisconsin 44, 39–54 (1990)
7. DeSalvo, J.: Reforming rent control in new york city: Analysis of housing expenditures and market rentals. Regional Science Association Papers and Proceedings 27, 195–227 (1971)
8. Frankena, M.: Alternative models of rent control. Urban Studies 12, 303–308 (1975)
9. Arnott, R., Johnston, N.: Rent Control and Options for Decontrol in Ontario, Ontario Economic Council, Toronto, Canada (1981)
10. Iwata, S.: He japanese tenant protection law and asymmetric information on tenure length. Journal of Housing Economics 11, 125–151 (2002)
11. Gilderbloom, I., Ye, L., Hanka, J., Usher, M.: Inter-city Rent Differentials in the U.S. Housing Market 2000: Understanding Rent Variations as a Sociological Phenomenon. School of Urban and Public Affairs. University of Louisville, USA (2008)
12. Mitra, S.: Applicability of Artificial Neural Network in Predicting House Rent, MURP Thesis, Dept. of Urban and Regional Planning, Bangladesh University of Engineering and Technology (2008)
13. The Daily Star: House Rent Legislation: Some issues,
    http://www.thedailystar.net/law/2008/09/04/index.htm

Integrated GIS and Remote Sensing Techniques to Support PV Potential Assessment of Roofs in Urban Areas

Flavio Borfecchia[1], Maurizio Pollino[1], Luigi De Cecco[1], Sandro Martini[1],
Luigi La Porta[1], Alessandro Marucci[2], and Emanuela Caiaffa[1]

[1] ENEA National Agency for New Technologies,
Energy and Sustainable Economic Development,
UTMEA Technical Unit "Energy and Environmental Modeling",
C.R. Casaccia, Via Anguillarese, 301 – 00123 Rome,
[2] Abruzzo Ambiente Srl
{flavio.borfecchia,luigi.dececco,sandro.martini,luigi.laporta,
maurizio.pollino,emanuela,caiaffa}@enea.it

Abstract. The last guidelines approved by Italian government to financially support the solar Photovoltaic (PV) Energy production development (Fourth and Fifth feed-in-scheme, January 2012 and later), in order to avoid soil consumption in agricultural or naturals areas, include specific indications for more advantageously funding installations exploiting roofs or covers surfaces. In this context it becomes important, for a suitable PV planning and monitoring, the extensive mapping of the available surfaces extent, usually corresponding to covers and properly assessing their quality in term of PV potential. Since the covers are mainly located in urban or industrial areas, whose 3D heterogeneity, albedo, atmospheric turbidity and casting shadows significantly influence the local solar irradiance, it is necessary to suitably account for these distributed factors by means of GIS mapping and advanced modeling tools in order to provide realistic estimates of solar available radiance at roofs level. The implemented methodology, based on remote sensing techniques, has allowed to estimate and map the global solar radiance over all the roofs within Avellino (southern Italy) municipality. Starting from LIDAR data, DSM of the entire area of interest (~42 Km^2) has been firstly obtained; then the 3D model of each building and related cover has been derived. To account the atmospheric transparency and the related time-dependent diffuse/direct radiation percentage on the area, data and tools from EU PVGIS web application have been also used. The final processing to obtain the solar radiance maps has been carried out using specific software modules available within commercial and open-source GIS packages.

Keywords: GIS, Remote Sensing, Photovoltaic, Solar radiation, LIDAR, PV.

1 Introduction

With the entry into force of the May 5, 2011 Decree and later (liberalization decree of January 24, 2012) in Italy, PV incentives are currently normed by the provisions

B. Murgante et al. (Eds.): ICCSA 2013, Part III, LNCS 7973, pp. 422–437, 2013.
© Springer-Verlag Berlin Heidelberg 2013

contained in the so-called Fourth feed-in-scheme ("*Quarto Conto Energia*") [1] by which it's intended to continue to support the national growth of the PV market. The decree contains a clear orientation to support the development and the dissemination of PV systems built on the infrastructure coverages, while there aren't additional incentives for those to be installed on the ground, in order to avoid soil consumption, especially in rural land [2, 3, 4, 5], which could have implications for the food industry, strategic in the near future [6]. In this context it assumes considerable importance for the PV, the characterization of the available surface on the roofs of residential and industrial buildings, including those of urban settlements, where this type of "land use" is densely represented. In general, until now, in order to map the PV potential using GIS [7, 8, 9, 10], the assessment of the solar irradiance has been carried out on the basis of orientation and exposure parameters, derived from natural relief maps called DEM (Digital Elevation Model) or DTM (Digital Terrain Model); currently, to achieve a more realistic evaluation, especially for urban areas [11, 12, 13], it is necessary to take into account a more detailed representation of building covers and their own 3D geometric and radiometric parameters.

The case study here described has been located in the area of the Municipality of Avellino, in the Campania region (Italy). Avellino (40°54'55"N 14°47'23"E, 348 m a.s.l., 42 km NE of Naples, Total population: 52,700) is situated in a plain called "Conca di Avellino" and surrounded by mountains: Massiccio del Partenio (Monti di Avella, Montevergine and Pizzo d'Alvano) on NO and Monti Picentini on SE. Due to the Highway A16 and to other major roads, Avellino also represents an important hub on the road from Salerno to Benevento and from Naples.

Figure 1 shows an output obtained using the tools currently available at the European Photovoltaic Geographical Information System GIS web site (PVGIS, http://re.jrc.ec.europa.eu/pvgis) [14], implemented by the EU to support the photovoltaic development in Europe [15, 16]. In particular, the picture shows the distribution of average annual radiation derived from DTM parameters, while on the left are reported the values of radiation and the related PV potential for electricity production (referred to various inclination angles of a PV south facing unit surface), for the urban area of Avellino. In such DTM-based approach, applied to urban areas, the results are generally poor since they generally tend to overestimate the real area of covers, whose angular asset parameters may be sensibly different from those derived from DTM/DEM.

In order to improve the extensive evaluation of roofs solar PV-systems potential, it is necessary to take into account the real surface available as well as the orientation and inclination. In this perspective, the assessment of the covers potentially exploitable for PV production, represents a fundamental task, which effectiveness must be derived from a reliable 3D model of buildings and/or other man-made objects in the area of interest. The recent improvements of Remote Sensing (RS) techniques (sensors, platforms and systems) make them very attractive and capable to contribute to the above mentioned surveying tasks, in terms of geometric resolution, operability and accuracy. In particular, active LIDAR (*Light Detection and Ranging*) has emerged as an effective technology for the acquisition of high quality Digital Surface Models (DSM), due to its ability to generate 3D dense terrain point cloud data with high accuracy [17]. Recent advances of airborne laser scanning systems and techniques have opened a new way of directly measuring elevation or generating DSM with high vertical and horizontal accuracies. Compared with traditional methodologies (e.g., photogrammetry), LIDAR has advantages in measuring 3D objects on the Earth Surface, in terms of density of measured points, accuracy, effectiveness, processing automation and fast delivery time [18].

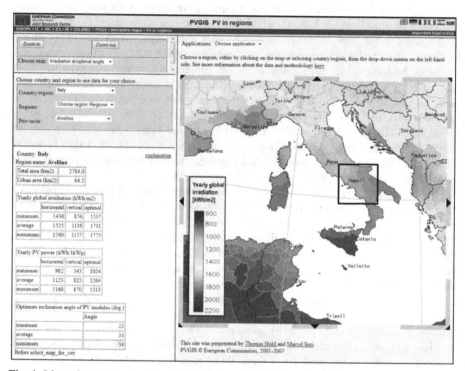

Fig. 1. Map of average annual solar radiation on inclined surface in an optimal way (source: PVGIS, http://re.jrc.ec.europa.eu/pvgis/). On left side, values of radiation and related PV potential for electricity production for the urban area of Avellino (contained within black box).

Referring to the case-study here described, LIDAR technology [19] has been exploited to produce a more detailed and effective characterization of the urban environment at building and coverage level, in order to support a reliable assessment of roofs PV potential. In this context, the structure and heterogeneity of the urban environment have been also considered in terms of albedo, atmospheric turbidity and shading, which in general have a strong influence on the local solar irradiance, as primary source for PV production. To reach the above mentioned goals, it is necessary to account, in an appropriate way, for these factors relating to the atmospheric transparency and to the contribution of reflected irradiance over the whole spectrum (albedo) due to the surrounding areas.

The description of atmospheric turbidity has been derived from time series (monthly distributions), coming from ground measurements and satellite data (thematic maps). This information allows to assess the direct and diffuse components, corresponding to a certain geographical location and reference month. To evaluate the reflective component arising from the local albedo, it is important to take into account reflectance, shape and the position of the objects placed in 3D space surrounding the PV surface. Their shape and relative position are both derivable from DSM, while optical RS may support the evaluation of their reflectance properties.

Starting from the above mentioned methodologies and techniques, this paper describes the development of an innovative approach, based on active and passive RS, to extensively assess a set of buildings cover parameters, in order to support evaluation and mapping of PV potential from roofs. In particular, from LIDAR data have been extracted specific geometric parameters of roofs (suitable surface, orientation, etc.) for every building in the area of interest; reflectance/albedo properties and atmosphere turbidity maps have been derived from RS data. Furthermore, to take account of the diffuse and reflected solar radiation, including cloud cover, have been also exploited products and features provided by the specific web-GIS application available at PVGIS [14] site.

2   Solar Irradiance

The solar radiation [20] that reaches the Earth, out of the atmosphere, on a plane perpendicular to the rays (normal extraterrestrial irradiance G_0), has an average amount of about 1360 W/m^2, with variations up to about 7%, mainly depending from the Earth-Sun distance. Its spectral distribution is that one of a black body at approximately 6000 °C, with a maximum in the visible range and absorption lines, due to the various elements diffused in space traveled from sunlight (Figure 2).

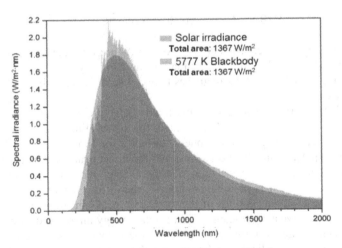

Fig. 2. Spectral distribution of solar radiation

By crossing the atmosphere, a portion of the incident energy is lost in several interactions, so that the part available on the ground is averagely about 900-1000 W/m^2, with a seasonal trend depending on the local height of the sun and, consequently, on the latitude of the site [21]. This value derives from the contribution of different components, whose variations depend on the local atmospheric transparency, the geometrical angular asset parameters of PV surface, as well on the reflectance proper-ties (albedo) of surrounding 3D objects. The percentage of atmospheric radiance that comes directly from the sun (beam/direct radiation) depends on the atmospheric transparency/turbidity value, which controls also the indirect component (diffuse radiation), coming from the different directions of the atmospheric hemisphere. This last increases with the decrease of atmospheric clearness, that in parallel reduces the direct component. The atmosphere clearness is characterized by the atmospheric Linke turbidity factor T_{LK} [22], defined as the ratio between the current atmospheric attenuation and that corresponding to a situation of "clear-sky" over the entire spec-trum of interest (mainly visible, NIR and SWIR). The diffuse component on a hori-zontal surface D_{hc} [W/m^2] is assessed as a product of G_0, of T_n, a diffuse transmission function (dependent only on the Linke turbidity factor T_{LK}) and of F_d, a diffuse solar altitude function (dependent only on the solar zenith angle θ_s) [23]:

$$D_{hc} = G_0 \, T_n(T_{LK}) \, F_d \, (\theta_s) \tag{1}$$

The estimate of the transmission function $T_n(T_{LK})$ gives a theoretical diffuse irradi-ance on a horizontal surface with the Sun vertically overhead for the air mass Linke turbidity factor. The following second order polynomial expression is used:

$$T_n(T_{LK}) = -0.015843 + 0.030543 \, T_{LK} + 0.0003797 \, T_{LK}^2 \tag{2}$$

The solar altitude function is evaluated using the expression:

$$F_d(\theta_s) = A_1 + A_2 \cos \theta_s + A_3 \cos^2\theta_s \tag{3}$$

where the values of the coefficients A_1, A_2 and A_3 are only depended on the Linke turbidity T_{LK}. The used model for estimating the clear-sky diffuse irradiance on an inclined surface D_{ic} [W.m^{-2}] distinguishes between sunlit, potentially sunlit and shad-owed surfaces [24].

  Figure 3, for example, shows the daily trends (left) of direct and diffuse irradiance on a 35° sloping surface, south (0°) oriented, for the Avellino site, calculated on the basis of the local average of atmospheric turbidity for June and derived from time series data [14]. The same diagram also shows the direct irradiance in a state of "clear sky" for the identical site: in the right side the average monthly trend of the diffuse fraction is shown. As it is possible to observe, the diffuse percentage consistence is relevant and up to about 35-55% of the total, with peaks in the winter months. The other factor, that have a significant effect on solar energy intercepted by a PV surface, is the geometric asset in terms of slope and aspect, that needs to be maintained as much as possible perpendicular to the sun, in order to maximize the global intercepted radiation.

  The graphs in Figure 4, obtained by means the PVGIS web application [14], clear-ly show how, for the studied site and south orientation, the optimal inclination angle varies during the year with the height of the sun (right side of Figure 4). For fixed PV installations, the optimal selection falls to a value of about 33°, which corresponds to the total irradiance trend of the black curve, in the graph on the left, resulting in a cumulative annual maximum value. In the same graph are also reported the specific values of irradiance on horizontal (blue) and vertical (violet) inclined PV surfaces.

Commonly, the photovoltaic potential is extensively assessed by GIS techniques, according to estimated irradiance on the ground and considering the natural altimetry parameters obtained from DEM or DTM [10]. The results coming from such a typology of 3D model are generally different from those obtained from detailed 3D models, capable to accurately describe roofs surfaces usable for PV-systems installation. In this context, therefore, a DSM obtained by LIDAR surveys can be more profitably used for covers identification and characterization, in terms of useful surface and related asset parameters, which are crucial for a more realistic estimate of the PV potential.

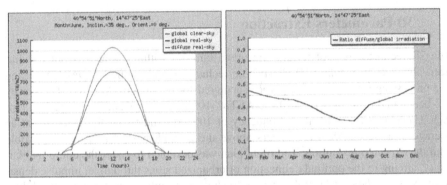

Fig. 3. Hourly solar irradiance and percentage of monthly average diffuse in the Avellino area in June, on the basis of monthly average atmospheric turbidity derived from recorded time series (source: PVGIS, http://re.jrc.ec.europa.eu/pvgis/)

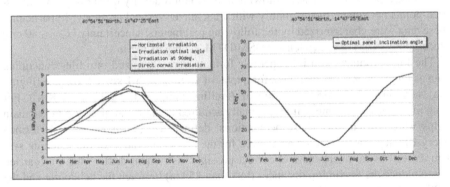

Fig. 4. Estimated irradiance on surfaces with different inclination and orientation to the south (left) and optimal angle (right) for the location of Avellino (source: PVGIS, http://re.jrc.ec.europa.eu/pvgis/)

Furthermore, the obtained 3D reconstruction, along with a broadband albedo map (derived from satellite RS multispectral data), allows the estimation of the irradiance residual contribution, in terms of shading and reflectance from the surrounding environment. In order to maximize the solar energy intercepted, it is important to consider slope and orientation of roof surfaces, potentially exploitable for PV-systems. These parameters usually vary for the different surfaces of the same roof. In fact, the direct component of solar irradiance on a surface I_s has especially a directional dependence:

$$I_s = I_{0s} \left[\sin(\theta_a)\cos(\theta_s) + \cos(\theta_a)\sin(\theta_s)\cos(\varphi_a - \varphi_s) \right] \quad (4)$$

where, using stereo goniometry, I_{0s} expresses the irradiance value on a sun perpendicular surface and θ_a, θ_s, φ_a, φ_s are respectively the zenith and azimuth angles of solar rays (s subscript) and of the normal to the PV surface (a subscript).

3   3D Parameters Extraction

The 3D reconstruction of terrain and urban built-up within the area of interest has been obtained by means LIDAR specific techniques and methodologies [25], using and processing a set of data on purpose acquired for the area of Avellino. The survey has been conducted by means of the ALTM3100 LIDAR system (developed by Optech, Canada), with acquisition frequency of 100 kHz and flying by helicopter at an average altitude of 900-1000 meters above the ground, having a speed compatible with the expected resolution. The flight plan was designed to cover the entire municipality, aiming to the optimal delineation of urbanized areas through an impulse density of the laser data (at least 4 return pulses/m²) and multi-returns handle capability, able to ensure the effective spatial identification of building cover features, contextually with vegetation discrimination. After the acquisition and pre-processing tasks, the obtained *points clouds* were showing a distribution density higher than four measured-points per square meter, jointly with a decimeter vertical accuracy. In particular, these data were characterized by the following parameters: Uncertainty $H_{Max} \leq 40$ cm; Uncertainty $V_{Max} \leq 2\,H_{Max}$; density DSM/DTM ≥ 4 points per m².

The final products, in a interoperable format, were organized into files containing data as *points clouds* (format .*las*), for a total of 48 tiles with 1 km x 1 km size (Figure 5-a), covering   the entire extension of Avellino municipality (about 42 Km²). For each tile at least two control lines normally acquired at the mean direction of the main strips were carried out. These acquisitions, defined as "tie-lines", were performed every 10 km circa and served as a check for the proper operation of the scanning system as well as to determine and correct possible instrument drifts. This was carried out to reduce the maximum distances (less than 25 km) between the helicopter and the GPS master-station on the ground. The measurements were performed only during periods of the day in which occurred simultaneously the following conditions: PDOP < 3 and number of seen satellites > 6, with an elevation mask of 15°. The UTM-WGS84 was the adopted geodetic reference system.

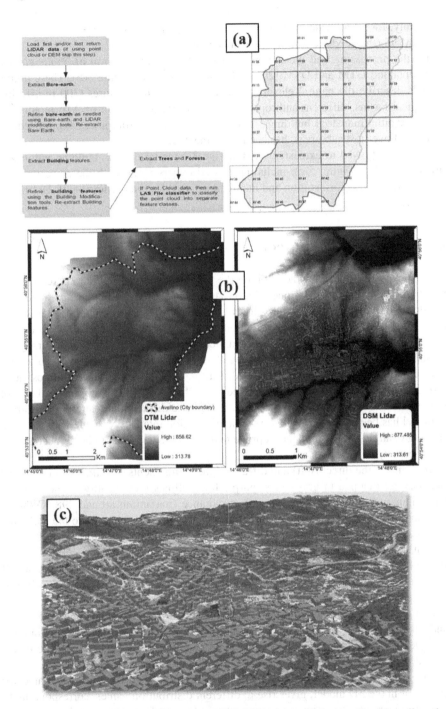

Fig. 5. LIDAR data processing: **a)** Tiles subdivision (48, 1 km x 1 km, outlined in yellow the dense urban area); **b)** DTM and DSM; **c)** Buildings 3D model of Avellino urban area

Finally, for all relevant operations, were used as Ground Reference Stations (GRS) the IGM 95 (the geodetic national network) cornerstones.

By the processing and mosaicking of LIDAR data, in the form of *points clouds*, were obtained DTM and DSM [26] of the area of interest (Figure 5-b). These digital models describe the natural elevation shape, including hill features (300 m asl), hydrological basins, different anthropic structures (e.g., the lay-out of A16 motorway) and residential/industrial buildings. For these latter, a specific processing procedure, calibrated over test sub-areas, has been implemented.

This has allowed to extensively extract the 3D model of each building within the urban area [27] (Figure 5-c), including related covers along with the corresponding surface parameters (orientation and inclination), used for the production of the irradiation map.

4   Radiation and Covers

The estimation of the solar radiation on the buildings roofs, using the 3D reconstruction based on the above mentioned DSM, has been accomplished by means some add-on tools available on commercial software packages (e.g., ESRI ArcGIS Solar Radiation Tool) [28]. The cumulative radiation, in terms of Wh/m^2, relative to June and July (that generally provide the highest annual PV contribution), has been firstly assessed. Thanks to the implemented procedure, both the direct and the diffuse irradiance components have been calculated, starting from a fixed value of atmospheric turbidity for the reference month, derived from the PVGIS [14] site, on the basis of time series measurements. To assess the atmospheric diffusion, a scattering anisotropic model has been adopted, with the incident radiation intensity variable with the local sun zenith angle, that is more appropriate than isotropic approach to better account the heterogeneity of the urban area.

In Figure 6 the direct (left) and diffuse (right) components of estimated cumulative radiance are mapped in gray tones (purple line outlines municipality boundaries). In particular, in the map of direct cumulate radiance (maximum around 298 KWh/m^2) the shading effects of 3D structures (buildings) are more evident, due to the daily and seasonal solar path as darker streaks. On the other hand, the map of the diffuse component (maximum around 85 KWh/m^2) shows a less marked shading effect, with a strong concentration around 3D structures (buildings). Subsequently, the evaluation of total radiance has been carried out exclusively on the covers of 3D built-up objects, previously extracted using the DSM.

In Figure 7, the total cumulated radiance incident on the roofs, estimated for June and July 2012, is represented in false color, with maxima on red and minima in blue. The thematic representation has been overlaid on Quick-Bird Avellino satellite image (true colors, 2.5 m ground resolution).

The solar radiation for June and July 2012, expressed in Kwh/m^2, calculated for each roof in the 3D model, gives rise to different distribution values, corresponding to the various orientations, slopes and geometries of the surfaces considered.

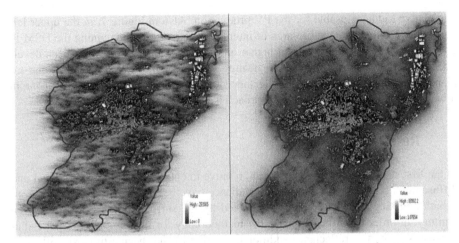

Fig. 6. Maps of the assessed distribution of cumulative solar radiance (Wh/m²): direct components (left) and diffuse (right), for June and July 2012

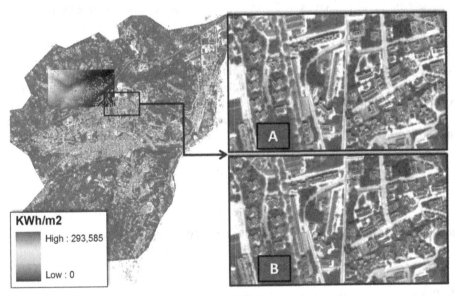

Fig. 7. Total solar cumulative radiation in kWh/m² calculated for the roofs of the buildings and infrastructure of the Avellino town with specific areas during June and July 2012

This can be observed in the images shown as detail-view in the right side of Figure 7, related to the sub-area outlined with an unfilled box. In the first image of detail-view (indicated with "A") it is possible to detect changes in cumulate radiance calculated for each roof with different orientation and inclination, characterized by uniform values for flat roofs and by mutual shading effects. In the detail-view "B" of Figure 7, instead, are reported the different geometric elements and surfaces of roofs, obtained from the DSM, for each of which has been estimated (by GIS techniques) the cumulative radiance usable for the PV production. Also in Figure 7, in the upper left part is represented another sub-area of interest (1 km x 2 km), containing the DSM in gray tones (lighter tones represent higher altitudes). On this sub-area has been developed and tested a refined procedure, dealing with a complete assessment of solar radiance including the additional reflectance component, that may be significant in urban environment. This improved approach is described in the next paragraph.

5   Results

In the first step of the pursued approach, relative to entire municipality, the direct and diffuse main components of the solar irradiance have been considered, while the contribution arising from the radiation reflected by the surroundings surfaces has been omitted. In order to further improve the methodology for the cumulative solar radiation assessment, this reflectance additional contribution has been estimated, still using 3D model from DSM, to describe the position of the PV surfaces on roofs and the reflecting surroundings.

The reflectance properties of the objects in the area of interest have been derived from wide band hemispherical map (Figure 8), usually called albedo, at suitable geometrical resolution of 30 m on ground. In the same figure it is possible to observe that the highest values are in correspondence to urban and industrial areas, while densely vegetated areas and roads present lower reflectance values. The preliminary estimation of the clear-sky ground reflected irradiance for inclined surfaces (R_i), relies on an isotropic assumption. The ground reflected clear-sky irradiance received on an inclined surface [W/m^2] is proportional to the global horizontal irradiance G_{hc}, to the mean ground albedo ρ_g and to a fraction of the ground viewed by an inclined surface $r_g(\gamma_N)$ [24]:

$$R_i = \rho_g \, G_{hc} \, r_g(\gamma_N) \tag{5}$$

where:

$$r_g(\gamma_N) = (1 - \cos \gamma_N)/2 \tag{6}$$

and global irradiance on a horizontal surface G_{hc} [W/m^2] is given as a sum of its beam and diffuse components:

$$G_{hc} = B_{hc} + D_{hc} \tag{7}$$

In Scharmer and Greif [23] typical albedo mean values for a variety of ground surfaces are listed. In general, the values of 0.2 or 0.15 are mostly used.

In order to evaluate the local albedo map, a multispectral Landsat ETM+ satellite image (August 2012) has been processed, using [29] the reflectance channels in the visible, NIR and SWIR ranges. The atmospheric correction has been performed on the basis of atmospheric visibility value of 30 km, in accordance with Linke parameter of atmospheric turbidity described in Paragraph 2.

Fig. 8. Avellino albedo map

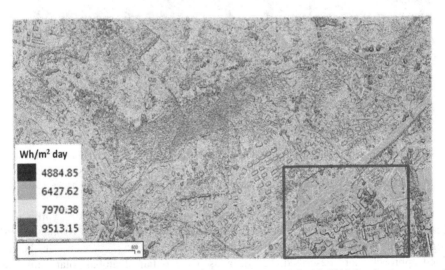

Fig. 9. Raster map of the estimated global solar radiation on July 14,.2012 for the area covered by the DSM (gray shades in the upper-left part of Figure 7).

In the assessment of cumulative sun radiance, previously described, tools and data available on PVGIS web site, have been used to estimate the atmospheric Linke turbidity for the reference month, related to Avellino location. The monthly average turbidity, in terms of Linke factor, is provided by PVGIS as maps with spatial accuracy of 1 km and is estimated from the global atlas implemented by Remund et al. [30], integrated and updated with data collected from satellite platforms (typically on geostationary orbit) and from ground-based measurements acquired by meteorological stations.

Fig. 10. Detailed map of subarea included in Figure 9 (red box) with the daily (July) cumulated sun specific radiance indication (values in Wh/m^2 day) for each roof surface extracted from DSM and characterized as GIS polygonal data. Aerial true-color orthophoto on background

Considering the spatial extension of Avellino study area (about 42 km^2, with a rectangular size of about 7x6 km), in the first approach was reasonable to assume that its atmospheric transparency was well characterized by a single point value. In the refined approach, here described, has been used the obtained global solar cumulate radiation map (Wh/m^2 day), over the interest area (Figure 9). The values have been estimated for June 14, 2012, by means of the specific software tool available within the open source package GRASS [31], using the DSM and the albedo data previously obtained with the Linke turbidity map related to July. The assessment has been carried out for each one of the three components: direct, diffuse and reflected sunlight irradiance values. The reflected component, even if not previously considered because lower than other two, was in this case still significant and approximately resulted equal to 1/3 of the diffuse component.

Figure 10 shows an example of a thematic final product, coming from the developed methodology, in which all the cover surfaces are categorized in terms of available sun radiance. Such a thematic product can be used, for example, in the framework of a specific web-GIS application, jointly with cadastral data of urban area of interest, in order to usefully support the PV inventory and planning.

6   Conclusions

The implemented methodology is characterized by the exploitation, in an integrated way, of active LIDAR and passive RS techniques, in combination with GIS procedures and using specific web-applications and data sources, available at European level for PV sector development. On the whole, such approach has allowed to extensively and effectively estimate, for the Avellino municipality, the solar cumulative radiance, available over the covers of 3D features and exploitable for PV production, on the basis of direct and diffuse local sun radiance components.

Subsequently, through a procedure refinement, the reflected component arising from local albedo has been also assessed for the previously detected and characterized cover surfaces. The related albedo map has been derived from an atmospherically corrected multispectral Landsat ETM+ image, acquired in a date within the reference months. In this way, the assessment of global irradiance from the three components (direct, diffuse and reflected) has been more reliably carried out for each roof in the urban area, potentially exploitable for PV production.

In this specific PV application sector, the aim of future researches will be to increase both the use of thematic products coming from different satellite missions, and the use of specific ground measurements, already available/accessible or to be on purpose acquired. Furthermore, it will be possible to better take into account the high heterogeneity of the urban environment (due to anthropogenic contributions as traffic, transports, heating, etc.) that affect the atmospheric transparency, also influenced by climate change factors.

These thematic products, currently available or in the next future obtainable, will allow to improve the spatial and temporal characterization of broadband turbidity (sensu Linke), in supporting the photovoltaic planning and production. In this context, to gather a better albedo estimation, it is also fundamental taking into account the anisotropic effects of non-Lambertian reflection, a typical phenomenon of the urban areas, eventually assessable by some Earth Observation products currently available (e.g., BRDF /albedo MODIS products).

References

1. Italian, Gestore dei Sistemi Energetici - GSE SpA, Fourth feed-in-scheme,
   http://www.gse.it/en/feedintariff/Photovoltaic/Fourth%20feed
   -in%20tariff/Pages/default.aspx (retrieved on February 2013)
2. Murgante, B., Danese, M.: Urban versus rural: The decrease of agricultural areas and the
   development of urban zones analyzed with spatial statistics. International Journal of Agricultural and Environmental Information Systems 2(2), 16–28 (2011)
3. Fichera, C., Modica, G., Pollino, M.: Land Cover classification and change-detection analysis using multi-temporal remote sensed imagery and landscape me-trics. European Journal of Remote Sensing 45, 1–18 (2012)
4. Di Fazio, S., Modica, G., Zoccali, P.: Evolution trends of land use/land cover in a Mediterranean forest landscape in Italy. In: Murgante, B., Gervasi, O., Iglesias, A., Taniar, D.,
   Apduhan, B.O. (eds.) ICCSA 2011, Part I. LNCS, vol. 6782, pp. 284–299. Springer,
   Heidelberg (2011)
5. Modica, G., Vizzari, M., Pollino, M., Fichera, C.R., Zoccali, P., Di Fazio, S.: Spatio-
   temporal analysis of the urban-rural gradient structure: an application in a Mediterranean
   mountainous landscape (Serra San Bruno, Italy). Earth Syst. Dynam. 3, 263–279 (2012)
6. Murgante, B., Borruso, G., Lapucci, A.: Sustainable development: Concepts and methods
   for its application in urban and environmental planning. In: Murgante, B., Borruso, G.,
   Lapucci, A. (eds.) Geocomputation, Sustainability and Environmental Planning. SCI,
   vol. 348, pp. 1–15. Springer, Heidelberg (2011)
7. Caiaffa, E.: Geographic Information Science in Planning and in Forecasting. In: Institute
   for Prospective Technological Studies (eds.) In Cooperation with the European S&T Observatory Network. The IPTS Report, vol. 76, pp.36–41. European Commission JRC-
   Seville (2003)
8. Huld, T.A., Šúri, M., Kenny, R.P.: Estimating PV performance over large geographical regions. In: Conference Record of the IEEE Photovoltaic Specialists Conference (2005)
9. Šúri, M., Hofierka, J.: A new GIS-based solar radiation model and its application to photovoltaic assessments. Transactions in GIS 8(2), 175–190 (2004)
10. Caiaffa, E., Marucci, A., Pollino, M.: Study of sustainability of renewable energy sources
    through GIS analysis techniques. In: Murgante, B., Gervasi, O., Misra, S., Nedjah, N., Ro-
    cha, A.M.A.C., Taniar, D., Apduhan, B.O. (eds.) ICCSA 2012, Part II. LNCS, vol. 7334,
    pp. 532–547. Springer, Heidelberg (2012)
11. Pellegrino, M., Caiaffa, E., Grassi, A., Pollino, M.: GIS as a tool for solar urban planning.
    In: Proceedings of 3rd International Solar Energy Society Conference-Asia Pacific Region
    (ISES-AP 2008), Sydney, Australia, November 25-28 (2008)
12. Cebecauer, T., Huld, T., Šúri, M.: Using high-resolution digital elevation model for im-
    proved PV yield estimates. In: Proceedings of the 22nd European Photovoltaic Solar Ener-
    gy Conference, Italy, pp. 3553–3557 (2007)
13. Balena, P., Mangialardi, G., Torre, C.M.: A BEP analysis of energy supply for sustainable ur-
    ban microgrids. In: Murgante, B., Gervasi, O., Misra, S., Nedjah, N., Rocha, A.M.A.C.,
    Taniar, D., Apduhan, B.O., et al. (eds.) ICCSA 2012, Part II. LNCS, vol. 7334, pp. 116–127.
    Springer, Heidelberg (2012)
14. Joint Research Centre (JRC) - Institute for Energy and Transport (IET): Photovoltaic Geo-
    graphical Information System-PVGIS, http://re.jrc.ec.europa.eu/pvgis/
    (retrieved on February 2013)

15. Šúri, M., Huld, T.A., Dunlop, E.D.: PV-GIS: A web-based solar radiation database for the calculation of PV potential in Europe. International Journal of Sustainable Energy 24(2), 55–67 (2005)
16. Šúri, M., Huld, T.A., Dunlop, E.D., Ossenbrink, H.A.: Potential of solar electricity generation in the European Union member states and candidate countries. Solar Energy 81, 1295–1305 (2007)
17. Ackermann, F.: Airborne Laser Scanning for Elevation Models. GIM - Geomatics Info. Magazine 10, 24–25 (1996)
18. Baltsavias, E.: A comparison between photogrammetry and laser scanning. ISPRS J. Photogramm. 54, 83–94 (1999)
19. Axelsson, P.: Processing of laser scanner data: algorithms and applications. ISPRS J. Photogramm. 54, 138–147 (1999)
20. Zaksek, K., Podobnikar, T., Ostir, K.: Solar radiation modeling. Computers & Geosciences 31, 233–240 (2005)
21. Duffie, J.A., Beckman, W.A.: Solar Engineering of Thermal processes, 2nd edn. John Wiley & Sons, USA (1991)
22. Linke, F.: Transmissions-Koeffizient und Trübungsfaktor. Beitr. Phys. fr. Atmos. 10, 91–103 (1922)
23. Scharmer, K., Greif, J. (eds.): The European solar radiation atlas. Database and exploitation software, vol. 2. Les Presses de l'École des Mines, Paris (2000)
24. Muneer, T.: Solar radiation model for Europe. Building Services Engineering Research and Technology 11(4), 153–163 (1990)
25. Gabet, L., Giraudon, G., Renouard, L.: Automatic generation of high resolution urban zone digital elevation models. ISPRS J. Photogramm. 52, 33–47 (1996)
26. Borfecchia, F., De Cecco, L., Pollino, M., La Porta, L., Lugari, A., Martini, S., Ristoratore, E., Pascale, C.: Active and passive remote sensing for supporting the evaluation of the urban seismic vulnerability. European Journal of Remote Sensing 42(3), 129–141 (2010)
27. Ricci, P., Verderame, G. M., Manfredi, G., Pollino, M., Borfecchia, F., De Cecco, L., Martini, S., Pascale, C., Ristoratore, E., James, V.: Seismic Vulnerability Assessment Using Field Survey and Remote Sensing Techniques. In: Murgante, B., Gervasi, O., Iglesias, A., Taniar, D., Apduhan, B.O. (eds.) ICCSA 2011, Part II. LNCS, vol. 6783, pp. 109–124. Springer, Heidelberg (2011)
28. ESRI ArcGIS Resource Center, http://resources.arcgis.com/
29. Liang, S.: Narrowband to broadband conversions of land surface albedo I – Algorithms. Remote Sensing of Environment 76, 213–238 (2000)
30. Remund, J., Wald, L., Lefèvre, M., Ranchin, T., Page, J.: Worldwide Linke turbidity information. In: Proceedings of the ISES Solar World Congress 2003, Göteborg, Sweden (2003)
31. Hofierka, J., Šúri, M.: The solar radiation model for Open source GIS: implementation and applications. In: Proc. of the Open Source GIS GRASS Users Conference, Italy (2002)

Deriving Mobility Practices and Patterns
from Mobile Phone Data

Fabio Manfredini, Paola Pucci, and Paolo Tagliolato

Dipartimento di Architettura e Studi Urbani, Politecnico di Milano, via Bonardi 3,
20133 Milano, Italy
{fabio.manfredini,paola.pucci,paolo.tagliolato}@polimi.it

Abstract. The paper addresses the issue of analyzing and mapping mobility
practices by using different kinds of mobile phone network data, which provide
geo-located information on mobile phone activity at a high spatial and temporal
resolution. We will present and discuss major findings and drawbacks, based on
an application carried out on the Milan urban region (Lombardy, Northern
Italy).

Keywords: Mobile phone data, Density of space usage, Mapping urban
domains, Milan urban region, mobility.

1   Introduction

Interpretative tools for the identification of mobility practices in the contemporary
metropolis are needed, not only for the some known limitations of traditional data
sources but also because new forms of mobility are emerging, describing new city
dynamics and time-variations in the use of urban spaces by temporary populations. In
Italy, the traditional data sources for urban and mobility investigations (ie surveys,
census) have known limitations, including the high cost of surveys, the difficulty of
data updating, the difficulty of describing city dynamics and time dependent
variations in intensity of urban spaces usages by temporary populations at different
scales.

New forms of mobility, close to daily mobility, are changing the way in which
urban spaces are used. They are characterized both by being based on the use of
transportation system, and by the efficient appropriation of information technologies
(internet, mobile phone). They intensified the density of the moves with which we can
read diversified uses of the city, that traditional sources of analysis are unable to
return with continuity. In Lombardy region, the systematic mobility represents only
the 29% of daily travels, which are attested on 2,65 travel/day in average, with a
propensity to mobility that changes in relationship to the professional condition.

As underlined by some authors [1], [2], [3], [4], [5], [6], the changes in the
management of mobility in contemporary cities are a useful key for understanding the
transformations of times, places and modes of social life and work programs,
structuring the metropolitan areas.

B. Murgante et al. (Eds.): ICCSA 2013, Part III, LNCS 7973, pp. 438–451, 2013.
© Springer-Verlag Berlin Heidelberg 2013

In this perspective, mobility may represent both a tool of knowledge and a project for urban planners, provided that a better understanding of different patterns of mobility in the form of "active biographies", which increase the range of "post-fordist living and labor styles" [7], is available.

Considering the role of mobility practices in social and spatial differentiation, it becomes important to formulate pertinent analytical approaches, aimed at describing the different densities of use of the city as a new challenge and a prerequisite for understanding the city and its dynamics.

Hence, from an analytical point of view, it becomes important to accompany the traditional quantitative approaches referred to a geographic displacement that tends to focus on movement in space and time, in an aggregate way and for limited periods, with data sources able to describe fine grain over-time variation in urban movements.

In this direction, an interesting contribution may come from mobile phone network data as a potential tool for the development of real-time monitoring, useful to describe urban dynamics, as it has been tested in several experimental studies [8], [9], [10].

The application researches were focused on two different products. Some studies dealt with aspects of representation of the data, emphasizing the most directly evocative aspects, to highlight how these data may represent the "Mobile landscapes" [9]. Other studies focused both on data-mining analysis and on the construction of instruments capable of deriving summary information and relevant data about the urban dynamics from cell-phone [8].

As opposed to the more traditional methods of urban surveys, the use of aggregated and anonymous mobile phone network data has shown promise for large-scale surveys with notably smaller efforts and costs [11].

If we consider the observed and aggregated telephone traffic as the result of individual behaviors and habits, we can treat mobile phone data as a functional source on the *real* use of the cities, capturing, for example, traces of temporary populations, which are difficult to intercept by traditional data source, but which, at the same time, increasingly affect urban practices both quantitatively and qualitatively.

In this direction an increasing number of studies concerns the exploitation of mobile phone data in urban analysis and planning [12]. In particular an interesting issue regards the classification of urban spaces according to their users' practices and behaviors [11], [13]. In [14] the authors outline the fact that city areas are generally not characterized by just one specific use, and for this reason they introduce the use of c-means, a fuzzy unsupervised clustering technique for land use classification, which returns for each area a certain grade of membership to each class.

In this general context, we used mobile phone data provided by Telecom Italia, the main Italian operator, in order to test the potentialities of this information in describing the density of use of urban spaces at different temporal and spatial scales as a precondition:

— to identify temporary populations and different forms of mobility that structure the relationships in the contemporary city;

— to propose diversified management policies and mobility services that city users require, increasing the efficiency of the supply of public services.

2   Mobile Phone Data

Milan is placed in an urban region which goes far beyond its administrative boundaries (see fig.1). The core city and the whole urban area have been affected in the last 20 years by changes in their spatial structures and have generated new relationships between the centre and suburbs. At the moment, the urban region of Milan is a densely populated, integrated area where 4.000.000 inhabitants live, where there are 370.000 firms and large flows of people moving daily in this wide area [15].

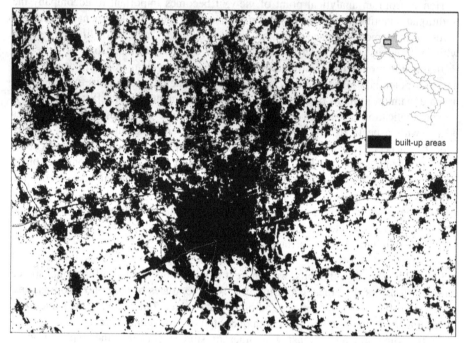

built-up areas

Fig. 1. Map of built-up areas in the Milan urban region (2007). Source: DiAP elaboration of DUSAF 2.1 data.

In order to analyze the complex temporal and spatial patterns emerging from mobile phone data, we used two different types of data provided by Telecom Italia.

2.1   Telephone Traffic Intensity – Erlang

The first data type concerns the mobile phone traffic registered by the network over the whole Milan urban region (Northern Italy).

Data are expressed in Erlang, namely the average number of concurrent contacts in a time unit. In the present case, the data represent the telephone traffic intensity every 15 minutes and was supplied by Telecom Italia in a spatialized form. From the telephone traffic recorded by each cell of the network, the telephone provider distributed the measurements, by means of weighted interpolations, throughout a tessellation of the territory in 250 meters x 250 meters squared areas (pixels) [1].

We performed time series analysis on this data along a period of 14 days in March 2009 (March 18th till March 31st), in order to evaluate specific characteristics of population behaviors at an hourly and daily base. We then applied a novel geo-statistical unsupervised learning technique aimed at identifying useful information on hidden patterns of mobile phone use. We will show that these hidden patterns regard different usages of the city in time and in space and that they are related to the mobility of individuals. The results return new maps of the region, each describing the intensity of one of the identified mobility pattern on the territory. This highlights, in our opinion, the potentials of this data for urban planning and transport research studies.

2.2   Aggregated Tracks of Users – Hourly Origin Destination Matrices

The second typology of data consists in localized and aggregated tracks of anonymized mobile phone users. It is an origin-destination datum derived from the Call Detail Record database. Italian privacy policies severely constrain the use of these data, even for research purposes. In the framework of a collaboration with Telecom Italia (T-Lab), we arrived at the definition of a datum which was free from privacy constraints, consisting in an aggregation of users' displacements based on CDR records. Telecom engineers set up a system for the automatic and blind extraction of data of this kind.

The system is fed with the CDR and a tessellation of a certain geographical region (in the present case the Lombardy Region). The output consists in time series of Origin-Destination matrices (where origin and destination zones are the tessellation's tiles), equivalent to a function $F(o,d,t) \rightarrow n$ which, at time t (t varying in within the 24 hours of a given day), assigns to origin o and destination d the number n of distinct users that performed some mobile phone activity[2] within o at time t-1 and a subsequent activity within d at time t.

The CDR's raw informations are available at the level of the antennas which handled the activity. The distribution of antennas in space depends on the amount of mobile phone traffic that needs to be managed. In dense urban areas we therefore observe a high density of antennas while in the suburbs the density of antennas may be very low. For the positioning of a user within a certain tile of a tessellation to be reliable, a technical constraint imposed that the tiles contained at least 13 antennas.

We defined two distinct tessellations which could gave us the possibility to map and to interpret main spatial patterns of mobile phone users' mobility: the first is more

---

[1] More formally we can define the Erlang E_{xj} relevant to the pixel x and to the j-th quarter of an hour as:

$$E_{xj} = \frac{1}{15} \int_{15(j-1)}^{15j} N_x(t)\, dt$$

where $N_x(x)$ is the number of mobile phone using the network within pixel x at time t, hence E_{xj} is the temporal mean over the jth quarter of an hour of the number of mobile phones using the network within pixel x.

[2] With mobile phone activity we intend each interaction of the device with the mobile phone network (i.e. calls received or made, SMSs sent or received, internet connections, etc..).

fine-grained, and it was obtained by a data-driven process taking into account the spatial distribution of antennas; the second, even if more coarse, was more directly related to administrative boundaries, consisting in the aggregation of adjacent municipalities' polygons.

We obtained the two tessellations as follows:

— Automatic clustering of antennas (526 polygons). Each zone is an aggregation of Voronoi cells obtained from the points of location of the antennas. More precisely we proceeded by clustering the positions of the antennas by means of an agglomerative hierarchical clustering algorithm (complete linkage, euclidean distance): we firstly cut the hierarchical tree in order to obtain 100 clusters; we then selected the groups with more than 500 antennas and we cut the corresponding subtrees according to an *inconsistency coefficient* less than a given threshold (see e.g. [18]), obtaining a final sufficiently balanced partition (i.e. with an homogeneous number of antennas per cluster). Finally, for each cluster, we calculated the polygon corresponding to the union of the voronoi cells of its antennas. Hence we obtained the tessellation having these polygons as tiles;

— Automatic aggregation of municipalities (313 polygons): Each zone is an aggregation of municipalities. Each zone contains not less than 13 antennas. An automatic procedure has been created in order to build new zones in an iterative manner.

On these tessellations it was possible to map the direction and the intensity of mobile phone users' movements at an hourly basis. The data set was collected in different working days: five Wednesday respectively in July, August, September, October and November 2011. Using this data, we performed an analysis aimed at evaluating the overall mobility of cell phone users in the Lombardy region.

3   Analysis

3.1   Treelet Decomposition of Erlang

Erlang measures can give insights on different aspects of the urban area to which they refer, and their analysis can be developed with various scopes: the segmentation of the area into districts characterized by homogeneous telephonic patterns; the identification of a set of reference signals able to describe the different patterns of utilization of the mobile phone network in time.

Treelet decomposition is an effective dimension reduction technique for Erlang profiles and, more generally, for data with peculiar functional features, like spikes, periodicity, outliers.

The methodology of Treelet decomposition [16], [19] allowed us to obtain: a reference basis reporting the specific effect of some activities on Erlang data; a set of maps showing the contribution of each activity to the local Erlang signal. The idea behind our approach is that different basic profiles (each being one element of the treelet decomposition basis) of city usages can concur in the same place and that the overall observed usage of a certain place is the superimposition of layers of these profiles.

We selected some results as significant for explaining specific patterns both of mobility and of city usages (commuting, nightly activities, distribution of residences, non systematic mobility). We tested their significance and their interpretation from an urban analysis and planning perspective at the Milan urban region scale.

Each of the following figures represents one of the extracted city usage profile and is organized as follows:

— Top panel depicts the considered basic profile: x-axis represents time, spanning 7 days from Wednesday to Tuesday at an hourly rate. The dotted lines correspond to 2 hours while the continuous lines separate the different days of the week; y-axis: Erlang values;

— Bottom panel depicts a map of intensity values: colors show how much the upper profile concur to explain one place's overall (telephone) use pattern.

Figure 2 is about the density of mobile phone activity late at night (in particular from midnight until 8 am). We can observe here some interesting hot spots where values are very high. For example, the exhibition district in the Northern Western side of the map. In the considered period an important Fair (the 2009 International Milan Design Week) was held and the peak fits well with the nightly activities necessary for the mounting and the organization of the site. Another point of interest is the Fruit and Vegetable Wholesale market in the South Eastern part of the region where consistent night work happens for delivering and distributing products that come from whole Italy and abroad. The city centre is characterized by a relative low value, according to the absence of relevant nightly activity inside it.

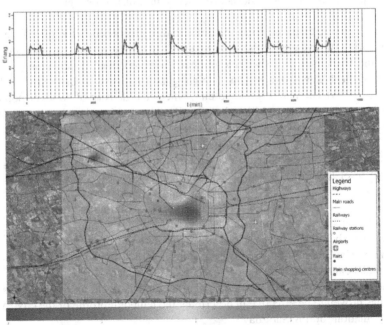

Fig. 2. Nightly activity. Hot spots highlight the presence of night work.

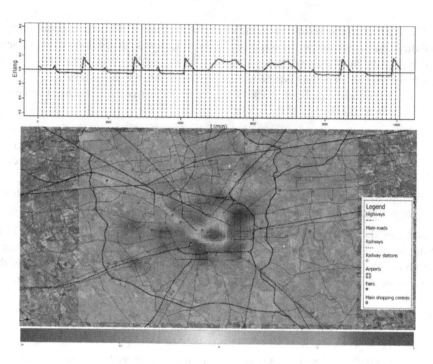

Fig. 3. Concentration of activities during evenings of working days and during daytime (from 8 am until 8 pm) of the week end: residential districts of the Milan urban region

Figure 3 puts in evidence some locations with high concentration of mobile phone activity during the evening of the working days and during daytime (from 8 am until 8 pm) of the week end. It shows a significant correspondence with main residential districts of the Milan urban region. It highlights a relevant concentration of homes along the second circular ring of the city, where the density of resident population reaches the highest value of Milan, but also in some municipalities with a residential profile and social housing in the south, south-west and in the north of the metropolitan area (Corsico, Rozzano, Sesto S.G). The city centre of Milan appears as a void, a fact that is consistent with the changes that occurred in the last decades, namely a gradual replacement of the residents with activities mainly related to the services sector and the commercial sector.

Figure 4 shows places with high density of activity during Saturday evening, from 8 pm until midnight. Focusing on the core city area, we notice several interesting patterns: a high activity in some places where there are many pubs and restaurants near the Milan Central Station, in the Navigli District, in the Isola Quarter and in other ambits characterized by the presence of leisure spaces (Filaforum Assago in the south of Milan), but also of activities in a continuous cycle as the hospitals. This treelet has proven to be effective in describing the temporal profile of the city lived by night populations during Saturday.

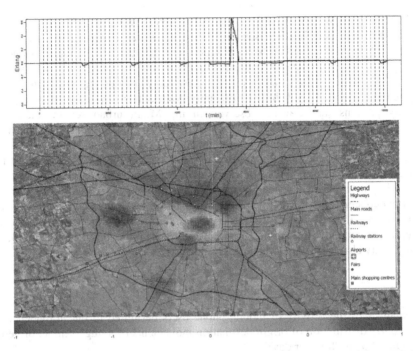

Fig. 4. Density of activity during Saturday evening (8pm-midnight). Saturday night population: leisure and hospitals

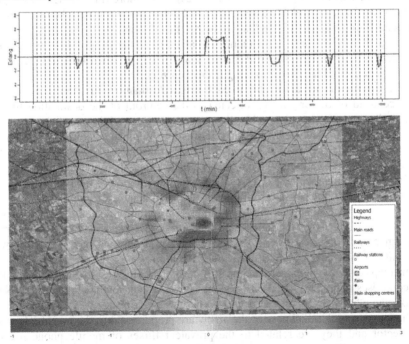

Fig. 5. Mobility practices. Saturday (10am- 8pm), shopping and leisure activity.

Figure 5 highlights another relevant mobility pattern, which is difficult to intercept through database traditionally used in urban studies: the shopping activity and, in general, the leisure activity. The map represents the density of mobile phone use during Saturday, from 10 am to 8 pm. Shopping and leisure are two of the main reasons of mobility in contemporary cities: they belong to the category of unsystematic mobility, and they significantly contribute to the even more complex mobility patterns that can be observed in the Milan urban region due to the distribution of commercial centers, commercial streets and, in general, of activities (museums, touristic sites, cinemas, just to cite some) inside and outside the city. These places attract, especially in certain days of the week, a huge amount of population coming from a vast territory that goes far beyond the administrative boundaries of the city. The map is the result of this spatial pattern and shows an important concentration of mobile phone traffic in the city centre and in other several places outside the city (most of them corresponding to the presence of commercial centers). The mainly residential areas, recognized in the previous Figure 3, are consequently characterized by the lowest value.

3.2   Aggregated Tracks of Mobile Phone Users

The analysis of the activity of mobile phone users permitted to put in evidence the main hourly distribution of origin destination movements of a huge sample of people (more than one million per day).

We started from the hourly origin destination matrices (October, 19th 2011) of mobile phone users among the 526 zones (tiles) of the more fine grained tessellation. For each zone it was available a set of directed connections towards the other and for each connection it was available the number of traced users.

Our goal was to find a synthetic visualization. Our proposal is to consider "prevalent fluxes" of mobility at different hours of a typical working day defined as the sum vector of all the fluxes moving from each zone.

More specifically we associated to each origin destination flux, the vector applied to the centroid of the origin tile, directed to the centroid of the destination tile and having the magnitude given by the value in the origin destination matrix corresponding to the selected origin and destination. We then calculated, for each tile, the sum vector of all the vectors applied to their centroid. It is characterized by two dimensions: the magnitude, which is function of the magnitudes of the original vectors, and the angle which expresses the prevalent direction of the flux.

A set of maps of the sum vector moving from each zone at different hours has been produced in order to highlight the main patterns of mobility during a typical working day (October 19th, 2011). The interested reader can visit our interactive web version of the map showing prevalent fluxes of mobility in a working day at the following URL: http://www.ladec.polimi.it/maps/od/fluxes.html .

The maps cover a wide area that goes from Milan in the West to Brescia in the East and comprises many populated city regions (Pavia in the South, Monza in the North and Bergamo between Milan and Brescia).

The length and the thickness of the arrow are proportional to the magnitude of the sum vector and they are related to the prevalent fluxes of mobile phone users at specific hours. The convergence of travels toward the main centers during the morning, the more complex direction of movements during the afternoon, are some interesting phenomena emerging from our analysis. The maps represent also the main infrastructures (railways and highways) and the zones of the tessellation, colored according to the intensity of the outgoing fluxes (light red color corresponds to lower intensity while dark red corresponds to higher intensity).

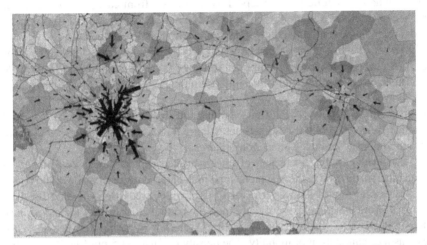

Fig. 6. Aggregated flows of mobile phone users: 9 am – 2011-10-19

Fig. 7. Aggregated flows of mobile phone users: 5 pm – 2011-10-19

The maps show *the variability* of phenomena that conventional data sources, such as census data, cannot give for a typical day. A broad use of the territory and an articulation of daily moves are visible every hour. The maps can be used as meaningful tools for monitoring the use of the infrastructural networks and of the urban spaces. On the one hand, the morning map (9pm; fig 6) confirms a polarization of movements towards the main centers offering job opportunities and highlights also the most commonly used infrastructures. On the other hand, the aggregated flows of mobile phone users in the afternoon (at 5 pm; fig 7) allow to recognize significant places for shopping and leisure, that are attended after work. This type of information is difficult, if not impossible, to monitor through conventional data at a comparable spatial and temporal resolution.

The automatic aggregation of municipalities (313 polygons) has been used for producing maps of mobile phone users' fluxes directed from each zone towards the others. We used this tessellation in order to have a better visualization of the complex and more relevant outcoming and incoming connections between the zones.

Only flows of more than 100 users are shown by oriented lines that connect the centroid of each origin zone with the centroid of the destination zones. In order to better visualize the overall patterns of mobility in the region, in and out flows from/to Milan city were excluded.

The map (fig. 8) evidences relevant relations and fluxes towards the main cities of the Lombardy Region, but also some interesting patterns in the Northern Milan area where a high density of huge connections emerges.

It is also evident a linear element of interconnected centers along some important infrastructural corridors (i.e. the Sempione road in the northern western side of the region and along the highway in the western side of the map between Bergamo and Brescia). These maps show the complexity of daily mobility patterns that modify the hierarchical structure of the cities where traditionally the physical relationship between jobs and homes was the main reason of mobility. The density of fluxes at 5 pm (fig. 9) describes not only the return home, but also the unsystematic mobility related to individual habits, as an effect of the diversified uses of the Milan urban region.

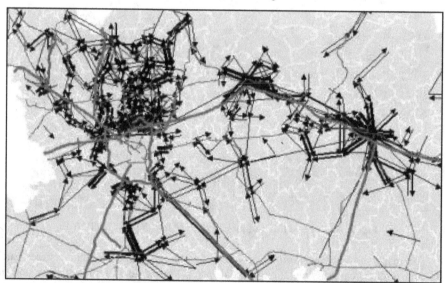

Fig. 8. Origin destination fluxes of mobile phone users: 8 am, 2011–10-19

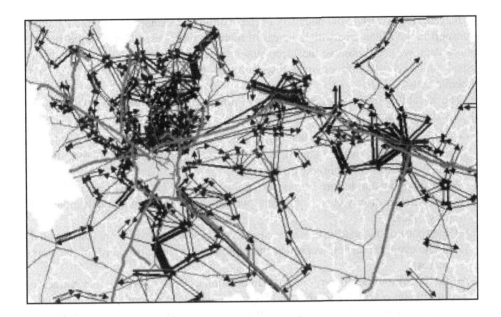

Fig. 9. Origin destination fluxes of mobile phone users: 5 pm, 2011-10-19

4   Implications for Policies

The research allowed us to test the potential of mobile phone data in explaining relevant urban usage and mobility patterns at the Milan urban region scale and in understanding the dynamic of temporary populations, two important topics that can be hardly intercepted through traditional data sources. This opens new implications for the urban research community which needs to elaborate new strategies to integrate traditional data with user generated data, such as mobile phone activity, in order to achieve a better comprehension of urban usages, in time and in space.

The presented data and methodology let the recognition of effective mobile populations in the urban environment. This knowledge can be exploited by decision makers for the definition of specific policies directed to temporary populations, which are more and more important in contemporary cities, otherwise ignored.

Describing the trends of use of urban spaces, the maps of mobile phone data give important information for mobility policies: the lack of coincidence between the mobility practices in the peak hours in the morning and in the afternoon when the chains of displacements are very articulate and complex, allows to recognize not only the variability in mobility practices, but also the places where these practices are occurring.

The commuters between 8 am and 9 am, become city users between 5 pm and 7 pm. This phenomenon strictly affects land use and can pose new questions and indications for transport policy.

Indeed, if we overlaid the boundary of the institutional management of local public transport in the Milan area with the areas of mobility practices, taken from the mobile phone data, we could observe the "deep structural effects of the mobility of people on urban policies" and the obvious disconnection between fixed jurisdictions and "mobile factors" [17].

The variability in the space-time of use of urban spaces resulting by mobile phone data is also revealed by:

— the spaces of night leisure that define a geography of places densely crowded at Saturday night, that is quite different from the territories of night work during the week (Monday to Friday night);

— the shopping and leisure spaces during the weekend (between 10 am and 8 pm) show the inner city center of Milan, but also some commercial malls along the ring roads;

— the space of the residence, where most significant call-densities are concentrated in the evenings and on weekends and represent the "negative copy" of the work places;

— the spaces of temporary events (International Design Week) that attract a significant portion of tourists and city users who visit several places in the city that are not identified by traditional sources and are not limited to traditional exhibition spaces.

The same data helps us to question some interpretations in the literature on the erratic behaviors of metropolitan populations, on the nomadism that characterizes the contemporary practices, that surveys on mobile phone data have already undertaken [10]. Some research about a significant sample of mobile phone data have, in fact, contested interpretations of nomadism of contemporary populations.

If they confirm the high density of commuting, they also show the strong recursion of the paths. In other words we move more during the day, but according to the known and usual paths.

References

1. Ehrenberg, A.: L'individu incertain, Paris, Calmann-Lévy (1995)
2. Urry, J.: Sociology Beyond Societies. Routledge, London (2000)
3. Kaufmann, V.: Re-thinking mobility. Ashgate, Aldershot (2000)
4. Ascher, F.: Les sens du mouvement: modernités et mobilités. In: Allemand, S., Ascher, F., Lévy, J. (eds.) Le Sens du Mouvement, Belin Bourdin, pp. 21–34 (2004)
5. Bourdin, A.: Les mobilités et le programme de la sociologie. Cahiers Internationaux de Sociologie 1(118), 5–21 (2005)
6. Scheller, M., Urry, J.: The new mobilities paradigm. Environment and Planning A 38, 207–226 (2006)
7. Nuvolati, G.: Resident and Non-resident Populations: Quality of Life, Mobility and Time Policies. The Journal of Regional Analysis and Policy 33(2), 67–83 (2003)
8. Ahas, R., Mark, Ü.: Location based services–new challenges for planning and public administration? Futures 37(6), 547–561 (2005)

9. Ratti, C., Pulselli, R.M., Williams, S., Frenchman, D.: Mobile landscapes: using location data from cell phones for urban analysis. Environment and Planning B: Planning and Design 33(5), 727–748 (2006)

10. Gonzalez, M.C., Hidalgo, C.A., Barabási, A.-L.: Understanding individual human mobility patterns. Nature 453(7196), 779–782 (2008)

11. Reades, J., Calabrese, F., Sevtsuk, A., Ratti, C.: Cellular census: Explorations in urban data collection. IEEE Pervasive Computing 6(3), 30–38 (2007)

12. Becker, R.A., Caceres, R., Hanson, K., Loh, J.M., Urbanek, S., Varshavsky, A., Volinsky, C.: A Tale of One City: Using Cellular Network Data for Urban Planning. IEEE - Pervasive Computing 10, 18–26 (2011)

13. Soto, Frias-Martinez: Automated land use identification using cell-phone records. In: Proceedings of the 3rd ACM International Workshop on MobiArch, pp. 17–22. ACM (2011)

14. Soto, Frias-Martinez: Robust land use characterization of urban landscapes using cell phone data. In: The First Workshop on Pervasive Urban Applications (PURBA) (2011b)

15. Balducci, A., Fedeli, V., Pasqui, G.: Strategic planning for contemporary urban regions: city of cities: a project for Milan. Ashgate, Burlington (2010)

16. Manfredini, F., Pucci, P., Secchi, P., Tagliolato, P., Vantini, S., Vitelli, V.: Treelet decomposition of mobile phone data for deriving city usage and mobility pattern in the Milan urban region. MOX Report 25/2012 (2012)

17. Estebe, P.: Gouverner la ville mobile. PUF, Paris (2008)

18. Jain, A., Dubes, R.: Algorithms for Clustering Data. Prentice-Hall, Upper Saddle River (1988)

19. Vantini, S., Vitelli, V., Zanini, P.: Treelet Analysis and Independent Component Analysis of Milan Mobile-Network Data: Investigating Population Mobility and Behavior. In: Analysis and Modeling of Complex Data in Behavioural and Social Sciences - Joint Meeting of of the Italian and the Japanese Statistical Societies (2012)

GIS Based Urban Design for Sustainable Transport and Sustainable Growth for Two-Wheeler Related Mega Cities like HANOI

Martin Ruhé[1], Hans-Peter Thamm[2], Leif Fornauf[3], and Matias Ruiz Lorbacher[4]

[1] German Aerospace Center, Rutherfordstrasse 2, 12489 Berlin, Germany
martin.ruhe@dlr.de
[2] Freie Universität Berlin, Malteserstrasse 74-100, 12249 Berlin, Germany
hpthamm@zedat.fu-berlin.de
[3] Technische Universität Darmstadt, Petersenstrasse 30, 64287 Darmstadt, Germany
fornauf@verkehr.tu-darmstadt.de
[4] International Academy (INA) gGmbH at the Freie Universität Berlin,
Malteserstrasse 74-100, 12249 Berlin, Germany
mrl@zedat.fu-berlin.de

Abstract. Growing urbanization and the increasing size of metropolitan regions is a challenge as well as an opportunity for the economic development and social balance of societies particular in developing countries like Vietnam. The very dynamical evolution of the cities implies special requirements for mobility concepts. These need to be adapted to possible future developments both with respect to their effects and to their implementation. Special challenges of sustainable transport in mega cities today have to address environmental problems and their social implications (e.g. air quality, noise, land use, fragmentation, and security), accessibility and ways to manage existing infrastructure capacities through smart engineering. As an example REMON (Real Time Monitoring of Urban Transport) is one of the research projects on mega cities supported by the Federal Ministry of Education and Research of Germany in cooperation with Vietnamese Partners like "Transport Development and Strategy Institute (TDSI)", "University of Transport and Communication (UTC)" and "Vietnamese-German Transport Research Centre (VGTRC), Vietnamese-German University, Ho Chi Minh City, Vietnam". The main objective of the project is to reduce the energy consumption and thereby the CO_2 production of urban traffic in mega cities. A comprehensive traffic monitoring system will be developed, implemented and applied. Based on real data, specific scenarios, measurements and immission models, the team will develop analysis tools to check the efficiency of the system and for calculating and optimizing the emission reduction additionally.

The project follows an interdisciplinary approach integrating land-use planning, transport science, engineering and political science in order to contribute to the mitigation of climate change. An important part of REMON is to assess the environmental impacts of a traffic management system and planned urban traffic developments based on real measured traffic data. Linking technological developments with urban and transport-planning policies is one of the foci of

B. Murgante et al. (Eds.): ICCSA 2013, Part III, LNCS 7973, pp. 452–465, 2013.
© Springer-Verlag Berlin Heidelberg 2013

the project: The identification of suitable strategies and policies is a prerequisite to ensure permanent and effective implementation of technological measures. The development of the technical system for traffic management to be installed in Hanoi is reported. All information and results of the traffic management system are open for the public.

The paper will give an overview about the specialties of the project and will point out the findings and goals of the described showcases.

Keywords: GIS, Sustainable Mobility, Sustainable Transport, Mega Cities, Hanoi (Vietnam).

1   Aims

The REMON project (Real Time Monitoring of Urban Transport - Solutions for Transport Management and Urban Planning in Hanoi) is a three-year research project on traffic related emissions and energy consumption. The key objectives of the project are the reduction of traffic induced air pollutants and emissions as well as the reduction of energy consumption within the Hanoi's urban transport sector. In order to achieve its goals, the project relies explicitly on geo-referenced information and spatial data. Crucial parts within the project are the traffic information system based on Floating Car Data (FCD) and Floating Phone Data (FPD), a precise digital street map, land use detection and an urban growth model as well as a transport demand modeling.

Hanoi is one of the cities in South East Asia with a strong urbanization and motorization rate. As a consequence, Hanoi suffers from severe traffic congestion. As described by Almec [1], Schipper [2], and Hai [3], the rapid urban growth of Hanoi does not only cause traffic related problems like air pollution, noise, and traffic jams, but also difficulties in urban development, urban planning and urban management.

The basic idea of the REMON project is to detect and track traffic conditions in real time via two methods: Floating Car Data (FCD) and Floating Phone Data (FPD). FCD and FPD base on GPS technology. More details can be found at Barceló and Kuwahara [4], Leduc [5], Schäfer [6], [7], Kühne [8], Listl and Dammann [9], Demir [10], and Messelodi [11]. The GPS data stems from on-board units in vehicles (cars, buses, taxis, motorcycles) as well as smartphones of motorbike drivers. FPD is due to Hanoi's characteristic as "motorcycle dependent city", as described by Barter [12], Khuat Viet [13], or Hai [14]. The term and characterization stem from the high share of motorbikes and motorized two-wheelers.

The FCD and FPD raw data will be converted into information for various applications: from informing road users of the current traffic situation on each street to controlling and managing traffic as well as long-term planning efforts and measures to solve traffic problems. Once set up and running, the real-time traffic information FCD and FPD system with its taxis and on-board units as well as smartphones equals a "distributed network of sensors" as Messelodi says [11] for traffic conditions, vehicular emissions and transport related issues like spatial accessibility and travel patterns.

The REMON project is informed by the central assumption and insight that traffic and urban transport are not separate and detached phenomena, but interwoven and interrelated with urban development (see among others Banister [15]; Te Brömmelstroet [16];

Wegener [17]). In order to reduce traffic, urban patterns of housing, business and production locations need to be considered in transport management as well as in urban and transport planning. To do so, the project team sets up a Geographical Information System (GIS), an expert system as well as an urban growth model and a traffic demand model. In close cooperation with the local stakeholders, these tools will help the project team to develop scenarios, recommendations and solutions for sustainable urban transport and urban development of Hanoi.

Setting up a GIS and relying on geo-referenced data is not only indispensable for developing scenarios. It is also crucial for displaying FCD and FPD and analyzing those data. A precise digital street map is a prerequisite for FCD and FPD and the visualization of traffic conditions. By analyzing the data, it is possible to identify bottlenecks and to implement traffic management measures as well as traffic information. In addition to those short-term measures, the project also aims at long-term approaches in urban and transport planning. The project team will use the data for the analysis of urban and transport planning related measures. In particular, the analysis of historical or long-term FCD and FPD data is meant to indicate potential for the better use of resources in transport and urban planning.

In the following, more details to FCD and FPD (2), the mobility demand model (3) and the urban growth model (4) will be given.

2   Vehicle Probes Data

Floating Car Data (FCD), also known as Probe Data, is by now a state-of-the-art technology for traffic data collection for traffic information systems. In various cities, FCD systems are running, helping to detect speed and location of the equipped vehicles. Among others, those cities are Berlin, Hamburg, Nuremberg, Munich, Stuttgart, Graz, Vienna, Gothenburg, Stockholm, Beijing, Ningbo, Chengdu, Hefei, Hangzhou, and Shanghai (see also Listel [9] and Demir [10]). In many cases, the FCD system runs on taxi fleets, but also on bus fleets and special fleets like parcel services. Starting in the early 2000s, DLR (German Aerospace Center – Deutsches Zentrum für Luft- und Raumfahrt) applied FCD technology in several German cities see Schäfer [6], Kühne [8], Steinauer [18], Leduc [5] for examples of use. The majority of traffic information systems based on FCD operating in urban areas, as a small vehicle fleet can already provide the coverage for the entire urban area. Yet, some companies provide traffic information systems on the national level, among others the car manufacturer BMW. Similarly, companies like TomTom and Navteq provide traffic information, combining traffic detection via GPS and mobile communication systems (GSM/3G) as well as data from stationary traffic detection devices (TomTom [19] and Leduc [5]).

Traffic detection systems relying on FCD approach usually receive the information from driving four-wheel cars. Using GPS technology, the location, times, as well as speed of the cars are detected in regular time intervals. The speed of the cars is the indicator for the traffic flow and traffic density on the streets and thus for traffic problems like traffic jams. Usually the FCD are not used for intersection management by traffic lights. The application of inner city traffic management is based mostly on inductive loops and sometimes also based on video cameras. Nevertheless, FCD based on taxi fleets are the most cost effective way to elevate data about the current situation of the car-based, individual traffic in a city. Using local public buses for collecting FCD in addition, gives a powerful tool for transit planning. The project REMON will show this in detail.

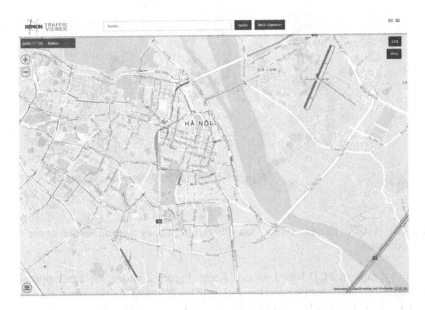

Fig. 1. Hanoi, FCD data coverage: Sunday 18:00 – 18:20

In contrast to European and North-American cities, motorcycles dominate urban transport in Hanoi. Many South East Asian cities like Hanoi, Ho Chi Minh City, Bangkok, or Jakarta, which all have a high share of two-wheelers, are referred as "motorcycle dependent cities" (see Barter [12], or Khuat Viet [13]). A challenging task in developing the traffic information system in Hanoi thus is the significant proportion of two-wheelers. The motorbikes are much more flexible and able to drive even if passenger cars are already stopped by traffic congestion. The high agility of the two-wheelers causes a challenge for traffic detection. The REMON project will develop and apply an approach similar to Floating Car Data, the Floating Phone Data (FPD). It is based on phones/smartphones equipped with GPS receivers. As the application of standard FCD is not applicable to the traffic flows in Hanoi, a comprehensive mathematical and statistical analysis has to be accomplished to determine the correlation between speed, traffic flow and traffic density for different street types in Hanoi. The relation between speed, traffic flow and traffic density is empirically tested and described in the so-called fundamental diagrams (described in FGVS [20]). But these results apply only for car-dominated traffic flows. It is one of the main scientific research tasks of the project to find new mathematical relations.

Figure 1 and 2 show the graphical user interface of the system of data collection and traffic information for trip planning and dynamic navigation. Figure 1 shows the

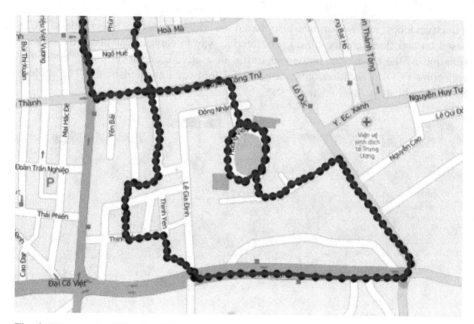

Fig. 2. First result of Floating Phone Data (FPD) collection by the beta version of the smart-phone application

current coverage of FCD based on taxi fleets. Figure 2 shows the first approach of FPD based on the smartphone application. The different colors show the service quality of the current traffic in three different levels on a digital street map of wider Hanoi area.

3   Mobility Demand Model

Providing real-time traffic information is crucial for influencing traffic, but also for transport planning and developing appropriate traffic management strategies. Howev-er, for transport planning it is indispensable to know why people move from one place to another. The analysis of travel demand is one of the most important tasks in trans-portation planning. For conducting a regional transportation planning study as well as for examining the transportation impacts of a new development site, the estimation of the expected travel demand is always a critical, but also very important point for transportation planning (see Meyer [21]). In order to analyze the transportation sys-tem, planners often use models. A model can be defined as an abstraction and a sim-plification of a 'real world' system to make conditional forecasts about what might occur within the system if certain changes to the system are introduced [21].

Transport demand modeling for passenger traffic is already very sophisticated. Years of experimentation and development have resulted in a general structure which has been called the classic transport model (see Ortuzar [22]). The general form of the model is illustrated in the following Figure 3. It consists of four major stages and hence is named the four-stage or four-step model.

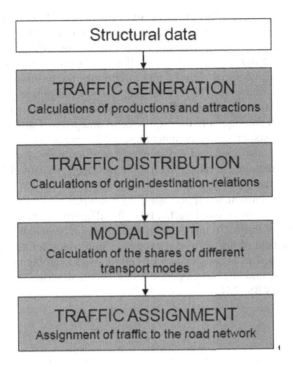

Fig. 3. The classic four-step transport model

1. Trip generation is the prediction of the number of trips produced by and attracted to each zone. The trip generation analysis predicts total flows into and out of each zone in a defined study area, but it does not predict where these flows are coming from or are going to.
2. Trip distribution is the allocation of these trips to particular destinations, in other words their distribution over space. The results are shown as origin-destination (O-D) flows between zones stored in relevant matrices.
3. The modal split predicts the percentage of travel flows that are dedicated to the available modes of transport (car, bicycle, bus, train, walk etc.) between each origin-destination pair.
4. Trip assignment is also referred to as route choice. It allocates trips by each mode on specific routes of travel to the respective modal networks, considering the impedance on each single route.

In general, the four steps of this transport model correspond to a sequential decision process in which people decide to make a trip (generation), decide where to go (distribution), decide what mode to take (modal split), and decide what route to use (assignment). The four step model, however, represents a pragmatic approach to reduce the extremely complex phenomenon of travel behavior into analytically manageable components that can be dealt with by using state of the art simulation tool and reasonable amounts of data.

The objective of the REMON project is the reduction of traffic jams by 5 to 10 per cent and the reduction of transport related emissions by 10 to 15 per cent by appropriate (dynamic) traffic management strategies. Traffic management influences the supply of traffic and transport systems as well as the demand for travel and transport through a bundle of measures with the aim to optimize the positive and negative impacts of traffic and transport. This implies that traffic management needs to consider and to weight different aspects (traffic safety, traffic quality, economic and environmental aspects) and consequently the requirements and goals of different stakeholders (motorists and cyclists, cities and communities, traffic operators).

Unexpected, but also expected deviations from a defined basic traffic condition need pre-defined measures to handle the problems effectively and efficiently. Hence, (dynamic) traffic management consists of influencing the current transport demand and influencing the current transport supply by coordinated situation-adaptive measures. Measures are taken to avoid traffic, to shift traffic (in mode, in destination, or in time), and to control traffic. In this context, a strategy can be defined as an action plan that includes a bundle of pre-defined measures to improve a defined initial situation. A situation is defined as a combination of certain events, problems, and conditions. Both, the situation and the developed strategy for it, describe a specific scenario (Figure 4) (see BMVBS [23]).

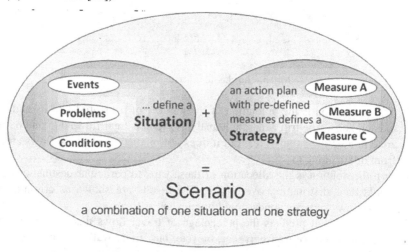

Fig. 4. Situation, Strategy, Scenario (Source: ZIV et al [24])

Taking the general approach of 'four step model' as the basis, many commercial products have been developed and are established now. These products allow applying different theoretical assumptions and approaches in the different modeling steps. After assessing of different software tools regarding their applicability, the software VISUM of the German company PTV AG has been chosen as the modeling tool to be used within the project. VISUM is a software system that allows modeling all private and public transport types in one single integrated model (see PTV [24]). The transport demand modeling is a crucial element of the REMON project. It will be used for the assessment of suitable traffic management strategies as well as for the scenarios on urban and transport development. Within the modeling software all 4 stages of the classical traffic model (4-step model) can be calculated. Furthermore, modules can be developed within the software, which is able to estimate transport-related emissions (noise, CO_2, PM, NOx).

Transport models can depict the effects on traffic flow and traffic congestions, respectively. The emissions can either be measured directly or estimated indirectly by using the demand model. To estimate the transport related emissions, transport models facilitate emission calculations. The emissions can be calculated by the equation

'emissions = traffic volumes * emissions factors'.

The HBEFA (Handbook Emission Factor for Road Transport) catalogue, an internationally agreed catalogue, provides the figures of the emission factors. Alternatively, they can be adjusted to the specific context. In the case of the REMON project, an adaption to the conditions in Hanoi will be necessary. One advantage of the REMON project consists in evaluating the emission calculations by crosschecking them with data generated in the FCD and FPD system.

For the calculation of transport-related emissions (among others CO_2, NOx, PM) and energy consumption, basically the amount of vehicles on the streets in a time interval (traffic density), the percentages of different motorized transport modes (fleet composition) and the emission factors, a representative value for quantity of a pollutant released by different types of vehicles, are needed. The traffic density and traffic flow will be derived from the FCD/FPD system; fleet composition and emission factors are still to be identified and determined. The demand model then calculates the emissions (among others CO_2, NOx, PM) per street section and aggregates them for the whole city.

After modeling the transport demand, the impact of traffic can be shown and analyzed through the results of the demand model. The results of demand modeling are to some extent also the basis for the scenario building in integrated transport and urban development. A transport supply system has diverse impacts, which may vary because of the application of different measures. The following impacts have to be considered:

— Impacts on the user of the transport system
— Impacts on the operators who have to 'produce' a transport service
— Impacts on the general public who benefits from the transport infrastructure
— Impacts on the environment which is harmed by pollution

In the case of the proposed project in Hanoi as a highly dynamic, motorcycle dependent city, the main challenges in transport demand modeling will be the adaption to the local conditions. The research questions that the project will investigate on are among:

— Efficient methods of data gathering for the transport demand modeling
— The modeling of local trip patterns
— The composition of the vehicle fleet in Hanoi

Considering the time limitation as well as the high cost of data collection, an aggregated model will be established in the REMON project. Here, aggregation is performed in three dimensions: the spatial, the temporal, and the socioeconomic (see also Meyer [21]).

- Spatial aggregation is performed by dividing the study area into a set of zones and then treating these zones as the basic units of analysis. Thus, rather than dealing with trips made by individuals from point to point, the analysis is concerned with total flows of people from zone to zone.
- Temporal aggregation is performed by grouping travel flows that will tend to occur at varying levels over time into discrete time periods. Thus, one might analyze peak period or off-peak period flows between zones in order to identify deficiencies in the network capacity.
- Socioeconomic aggregation occurs whenever individuals are categorized into 'homogeneous' groups. In such cases, the explicit assumption is that all members of a given group behave in the same way, or at least the variance in their behavior is small relative to the differences in behavior observed between their group and other groups.

After setting up a transport demand model for Hanoi, the scenario of the traffic situation will be visualized. Then as a second step, the most important traffic links and roads of Hanoi (5 to 10 routes or axes) will be selected in cooperation with transport operators. For those links and routes, there will be a continuous automated monitoring. Deviations from the normal travel times will be displayed. Professional users like operators and traffic police will receive primarily this information in order to issue recommendations and warnings to avoid congested areas.

Another potential application for traffic management in Hanoi is described as following: The visualization of the current traffic situation will include journey times and the monitoring of travel times. For the monitoring of the current travel times, a multi-stage concept has been developed which will be implemented within this project as well as other projects. First, users will be able to query the current, statistically proven and most probable travel time for a route between a defined start and destination. This pre-trip information will be given to the users. Additionally, the user will be able to select different routes ranging from the shortest to the fastest route. Together with the dynamic navigation, the system will display areas with high traffic density and strongly reduced traffic speeds. Thus, the user can decide to avoid those areas or to be guided to the route of least traffic load by using the function of re-routing.

This working area (mobility demand modeling) encompasses the set-up of a goal system, the development of an assessment method for traffic management strategies, and finally the assessment and selection of specific traffic management strategies for Hanoi. Another major task is the set-up of the transport demand modeling, which will be used for the assessment of suitable traffic management strategies, the calculation of energy consumption and emissions in the urban transport sector in Hanoi as well as for the scenarios on urban and transport development. Here it is essential to point out that in this working area there are much cooperation with other partners. Transport demand modeling is not an isolated issue in the project. The network modeling at the beginning relies more or less on the visible and reliable network data that is collected and provided through GIS. Due to the fact that the transport has finally a long-term impact on the land use and the urban development, the cooperation with correspond partners specialized in that field is very meaningful.

4   Urban Growth Model

Sustainable urban planning depends on sound information about the future development of the city under different boundary conditions. Especially in rapidly growing cities like Hanoi, reliable future projections are important to bridge the time gap between the building of residential and industrial areas (one to two years) and planning and realizing of infrastructure like roads (sometimes more than 5 years). Against this background, an urban growth model for Hanoi City will be developed within the REMON project. It will model spatial explicit scenarios about future urban development under different socio-economic and technical boundary conditions. In the interdisciplinary approach of the REMON project, the results of the modeling can be used for sustainable traffic planning as well as the impact of various planning scenarios on the CO_2 output. It will be challenging to couple the urban growth model with the traffic model due to the different spatial representation.

There are different approaches for modeling land use change in general and especially urban growth (Briassoulis [27],and Mulligan [28]), from simple statistical models to very complex models which try to describe the complexity of human behavior in detail (Silva [29]). Verburg [30] gives a good overview over the different model approaches. Every model type has advantages and disadvantages. So it is very important that on one hand the functional relations which are causing the urban growth are assessed with a sufficient accuracy (Wegener [31]) and that a reliable database for calibration and validation of the model is available (Pontius [32]).

In respect of the availability of the data and manageable complexity of the model within the REMON project, a statistical dynamical model approach, which can as well represent human decision making, will be chosen for Hanoi like it is described by Verburg [33].

To understand the underlying processes causing the urban dynamics of Hanoi, historical spatio-temporal change patterns will be derived from historical satellite data and analyzed with geo-statistics using additional socio-economic data (Ward [34], Yang [35], and Xian [36]). If available, other data explaining underlying causes for the change, like human decision making, will be incorporated in this analysis. As a result, the driving forces for the urban dynamics can be identified and their influence can be quantified. Hereby the traffic model can be a tool to investigate the interrelation between traffic and urban dynamics. The functional relationships of urban dynamics will be described with respective formulas in the urban growth model. There will be numerous sensitivity analyses (see Pontius [37]) of the model to understand the influence changes in the input parameter and functional relationships estimated (Lambin [38] and Pontius et al [39]). The model will be no "black box". Its design is modular so that all intermediate results can be evaluated. As well detailed sensitivity analysis for very parameter can be performed. The model will be integrated in an advanced information system with a user friendly interface. Therefore urban planners are enabled to compute possible states of future urban development under different boundary conditions without losing too much time for parameterization. It is planned to realize the model in the Java based XULU framework (eXtendable Unified Land Use and land cover platform) (Schmitz [26]), which provides the necessary flexibility, good interoperability with the databases. As well the realization of additional modules is comparatively efficient.

Urban land use and transport is closely interlinked especially in terms of social and economic activities. Consequently, land use policy has a significant impact on transport planning and traffic. In turn, the existing or planned transport system has a strong impact on the accessibility and thus on the attractiveness of a specific urban area (Wegener [31]). For that reason, it is obvious to combine and moreover to integrate the land use and the transport demand model into an integrated land use transport interaction model. In this context, several land use interaction models have been developed and have been in use for many years, such as

— static models (e.g. IMREL, DSCMOD),
— entropy-based models (e.g. LILT),
— spatial-economic models (e.g. MEPLAN, TRANUS, MENTOR) and
— activity models (e.g. IRPUD, DELTA)
  (see Department for Transport [40], pp. 12-16.].

The work in the described project focuses on the present traffic and land use conditions and tries to develop and to implement dynamic strategies for real-time operations. The development and the use of a land use interaction model would be useful for mid- and long-term planning processes and thus could be the objective for the next step after first approaches and a stable foundation have been elaborated within the presented project. Although the development of a land use transport interaction model is not part of project, an interaction between the urban land use model and the transport demand model can be realized to estimate the mid- and long-term perspectives for Hanoi. This could be achieved by utilizing the results of the land use model as part of the input data of the transport demand model and vice versa, for example by analyzing the accessibility of urban areas.

5   Outlook

During the next 30 months the project team will work on several tasks. First of all the team of Vietnamese and German experts is looking for smartphone users bringing the system on a wider basis of data collectors. The modeling tasks for transport and land use have to be done.

The final outputs of the REMON project will be a running real-time traffic information system, well-adapted traffic management strategies, policy recommendations on urban development as well as showcases of energy-efficient urban and transport planning in Hanoi.

One result of the project will be the digitization and visualization of traffic conditions. In Hanoi, a new source of information about the current (real time) situation on the streets of Hanoi will be available with the REMON project. The local government, transport authorities, urban and transport planners will be able to measure traffic conditions and conduct an efficient monitoring of Hanoi's transport infrastructure.

The possibilities to use of the generated traffic data are ample, thus the data can be used for:

— Information on current traffic conditions to road users (via Internet, mobile phones, radio, television) and to the transport authorities
— Data analysis
— Identification of capacity overloads, bottlenecks and hot spots of traffic congestion
— Intersection Traffic Monitoring and Local Net Assessment
— Accessibility analysis
— Traffic management and traffic quality
— Managing and monitoring of traffic conditions
— Optimization of existing transport infrastructure by applying traffic management measures and monitoring their effectiveness with the FCD/FPD system
— Emission modeling and energy consumption calculations based on FCD and FPD

References

1. ALMEC Corporation, Nippon Koei Co. Ltd., Yachiyo Engineering Co. Ltd.: The Comprehensive Urban Development Programme in Hanoi Capital City of the socialist Republic of Vietnam (HAIDEP): Final Report. Japan International Cooperation Agency (JICA) and Hanoi People's Committee, Hanoi (2007)
2. Schipper, L., et al.: Measuring the Invisible: Quantifying Emission Reductions from Transport Solutions – Hanoi Case Study. World Resource Institute, Washington, DC (2008), http://pdf.wri.org/measuringtheinvisible_hanoi-508c_eng.pdf
3. Hai, L.D.: Influence of Asian Transport on Urban Transport Policy and Planning in Ha Noi, Viet Nam. In: Eastern Asia Society for Transportation Studies (ed.) Proceedings of the Eastern Asia Society for Transportation Studies, 5th International Conference of Eastern Asia Society for Transportation Studies, Fukuoka, Japan, October 29-November 01 (2003), http://www.easts.info/2003proceedings/papers/1654.pdf
4. Barceló, J., et al. (eds.): Traffic Data Collection and its Standardization. Springer, New York (2010)
5. Leduc, G.: Road Traffic Data: Collection Methods and Applications. In: Working Papers on Energy, Transport, and Climate Change N. 1. European Commission, Joint Research Center, Institute for Prospective Technological Studies, Brussels (2008), http://ftp.jrc.es/eurdoc/jrc47967.tn.pdf
6. Schäfer, R.-P., et al.: Neue Ansätze im Verkehrsmonitoring durch Floating Car Data. In: Proceedings 19. Verkehrswissenschaftliche Tage Dresden (2003)
7. Schäfer, R.-P., et al.: Analysis of Travel Times and Routes on Urban Roads by Means of Floating Car Data. In: Proceedings of the ITS World Congress 2002, Chicago, USA (2002)
8. Kühne, R., et al.: New Approaches for Traffic Management in Metropolitan Areas. In: 10th International Federation of Automatic Control, Tokyo, Japan (2003)
9. Listel, G., et al.: Untersuchungen zum Einsatz von Taxi-Floating Car Data im Ballungsraum Rhein-Main. Straßenverkehrstechnik 53(3), 147–151 (2009)
10. Demir, C., et al.: FCD for Urban Areas: Method and Analysis of Practical Realisations. In: Proceedings of the 8th ITS World Congress, Madrid, Spain (2003)
11. Messelodi, S., et al.: Intelligent extended floating car data collection. Expert Systems with Applications 36(3), Part 1, 4213–4227 (2009)
12. Barter, R.P.: An international Comparative Perspective on Urban Transport and Urban Form in Pacific Asia: The Challenge of Rapid Motorization in Dense Cities. Perth (1999)
13. Khuat Viet, H.: Traffic Management in Motorcycle Dependent Cities. Darmstadt (2006)

14. Hai, P.H., et al.: The motorcycle ownership behaviour in Hanoi city, Vietnam: How unique they are compared to other countries? In: Eastern Asia Society for Transportation Studies (ed.) Proceedings of the Eastern Asia Society for Transportation Studies, 7th International Conference of Eastern Asia Society for Transportation Studies, Dalian, China, September 24-27 (2009), http://www.jstage.jst.go.jp/article/eastpro/2009/0/97/_pdf/-char/ja/

15. Banister, D.: Unsustainable Transport: City transport in the new century. Routledge, London (2005)

16. Te Brömmelstroet, M., et al.: Developing land use and transport PSS: Meaningful information through a dialogue between modelers and planners. Transport Policy 15(4), 251–259 (2008)

17. Wegener, M.: Modelle der räumlichen Stadtentwicklung - alte und neue Herausforderungen. Schriftenreihe Stadt Region Land 87, 73–81 (2010)

18. Steinauer, B., et al.: Integration mobil erfasster Verkehrsdaten (FCD) in die Steuerverfahren der kollektiven Verkehrsbeeinflussung. Bundesministerium für Verkehr, Bau und Stadtentwicklung, Abteilung Straßenbau, Straßenverkehr, Bonn (2006)

19. TomTom: TomTom Traffic Homepage (2013), http://www.tomtom.com/en_gb/licensing/products/traffic/

20. FGSV: Hinweise zum Fundamentaldiagramm – Grundlagen und Anwendungen, Forschungsgesellschaft für Straßen und Verkehrswesen, Köln (2005)

21. Meyer, M.D., et al.: Urban Transportation Planning, New York. McGrall-Hill Series in Transportation (2001)

22. Ortuzar, J., de, D., et al.: Modelling Transport. John Wiley & Sons, New York (2006)

23. BMVBS (Bundesministerium für Verkehr, Bau und Stadtentwicklung): Leitfaden Verkehrstelematik - Hinweise zur Planung und Nutzung in Kommunen und Kreisen (ITS Manual – Recommendations for the Planning and Application in Municipalities and Counties), Berlin (2006)

24. PTV Groups: VISUM 12.5 Fundamentals Karlsruhe: PTV Planung Transport Verkehr AG (2012)

25. ZIV (Zentrum für integrierte Verkehrssysteme) et al: Verknüpfung von Strategien, Maßnahmen und Systemen des regionalen und städtischen Verkehrsmanagements (Linking Strategies, Measures and Systems for Regional and Urban Traffic Management). Project Report. Darmstadt, Frankfurt a. M., Aachen (2000)

26. Schmitz, M., Bode, T., Thamm, H., Cremers, A.B.: XULU - A generic JAVA-based platform to simulate land use and land cover change (LUCC). In: Oxley, L., Kulasiri, D. (eds.) Proceedings of MODSIM 2007 International Congress on Modelling and Simulation (Modelling and Simulation Society of Australia and New Zealand), pp. 2645–2649 (2007)

27. Briassoulis, H.: Analysis of Land Use Change: Theoretical and Modeling Approaches (Regional Research Institute, University of West Virginia) (2000), http://www.rri.wvu.edu/WebBook/Briassoulis/contents.htm

28. Mulligan, M., Wainright, J.: Modelling and model building. In: Wainright, J., Mulligan, M. (eds.) Environmental Modelling. Finding Simplicity in Complexity, pp. 5–74. John Wiley & Sons, Chichester (2004)

29. Silva, E.A., Clarke, K.C.: Complexity, Emergence and Cellular Urban Models: Lessons Learned from Applying Sleuth to Two Portuguese Metropolitan Areas. European Planning Studies 13(1), 93–115 (2005)

30. Verburg, P.H., et al.: Modelling the Spatial Dynamics of Regional Land Use: The CLUE-s Model. Environmental Management 30(3), 391–405 (2002)

31. Wegener, M., et al.: Land-Use Transport Interaction: State of the Art. IRPUD, Dortmund (1999)
32. Pontius J.R, R.G., Huffaker, D., Denman, K.: Useful techniques of validation for spatially explicit land-change models. Ecological Modelling, 179 (4), 445-461 (2004)
33. Verburg, P.H., et al.: Modelling Land-Use and Land-Cover Change. In: Lambin, E., Geist, H. (eds.) Land-Use and Land-Cover Change. Local Processes and Global Impacts, pp. 117–136. Springer, Berlin (2006)
34. Ward, D., et al.: Monitoring Growth in Rapidly Urbanizing Areas Using Remotely Sensed Data. Professional Geographer 52(3), 371–386 (2000)
35. Yang, X., Lo, C.P.: Using a time series of satellite imagery to detect land use and land cover change in the Atlanta, Georgia metropolitan area. International Journal of Remote Sensing 23(9), 1775–1798 (2002)
36. Xian, G., Crane, M.: Assessment of urban growth in the Tampa Bay watershed using remote sensing data. Remote Sensing of Environment 97, 203–215 (2005)
37. Pontius, J.R., Huffaker, R.G., Denman, D., Useful, K.: techniques of validation for spatially explicit land-change models. Ecological Modelling 179(4), 445–461 (2004)
38. Lambin, E.F., Meyfroidt, P.: Land use transitions: Socio-ecological feedback versus socioeconomic change. Land Use Policy 27(2), 108–118 (2010)
39. Pontius, J.R, R.G., et al.: Comparing the input, output, and validation maps for several models of land change. Annals of Regional Science 42, 11–37 (2008)
40. Department for Transport: Transport Analysis Guidance. Department of Transport, London (2005)

New Concepts for Structuring 3D City Models – An Extended Level of Detail Concept for CityGML Buildings

Marc-O. Löwner[1], Joachim Benner[2], Gerhard Gröger[3], and Karl-Heinz Häfele[2]

[1] Institute for Geodesy and Photogrammetry, Technische Universität Braunschweig, Germany
m-o.loewner@tu-bs.de
[2] Institute for Applied Computer Science, Karlsruhe Institute of Technology, Germany
{joachim.benner,karl-heinz.haefele}@kit.edu
[3] Institute for Geodesy und Geoinformation, University Bonn, Germany
groeger@igg.uni-bonn.de

Abstract. We propose a new Level of Detail (LoD) concept for CityGML buildings that differentiates a Geometrical Level of Detail (GLoD) and a Semantical Level of Detail (SLoD). These two LoD concepts are separately defined for the interior characteristics and the outer shell of a building, respectively. The City Geography Markup Language (CityGML) is an open and application independent information model for the representation, storage, and exchange of virtual 3D city models. It covers geometric representations of 3D objects as well as their semantics and their interrelation. The CityGML Level of Detail concept in general offers the possibility to generalize CityGML features from very detailed to a less detailed description. The current LoD concept suffers from strictly coupling geometry and semantics. In addition it provides only one LoD (LoD4) for the description of the interior of a building. The benefits of our new LoD concept are first, a substantially higher informative value for the Level of Detail, second, a better description of the interior Level of Detail, third, a broadening of the opportunities for indoor modelling, and last, a better assignability to all other modules represented in CityGML. Due to more combinations of GLoD and SLoD, the Level of Detail definition for every module in CityGML can be defined according to the nature of modelled real world phenomenon.

Keywords: 3D City Models, CityGML, Level of Detail, Geometrical Level of Detail, Semantical Level of Detail.

1   Introduction

Semantically enriched virtual 3D city models support urban modelling in many ways. Possible applications are environmental and energy planning, disaster management, noise simulation, urban planning, and public participation in planning processes. In order to fully exploit virtual 3D city models, a commonly accepted data model for storage and exchange of geometry, semantics and relations of the modelled features is needed.

CityGML [1] is such an interoperable data model. It has been issued by the Open Geospatial Consortium (OGC), which is – besides the official International Organization for Standardization (ISO) – the most important standardization

B. Murgante et al. (Eds.): ICCSA 2013, Part III, LNCS 7973, pp. 466–480, 2013.
© Springer-Verlag Berlin Heidelberg 2013

organization in the field of geospatial information technologies. CityGML is a common information model and encoding standard for the representation, storage, and exchange of virtual 3D city and landscape models. In addition to 3D geometric representations, it provides concepts to represent their semantics and their relations. CityGML is commonly accepted in the field of 3D city models; the number of available city models and their applications has increased significantly in the last ten years. Applications that rely on CityGML are e.g. the Energy Atlas of Berlin that supports investigations on energy consumption, energy demand, and energy saving potentials. Noise simulation and mapping has been performed for North Rhine-Westphalia in Germany using an extended CityGML data model [2], [3]. A fragmentary overview of CityGML applications in Germany is given in [4].

An advantage of CityGML is its scalability to the requirements of the user and the data available. The functionality of CityGML can be extended by applying the Application Domain Extension (ADE) mechanism. This mechanism extends CityGML classes by additional attributes and relations. Another way to extend CityGML is the definition of generic classes and attributes, which is more flexible, but hampers interoperability, since there is no common schema for the extension.

However, confining the functional range of CityGML is especially important in practice. On the one hand CityGML is organized in thematic modules that support a valid creation of tailored CityGML instance models without implementing the whole standard. On the other hand, almost every thematic class may be represented in different Levels of Detail (LoD). The LoD concept enables first, a gradual refinement of the geometrical characteristic, and second, the adjunction of semantic properties. Therefore it supports gradual data collection with respect to different application requirements and efficient data visualization and analysis.

Different LoDs first, serve different applications and, second, provide information about the quality of a modelled feature. The LoD concept has been developed first for the *Building* module and has been adapted to the other modules afterwards. Although the LoD concept of CityGML is used in practice and is subject to scientific research, from today's perspective it has considerable disadvantages as to informational content as well as to clearness of definition.

In this paper we propose a new approach to define the Levels of Detail for CityGML features. We start with a short description of CityGML to represent geometrically and semantically virtual 3D city models. In particular, we will focus on the *Building* module and will describe the current Level of Detail concept afterwards. In section 3 we will carve out the main deficits of this concept and develop a new approach that distinguishes between a geometrical and a semantical Level of Detail. We follow up with a discussion on the benefits of this new approach.

In this paper, names of classes and attributes used in CityGML and are written in *italics*.

2   CityGML – An International Standard for Virtual 3D City Models

The City Geography Markup Language (CityGML) is an open and application independent information model for the representation, storage, and exchange of virtual 3D city models. In addition to geometric representations of 3D objects it provides concepts to store their semantics and their interrelation. In addition, it covers the generalization and aggregation of semantically defined features. Therefore, it supports 3D content for visualization, but goes far beyond that point to support manifold analytical capacity. Unlike the Keyhole Markup Language (KML) used in the context of Google Earth, Collada or X3D, for instance, it distinguishes real world features providing 98 classes with 372 well defined attributes in total. These classes may have geometrical properties or not. Thus, in addition to visualisation application it supports the exchange of 3D city models for environmental simulations, energy estimations, disaster protection and others.

CityGML is an Open Geospatial Consortium (OGC) encoding standard and was released as version 1.0 in 2008 [5] and as version 2.0 in March 2012 [1]. Besides, the International Organization for Standardization (ISO), the OGC is seen to be the most important organization in the field of geospatial technologies.

CityGML is implemented as an application schema of the extensible Geography Markup Language (GML 3.1.1) [6] which is itself based on the Extensible Markup Language (XML). Hence, the exchange of CityGML benefits from all GML-intermateable techniques for data exchange, processing, and cataloguing, provided by the OGC. These include the Web Feature Service (WFS), the Web Processing Service (WPS), and the OGC Catalogue Service, for instance.

CityGML is organized in 13 thematic modules that enable a vertical scaling of a city model. This modularization is carried out by different XML-Schemas with different namespaces. The benefit of vertical modularization is the valid creation of thin CityGML instance models without implementing the whole standard. The most important of these thematic modules are the *Building* module containing semantic classes to represent buildings, i.e. houses or garages (cf. sec. 2.1) and the fundamental *Core* module. While the *Bridge* module and the *Tunnel* module are modelled as the *Building* module, the others are less detailed.

Besides offering an opportunity to confine CityGML by using only selective modules and the Level of Detail concept (rf. sec. 2.2) it is expandable, also. For this the Application Domain Extension (ADE) concept was developed. It allows the user, first, to add attributes or relations to CityGML classes and, second, to define new classes by generalization from CityGML classes. All attributes and classes then have to be defined in an own ADE namespace.

2.1   The CityGML *Building* Module

In CityGML the most important thematically module is the *Building* module. The central class is the abstract class _*AbstractBuilding* that is specialized to a *Building* or a *BuildingPart*, respectively. Both, *Building* and *BuildingPart* inherit the attributes *yearOfConstruction*, *yearOfDemolition*, *roofType*, *storeysAboveGround*, *storeysBelowGround*, *storeyHeightsAboveGround* und *storeyHeightsBelowGround* from _*AbstractBuilding* as well as *class*, *function*, and *usage*. As in all other thematic classes in CityGML the attribute *class* represents a classification of the *Building*, e.g. 'habitation' or 'business'. The attributes *function* and *usage* contain information about planned and actual utilisation of the building, e.g. 'holiday house' or 'public building'. All values of the attributes may be defined in external code lists and therefore can be adapted to national standards or project oriented needs. The same accounts for the attribute *roofType*.

_AbstractBuilding is specialized either to *Building* or to *BuildingPart*, allowing the representation of an aggregation hierarchy of arbitrary depth of connected buildings and parts of buildings. Disconnected groups of buildings sharing the same semantics, e.g. industrial complexes may be modelled as *CityObjectGroups*.

The building's geometric representation as well as the improvement of semantical description may be gradually refined, applying the CityGML concept of Levels of Detail (LoD) that is outlined in the next section.

2.2   The CityGML Level of Detail Concept

The Level of Detail concept (LoD) is a characteristic quality of CityGML. Next to the horizontal modularization the LoD concept offers the possibility to generalize CityGML features from very detailed to a less detailed description. In CityGML the LoD concept enables first, a gradual refinement of the geometrical characteristic, and second, the adjunction of semantic properties. Therefore it supports gradual data collection with respect to different application requirements as well as efficient data visualization and analysis. The LoD concept is a prerequisite to model buildings in the context of cities or even regions as well as detailed buildings with interior structures. Thereby, CityGML is different to the Industry Foundation Classes (IFC) developed by·buildingSMART for the representation of highly detailed building models [7].

The Level of Detail concept can be applied to all main thematic classes representing the most important types of objects within virtual 3D city models, i.e. *Bridge*, *Building*, *CityFurniture*, *CityObjectGroup*, *Generics*, *LandUse*, *Relief*, *Transportation*, *Tunnel*, *Vegetation*, and *WaterBody*. In a CityGML instance document the coexisting representation of one and the same object in different Levels of Detail is possible. The current LoD concept was developed primarily for the *Building* module and adopted for other modules, afterwards. That does not apply for the building's LoD0 representation that was added to the *Building* module in version 2.0 to handle 2D map representations for buildings. All in all there are 5 Levels of Detail in the *Building* module that are depicted in figure 1.

Level of Detail 0 (LoD0): The building is represented by the building footprint or the roof outline. Both are horizontal surfaces with a constant height value. This representation corresponds to a city map representation and enables the integration of 2D data coming from cadastral map excerpts, for instance. A possible application for a LoD0 building representation might be density or distance calculations for fire precautions or just land tenure visualization.

Level of Detail 1 (LoD1): The building is represented by a block model, i.e. a vertical extrusion solid without any semantically structuring. The geometric representation is realized by a *gml:Solid* or a *gml:MultiSurface*.

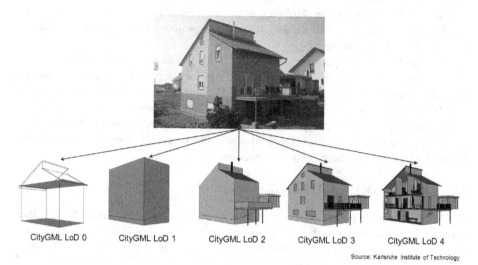

Source: Karlsruhe Institute of Technology

Fig. 1. Representation of a building using LoD0 - LoD4 [8]

Possible applications for LoD1 building models are noise mapping approaches [2], [3] or the estimation of real volume in flood planes for flood prevention. Even for modelling mobile communications networks, a LoD1 city model would be sufficient as long as no reflection properties are needed. Another application could be a multiple line of sight analysis to optimize the deployment of WLAN routers or checking blind spots for closed circuit television monitoring systems [9].

Level of Detail 2 (LoD2): The building is represented by a geometrically simplified exterior shell. The outer facade of a building may be differentiated semantically by the class *_BoundarySurface* as a part of the building's exterior shell apportioned a special function. This can be a wall (represented by the class *WallSurface*), roof (*RoofSurface*), ground plate (*GroundSurface*), outer floor (*OuterFloorSurface*), outer ceiling (*OuterCeilingSurface*) or a *ClosureSurface*. A *ClosureSurface* does not correspond to an object in the real world but is introduced to support the generation of closed volumes. Figure 2 depicts a LoD2 building representation using the boundary surfaces with the exception of a *ClosureSurface*. Further, additional building elements like chimneys, dormers, and balconies may be associated to a building in LoD2 using the class *BuildingInstallation*.

Compared to LoD1, the outer shell of a LoD2 building is differentiated both semantically and geometrically. Hence, more applications are possible in LoD2. Analysing the roof surfaces of a building leads to an estimation of solar energy potential [10]. Comprise building installations like dormers and chimneys can even improve this estimation if shadowing effects are considered [11]. Analysing the total surface of a building's wall surfaces could help to estimate thermal insulation effort.

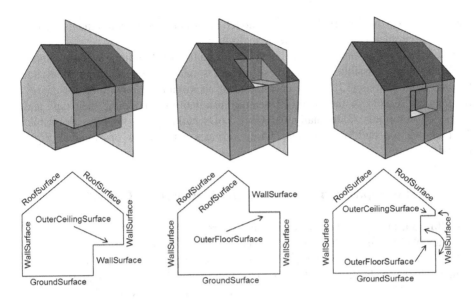

Fig. 2. Examples of boundary surfaces for a building in LoD2 ([8], according to [1] p. 70)

Level of Detail 3 (LoD3): The building is represented by a geometrically exact outer shell. Semantically this representation may be enriched by two features, *Door* and *Window*, as a specialization of the class *_Opening*. Since address information can be associated with both, a *Building* and a *Door*, the address information can be applied in much more spatial detail in LoD3, i.e. larger building complexes may get more than just one address.

With LoD3 city models, again the number of applications increases. Since doors and windows are represented, access ways to buildings can be analysed for evacuation scenarios or police operations. Windows may be sampled for coefficient of heat transmission and area to estimate restructuring requirements. In [12] an absolute vehicle positioning architecture in urban environments was proposed, realized by combining 3D city data and car side laser scanner. In particular, the geometries of window ledges could increase the number of observations for positioning calculations.

Level of Detail 4 (LoD4): In addition to the LoD3 representation of the building's outer shell, interior structures are represented in LoD4 by the class *Room* that again may be semantically enhanced by the attributes *class, function,* and *usage*. *Rooms* are bounded by one or many *InteriorWallSurface, FloorSurface,* and *CeilingSurface*. Installations within a room that are not movable, i.e. radiators or fireplaces, are represented using the *IntBuildingInstallation* class. Furniture, like tables and chairs, can be represented with the class *BuildingFurniture*. Since CityGML version 2.0 the geometry of *IntBuildingInstallation and BuildingFurniture* may be represented using an *ImplicitGeometry*. That is a prototypal geometry that is, on grounds of costs, used for more than one object. Even VRML-, DXF- or 3D Studio MAX files or a suitable web service may be imported to the city model. Next to memory requirements the use of implicit geometries enables a faster visualisation.

Since LoD4 city models hold interior structures of building available more applications are possible. Among these are the semiautomatic checking of digital building applications [13], or the calculation of air volume of a building for energy requirement reasons or the query for heating installations concerning their type and energy consumption. Additionally, CityGML LoD4 buildings are virtually assessable. That allows for build-up locations based services for visitors or simulate flight behaviour.

3   A New Separated Level of Detail Concept for CityGML

CityGML does not provide a consistent and superordinate Level of Detail concept for all available modules. There are separated concepts and descriptions, rather than definitions that may be suitable for the modules representing buildings, tunnel, bridges, land use, furniture and so on. To some extent these concepts appear to be transferable. However, this is complicated because the entities modelled are quite different in nature. Hence, a single view of all the entities modelled in CityGML must result in an erroneous interpretation and application of the Level of Detail concept. As a result, concepts for Levels of Detail must be discussed for each module separately.

3.1   Deficits of the CityGML 2.0 LoD Concept

The Level of Detail concept in CityGML was first developed for the *Building* module. Since buildings represent the most important entities of a city, a well-defined LoD concept for the corresponding CityGML module is important. It distinguishes between very rudimental block models (LoD1), a representation containing typical roof forms and a generalized shape of the building's facade (LoD2), a geometrically exact representation of the exterior shell (LoD3) and a representation of internal structures (LoD4). But this cannot be simply adapted to all other entities of an urban landscape. There is no clear reason to define an interior of a *SolitaryVegetationObject* that is used to model single vegetation objects like plants and trees. What exactly has to be modelled to represent the interior of a *WaterBody*. Further, taking into account the fact that the LoD concept accounts for both, geometry definition and semantically depth, what is the meaning of different LoDs for land use or relief?

Even for the *Building* module, the current Level of Detail concept seems to be insufficient. Take the interior structure of a building, e.g. rooms, interior boundary surfaces, installations etc., as an example. These features can only be modelled in one geometrical LoD, LoD4 that requires a geometrically exact representation. However, often this information is not available and is not needed for various use cases. An example is the Spanish cadastre, which often provides a coarse outer building shell (LoD1) and information on a story and interiors and its spatial structure. Currently, these models cannot be represented adequately in CityGML. Further, the interior structure of a building can only be modelled when, simultaneously, there is a geometrically exact model for the exterior shell of the building is available. This limits the application of indoor models.

For Level of Detail representations higher than 2, this attribute does not provide much information on the actual semantically content of a building model. For a LoD3 building, the range of valid representations goes from a pure *gml:MultiSurface* geometry for the exterior shell to semantically structured representation with wall surfaces, roof surfaces, outer installations, doors and windows (rf. [14]).

3.2   General Consideration for a New Level of Detail Concept for CityGML

General considerations must be addressed to overcome the aforementioned problems in defining, assigning and applying the Level of Detail concept to CityGML classes and instances. This involves also a readiness to reassess the building as the constructional drawing for LoD concepts for other city objects and modules. Below we suggest some of these general considerations.

The Level of Detail attribute is a sign of quality for every single *CityObject*, i.e. a feature of the urban environment that is represented in CityGML. In other words, the LoD is a measure of the consistency between real world feature and modelled feature, both geometrically and semantically. Geometrically a LoD declaration should be a sign of maximum geometrical deviation of points in the real world and the model. This maximal deviation is required to be more precise the smaller the modelled features are. Semantically the value of a LoD attribute should indicate how many of important subcomponents are modelled from real world features. That should also include the question whether values are assigned to relevant attributes of a thematic class. Since almost all of the attributes in CityGML are optional, this is important information. Finally, the LoD attribute could contain information on the degree of conformity of the feature's appearance both in the real world and the model representation.

As a consequence, for almost all semantic classes in CityGML, the LoD cannot be expressed by simply one number between 0 and 4 but rather as an explicit attribute expressing all supported Levels of Detail. Since at least geometry and semantical depth have to be considered, this attribute needs to be one of a complex type.

We propose the differentiation of a Geometrical Level of Detail (GLoD) and a Semantical Level of Detail (SLoD). These two LoDs are separately defined for the interior characteristics and the outer shell of a building, respectively. For the sake of completeness, however, an additionally Level of Detail on the appearance (ALoD) of a model has to be considered. It should allow statements whether there is colour or texture information either are available, partly available or not available. However, this ALoD is not discussed here.

3.3   Geometrical Level of Detail (GLoD)

The Geometrically Level of Detail (GLoD) denotes the geometrical resolution and the deviation of the modelled feature from the real world phenomenon. Since more information on Level of Detail for interior building structures is embedded, the GLoD is divided into an outer shell GLoD and an interior GLoD.

The highest number of our new GLoD concept is 3 resulting in 4 GLoDs. The former concept comprises 5 Levels of Detail, where the LoD4 just indicates that interior structures of a building are modelled. Since the new concept distinguishes between outer shell and interior structures of a building, the fifth LoD, i.e. LoD4, can be omitted. Even more detailed characteristics of the model's quality can be expressed when exterior GLoD and interior GLoD are combined. For the outer shell of a building the GLoD has the following encoding (rf. fig. 3):

Geometrical Level of Detail 0 (GLoD0): 2D / 2.5D geometry. The outer building is represented by a two-dimensional surface, a planar surface embedded into the 3D space, or a non-vertical surface. The latter means that within the extent of the surface the vertical height z of a surface point is a unique function $z=f(x,y)$ of the horizontal point position. Attributes specify the meaning of the surface, i.e. whether it represents the building outline or the roof outline.

Geometrical Level of Detail 1 (GLoD1): Vertical extrusion body. The building is represented by a block model. An attribute specifies the meaning of the surface used for the extrusion, i.e. whether it represents the building footprint or the roof outline.

Geometrical Level of Detail 2 (GLoD2): Generalized geometry. The building is represented by a geometrically simplified outline contour with a well-defined maximum geometrically deviation between model and real world feature.

Geometrical Level of Detail 3 (GLoD3): Exact geometry. The building is represented by a geometrically exact outer shell. Again, there is a well-defined maximum geometrically deviation between model and real world feature much smaller than in GLoD2.

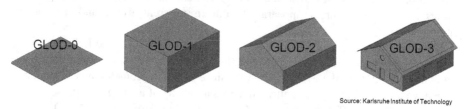

Source: Karlsruhe Institute of Technology

Fig. 3. Representation of the outer shell with the new GLoD concept. Here, no additional semantics are represented, i.e. the Semantically LoD (SLoD) is 0 (see section 3.4)

The interior GLoD solely characterises the Level of Detail for inner structures of a modelled building. A priory, it is not bound to a specific GLoD for exterior feature granularity. Therefore, this new concept allows modelling highly detailed interior structures of a building without providing exact outer building shell at the same time. Since the current LoD4 concept needs to have exact outer shell geometry, this representation is not possible that concept. The numbering of the interior GLoD follows that of the outer shell GLoD (rf. fig. 4).

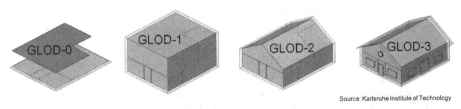

Source: Karlsruhe Institute of Technology

Fig. 4. Representation of the interior of a building, i.e. rooms, using the new GLoD concept. Here, no additional semantics are represented, i.e. the Semantical LoD (SLoD) is 0 (see section 3.4)

3.4   Semantical Level of Detail (SLoD)

The new Level of Detail concept proposed here separates clearly between geometrical and semantical accurateness of a modelled feature. Besides, since the separation of geometry and semantic is a basic principle of the OGC Abstract Specification [15], our approach is more compliant to OGC policies than the current one. The Semantical Level of Detail (SLoD) denotes the degree to which the semantical structure of a real world phenomenon is reflected in the modelled feature. Next to geometrical integrity it describes the semantical depth that is expressed in a model, i.e. whether the boundary surfaces have a type and additionally attributes or are just surfaces without any meaning. The SLoD is represented using the following graduations:

Semantical Level of Detail 0 (SLoD0): There is no semantical structuring of the building's outer shell or a room, only the building is represented semantically. Hence, no differentiation of the boundary surface is available.

Semantical Level of Detail 1 (SLoD1): The geometry representing the outer shell is completely structured by boundary surfaces. As mentioned in sec. 2.2 a *BoundarySurface* is specialized in several classes representing semantics of these surfaces, e.g. an *OuterCeilingSurface*.

Semantical Level of Detail 2 (SLoD2): In addition to the boundary surfaces in SLoD2, particular parts of the outer shell are modelled by *BuildingInstallation*. Building installations themselves are represented by boundary surfaces as well.

Semantical Level of Detail 3 (SLoD3): Openings, i.e. *Doors* and *Windows* are represented.
    Like the GLoD, the SLoD can be applied for the outer representation of a building as well as for the interior. Due to the combination of different SLoDs with GLoDs, there are substantially more ways to describe a *CityObject's* model correctness than applying the current LoD concept.

3.5   Possible Combinations of GLoDs and SLoDs and Their Relationship to the Current Concept

The main advantage of the new LoD concept, i.e. first, separating information about geometry and semantics, and second, separating the outer shell from the interior of a building, is an enhanced description of the deviation between the real world feature and the modelled object. It enables more information depth on the modelled feature by combining different geometrical and semantical Levels of Detail. However, not every combination of GLoD and SLoD is valid. In GLoD0, where the building's outer shell geometrically is represented by an (almost) planar surface, it does not make sense to identify _BoundarySurface or BuildingInstallation, which would have to be modelled as lines. For the extrusion volume used in GLoD1, it is implicitly obvious that the lower horizontal plane corresponds to a GroundSurface, the upper plane corresponds to a RoofSurface, and the vertical planes correspond to WallSurface. An additional modelling of these structures as BoundarySurface therefore would not generate any additional information and is prohibited. Table 1 shows valid combinations of GLoDs and SLoDs for the exterior shell of a building. It can also be seen, that this combinations are capable to represent the current Level of Detail concept.

Table 1. Possible combinations of GLoD and SLoD representing the exterior shell of a Building. Every LoD of the former model can be represented.

	SLoD0	SLoD1	SLoD2	SLoD3
GLoD0	LoD0	prohibited	prohibited	prohibited
GLoD1	LoD1	prohibited	prohibited	prohibited
GLoD2	LoD2	LoD2	LoD2	new
GLoD3	LoD3	LoD3	LoD3	LoD3

Since the current LoD concept offers only one LoD for the interior of buildings, i.e. the LoD4, the new concept gives more information. Further, the combination of different GLoDs and SLoDs allows the representation of more semantical information, even if geometry is not available in the best resolution, e.g. GLoD3. Table 2 shows possible combinations of GLoD and SLoD for the interior of a building as well as the relationship between the current LoD concept and the new one.

Table 2. Possible combinations of GLoD and SLoD representing the interior of a building. Every LoD of the former model can be represented.

	SLoD0	SLoD1	SLoD2	SLoD3
GLoD0	new	prohibited	prohibited	prohibited
GLoD1	new	prohibited	prohibited	prohibited
GLoD2	new	new	new	new
GLoD3	LoD4	LoD4	LoD4	LoD4

Since GLoD and SLoD can be combined once for the outer shell of a building and the interior of a building, the overall Level of Detail is given by a combination of interior and outside. While allowance of valid combinations is ambiguous, possible combinations need to be discussed. Here, we present a more restrictive set of combinations that is expected to fit the requirements of users quite well (rf. fig. 5).

		outer shell									
		not modelled	GLoD0 / SLoD0	GLoD1 / SLoD0	GLoD2 / SLoD0	GLoD2 / SLoD1	GLoD2 / SLoD2	GLoD2 / SLoD3	GLoD3 / SLoD0	GLoD3 / SLoD1	GLoD3 / SLoD2	GLoD3 / SLoD3
building interior	not modelled	X	LoD0	LoD1	LoD2	LoD2	LoD2	new	LoD3	LoD3	LoD3	LoD3
	GLoD0 / SLoD0	X	new	X	X	X	X	X	X	X	X	X
	GLoD1 / SLoD0	X	X	new	X	X	X	X	X	X	X	X
	GLoD2 / SLoD0	X	X	X	new	new	new	new	X	X	X	X
	GLoD2 / SLoD1	X	X	X	new	new	new	new	X	X	X	X
	GLoD2 / SLoD2	X	X	X	new	new	new	new	X	X	X	X
	GLoD2 / SLoD3	X	X	X	new	new	new	new	X	X	X	X
	GLoD3 / SLoD0	X	X	X	X	X	X	X	LoD4	LoD4	LoD4	LoD4
	GLoD3 / SLoD1	X	X	X	X	X	X	X	LoD4	LoD4	LoD4	LoD4
	GLoD3 / SLoD2	X	X	X	X	X	X	X	LoD4	LoD4	LoD4	LoD4
	GLoD3 / SLoD3	X	X	X	X	X	X	X	LoD4	LoD4	LoD4	LoD4

Fig. 5. Possible combinations of GLoD and SLoD representations for the outer shell and the interior of a building characterising its total Level of Detail. A 'X' indicates an invalid combination and LoDx the correspondent to the current LoD concept.

The above proposed combinations fit well for the *Building* module. One of the major sources of criticism on the current LoD concept is that it was simply transferred from a building LoD definition to all other modules in CityGML. In general, this accounts for this approach, also. But the main improvement is the intricacy of design that enables a balanced design of the table given in fig. 5 for every single module. As a result, for some modules the interior GLoD and SLoD will be omitted, e.g. for the *Vegetation* module or the *LandUse* module.

An example of improved application using the new LoD concept is given in fig. 6. Here, the exterior shell of a building is represented only in GLoD1 and SLod0. That is a block model with no semantical differentiation of boundary surfaces. It can be generated by extruding municipal maps. Nevertheless, the interior of the building, i.e. the rooms and interior building installations, are modelled in the highest GLoD and SLoD, i.e. GLoD3 and SLoD3, respectively. This option is important for interior designers that are not interested in the outer shape of a building.

Fig. 7 depicts a reverse situation. Here, the outer shell of a building is represented using a combination of the most precise GLoD3 and SLoD3. This reflects the current LoD3 or LoD4 representation, depending on whether the interior of a building is modelled or not. Because the interior of a building was only representable in LoD4 in the current concept, maximum precision and semantic information were required. Here, a less detailed interior Level of Detail may be applied, i.e. a GLoD0 and SLoD0 representing just floor plans. This building representation might be useful for sales conversations.

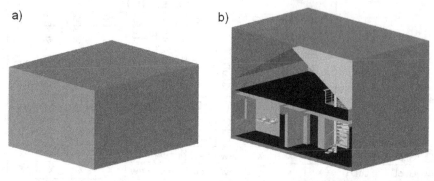

Source: Karlsruhe Institute of Technology

Fig. 6. Representation of the buildings a) outer shell in GLoD1 and SLoD0 and b) the interior of the same building in GLoD3 and SLoD3

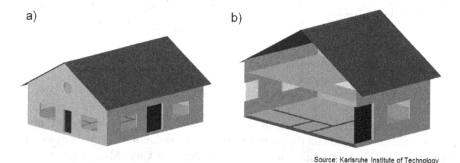

Source: Karlsruhe Institute of Technology

Fig. 7. Representation of the buildings a) outer shell in GLoD3 and SLoD3 and b) the interior of the same building in GLoD0 and SLoD0

4   Conclusions and Discussion

We proposed a new Level of Detail concept for the international Open Geospatial Consortium encoding standard CityGML, a common information model for representing, storing, and exchanging virtual 3D city models. The concept involves the separation of a Geometrical LoD (GLoD) from that of a Semantical LoD (SLoD) as well as the separation of exterior and interior properties of a *CityObject*.

Since August 2008, CityGML 1.0 is an Open Geospatial Consortium encoding standard. It was republished in March 2012 in the version 2.0. Next to the quality of model concepts and certain flexibility, continuity is a major reason for a standard to be accepted. The version change from 1.0 to 2.0 was characterized by only small changes. A new major release was necessary only because of OGC policies, disallowing additions on existing classes (rf. [16]). Since a new release offers just extensions to the old one, i.e. additional features and attributes, instance models based on the former encoding standard are still valid or can easily be converted. Here, we state that our new LoD concept is an extension that first, enables more information depth on the modelled feature and, second, is easily be applied to instance models of CityGML version 2.0. Compared to the current concept, the new approach offers the following advantages:

- A substantially higher informative value for the Level of Detail of *CityObjects* due to manifold combination of GLoD and SLoD for both, the interior and exterior of a *CityObject*.
- A better description of the interior Level of Detail. Since the current concept offers only one LoD (LoD4), ten different quality classes are available now.
- A broadening of the opportunities for indoor modelling. Since the current LoD concept demands on high resolution geometry, the new approach allows for the representation even of 2.5D data for interior objects. That enables novel applications, for example the representation of room or floor plans for indoor navigation.
- A better assignability to all other modules represented in CityGML. Due to more combinations of GLoD and SLoD, the Level of Detail definition for every module in CityGML can be defined according to the nature of modelled real world phenomenon.

Further work needs to be done to develop this approach of GLoD and SLoD to a comprehensive metadata model. This metadata model should allow the testing and evaluating of instance documents for certain applications. This metadata model has to provide answers to the following questions:

- Does the model represent an explicit building volume, directly or indirectly?
- Is there explicit information about the appearance available, i.e. colour or textures?
- If so, is this information related to visual nature or to special information, e.g. to a thermographic image of the surface?
- For which subset of the set of (mostly optional) attributes are values provided?

The work on the new Level of Detail concept is still in progress and under discussion in the modelling working group of the Special Interest Group 3D (SIG 3D, see www.sig3d.org) of the initiative Spatial Data Infrastructure Germany (GDI-DE). After the group has agreed upon the new concept, it will be forwarded as an official change request to the OCG, particularly to the CityGML Standards Working Group (CityGML SWG) of the OGC.

References

1. Gröger, G., Kolbe, T.H., Nagel, C., Häfele, K.-H. (eds.): OGC City Geography Markup Language (CityGML) Encoding Standard, Version 2.0, OGC Doc No. 12-019, Open Geospatial Consortium (2012)
2. Czerwinski, A., Sandmann, S., Stöcker-Meier, E., Plümer, L.: Sustainable SDI for EU noise mapping in NRW - best practice for INSPIRE. International Journal for Spatial Data Infrastructure Research 2(1), 90–111 (2007)

3. Czerwinski, A., Gröger, G., Reichert, S., Plümer, L.: Qualitätssicherung einer 3D-GDI - EU-Umgebungslärmkartierung Stufe 2 in NRW. Quality management of a 3D-SDI – phase 2 of the EU ambient noise mapping in NRW. Zeitschrift für Geodäsie. Geoinformation und Landmanagement (2013)

4. Löwner, M.-O., Casper, E., Becker, T., Benner, J., Gröger, G., Gruber, U., Häfele, K.-H., Kaden, R., Schlüter, S.: CityGML 2.0 – ein internationaler Standard für 3D-Stadtmodelle, Teil 2: CityGML in der Praxis. CityGML 2.0 – an international standard for 3D city models, part 2: CityGML in practice. Zeitschrift für Geodäsie, Geoinformation und Landmanagement 2, 131–143 (2013)

5. Gröger, G., Kolbe, T.H., Czerwinski, A., Nagel, C. (eds.): OpenGIS® City Geography Markup Language (CityGML) Encoding Standard, Version 1.0.0, OGC Doc No. 08-007r1, Open Geospatial Consortium (2008)

6. Cox, S., Daisey, P., Lake, R., Portele, C., Whiteside, A. (eds.): OpenGIS® Geography Markup Language Implementation Specification, Version 3.1.1, OGC Doc No. 03-105r1, Open Geospatial Consortium (2004)

7. Eastman, C.: Building Product Models. CRC Press LLC (1999)

8. Löwner, M.-O., Benner, J., Gröger, G., Gruber, U., Häfele, K.-H., Schlüter, S.: CityGML 2.0 – ein internationaler Standard für 3D-Stadtmodelle, Teil 1: Datenmodell. CityGML 2.0 – an international standard for 3D city models, part 1: Data model. Zeitschrift für Geodäsie, Geoinformation und Landmanagement 6, 340–349 (2012)

9. Ying, M., Jingjue, J., Fulin, B.: 3D-City model supporting for CCTV monitoring systems. In: Symposium on Geospatial Theory, Processing and Application, Ottawa (2002)

10. Baumanns, K., Löwner, M.-O.: Refined estimation of solar energy potential on roof areas using decision trees on CityGML-data. Geophysical Research Abstracts 11 (2009)

11. Ben Fekih Fradj, N., Löwner, M.-O.: Abschätzung des nutzbaren Dachflächenanteils für Solarenergie mit CityGML-Gebäudemodellen und Luftbildern. Estimation of usable roof surface area for solar energy using CityGML building models and aerial images. In: Löwner, M.-O., Hillen, F., Wohlfahrt, R. (eds.) Geoinformatik 2012 Mobilität und Umwelt. Conference Proceedings of the Geoinformatik 2012, pp. 171–177 (2012)

12. Löwner, M.-O., Sasse, A., Hecker, P.: Needs and potential of 3D city information and sensor fusion technologies for vehicle positioning in urban environments. In: Neutens, T., De Maeyer, P. (eds.) Developments in 3D Geo-Information Sciences. Lecture Notes in Geoinformation and Cartography, vol. 27, pp. 143–156 (2010)

13. Benner, J., Geiger, A., Häfele, K.-H.: Concept for Building Licensing based on standardized 3D Geo Information. In: Proc. 5th International 3D Geoinfo. Conference, November 3-4, pp. 9–12 (2010)

14. Stadler, A., Kolbe, T.H.: Spatio-semantic Coherence in the Integration of 3D City Models. In: Proceedings of the 5th International Symposium on Spatial Data Quality, Enschede (2007)

15. Kottman, C. (ed.): The OpenGIS Abstract Specification. Topic 8: Relationships between Features. OpenGIS Project Doc No. 99-108r2, OGC (1999)

16. Reed, C. (ed.): Policy Directives for Writing and Publishing OGC Standards: TC Decisions, OGC Doc 06-135r11, Open Geospatial Consortium (2011)

Walking into the Past: Design Mobile App for the Geo-referred and the Multimodal User Experience in the Context of Cultural Heritage

Letizia Bollini[1], Rinaldo De Palma[2], and Rossella Nota[1]

[1] Department of Psychology, University of Milano-Bicocca
Piazza dell'Ateneo Nuovo 1, 20126 Milano, Italy
letizia.bollini@unimib.it
[2] Bitmama SRL,
Corso Francia, 110, 10143 Torino, Italy

Abstract. Information technology and new mobile revolution could be a strategic resource in the field of humanities studies and a powerful tool to bring culture in the people's everyday life.

The research explores the potentialities of the Web 3.0 applied to Cultural Heritage and tests three of its specificities: the semantic information architecture and the associative organization of contents, the geo-location of information in the physical environment and the social dynamics between people and places in geo-social networks.

Starting from a theoretical analysis the work maps and classifies the existing best practices both for semantic knowledge-based platforms and mobile apps developed by the most important international museums, historical house-museums and private galleries.

On the basis of these theoretical statements, of a literature analysis and a benchmark, the research explores and experiments – through a case study – the potentiality of a hybrid design process – both user-experience and technology driven – to revive a masterpiece of Italian literature – *The Betrothed* by Alessandro Manzoni – through a multi-layered historical storytelling approach to contents and a geo-referred experience of the places, monuments and routes described in the novel (XVII sec.), lived by the author (XIX sec.) and today thanks to a mobile app.

Using the technical repertory of geo-referred data, AR, QR-Code and multimodal contents – videos/audio, interactive documents, shared dairies and UGC – the pilot of *The Betrothed 3.0* allows the users to discover Milan.

The user can navigate through the timeline and decide to visit the city experienced by the protagonists of the novel or by the author who lived and dwelt in the places described, but in another time period – his house-museum is one of the main touch point of the app – making comparisons with the contemporary reality.

The project tries to set a design pattern for engaging in an effective way people facing the Humanities through technology.

B. Murgante et al. (Eds.): ICCSA 2013, Part III, LNCS 7973, pp. 481–492, 2013.
© Springer-Verlag Berlin Heidelberg 2013

Keywords: Geo-localization, geo-localized interaction, user experience design, mobile application design, digital story-telling.

1   Cultural Heritage and the Web 3.0

Information technology and above all the new *mobile revolution* could be a strategic resource in the field of humanities studies and a powerful tool to bring culture in the people's everyday life.

The research explores the potentialities of the so-called *Web 3.0* applied to the context of Cultural Heritage and tests three of the specificity of the third generation of Internet: the *semantic* information architecture and the associative organization of contents, the possibility to *geo-locate* information in the physical environment thanks to mobile devices and the *social dynamics* between people and between people and places in geo-social networks.

Starting from a theoretical analysis, the work maps and classifies the existing best practices both for semantic knowledge-based platforms – i.e. *Europeana, Culture-Sampo* – and virtual tour or mobile apps developed by the most world-wide known museums, historical house-museums and private galleries in the field of Science, Art and Cultural Heritage.

1.1   Archive vs. *Museums*

The museum sector was one of the first places where the Internet was early adopted thanks to the intrinsic nature of repository which is very similar to the structure of the web: a large world-wide archive enriched by increasing new entries, constantly updated and cataloged [1], as well as the museums are the places of the collection and preservation of memory, relics of the history and evolution of human culture in its various fields.

Since its inception the web has *enabled* to break down one of the most narrow limits of documentary sharing of historical memory: the *space-time* of the museum, in its physicality.

Internet has opened the *closed box* of the Wunderkammer culture letting it accessible, usable, *surfable* to users and public that would otherwise be left out in a sort of – to paraphrase Rifkin [2] – to *physical-divide*. Now, let us go back to the second half of the nineties – to coincide with the bursting of the new economy and the internet phenomenon – the web-sites of major museums such as the Louvre, the Hermitage or, in Italy, the Uffizi in Florence.

In this first Internet *era* web-sites were seen as a *reproduction* or *duplication* of the museum in its physical dimension shaped along the spatial paths, the   rooms succession and exhibition halls.

In fact, technologies – such as QuickTime VR or VRML – already made possible to explore real/virtual environments through spatial movements to 180-360° based on a physical and photographic mapping of environment, but the poor performance of the connections to put still-massive adoption of these instruments while already ripe for exploitation.

1.2   *Mimesis* **and Showing: The Museum 1.0**

The Net – according to this first model 1.0 [3], which we can define *hypertext archival*– uses the medium according to a spatial metaphor of the real domain of the object represented. But the cyberspace virtual nature of the network is a world without limits or boundaries in which – as pointed out by Cappelli [4] – the culture is no longer something given and fixed, but a *continuum* fluid and always changing, instead to be forced to the physical dimension which reproduces the rooms, the fittings, the provision and exploring sequentially, with a re-current tendency to overlap between reality and digital representation.

The museum in the web isn't a *representation* limited and constricted to the space of a page – in the two-dimensionality of the monitor – the museum and its collections were translated – from a linear sequence/succession of rooms – into static maps slavishly pointing out the location of the masterpieces with a correspondent page, sometimes enriched by re-call context shaped on the model of the encyclopedia's *lemma*.

The act of communication meant to be the replacement and equal to the enjoyment of art, the science of historical memory or ephemeral of an exhibition and *un-mediated* personal experience of the viewer/user and his relationship with the material shown and the aroused emotions of memory, aesthetics and knowledge.

The great revolution introduced by digital technology – instead – is to open places and archives to the collective and massive dimension, and sharing of knowledge itself to make knowledge diachronic and distributed beyond the four dimensions (3D and time) in a kind of rearticulated *space-time* and displacement where the "here and now" are just a *click*.

1.3   **Fluid Culture: Mounting and Nomadism**

The natural evolution of the spatial hypertext network has enabled us to deconstruct real-mind, not just the physical space of the museum, but also the conceptual space of the collection and of the belonging. The works are primordial *cells*, atoms of culture and memory that aggregate and/or dissociated are re-framed in conceptual links other than a static exhibition. The concepts are *nomadic* and transverse intertwine in [a] continuous movement that stratify of interpretations, documentazions and contamination. And utopistic space in which the same pattern, the same work, the same piece of history can simultaneously belong to a physical place and multiple virtual places that reunite it in a dialogical relationship between simility and diversity, the *if* and the *else*. Moreover, the metadata, ontology and semantics, applied to a single object, build infinite levels of meanings and interpretations: the chimera of multidimensional cataloging, which generates – without ever exhausted – the Escher architecture, the spatial-diplayed key-images and catalogation of the Mnemosyne by Warburg [5] or the universe of multi-faceted classification of Ranganathan [6].

The knowledge becomes sedimented/interconnected link, questioned according to specific logic concentric which expands its orbits of knowledge incremental up to exo-disciplinary fields – otherwise unsearchable – for quality, quantity and multidimensionality of content accessible.

1.4   Digital Media: *Medium* and Subject of Exhibiting

If the first use of the Internet was a sort of *hyper-propagation* of information, a second area arose opening and experimenting *digital* and *interaction* in various forms. Technology becomes both a medium and a subject itself of exhibition. That's the case of the *meta*-museum in Karlsruhe, built in 1989 become a physical exhibition place – hard to find an appropriate definition – for this act of *experinence* of such a museum where masterpieces are immaterial and *alive* just when performed by the visitors.

The museum presents a wide range of pieces: from the first concept aimed to *reificate* – making real whatever *real* means – the Gesamtkunstwerk concept that means the multimedia-like *Licht Raum Modulator* of Laszlo Moholy-Nagy – kinetic sculpture made by the temporal unfolding of light, form and space – to the most recent virtual artificial intelligent or interactive environments as *Lorna* by Lynn Hersham which dialogs in a game of reflecting mirrors between *human* and *artificial* in a conceptual research of common language in the balance between *identity* and *alienation*, *real* and *digital* [7].

The museum is a kind of evolutionary path between the two extremes: on the one multimediality physical or instrumental – as suggested Donald Preziosi [8] – a virtualization more and more thrust. In this extreme, the digital becomes the pure form of a dialogue/interaction between human beings and himself and the human beings and the art.

Another facet of the open world of technology, the exhibition space starts to have more and more their forms and specifications. The *camouflage* pattern of the new medium and a metaphor hyper or – vice versa – slavishly realistic, in an attempt to become transparent to the user within the exhibition. And the research and experimentation of new languages will increasingly close to its potential, which divides into two prevailing models: the 2.0 archive and interaction proxemics.

1.5   Archive and Spatial Interaction: The Multimodal Performance

The first model – the archival organization – adopts an instrumental approach based on the structure of databases: the objects on display are cataloged and exposed according to research carried out by the user and the criteria identified to implement a strategy exploration of the underlying material – but at the same time to reinterpret what concerns the mode and dynamics of data mining intend.

Thah's the case of the Dutch group LUST, author of projects such as the *Grapich Design Museum* or *Random Generation*. The metaphorical model of reference is the *stack* – the archive of the earliest forms of hypertext of the highties – repeated on an interactive-table queried dynamically with gestural interfaces (NUI) in a physical space. The two-dimensionality of the surface that runs horizontally along the lines of the *infinite plane* is given depth, both conceptual and physical, through the ability to zoom in and out deeply diving into the content presented in a gesture that brings to the foreground or background of the individual units textual and multimedia information to the user.

The second one uses the full potentiality of the medium in its multi-dimension-modality and interactivity.

A good example is carried out by the research group Art + Com based on conceptual stages such as that implemented the installation Zerseher/De-Viewer (1992), and in which the gaze of the viewer distorts the picture itself – a reproduction of an oil painting by Giovanni Francesco Caroto – *reflecting*over not only the relationship between the work and the visitor, but above all – quoting McLuhan –the medium and the message. Concept which is resolved in the performative act in a co-authorial involvement of the user.

The approach of immersive physical and virtual museum scenario is extended and structure in the work of Studio Azzurro or in the info-productions of IO where physical interaction, proxemics and proprio-ceptive relationships with the environment, come out of the 2D space of the monitor and became pervasive as an overlapping layer of knowledge displayed on reality.

And the direct manipulation of *non-existent* – which query the physical object – interacts with a virtual space and the conceptual load of information [9].

If the traditional models of interaction permit to stage object in a fixed spaces, pre-determined, strictly separated from the place where the user is located and from which orchestrates the interaction, the experience is rather symbolic and organized according to a metaphor based on the principles that interaction must according to.

The current boundaries digital design exhibition, however, operating according to a logic similar to the real world, in which place of inter-opration and location of the staging merge and where the user's gesture is no longer projective and metaphorical, but natural – or at least – simultaive of the physicality.

The exhibition then becomes a layering of knowledge and languages, ranging from pyshical to digital, from the real to the cognitive *continuum* in a modulation of communication models coordinated by an act of *directing*: the author – on one hand, – and the user as a co-author – on the other.

To this new paradigm corresponds, therefore, a change in perspective – if you do not plan a second contact surface, but an interactive event, a performace that will *re-frame* the space-time dimension of the user.

The task of the designer, the multidisciplinarity skills brought into play, it seems so similar, not so much to the architecure planning, but rather to the director culture. This opens new perspectives in the research of new languages, that should be capable of finding a dimension that exploits the potential of the media, as original and experimental research of new and appropriate expressive solutions.

2   From the Wunderkammern to the Multimodal Storytelling

The concept of *Wunderkammern* (from the German *wunder = kammer = bed* and *wonders* and *chamber of wonders*) is the predecessor to the modern museum and the steps that led to this evolution are three:

- *collecting:* all kinds of European encyclopedic collection from the sixteenth to the late eighteenth century, a varied set of objects, which aims a project collecting. So we have the concept of pure wunderkammer;

- *cataloging* in the "room of wonders" were preserved natural objects (dried plants, stuffed animals, minerals) next to art objects, scientific instruments, objects of ethnographic interest, apparently amassed without any criteria. Subsequently, the objects begin to be arranged in a specific order, although not immediately obvious to the eyes of the modern observer, it is particularly famous case Quicchenberg Samuel, who published Inscriptiones speed tituli theatri very broad (1565). The purpose of these exhibitions was essentially autocelebrare rich magnificence of her house through the vastness of the apparatus and the rarity of the objects collected, and especially to arouse the wonder of the few noble and cultured lucky visitors. This transitional phase allows to define taxonomies then serve the next stage;

- *exhibiting*: initially around these places were created of real networks of intellectuals who exchanged their knowledge and their "wonders", but they also begin to feel the need to catalog the objects with the ultimate goal of showing their exposures to a wider audience outside their circle.

The next step is the analysis of the meaning of the two words that make it up: *house-museum*. First, we must break down the word: house and museum, the first we mean everything that is life, the history and experience of a person who can be summarized with the word user experience. Museum, whose birth the meaning has been explained previously, is the set of objects that are shown to the public.

In particular, the case study considered is the house-museum of Manzoni in Milan, which is located in the house on Via Morone 1, where the author lived with his family from 1814 to 1873, the year of his death.

The report writer is very close to Milan in 1848, revolution broke out of the Five Days of Milan, makes the three sons to take part and although one of them was taken prisoner and hostage of the Austrians, signed an appeal to all peoples and Italian princes because they help the people of Milan.

The tie that binds the writer to his city can also be found in his works: *History of the pillory* and *The Betrothed*.

The project was focused on the latter novel, set from 1628 to 1630 in England, especially in Milan, during the Spanish occupation, was the first example of the historical novel in Italian literature. The roman is based on a rigorous historical research and episodes of the seventeenth century, such as the story of the Nun of Monza and the Great Plague of 1629-1631, all are based on archive documents and chronicles.

3   The Betrothed 3.0: A Mobile the Geo-referred and the Multimodal User Experience

The project, The Betrothed 3.0, describes the second journey made by Renzo in Milan as he goes in search of Lucia, complete with contents that show the house-museum Manzoni.

The path is analyzed according to three historical periods: 1500-1600, a period in which the novel, 1700-1800, a period in which he lived the author, and since 1900, which is the most recent history of Milan, illustrated not only from a perspective of cultural history, but it is also full of ideas and suggestions of tourist in order to allow the visitor to the city even further from the gastronomic point of view, traditions and commercial.

To understand the concept of the mode and multi-mode you can insert content related to the example of Milan's Porta Nuova (because the story board refers to that place), divided according to the historical period.

The Betrothed 3.0 is a research project that concluded with the design and prototype development of an iPhone application.

Starting from the path taken by Renzo Tramaglino the streets of Milan in the seventeenth century, were drawn 3 historical, mentioned earlier, that trace the steps of the protagonist of the novel. Each route has its own historical setting and are shown for each of the points of interest that are associated with multimedia content in context at the time of the itinerary.

Fig. 1. Wireframe showing the funcionalities and Use-cases

The Betrothed 3.0 uses the technique of story telling # to propose such content. This allows for a more dynamic and less tedious in the aspect of information presentation.

The goals that you place the research team for the design of the app are:

- Create a link between the novel The Betrothed by Alessandro Manzoni and the city of Milan, where he set a part of the book
- Create a link between the home and museum of Manzoni Milan
- Provide a tool that allows to follow the path by Renzo Tramaglino been going through the streets of the city
- Create a series of tours starting from the location mentioned above
- Associate a historical context for each route
- Show a series of multimodal content associated with points of interest
- Use the storytelling to offer content
- Use the geolocation and augmented reality for a better user experience
- Give the user a chance to co-author the content offered

- Create a guide historical alternative to the city
- steal the best from the applications on the market trying to improve the limits.

You once chosen your route, is guided from the streets of Milan the Map shows the route with the relevant points of interest and directions on how to reach them. For each there is a tab-depth where are shown the multimedia contents mentioned above. The user not only has a role of spectator but was invited to co-author the content offered. In fact on the cards, it is possible to add photographic material, video and text. Furthermore, there are buttons for sharing on social networks most used (Facebook, Twitter and Google+). This approach is closely linked to the world of social networking and UGC (User Genereted Content) upon today is based on the operation of the Internet. Users play a dual role, both active and passive at the same time, they are called to increase and share knowledge.

The application has a function (*Look Around*) that uses Augmented Reality and that shows on the screen of your smartphone multimedia content at the points of interest around the city. In addition, small arrows show the direction to follow to reach the next stop on the itinerary. The user then has two ways to follow the routes proposed by the APA and the Map and Look Around. The first allows him to have an overview of the data path can also display the tabs of the various stages, while the second allows him to directly access to multimedia content if it is located in proximity of one of the stages. The application, therefore, acts as a guide and a navigator for the routes. It is also proposed as an alternative historical guide to the city of Milan.

4   *Open-Air* Cultural Heritage: Gelocalization and AR

In recent years we have seen the increase of the use of the internet, with a consequent increase in its importance in the process of renewal and cultural diffusion, facilitated the development of a user without boundaries. A large virtual community of people made real and virtual interactions, that is not bound to a place but only the accessibility to the network, able to communicate instantaneously with the rest of the world.

This connectivity together limitless virtual evolution of relational and social gave birth, and the subsequent huge growth of social networks that have misinterpreted the concept of privacy on the Internet so that now there is a widespread sharing life mania, ie 'irrepressible desire to share anything of our lives, including their geographic location, with all our contacts even if unknown.

This trend has been ridden by social networks, seeing this trend in the ability to reach more and more profit, have equipped their applications Mobile Location Based Service that services that allow you to geographically locate a mobile device by means of a GPS receiver or by the Cellular networks, thus enabling geolocalize an individual in real time.

Applications geo mainly do two things: communicate their location to other users and connect you to a real reference. This enriches the user experience in the information processed by the APA change depending on the context that changes depending on where you are located.

Today all smartphones are equipped with a GPS receiver which, by communicating with a series of satellites, calculates the exact position in which there is located, returning of the coordinates that correspond to a precise point on a map. When the GPS signal is not available, the smartphone using the information from the cell phone with which you can determine approximately its position.

4.1   Augmented Reality and Georeferred Interaction

The For Augmented Reality refers to the enhancement of human sensory perception through information, generally handled and conveyed electronically, that would not be perceived with the five senses."

Unlike virtual reality in which you are immersed in a world composed entirely of sequences of bits, Augmented Reality combines real world with computer graphics. The means by which you live the two sensory experiences are very different. AR is sufficient for a display and an optical sensor, components that are supplied to all smartphones on the market.

Technically Augmented Reality is a technology that:

- Combines the physical world with computer graphics
- Allows interaction with objects in real time
- Trace the environment in real-time
- Allows the recognition of images and objects
- Provides an environment that changes in real time

The software that generate Augmented Reality, show on a monitor of the virtual objects at particular points that can be defined through the coordinates of the markers or visual as for example the QR code. Applications using Augmented Reality without any kind of marker are called markerless show of virtual objects by simply using the position of the smartphone and its motion sensors such as the gyroscope and accelerometer. The optical sensor is thus superfluous if not to show on the display the real world that surrounds the user.

4.2   Maps and *Look around:* An Immersive Geo-located Ux

The Starting from the path taken by Renzo Tramaglino the streets of Milan in the seventeenth century, were drawn three historical routes that follow the steps of the protagonist of the novel. Each route has its own historical setting and are shown for each of the points of interest that are associated with multimedia content in context at the time of the itinerary. The Betrothed 2.0 uses the technique of story telling to propose such content. This allows for a more dynamic and less tedious in the aspect of information presentation.

You once chosen your route, is guided from the streets of Milan using the geolocizzazione. On The Map shows the route with the relevant points of interest and directions on how to reach them. For each there is a tab-depth where are shown the multimedia contents mentioned above. The user not only has a role of spectator but was invited to co-author the content offered. In fact on the cards, it is possible to add photographic material, video and text. Furthermore, there are buttons for sharing on social networks most used (Facebook, Twitter and Google+). This approach is closely linked to the world of social networking and UGC (User Genereted Content) upon today is based on the operation of the Internet. Users play a dual role, both active and passive at the same time, they are called to increase and share knowledge.

The application has a function (a Look Around) that exploits the Augmented Reality markerless ie without the use of external markers. At the points of interest that make up the route is shown on the display of the device, a label with the name of the place, and a button to view the data in-depth. Are also proposed a series of comics that indicate the existence of one or more multimedia content associated with the area that you are framing with the camera of your smartphone. The display also shows the direction to follow to reach the next stage of the route.

Fig. 2. Map view, look around and social features of the app

The user then has two ways to follow the routes proposed by the APA:

- *Map* that allows to have an overall view of the path giving the possibility to access the cards attached to the individual stages.

- the *Look Around* allows direct access to multimedia content in case you are standing near one of the stages.

The application, therefore, acts as a guide and a navigator for the routes. It is also proposed as an alternative historical guide to the city of Milan. The geolocation is the technology upon which the operation of The Betrothed 3.0 as it allows the user to be located on the map of the city in real time and to be guided along the route proposed. In particular, the application identifies its location and shows the path to follow to follow the selected route through the stages that compose it.

When you start the app, it shows an alert that asks the user if he wants to be geolocated and if the answer is positive, the indicator is activated GPS iPhone calculated the exact position and indicated on the map. The user must then select one of the routes and after having explored the information, the app will ask to be guided to the first stop on the route chosen. The application does nothing more than offer the fastest path to reach point B (the first stage of the path) from point A (current position). All the movements are monitored and displayed on the map. The directions change in the event that the user reaches a check point that is a point where there is a stage or a change of direction. Once you reach the first stop, start your tour itself and the user is guided sequentially through all the points of interest that make up the route in the same manner as described earlier.

The geolocation is crucial to the operation of the Look Around that exploits the Augmented Reality, as the app takes as a reference point, the position of the user and displays the contents, previously georeferenced, which are closest to him.

A final use of this technology is to signal to the user, through the notifications, the presence of multimedia content closer to its position in the case in which the application was in the background ie both remained running despite the user has closed . The notifications are displayed through the alert accompanied by a beep and invite you to look at the content or proposed to deepen the knowledge of a stage.

All resources that form the basis of the application is georeferenced information that are assigned to each of the coordinates and placed on the map. When the user is close to the stage and its content, the app activates all the functions described above to enable you to explore, condivderli and enrich them.

The application, The Betrothed 3.0 provides for the use of Augmented Reality markerless ie without the use of external markers. At the points of interest that make routes through the Look Around, is shown on the display of the device, a label with the name of the place, and a button to view the data in-depth. Are also proposed a series of comics that indicate the existence of one or more multimedia content associated with the area that you are framing with the camera of your smartphone.

The display also shows the direction to follow to reach the next stage of the route.

5   Conclusions

The research project is based on a specific case history; however the project can be generalized – easily extended – and represents a reproducible experience for different disciplines focusing on the problem of knowledge sharing. The stratified/multi step approach to the problem of the cultural heritage allows actually to realize in many phases the translation in digital form and the consequent sharing of the cultural heritage avoiding the risk that the historical documents will remain unknown and on the worst case will get forgiven and lost and/or relegated in specific *containers* and places apart from every day life.

A first level is the storytelling approach that allows to disseminate specialized information to the public at large crossing the border to different disciplines and cultural experiences.

The second level – that of the implementation of search tools specific of the data basis contents and mainly focussed on the georefenced search and cultural domain the historical documents pertains to – the content distributed on the mobile platform in different/multimodal ways become then more an efficient *cultural tool* than a simple low-level digitalization.

The third refers to the creation of a social community of interest and experience that starts from the distributed editing and produces links and exchanges extending to a wider field and longer lasting than the original project.

The actions of exploiting innovation technologies and of designing user experience will be carriers of innovation and winning strategy to overcome the chronic lack of grants for the public projects fostering the sharing and the exploitation of the treasures of a cultural heritage.

References

1. Borsotti, M.: Fruttiere: il progetto di allestimento tra luogo e arte. In: Borsotti, M., Satori, G. (eds.) Il Progetto di Allestimento Ela Sua Officina. Skirà, Milano (2009)
2. Rifkin, J.: L'era delll'accesso. Mondadori, Milano (2002)
3. Carlini, F.: Lo stile del web. Einaudi, Torino (2006)
4. Cappelli, O. (a cura di): Mezzo mondo in rete. Laterza, Bari (2003)
5. Warburg, A.: cit. In: Gombrich, E.H., Eribon, D. (eds.) Il linguaggio delle immagini. Einaudi, Torino (1994)
6. Ranganathan, S.R.: Prolegomena to Library Classification. Asia Publishing House, New York (1967)
7. Maldonado, T.: Dal reale al virtuale. Feltrinelli, Milano (2001)
8. Preziosi, D.: Advantages and limitations of visual communication. In: Krampen, M. (a cura di) Visuelle Kommunikation und/oder Verbale-Kommunikation? pp. 25–35. Georg Holm Verlag, Hildesheim (1983)
9. Bollini, L., Borsotti, M.: Reshaping exhibition & museum design through digital technologies: a multimodal approach. The International Journal of Virtual Reality 8(3), 25–31 (2009)
10. Jenkins, H.: Convergence culture. New York University Press, New York (2006)
11. Sterling, B.: La forma delle cose. Apogeo, Milano (2005)
12. Bleecker, J., Knowlton, J.: Locative Media: A Brief Bibliography And Taxonomy of Gps-Enabled Locative Media. Leonardo Electronic Almanac 14(3) (2006)

Building Investments for the Revitalization of the Territory: A Multisectoral Model of Economic Analysis*

Gianluigi De Mare, Antonio Nesticò, and Francesco Tajani

Faculty of Engineering, University of Salerno, Italy
{gdemare,anestico,ftajani}@unisa.it

Abstract. Following the crisis generated by the financialization of private real-estate, construction prices have gradually decreased depriving the housing market of the necessary growth stimuli. Many countries have set up measures to revive this highly strategic area for the national economy. With reference to the Campania Region Law n. 19 dated 28 December 2009, known Housing Plan, this work has two objectives: to recognize the fundamental estimation problems that need to be solved in the implementation of the Campania Housing Plan; in addition, predict the effects of the regulations on the regional economy, both in overall terms as well as for each production sector, with particular attention being given to the construction industry. Regarding the first objective, the contents of the law are analysed on the basis of the principles that govern the appraisal. The consequences of the Campania Housing Plan on the economic system are then evaluated using *input-output* matrices, which are able to capture the structural relationships that exist among the various productive sectors. The numerical calculations require a preliminary investigation aimed at collecting a list of interventions approved by local governments in accordance to the Housing Plan. The cost of the works, as proposed in the applications submitted to the local administrations, is the *input* data for the implementation of the *Social Accounting Matrix* 2010 of the Campania Region.

Keywords: housing market, regional economy, inter-sectorial matrices.

1   Introduction

With the Order dated April 1, 2009[1], the Italian government has encouraged the promulgation of regional regulations with the aim of revitalizing the construction industry. The initiatives promoted in this field have a twofold objective. On the one hand, to revitalize the national economic structure, by acting on a sector that is capable of a strong recovery[2]. While on the other, to respond to pressing housing needs of

---

* This paper is to be attributed in equal parts to the three authors.
[1] An agreement between the State, Regional and local authorities, under article 8, comma 6, of the Law dated 5 June 2003, n. 131, on the Act relating to the measures to revive the economy through construction (Acts archive n. 21/CU of April 1, 2009).
[2] «The building industry in Italy accounts for 10% of the GDP, with about 2 million workers, of which 65% are employees» (www.fenealuil.it).

B. Murgante et al. (Eds.): ICCSA 2013, Part III, LNCS 7973, pp. 493–508, 2013.
© Springer-Verlag Berlin Heidelberg 2013

the growing number of disadvantaged families through *social housing*[3] projects. The main initiatives allowed by the national regulations aim to a) improve the architectural quality and energy efficiency of buildings, and b) simplify the bureaucratic procedures in granting concessions.

Under Presidential Decree 616/1977[4], each Region has transposed the content of the Order into its own Regional Law (R.L.) (known as Housing Plan) for the governing of the territory. With Law n. 19 dated 28 December 2009[5], as amended by Law n. 1 of 5 January 2011, the Campania Region has four main categories of private intervention: 1) increase in the volume of existing assets[6], 2) demolition and reconstruction of buildings, not necessarily in ruins, with an increase in volume, 3) rehabilitation of degraded urban areas, and 4) change of use for residential purposes of existing buildings. The redevelopment of depressed urban areas aims to, in addition to exploiting the existing building and urban patrimony, solve the housing problems of young couples and disadvantaged families, by providing that a part of the changes made are dedicated to building social housing. The eligibility of the works included in the categories listed is subject to the submission of applications within a set time period and in accordance to the constraints and construction methods specified by the Regional Law for each type of intervention.

This paper proposes the estimation of the economic impact created by the Housing Plan Law in the Campania Region. Firstly, the study analyzes the evaluation issues arising from the Regulation. This is followed by a survey of the applications submitted and of those actually granted by the local authorities, measuring the effects that the realisation of the proposed works may have in different production sectors. Quantitative procedures based on the use of *inter-sectorial matrices* are adopted. The model, based on inferential mechanisms, is applied to a sample taken from a vast area in the province of Salerno. The results obtained characterize the effectiveness level of the provisions of the Law in the revitalization of the regional economy. The calculations carried out define an analysis process that can be easily exported to other regional contexts.

---

[3] According to the European Coordinating Committee for Social Housing (CECODHAS), *social housing* is to offer «accommodation and service with a strong social connotation, to those who fail to meet their housing needs in the market (due to either economic reasons or lack of an appropriate offer) in an attempt to improve their condition».

[4] This Decree has given exclusive power over urban planning to the Regional administrations, with the State having a role of guidance and coordination of the asset as well as protecting the territory, and specific tasks assigned by the legislation of the sector.

[5] "Urgent measures for the economy, the re-qualification of existing assets, the prevention of seismic risk and administrative simplification".

[6] It should be noted that, according to the amendments made by Law 1/2011, «existing volume means the gross volume already built or under construction or completed but not yet with a habitability certificate, or with the possibility to build under the current regulations». Therefore, any increase in volume is not allowed for both existing buildings as well as any building areas that have not yet expressed, in whole or in part, their intent to build.

2   Evaluation Issues Relating to the Housing Plan in Campania

Most of the initiatives allowed by the Regional Law cannot be carried out without ex ante evaluations on the cost effectiveness of the projects[7]. In fact, if the ordinary reasons that induce an owner to increase the volume of his home may relate to the direct use of the constructed volumes (for example, an extra room or an extension of the spaces available), any demolition, reconstruction and rehabilitation initiatives of degraded areas are mainly dictated by speculative aims. In these cases, it is necessary to evaluate the *highest and best use*[8] of the property in question, considering the valorisation of the different solutions, and identify the most profitable alternative. Thus, for example, for a building in ruins, the transformation value of the recuperated building[9] must be compared with the market value of the building obtained from the demolition and reconstruction with an 35% increase of the initial volume[10]. The rent value of a property with a production destination must be compared to the value obtained from the same market as a result of the conversion of the existing building[11]. The profitability value of the company in activity

---

[7] On the centrality of evaluation issues for the practical feasibility of the investment, see [13], [25], [27].

[8] «The *highest and best use* (HBU), which is the most convenient and best use, is the use that has the maximum transformation or market value of the planned uses for a property. The HBU therefore indicates a more profitable target. This can be the current one of the property if the market value (MVEU) is greater than the transformation values of the alternative uses. [...] The choice of the HBU refers to uses: physically and technically feasible (technical constraints); legally allowed (legally binding); financially viable (budget constraint); cost-effective (economic criterion)» [33]. «The most convenient and best use is defined as follows: The most likely, physically possible, appropriately justified, legally permissible and financially viable, use to induce the provision of a higher value of the object of evaluation» [2]. It is also worth referring to [5], [24].

[9] With any increase in volume, if permitted.

[10] «Notwithstanding the planning instruments in force, an increase, of up to a maximum of thirty five percent, of the volume of the existing residential building is allowed for the demolition and reconstruction, to be achieved within the existing building in which it is located, owned by the applicant», Law 1/2011, art. 5, co. 1. It is therefore worth noting that a series of measures aimed at cutting bureaucratic procedures in approving projects are currently being discussed, regardless of the Housing Plan. «The latest [...] is the transition from the field of building renovation of the interventions that need a permit to build [...] to the simplified, with the Scia (certified report of start of the work) being enough to start work without prior permission, and the local authority being able to intervene within 60 days. [...] Amendment to Article 10 of the Construction guidelines (Decree 380/2001) and will extend to the Scia tacit assent to the work that will lead "to a building organism in whole or in part different from the previous year and involving an increase in housing units, changes of volume, shape, or surface". This project is part of the so called "freedom of shape", which should extend to the demolition and reconstruction that can be rebuilt without necessarily having to meet the shape of the old demolished building» [15].

[11] «For abandoned buildings, notwithstanding the general town planning and building parameters, [...] reconstruction interventions with the same existing volume are allowed, even with a change of use, providing for the construction of no less than thirty per cent for social housing [...]. The volume resulting from the replacement housing may have the following destinations: housing, offices for no more than ten per cent, neighborhood stores, craftsmen's workshops. [...]», R.L. 1/2011, art. 7, co. 5.

must be compared to the sum of the income of the company outsourced to a suitable *landing area* and the transformation value of the *take-off area* destined for new functions[12].

The economic convenience of housing transformations allowed by the Housing Plan is significantly influenced by the volumetric consistencies bound to the social functions, especially those for *social housing*[13], as well as by their management methods[14]. In most cases, the Regional Law sets the rate to be reserved for social housing. In one case, however, states that the percentage is determined «in relation to the transformation value» of the area[15].

A further disciplinary note relates to article 11^{-a} of the new Regional Law 1/2011, concerning the relocation of residential units located in areas with a very high landslide risk as well as in the red zone at risk of eruption of Mt. Vesuvius. This article evidently refers to equalization issues[16]. In fact, it seems to be apodictic that the

---

[12] «For polluting industries or those that are not compatible with the surrounding residential activities, the replacement housing is allowed, subject to the prior relocation of the activity in the region, ensuring, with a suitable relocation plan, the increase of ten per cent in the following five years of current employment levels. [...]», R.L. 1/2011, art. 7, co. 5-a. For further details on the estimative aspects that the examples raise, see [6], [7], [8], [9], [11], [12], [14], [20], [34].

[13] Disciplinary references can be found in [32].

[14] Art. 1, co. 2 of Ministerial Decree 22.04.2008 states that «a social housing unit is defined as property used for residential use in a permanent location that acts as a general interest [...] to reduce the housing problems of individuals and families [...] who are not able to rent accommodation in the free market. [...]». Co. 3 also provides that «the definition in paragraph 2 also includes the housing built or retrieved from public and private parties [...] for the temporary renting of at least eight years and also to the property». Art. 2 states that «the regions, in consultation with the regional Anci, define the requirements for admission and permanence in the social accommodation [...] the regions, in consultation with the regional Anci, set out the requirements to benefit from easier access to the property and establish procedures, criteria for the determination of the selling price specified in the agreement with the local authority [...]».

[15] «[...] the local authorities have to conclude the proceedings, even on a proposal from the owners, individuals or grouped in a consortium, with a measure to be taken [...] notwithstanding its planning instruments applicable to the areas where urban renewal and construction is subject to the disposal by the owners, individual or grouped in a consortium, and in relation to the transformation value of areas or properties to be allocated to social housing, in addition to the mandatory minimum provision of public spaces, or reserved for collective activities in public parks or carparks with reference to Ministerial Decree No. 1444/1968. [...]», R.L. 1/2011, art. 7, co. 2.

[16] Article 11-a: «1. In order to prevent the landslide risk or that of the eruption of Mt. Vesuvius and protect the safety of persons and the security of inhabited settlements, relocation should be encouraged within the same municipality, or other surrounding municipalities through an agreement between them, of buildings containing residential housing units in the areas classified by the Basin Authority as in danger or under very high landslide risk [...]. 2. The owners of buildings under the condition of danger or very high risk [...] can ask to carry out, outside of the same areas and in areas used for residential urban planning, an additional increase in volume, as well as those permitted on the basis of the current planning instrument, [...] equal to the volume of the housing unit assigned as the first house increased up to a maximum of thirty five percent [...]. 3. The applicant, however, shall, after concluding a special agreement, demolish the building and restore the environmental areas pertaining thereto as well as transfer the same to the unavailable patrimony of the town, prior to the conclusion of the construction the new building».

increase in the volume of the housing units built in an "safe" area («up to a maximum of thirty five percent») is devoid of any economic considerations on the different positional value of the *areas of landing* and *takeoff*[17].

Finally, it should be also pointed out that the application of the Regional Law on the property market could lead in the short run to a reduction in the selling prices, resulting in an increased supply of homes in response to the demand. This is a matter of no small importance in the current economic contingency.

3   Effects of the Housing Plan in Campania: The Field Survey

In order to quantify the effects that the Law 19/2009 of Campania is able to generate on the regional economy[18], a survey was carried out to verify the implementation of the aforementioned regulation at a local level. The geographical area covered by the study includes the fourteen municipalities that make up the vast area of the Agro-Nocerino-Sarnese (SA)[19].

The following data were collected from the Technical Offices of all the Local Councils:

1. Local Council Ordinance, with definition of the urban context subject to the Housing Plan[20] (article 4 co. 6 of R.L. 19/2009);
2. number of applications received pursuant to R.L.19/2009;
3. number of applications approved in accordance to R.L. 19/2009;
followed by:
4. classification of applications according to the article and paragraph of the Law which each type of intervention refers to.

Table 1 summarizes the results of the survey. Of the fourteen municipalities of the wide area, seven have issued a Council Ordinance. 367 applications were submitted in total.

Most of the applications (97) regard the change of use from rural to residential (art. 4 co. 7). This type of intervention is followed by that of a 20% volumetric increase (art. 4 co. 1), the demolition and reconstruction with an increase in volume (art. 5), recovery of attics (art. 8 co. 2) and, ultimately, the redevelopment of urban areas (article 7 form. 5). Figures 1 and 2 report the applications received in each municipality

---

[17] On the issue of urban equalization, see [3], [23], [24], [30].

[18] The study described was carried out between November and December 2010, about a year after the promulgation of the R.L. 19/2009 and before the enactment of the amendments made by R.L. 1/2011.

[19] The area known as Agro-Nocerino-Sarnese is located in the valley of the Sarno River, halfway between Naples and Salerno. The municipalities that are part of it (San Valentino Torio, San Marzano sul Sarno, Sarno, Pagani, Nocera Inferiore, Nocera Superiore, Castel San Giorgio, Siano, Bracigliano, Corbara, Angri, Sant'Egidio del Monte Albino, Roccapiemonte, Scafati) are all in the Province of Salerno, covering a total area of 158 km^2 and over 285,000 inhabitants, with a population density equal to 1,807 inhabitants/km^2.

[20] R.L. 19/2009 provides that within sixty days from the date of entry into force of the same, the local Councils could identify, by means of specific Council Ordinances, the areas to be excluded from the application of the law.

in relation respectively to the total number of applications in the Agro-Nocerino-Sarnese territory and the number of homes in the same municipality.

Upon data collection, none of the applications had been approved.

The study was supplemented by surveys, carried out in the five main cities of the province of Campania, which have made it possible to confirm the number of approved applications equal to 37% of those presented.

Table 1. Data collected from the Local Councils of the Agro-Nocerino-Sarnese

	Local Council Ordinance	change of use (art.4 co.7)	20% extension (art.4 co.1)	demolition and reconstruction (art.5)	requalification of degraded urban areas (art.7 co.5)	recovery of attics (art.8 co.2)	total applications
Angri	YES	15	27	33	1	77	153
Bracigliano	NO	0	1	1	0	0	2
Castel S. Giorgio	YES	2	5	6	0	0	13
Corbara	YES	0	1	2	3	0	6
Nocera Inf.	YES	45	26	9	1	0	81
Nocera Sup.	NO	0	0	0	0	0	0
Pagani	YES	2	1	2	1	0	6
Roccapiemonte	NO	0	0	0	0	0	0
S. Marzano S.	NO	3	1	0	0	0	4
S. Egidio M.A.	YES	1	7	8	6	4	26
S. Valentino T.	NO	3	2	0	0	0	5
Sarno	NO	13	12	16	0	0	41
Scafati	YES	13	11	6	0	0	30
Siano	NO	0	0	0	0	0	0
total		97	94	83	12	81	367

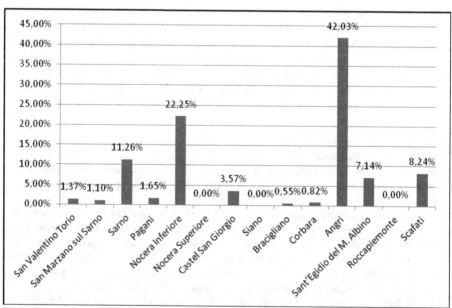

Fig. 1. Percentage of the applications presented, in relation to the total, per Council in the Agro-Nocerino-Sarnese

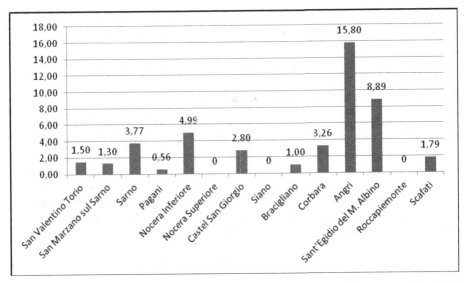

Fig. 2. Number of applications presented, in relation to the number of homes, per Council in the Agro-Nocerino-Sarnese (data multiplied by 10^3)

4   Estimation of the Effects of the Housing Plan on the Regional Economy of Campania

The sample is the starting point for predicting the economic effects of the Housing Plan Law in Campania. The computational tool is the inter-sectorial matrix. This makes it possible to determine the impacts (output) generated by a change in the aggregate demand (inputs, such as investment in the productive sector) on the economy of the territory where the matrix is associated[21].

The logic of the estimation is based on the assumption that every application has an associated implementation cost, i.e. the cost of the approved project. This spending generates an increase in investment in some sectors of aggregate demand (e.g., construction and professional activities), which in turn produces indirect effects on all the branches of the economy of a territory. Therefore, if it is possible to estimate the number of applications presented in Campania, based on the data found in the Agro-Nocerino-Sarnese, the sum of the costs of the interventions relating to the total number of applications is the *input* to identify the drag effect on the regional economy.

In this paper, the inter-sectorial matrix is the *Social Accounting Matrix* (SAM) of the Campania region, updated to 2010[22].

---

[21] The logic of the *input-output* system is structured in the inter-sectorial matrix, an accounting framework that synthesizes the flows arising from exchanges of goods and services that take place between the various productive sectors and between producers and end-use sectors. For further details, see [1], [17], [18], [19], [21].

[22] The preparation of the SAM Campania is a collaboration between the University of Rome Tor Vergata, Institute for Industrial Promotion (IPI) and the Ministry of Economic Development.

The implementation of the SAM Campania requires two preliminary steps:

1) estimation of the number of applications presented in the Region in accordance with R.L. 19/2009;

2) estimation of the costs of carrying out the works set out in the applications.

4.1   The Number of Applications in Campania

Since the Housing Plan Law has the main objective of increasing the number of homes, it is reasonable to assume that the number of applications in a defined geographical area depends largely on the number of houses in that area. Obviously omitting a number of other factors (population, income of the resident population, the prevailing building type, level of urbanization, social quality, educational level, presence of degraded areas, etc..) which are potentially influential on the number of applications presented in Campania, but in respect of which, to some extent, building density can be taken as a *proxy*. The estimate, therefore, is developed by weighting

TOTAL NUMBER OF APPLICATIONS IN CAMPANIA		9,018	
26.4%	2,381	art. 4 co. 7	change of use
25.6%	2,309	art. 4 co. 1	20% extension
22.6%	2,038	art. 5 co. 1	demolition and reconstruction of 35%
3.3%	298	art. 7 co. 5	requalification of abandoned areas
22.1%	1,993	art. 8 co. 2	recovery of attics

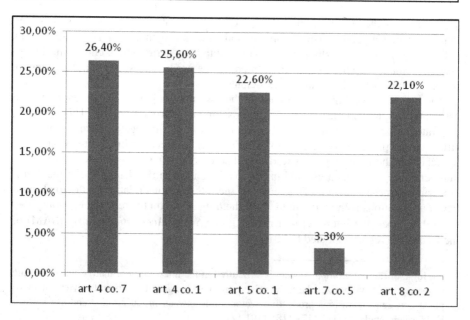

Fig. 3. Estimation of the number of applications presented in Campania and classification according to the Law

the number of applications presented in the Agro-Nocerino-Sarnese with the number of homes in the same territory. The assumed ratio is then extended to the entire region, with the patrimonial consistency being known. Since the year of assessment is 2011, from the calculations based on data from ISTAT relating to previous periods, there are, respectively, 105,725 housing units in the Agro-Nocerino-Sarnese and 2,598,039 in Campania. Taking into account that the number of applications in the Agro-Nocerino-Sarnes is 367, the probable number of applications presented in the region is equal to 9,018. In Figure 3, the estimated number is distributed among the intervention categories allowed by R.L. 19/2009, assuming that the percentage distribution coincides with that found in the study area.

4.2 Cost Analysis

The estimation of the costs for the implementation of the interventions described in the applications is developed by identifying an archetype for each of the project categories set out by R.L. 19/2009[23].

The total costs of these cases in relation to the most widespread building types and contexts in the Agro-Nocerino-Sarnese are quantified below[24].

20% volumetric Increase (art. 4 co. 1)

The typical case of a detached building with a volume of 700 m^3 is assumed. The application of article 4 comma 1 makes it possible to increase the volume by 20%, resulting in an overall cubic capacity of 840 m^3.

The total cost of the intervention is the sum of the construction costs, urbanisation costs and professional fees.

The construction costs are estimated by using a synthetic procedure, with reference to the prices indicated in [30]. Given that the unit cost of construction is 298 €/m^3, this results in:

$$\text{Construction costs} = €/m^3 \; 298 \times m^3 \; 140 = € \; 41,720 \; .$$

The urbanization costs and professional fees are assessed as a percentage of construction cost, respectively 10% and 7%. Therefore:

$$\text{Urbanisation costs} = 10\% \times € \; 41,720 = € \; 4,172 \; ;$$

$$\text{Professional fees} = 7\% \times € \; 41,720 = € \; 2,920 \; .$$

---

[23] On the procedures for estimating the construction costs, see among others: [4], [10], [16], [22], [28].

[24] It is worth noting that the cases in relation to the categories "change of use" (art. 4 co. 7) and "recovery of attics" (art. 8 co. 2) are not associated. In fact, these categories do not usually involve significant changes in volume or work, so that the corresponding *total cost* is given only by the technical expenses for the protocol procedures and the approval of the practices. Expenditure for the purposes of this study, are negligible in terms of contribution to the *overall cost*, the latter understood as the product of the total cost and number of applications presented.

The total cost for the 20% volumetric increase is therefore equal to:

$$\text{Total cost} = €\,41{,}720 + €\,4{,}172 + €\,2{,}920 = €\,48{,}812\,.$$

Demolition and Reconstruction (art. 5)

For this category, the typical case of a masonry building with a volume of 500 m³ is considered. Under art. 5, it is possible to demolish the ruins and build a building in place with a volume increase of 35% compared to the existing building, so that the final volume is 675 m³.

The total cost of the work includes the demolition costs, the fees for the disposal of the material, the construction costs of the new asset, the urbanization costs and professional fees.

The demolition costs and fees for the disposal of the material are based on [29]. The respective unit costs amount to € 15.24/m³ (full vacuum) and 4.20 €/m³ (actual volume of the material ≈ 130 m³). Thus:

$$\text{Demolition costs} = €/m^3\,15.24 \times m^3\,500 = €\,7{,}620\,;$$

$$\text{Waste material disposal fees} = €/m^3\,4.20 \times m^3\,130 = €\,536\,.$$

The construction costs are estimated by using a synthetic procedure, with reference to the prices indicated in [30]. Given that the unit cost of construction is 298 €/m³, this results in:

$$\text{Construction costs} = €/m^3\,298 \times m^3\,675 = €\,201{,}150\,.$$

The urbanization costs and professional fees are assessed as a percentage of the sum of the demolition costs, the waste material disposal fees and the construction costs, which are respectively 10% and 7%:

$$\text{Urbanisation costs} = 10\% \times (€\,7{,}620 + €\,536 + €\,201{,}150) = €\,20{,}931\,;$$

$$\text{Professional fees} = 7\% \times (€\,7{,}620 + €\,536 + €\,201{,}150) = €\,14{,}651\,.$$

Ultimately, the total cost for the demolition and reconstruction is:

$$\text{Total cost} = €\,7{,}620 + €\,536 + €\,201{,}150 + €\,20{,}931 + €\,14{,}651 = €\,244{,}888\,.$$

Requalification of Degraded Urban Areas (art. 7, co. 5)

For owners of abandoned buildings, R.L. 19/2009 makes it possible to convert the entire volume of the area for residential, commercial or tertiary use. The conversion is permitted in compliance with the minimum planning standards set out by DM 1444/1968. The typical case is given by an area of 8,000 m² which includes industrial factories with a volume of 24,000 m^3 (4,000 m^2 × 6 m in height)[25]. Within the constraints of Ministerial Decree 1444/1968, through the change of use, the result is a building complex of 15,000 m^3 (four tower buildings of 3,750 m^3 each) with 2,700 m^2 designed to urban standards[26] and 1,500 m^2 for parking[27]. Figure 4 shows how the space is distributed.

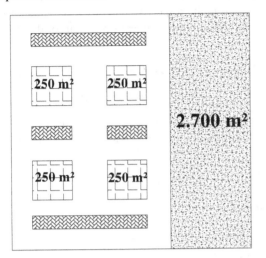

Legend

Parking areas according to the Tognoli Law

Standard public areas

Abutments of factories to be built

Fig. 4. Distribution of the areas in a typical case

---

[25] R.L. n. 14/1982 and s.m.i. (R.L. n. 7/1998 and R.L. 15/2005) provides that for new production facilities, "the coverage ratio, unless otherwise regulated by the Industrial Development Area Plans should be contained within the 1:2 ratio of the surface used for the production plant". It is worth noting that most of the currently disused factories in the Agro-Nocerino-Sarnese are characterized by a higher coverage ratio of 1:2, due to them being realized prior to the above mentioned law. In order to take into account the current state, in the typical case described, a 1:2 ratio between the area of the abutments of the assets and the land area (not land) of the lot is considered.

[26] Art. 3 of Ministerial Decree 1444/1968 states that the maximum ratio between the spaces allocated to residential and public spaces, or reserved for collective activities in public parks or parking lots "are set to such an extent as to ensure for each inhabitant, established or to be set up, the minimum imperative equipment, of 18 m^2 for public or reserved for collective activities in public parks or parking, with the exception of road space available for offices. [...] For the purposes of compliance with the aforementioned relationships in the training of planning instruments, it is assumed that, unless otherwise shown, for each inhabitant installed or set up there is an average of 25 m^2 of gross floor area (approximately 80 m^3 empty for full), plus possibly a share not exceeding 5 m^2 (approximately 20 m^3 for full vacuum) to destinations not specifically residential but closely associated with the residences (shops for basic needs, community services for homes, professional offices, etc.)".

[27] Art. 2 of Law 122/1989 (Law Tognoli) states that "in new buildings and also in areas belonging to the construction no less than one square metre for every ten cubic metres of the construction must be reserved for parking spaces".

The total cost is the sum of the demolition costs of the existing assets, the costs of disposing of the waste material, the construction costs of the new buildings, the cost to create outdoor areas of the new building complex, the urbanization costs and professional fees.

The demolition costs of the existing buildings and the disposal costs of the waste material are estimated according to [29]. From the respective unit costs, equivalent to € 13.08/m³ (full vacuum) and € 15.76/m³ (the actual volume of 3,500 m³ of debris), thus resulting:

$$\text{Demolition costs} = €/m^3\ 13.08 \times m^3\ 24,000 = €\ 313,920;$$

$$\text{Waste material disposal fees} = €/m^3\ 15.76 \times m^3\ 3,500 = €\ 55,160.$$

The construction costs and the cost of creating external spaces are estimated with reference to the prices indicated in [30]. The unit construction cost of a tower building is 283 €/m³, which gives:

$$\text{Construction costs} = €/m^3\ 283 \times m^3\ 15,000 = €\ 4,245,000.$$

The unit cost of creating external spaces is 43 €/m², thus:

$$\text{External spaces cost} = €/m^2\ 43 \times m^2\ [8,000 - 2,700 - (4 \times 250)] = €\ 184,900.$$

The urbanization costs and professional fees are respectively 10% and 6% of the sum of the demolition costs, the disposal costs, construction costs and the costs to create external spaces:

Urbanisation costs = 10% × (€ 313,920 + € 55,160 + € 4,245,000 + € 184,900) = € 479,898;

Professional fees = 6% × (€ 313,920 + € 55,160 + € 4,245,000 + € 184,900) = € 287,939.

For the typical case of the requalification of urban areas, the total cost is:

Total cost = € 313,920 + € 55,160 + € 4,245,000 + € 184,900 + € 479,898 + € 287,939 = € 5,566,817.

4.3   Implementation of the SAM Campania

Table 2 shows, for each intervention category, both the *total cost* given in paragraph. 3.2, as well as the *overall cost* as the product of the total cost and number of applications presented.

Table 2. Total cost and overall cost per intervention category

intervention category	total cost [€]	overall cost [€]
20% volumetric increase	48,812	112,694,872
demolition and reconstruction	244,888	499,125,791.10
requalification of degraded urban area	5,566,817	1,656,742,074

The *input* data required to activate the Social Accounting Matrix of Campania relate to the Construction and Professional Activity sectors. They are obtained from the overall cost by subtracting the expenditure items for professional fees. These data are reported in Table 3.

Table 3. Input data of the SAM Campania

intervention category	Construction [€]	Professional activity [€]
20% volumetric extension	105,952,444	6,742,428
demolition and reconstruction	469,263,564	29,862,227
requalification of degraded urban area	1,571,048,518	85,693,556
total	2,146,264,526	122,298,211

The implementation of the SAM Campania gives the effects on the regional economy generated by investments in the construction and professional activities sectors. The result is expressed synthetically using three indicators: change in regional GDP ($\Delta GDP = 2.889\%$), increase in employment (48,021 units of work) and monetized environmental damage (€ 214,904,024).

The *output* expresses the potential impact, over a period estimated to be between three and five years, that the Housing Plan would have on the regional economy if all the applications submitted were approved (optimistic scenario). It is reasonable to assume that this last condition can hardly be satisfied. From the data obtained from the provincial capitals, only 37% of the applications presented are approved. Thus, in addition to the optimistic scenario, it is also worth considering the realistic scenario that the percentage of approved applications in the Campania region coincides with the averages in the cities of Naples, Avellino, Benevento, Caserta and Salerno. In such a case, the *input* for the SAM Campania is the 37% of the optimistic scenario. The *outputs* indicate a 1.069% rise in the GDP, 17,768 new jobs and an environmental impact of € 79,514,788.

Table 4 summarizes the results.

Table 4. Output of the realistic and optimistic scenarios

	ΔGDP [%]	economic impact [M€]	employment [units of work]	environmental impact [€]
realistic scenario	1.069	5,269.08	17,768	79,514,606
optimistic scenario	2.889	14,240.76	48,021	214,904,340

5   Conclusions

This study, the results of which were submitted to the judgment of experts in thematic conferences, estimates the effects of the Housing Plan in Campania (R.L. 19/2009) on the regional economic system. It first discusses the objectives of the Law and the contents of the various articles. The evaluation issues are therefore analyzed, indicating the constant support function carried out by estimations in relation to public and private investment decisions.

The regulatory framework is the starting point for the field survey, which was carried out in numerous technical offices of local councils. The information obtained relates to the administrative and technical aspects of the process initiated by Law 19 and the subsequently approved applications.

Appreciation of the private financial resources that the regulation is able to mobilize is carried out with a synthetic procedure, based on unit costs derived from current literature and practices. The data are used as an *input* for the activation of the inter-sectorial matrix in Campania that makes it possible to predict the probable impacts on the regional economy generated by the implementation of construction projects that the Law contemplates. Two different scenarios are evaluated in the analysis: one optimistic, assuming that all the applications are approved by the local councils; one realistic, taking a percentage of unapproved applications. The results of the realistic scenario (economic impact on the production sectors of € 2975.80 million and 17,768 new jobs), compared with the forecasts made by ANCE[28] (effects on the economy, € 19 billion and 40,000 units of work) show that the objectives are, to date, only partially satisfied.

The application of the inter-sectorial matrix also make it possible to make eco-sustainable considerations of the effects of the Housing Plan, through the monetary quantification of environmental damage, for which appropriate mitigation tools should be expected.

The logic defined in the study – from the actual retrieval of data processing through an economic analysis methodology – represents a practical assessment which can be used in other regional contexts.

It is also worth highlighting the different responses given by the administrative authorities responsible for issuing approvals in the cities and the provinces. The data obtained from the surveys carried out in the provinces clearly highlight that no applications had been authorized at the time of the study. While, in the main cities 37% of

---

[28] See: www.edilportale.it

the applications had been approved. This shows the inconsistency of the regulatory measures aimed at deregulation, when they are particularly complex to interpret. The *expertise* available in the technical offices in the suburbs is often inadequate when having to assume any form of responsibility that comes from the loosening of legal constraints, resulting in the stalemate of the bureaucratic machine and the failure of any investment initiative. Finally, it deals with verifying the effectiveness of the regulation as amended by Law 1/11.

References

1. Abbate, C.C., Bove, G.: Modelli multidimensionali per l'analisi input-output. Quaderno di Ricerca. ISTAT (1993)
2. Associazione Bancaria Italiana: Codice per la valutazione degli immobili in garanzia delle esposizioni creditizie, Roma (2009)
3. Curto, R.: Un approccio economico alla pianificazione. In: Mantini, P., Oliva, F. (eds.) La Riforma Urbanistica in Italia. Pirola, Milano (1996)
4. De Mare, G., Morano, P.: La stima del costo delle opere pubbliche. UTET, Torino (2002)
5. De Mare, G., Nesticò, A.: Il diritto di superficie nelle trasformazioni urbane: profili estimativi. Rivista SIEV valori e valutazioni. 4/5. DEI Tipografia del Genio Civile, Roma (2010)
6. De Mare, G., Lenza, T.L., Conte, R.: Economic evaluations using genetic algorithms to determine the territorial impact caused by high speed railways. World Academy of Science, Engineering and Technology (71) (2012); ICUPRD 2012
7. De Mare, G., Morano, P., Nesticò, A.: Multi-criteria spatial analysis for the localization of production structures. Analytic Hierarchy Process and Geographical Information Systems in the case of expanding an industrial area. World Academy of Science, Engineering and Technology (71) (2012); ICUPRD 2012
8. De Mare, G., Nesticò, A., Tajani, F.: The rational quantification of social housing. In: Murgante, B., Gervasi, O., Misra, S., Nedjah, N., Rocha, A.M.A.C., Taniar, D., Apduhan, B.O. (eds.) ICCSA 2012, Part II. LNCS, vol. 7334, pp. 27–43. Springer, Heidelberg (2012)
9. De Mare, G., Manganelli, B., Nesticó, A.: Dynamic Analysis of the Property Market in the City of Avellino (Italy):The Wheaton-Di Pasquale Model Applied to the Residential Segment. In: Murgante, B., et al. (eds.) ICCSA 2013, Part III. LNCS, vol. 7973, pp. 509–523. Springer, Heidelberg (2013)
10. De Mare, G., Manganelli, B., Nesticó, A.: The economic evaluation of investments in the energy sector: A model for the optimization of the scenario analyses. In: Murgante, B., et al. (eds.) ICCSA 2013, Part II. LNCS, vol. 7972, pp. 359–374. Springer, Heidelberg (2013)
11. Famularo, N.: Lezioni di Estimo Civile e Rurale. Edizioni Italiane, Roma (1945)
12. Ferrero, C. (ed.): La valutazione immobiliare. Principi e metodologie applicative. Egea, Milano (1996)
13. Florio, M.: La valutazione degli investimenti pubblici. Il Mulino, Bologna (1991)
14. Forte, C.: Elementi di Estimo Urbano. Etas Kompass, Milano (1968)
15. Frontera, M.: Segnalazione semplificata per far decollare il Piano Casa. Il Sole 24 ORE, Milano (Aprile 23, 2011)
16. Grillenzoni, M., Grittani, G.: Estimo, teoria, procedure di valutazione e casi applicativi. Calderini, Bologna (1994)
17. Guarini, R., Tassinari, F.: Statistica economica. Il Mulino, Bologna (1990)
18. Leontief, W.: Structure of American Economy, 1919-1939: An Empirical Application of Equilibrium Analysis. Oxford University Press, New York (1951)

19. Leontief, W.: Environmental Repercussions and the Economic Structure: An Input-Output Approach. The Review of Economics and Statistics 52(3) (1970)
20. Medici, G.: Principi di Estimo. Calderini, Bologna (1972)
21. Miller, R.E., Blair, P.D.: Input-Output Analysis, Foundations and Extensions. Prentice-Hall, Engelewood Cliffs (1985)
22. Mollica, E.: Principi e metodi della valutazione economica dei progetti. Rubettino, Catanzaro (1995)
23. Morano, P.: Un modello di perequazione urbanistico-estimativo. Graffiti, Napoli (1998)
24. Morano, P.: La stima degli indici di urbanizzazione nella perequazione urbanistica. Alinea, Firenze (2007)
25. Morano, P., Nesticò, A.: Definizione del piano economico-finanziario di un'opera infrastrutturale. Atti del XXXII Incontro di Studio Ce.S.E.T. Venezia (2002)
26. Morano, P., Nesticò, A.: Un'applicazione della programmazione lineare discreta alla definizione dei programmi di investimento. Aestimum, vol. 50. Firenze. University Press, Firenze (2007)
27. Nuti, F.: Analisi costi-benefici. Il Mulino, Bologna (1988)
28. Patrone, P.D., Piras, V.: Construction Management. Alinea, Bologna (1997)
29. Prezzario, O.P.: Regione Campania. DEI Tipografia del Genio Civile, Roma (2010)
30. Prezzi Tipologie Edilizie. DEI Tipografia del Genio Civile, Roma (2010)
31. Stanghellini, S.: Fattibilità ed equità: da requisiti del piano a dimensioni della valutazione. Urbanistica Informazioni 105 (1995)
32. Stanghellini, S.: I suoli urbani per le politiche abitative. Rivista SIEV valori e valutazioni. DEI Tipografia del Genio Civile, Roma (2009)
33. Tecnoborsa: Codice delle Valutazioni Immobiliari. Telligraf, Roma (2005)
34. Vaudetti, F.: La stima delle aree fabbricabili. Calderini, Bologna (1957)

Dynamic Analysis of the Property Market
in the City of Avellino (Italy)*

The Wheaton-Di Pasquale Model Applied
to the Residential Segment

Gianluigi De Mare, Benedetto Manganelli, and Antonio Nesticò

Faculty of Engineering, University of Salerno, Italy
{anestico,gdemare}@unisa.it, benedetto.manganelli@unibas.it

Abstract. The dynamics of the housing market have been the subject of study and modelling for several decades (including: Muth, 1963; Ozanne, Thibodeau, 1983; Stiglitz, 1993; Green, Malpezzi and Mayo, 2005). The contingent relationships between the property market and micro and macroeconomic situations are particularly interesting. Studies were carried out by Di Pasquale and Wheaton (1996), with reference to the office property market and the correlation with the construction industry. The model has been adopted in this study, but applied to the residential property market in a medium size city located in southern Italy. The aim includes both the reading of the historical past ten years, with a dynamic approach to the problem, as well as a predictive application of the model for the near future (2013-2015).

Keywords: Residential property market, Analysis estimation, Models for economic forecasting.

1   Introduction and Objectives

Problems within the housing market are becoming more closely associated to general economic issues, given the importance of the financial resources invested in property as well as those connected to the construction industry.

It is evident that the recent global recession has led, and still leads, to common depressive effects on the housing sector, albeit with different consequences according to specific local and particular market segments.

In 2010, the links between the economic crisis and developments in the property market [1] were already being discussed, highlighting the concatenation of financial, ethical and socio-political profiles in the debate on both national and international levels.

Etiologic connections have been considered since antiquity when Aristotle considered the study of economics and its mechanisms as a tool for the advancement of the community, with a close union between ethics and political philosophy. Thus, *Homo economicus* was an interpretative model of human behaviour, which was strongly

---

* This paper is to be attributed in equal parts to the three authors.

B. Murgante et al. (Eds.): ICCSA 2013, Part III, LNCS 7973, pp. 509–523, 2013.
© Springer-Verlag Berlin Heidelberg 2013

influenced by, as well as dependent on, environmental restrictions. However, man has been able to break the bond of subjection to nature thanks to applied sciences.

In the modern era, the relationship between ethics and economics was considered by scholars of the 1700 and 1800s. In particular, the Scot A. Smith (1723-1790) in *The Wealth of Nations*, but also *The Theory of Moral Sentiments*, highlighted the bond between self and collective interests, between private and public spheres.

In the 1900s, the two dimensions are progressively separated, leading to the numerous current inconsistencies and difficulties inherent in the global economic system. Thus, the recent financial crisis can be interpreted after acritical reading of the socio-economic models adopted over the last twenty years, which have been permeated by a boundless trust in engineering applications to the world of finance and their ability to self-limit their own distortions as well as those of contaminated economic areas. The most severe applications have resulted in the virtualization of business to the extent that the assets underlying the speculative operations are no longer recognizable by the end user to whom the securities derived from financial manipulations are transferred.

This decline has also affected the property market (consider the mortgage crisis and derivatives linked to them), which today is the main partner of the financial markets in developed economies, dissolving its historical characteristics of stability and geo-referencing into a frantic portability, caused by the implemented securitization arrangements. This is also due to the economic interests that revolve around the industry in western countries. It is worth considering that in Italy [2] the sector accounts for 20% of the GDP, 60% of the wealth of households and involves 30% of the loans made by financial institutes.

To date, the perception of the property market is extremely different depending on the point of view. In fact, for the buyer looking for either a residential or commercial property, the outlook remains restricted and limited to the asset as well as the characteristics that it must have in order to meet the daily needs. For both large financial groups as well as small investors, the property market has become over the last ten years a complex and articulated investment tool, similar to the traditional stock market mechanisms such as stocks, bonds and derivatives. In fact, the stock market has enshrined the collaboration that has virtualized the world of brick and mortar.

Thus, there is, and must always be considered, a micro and macro reading of the property market, with both requiring distinct and powerful interpretation keys, in terms of theoretical and mathematical-computational instrumentation needed for the modeling and interpretation of the structural phenomena of the context.

This study follows the experience of 2010 and will offset a local approach to commercial issues. It implements a model well known in current literature [3], with it being applied to the office property market for the interpretation of the trend of a specific local residential market in the city of Avellino, in Campania (Italy). The protocol reconstructs and explains the commercial trend for the housing segment between 2001 and 2012, drawing on an extensive and institutional database. It also proposes, in the concluding part, a scenario analysis for the three year period 2013-2015, in order to support the choices of expenditure for both the *property market* as well as that for the *asset market*.

2   The Wheaton-Di Pasquale Model Applied to the Residential Market in Avellino

In current literature, the analysis of the property market has prompted the search for models that adequately interpret the changes. These range from models based on long term observations [4, 5] to models based on the explanation of the dynamics of property prices [6, 7] in the short term; models that focus on the elasticity of the properties on offer [8, 9]; models that benefit from innovative protocols under the computational profile [10, 11, 12, 13, 14].

When considering the property market in its entirety, it is worth referring to a set of different submarkets according to a plurality of characteristics (localization, infrastructure, type, quality construction, etc..). In disciplinary terms, such peculiarities are the *extrinsic* and *intrinsic* characteristics of the properties. Although, it is easier to describe a property and distinguish it from others, it is particularly complex to carry out studies that have an absolute validity. The property market, in fact, tends to be divided into different segments whose identification can follow deductive and inductive procedures. Hence, the need for a consistent and detailed dataset, rigorous in the surveying processes, which   possibly refers to different periods when wanting to feel the dynamic dimension of the phenomena investigated.

An initial macroscopic distinction occurs when differentiating the *property market* from the *asset market*.

In the *property market*, the offer is compared with the demand linked to the direct use of the building or *occupancy*. In contrast, in the second market or *asset market*, an investment on the property market that has as the aim of the return on invested capital is analysed.

The DiPasquale-Wheaton model (or four quadrants model), published in 1996, and adopted by others, deals with the relationship between the two segments mentioned.

In particular, it illustrates the interaction of the two markets through four binary relations established between the *demand for living space*, the *net income* that the property is capable of delivering, the *selling price levels* of the property and the *construction of new buildings*.

Interest rates are exogenous to the model.

The functions considered, which appear in the representation (see figure 1) as oblique segments, identify a finite number of equilibrium states from which to derive the overall balance of the market. This is identified by the junction between the different lines through a connection line of the functions. It describes the *time t* in the market, which represents a state of equilibrium between the stock of buildings, the levels of the rent, property prices and the rate of replacement of existing buildings with new constructions.

The lines function may translate or rotate on themselves in relation to changes in the underlying logical components/operators in that sector, modifying the general equilibrium of the whole system.

The direct use of the property is orientated by the demand for usable increase in rent, along with the socio-economic conditions, leads to a natural decline in the demand for square metres per direct use (*occupancy*). The opposite is also true.

It should be noted that the model presented not only has a deterministic nature but also disregards the role of the state. In the case of an increase in income or purchasing power, or a more general improvement of the economic conditions of the area where the property is located, a *ceteris paribus* is generated, a shift of the demand curve in high, and in consequence, for a given level of rents, an increase of the request of square metres. Subsequently *the pipeline*, i.e. the annual consumption of square metres tends to increase.

The phenomenon described relates to the *property market*, but will also have an immediate influence on the *asset market*.

The increase in rents affects property values, linked to them, making them rise. Consequently, the more opportunities for profit from them may help to determine an increase in production of square metres with an increase of square metres available, aimed at satisfying both the need to remunerate the capital as well as the demand for housing. The figure shows what has been briefly described.

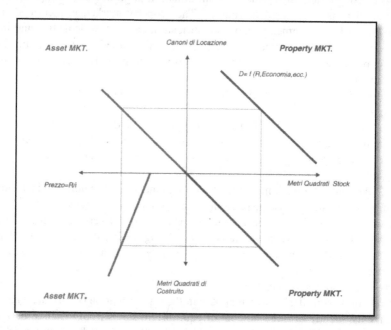

Fig. 1. The relationship between the Asset Market and Property Market [15]

The model presented is subject to further investigation. The relationship is dynamic and thus variable in the medium and long term. To better understand how it works, imagine a shift to the north-east of the demand curve, representing an increase in demand for offices in a given urban region (Figure 2). The phenomenon is usual in the case of targeted investments in the services sector, with a consequent increase in the number of employees and tertiary spaces to carry out productive activities. This increased demand will generate a shift of the demand curve that is to be found in the classic economic model of supply and demand, and that will influence the system resulting in an increase of the quantity of traded product, or an increase in rent level, or, depending on the responsiveness of the supply system, a combination of the two events. The increase in demand from D_0 to D_1, will in fact, in the short term (a year or two), only increase rents (and not the quantities), because the supply is unlikely to deliver the required volumes, given the *lag time* of the construction sector. This increase in demand will intercept the straight line of the evaluations (the one that refers to the relationship between prices and royalties and that then expresses the *cap rate* of the market for that product), and then fall down, intercepting the horizontal axis, thus defining a new market price for office properties and consequently a new convenient relationship between construction costs and selling prices. This will make it more profitable for investors operating in the sector. The synthesis line, which now comes to intercept P_1, will travel to the south, but it will not continue its course due to the inertia of the construction sector: the level of the amount of product will be (again in the short term) stable at point C*, to move in the long run, to C**. Only when the amount of new product C** is realized, will the dotted line be able to follow its own path and return to a balance that will catch, in turn, Q**, R** and then finally P**. This time interval between the change of the demand and the consequent reaction in the other three quadrants (and in particular in the southwest quadrant, the supply of new buildings), is exactly the *time lag* that generates cycles on the real estate market.

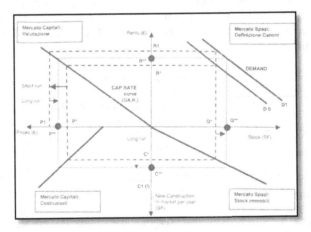

Fig. 2. Dynamic adjustments of the 4 quadrants model [Geltner, Miller, 2000]

2.1   The Reference Database

The data needed to implement the model between 2001 and 2012 in the city of Avellino are those required by the significant variables of the four quadrants. In addition to further predictors, they are important touniquely determine the functions representing the phenomena investigated.

In particular, for the prices and rents of the urban residential market, reference was made to the data from the Property Market Observatory managed by the Land Registry Office, supplemented – if necessary – with unit valuesavailable in *Property Directory* [16]. This results in the following:

Table 1. Average residential market prices in Avellino

AVERAGE MARKET PRICES AND RENTS OF RESIDENTIAL PROPERTY
(CITY OF AVELLINO)

YEAR	MARKET PRICE (P) $[\text{€}/m^2]$	RENT $[\text{€}/m^2\text{YEAR}]$	NET INCOME (R) $[\text{€}/m^2\text{YEAR}]$	CAPITALIZATION RATE $i = R/P$
2001	1,291	69.21	38.07	0.029
2002	1,433	76.90	42.30	0.030
2003	1,400	74.29	40.86	0.029
2004	1,633	85.69	47.13	0.029
2005	1,933	96.31	52.97	0.027
2006	1,933	96.28	52.95	0.027
2007	2,066	102.90	56.60	0.027
2008	2,066	101.46	55.80	0.027
2009	2,133	98.30	54.06	0.025
2010	2,166	94.57	52.02	0.024
2011	2,166	85.08	46.79	0.021
2012	2,200	81.16	44.64	0.020

A measurement of the maximum values reached by the location in the city is to be added to this information, due to it being useful in determining the straight lines representing the relationship between rent and price. The data are described in Table 2, adapting them from the sources mentioned above.

Table 2. Maximum rents and incomes from the renting of residential properties in Avellino

MAXIMUM RENT OF RESIDENTIAL PROPERTY
(CITY OF AVELLINO)

Year	Maximum Rent $[\text{€}/m^2\text{year}]$	Maximum rent income (R_{max}) $[\text{€}/m^2\text{year}]$
2001	91.20	50.16
2002	96.00	52.80
2003	105.60	58.08
2004	116.40	64.02
2005	122.40	67.32
2006	122.40	67.32
2007	122.40	67.32
2008	122.40	67.32
2009	122.40	67.32
2010	122.40	67.32
2011	124.20	68.31
2012	124.20	68.31

It is also necessary to have the property holdings on the residential sector during the period under study. These were obtained from the ISTAT data as well as data on new buildings provided by the Technical Department of the Municipality.From the analysis of more than 100 cards for year, the following distributions were obtained.

Table 3. Entire stock of residential properties in Avellino

Entire stock of residential properties (S) [m^2]	
2001	3,842,965
2002	3,857,415
2003	3,870,830
2004	3,890,995
2005	3,912,454
2006	3,941,751
2007	3,967,017
2008	3,983,702
2009	3,995,637
2010	4,009,179
2011	4,027,426
2012	4,051,837

Table 4. Annual volume of new residential properties in Avellino

Annual volume of new constructions (C) [m^2]	
2001	19,706
2002	14,450
2003	13,415
2004	20,165
2005	21,459
2006	29,297
2007	25,266
2008	16,685
2009	11,935
2010	13,542
2011	18,247
2012	24,411

The definition of the construction costs of residential buildings during the period of analysis is also required. This function is used to estimate the price-volume of new constructions in the third quadrant of the model.

Thus, in reference to ISTAT and databases from professional Orders, the following is obtained:

Table 5. Construction costs of houses in Avellino

DATE	CONSTRUCTION COSTS [€/m²]
2001	549.65
2002	568.61
2003	588.51
2004	598.16
2005	622.00
2006	641.28
2007	666.29
2008	690.28
2009	693.73
2010	704.28
2011	731.01
2012	757.73

Finally, in order to construct the representative function of the first quadrant, the data on the demographic trends in the municipality are essential.

The information, taken from ISTAT and the National Association of Italian Municipalities, are structured as shown in Table 6.

Table 6. Demographic data of the City of Avellino from 2001 to 2012

YEAR	Population (male)	Population (female)	Total Population	Number of families	Number of memebrs (E)
2001	25,247	27,456	52,703	19,716	2.67
2002	25,202	27,488	52,690	19,956	2.64
2003	26,003	28,274	54,277	20,261	2.68
2004	26,877	29,523	56,400	20,445	2.76
2005	27,121	29,872	56,993	20,615	2.76
2006	27,085	29,843	56,928	20,845	2.73
2007	27,022	29,886	56,908	21,114	2.70
2008	27,037	30,034	57,071	21,278	2.68
2009	26,965	29,974	56,939	21,422	2.66
2010	26,620	29,892	56,512	21,529	2.62
2011	26,510	29,829	56,339	21,512	2.62
2012	26,451	29,731	56,095	21,590	2.59

2.2   The Quadrants of the Model

The lines identifying functional relationships analyzed are shown for each quadrant and for each year of the analysis period.

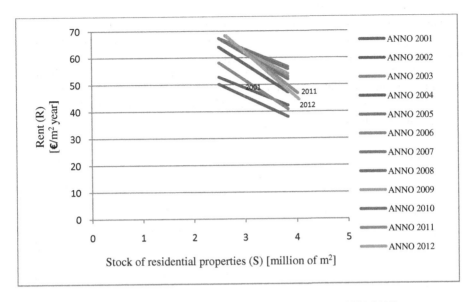

Fig. 3. Residential property demand curves in Avellino (2001-2012)

The equation is:

$$R = -\frac{S}{E*a} + \frac{b}{a} \tag{1}$$

where:
R = rental income net of expenses;
S = stock of residential properties available;
E = average number of members per household (see Table 6).

The straight line for each year of the survey is achieved by imposing the transition to the equilibrium point of the same year (see Table 1) and for the maximum point, which is obtained by combining the maximum income reported in Table 2 with the extent of the available stock in correspondence of the maximum income considered. This is obtained by dividing the stock that houses located in areas where the maximum income is measured by the total number of homes on the municipal territory.

It is interesting to note that the lines of the strip will go upward between 2001 and 2004, indicating a growing demand for living spaces, associated with an increase in the purchasing power of the population. The trend is stable between 2004 and 2010, due to the effects of the global economic crisis (it is worth remembering, in fact, that the demand curve is also a function of a number of exogenous variables in the model, such as the purchasing power, the income level and, more generally, the economic situation in which there is an area where the property under study falls). Between 2011 and 2012, there are signs of weakness (price go down). Although the slope of the straight line stresses the phenomenon, with a retreat of the absolute indicative position of a reduced availability of resources on the part of buyers, but with accentuated inclination that involves apical values for income. This phenomenon is usually at the end of regressive phases, in which the demand builds up and tends to produce signals in advance of recovery through housing   requests.

Further considerations can be made from the quadrants that follow, for which the straight lines and the related functional equations are represented. However, the essential elements deductible from the historical analysis are summarized in the next section, dedicated to the prediction of the scenario.

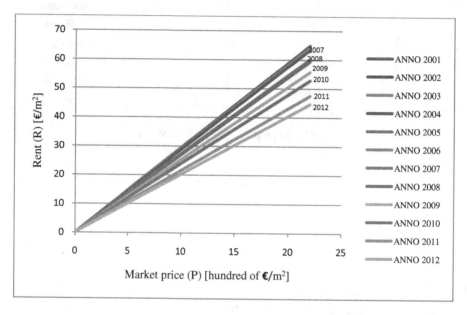

Fig. 4. Average price curves of the residential property market in Avellino (2001-2012)

The equation is:

$$P = \frac{R}{i} \tag{2}$$

where:
P = market price;
R = rental income net of expenses;
i = capitalization rate (taken from Table 1).

The equation is:

$$C = \frac{(P-\beta)}{\alpha} \tag{3}$$

where:
C = volume of new residential constructions;
P = market price.

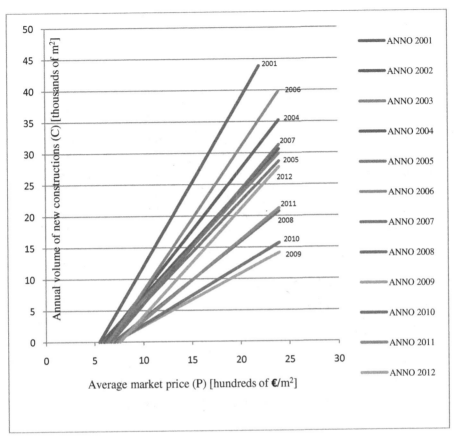

Fig. 5. Offer of new residential property curves in Avellino (2001-2012)

The equation is:

$$S = \frac{c}{d}$$ (4)

where:
d = fraction of the existing stock;
S = stock of residential properties available.

3   Forecast of Changes for the Period 2013-2015

The model described, as mentioned, can also be adopted for predictive purposes.

The historical sequence of the data is projected in the near future, with particular attention to the logic of the forecast and the constraints of the system. The number of required parameters is obtained and summarized below.

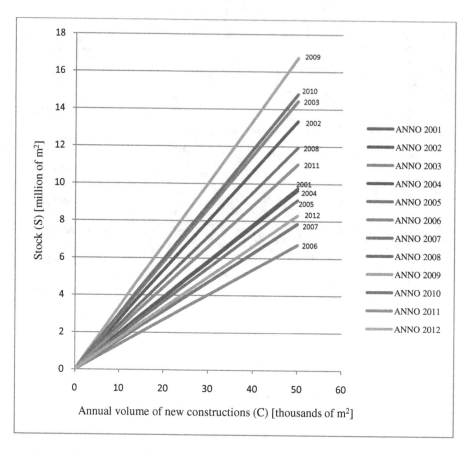

Fig. 6. Regulation of entire residential property stock curves in Avellino (2001 – 2012)

Table 7. Forecast data for the period   2013-2015

Year	Market Price (€/m²)	Net Rent Income (€/m² *anno)	Maximum Rent Income (€/m² *anno)	Construction costs (€/m²)	Annual volume of new constr. (m²)	Entire stock of residential properties (m²)	Total pop.	Number of families	Number of members per family (E)
2013	2,236	42.69	68.31	771	16,964	4,068,801	55,825	21,605	2.58
2014	2,266	42.69	68.31	790	16,964	4,085,765	55,570	21,635	2.57
2015	2,296	42.69	68.31	808	16,964	4,102,729	55,315	21,666	2.55

The completed Wheaton-Di Pasquale is the following:

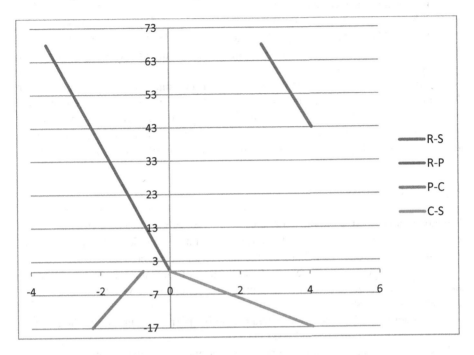

Fig. 7. Completed Wheaton-Di Pasquale diagram for the residential property market in Avellino (2013-2015)

It is assumed that the negative values on the axes are instrumental to a complete representation of the phenomenon through the use of the Excel software, but that they must be considered obviously positive.

Firstly, it is possible to observe that the prospective scenario is essentially static in the years concerned. This indication is plausible in light of the contradictory signals derived from the financial, stock and economic markets, since the three curves represented in each of the areas are almost completely overlapping.

The angular coefficients that typify the straight lines should also be appreciated.

Those of the first two quadrants are similar, which leads to comparable replacement rates. In other words, the rental and sales markets in Avellino have similar trends, suggesting an alignment of the dynamics despite the greater inertia to changes endemic to the renting sector. This indication is important for investors and thus the *asset market*, as it outlines a reactive scenario with respect to a possible economic recovery.

The straight line in the third quadrant is less inclined (about 85 degrees compared to 87 of the other two, but the tangent function tends to infinity toward 90 degrees, so a few degrees determine significant effects). This represents the damping effect associated with the so-called *time lag*, which is amplified even more when considering the slope of the line in the fourth quadrant (about 76 degrees). In fact, if the market prices and rents react flexibly to exogenous stress, it behaves differently to new construction due to the structural inertia at the time of execution of the construction work. The mechanism of replacing old units with new residences is even slower. This aspect is typical of all European markets than, in respect to the North American one, and depends mainly on building materials used, as well as the greater constraints on the preservation of historical heritage.

Due to the composite picture described, it can be concluded that any increase in demand for rental spaces (represented by a translational motion upward curve in the first quadrant) will cause an increase in net rent values, followed by an amplified percentage increase in the sale prices. While, the repercussions will be contained in terms of new buildings and even more from the viewpoint of building replacement.

4   Conclusions

The model developed, from the experience gained by Wheaton and Di Pasquale in the tertiary sector, allows to relate multiple aspects of interest in the study of the residential real estate market, testing static profiles and dynamic phenomenon.

Specific complexity is inherent in the collecting of the data necessary to build the interpretive functions, with them being particularly numerous where wanting to have a representative for several years.

The model is used for descriptive and predictive purposes, supporting the scenario analysis of both the property and asset markets.

It is applied to the residential property market, where as it was originally intended for the commercial property market, due to the reality of the city under study having a number of similarities between the two sectors, which are considered complementary, especially by investors.

The research results lead to numerous insights, which are explained in the text.

References

1. De Mare, G., Tajani, F.: Crisi economica e mercato immobiliare, nessi eziologici e prospettive nel futuro prossimo (2010), http://www.e-stimo.it (rivista online per le valutazioni e le stime)
2. Pennetta, F., et al.: L'andamento del mercato immobiliare italiano e i riflessi sul sistema finanziario, Banca d'Italia (Dicembre 2009)
3. Wheaton, W.C., Di Pasquale, D.: Urban Economics and Real Estate Markets. Prentice Hall, Englewood Cliffs (1996)
4. Ozanne, L., Thibodeau, T.: Explaining metropolitan housing price differences. Journal of Urban Economics 13(1) (1983)
5. Goodman, A.C.: An econometric model of housing price, permanent income, tenure choice and housing demand. Journal of Urban Economics 23 (1988)
6. Stiglitz, J.E.: Endogenous Growth and Cycles. NBER Working Paper No. 4286 (March 1993)
7. Malpezzi, S., Wachter, S.M.: The role of speculation in real estate cycles. Journal of Real Estate Literature, ARES (2005)

8. Muth, R.F., Bailey, M.J.: A regression method for real estate price index construction. Journal of the American Statistical Association (1963)
9. Green, R.K., Malpezzi, S., Mayo, S.K.: Metropolitan-specific estimates of the price elasticity of supply of housing, and their sources. American Economic Review, JSTOR (2005)
10. De Mare, G., Nestico', A., Morano, P.: Multi-criteria spatial analysis for the localization of production structures. Analytic hierarchy process and geographical information systems in the case of expanding an industrial area. World Academy of Science, Engineering and Technology (71) (2012); ICUPRD 2012
11. De Mare, G., Lenza, T.L., Conte, R.: Economic evaluations using genetic algorithms to determine the territorial impact caused by high speed railways. World Academy of Science, Engineering and Technology (71) (2012); ICUPRD 2012
12. De Mare, G., Nesticò, A., Tajani, F.: The rational quantification of social housing. In: Murgante, B., Gervasi, O., Misra, S., Nedjah, N., Rocha, A.M.A.C., Taniar, D., Apduhan, B.O. (eds.) ICCSA 2012, Part II. LNCS, vol. 7334, pp. 27–43. Springer, Heidelberg (2012)
13. De Mare, G., Nesticó, A., Tajani, F.: Building investments for the revitalization of the territory: A multisectoral model of economic analysis. In: Murgante, B., et al. (eds.) ICCSA 2013, Part III. LNCS, vol. 7973, pp. 493–508. Springer, Heidelberg (2013)
14. De Mare, G., Manganelli, B., Nesticó, A.: The economic evaluation of investments in the energy sector. A model for the optimization of the scenario analyses. In: ICCSA 2013. LNCS, vol. 7972, pp. 374–389. Springer, Heidelberg (2013)
15. Del Giudice, V., D'Amato, M.: Principi metodologici per la costruzione di indici dei prezzi nel mercato immobiliare. Maggioli Editore, Rimini (2008)
16. Tamborrino, M.: Come si stima il valore degli immobili, Gruppo24Ore (2012)

Spatial Representation: City and Digital Spaces

Gilberto Corso Pereira[1], Maria Célia Furtado Rocha[2], and Pablo Vieira Florentino[3]

[1] Federal University of Bahia, Salvador, Brazil
corso@ufba.br
[2] PRODEB, Salvador, Brazil
mariacelia.rocha@prodeb.ba.gov.br
[3] Federal Institute of Bahia, Salvador, Brazil
pablovf@ifba.edu.br

Abstract. This paper refers to forms of space representation, essentially derived from cartographic techniques, that bases many web applications which aid to promote the development of new forms of social interaction and influence the perception and use of public spaces. The paper discusses the relationship between spatial representation, spatial knowledge and technologies of representation and visualization of space and how developments and convergence of technologies may influence and shape the cultural practices and the use of contemporary urban space.

Keywords: Spatial Representation, Spatial Knowledge, Geographical Information.

1 Introduction

Representation is a key concept for understanding the social uses of technologies associated with manipulation, presentation, storage and maintenance of geographic information. Spatial representations have became a common characteristic present in mobile devices applications and social networks platforms. This paper refers to forms of space representation, essentially derived from cartographic techniques, that bases many web applications which aid to promote the development of new forms of social interaction and influence the perception and use of public spaces. It relates to previous and on going research on spatial representations and the role of geographic information for urban planning and management developed in LCAD (Laboratory of Applied Computer Graphic) at Faculty of Architecture, Federal University of Bahia, Brazil.

In the paper we address the relationship between spatial representation, knowledge and technologies of representation and visualization of space. We present examples to discuss how developments and convergence of technologies may influence and shape the cultural practices and the use of contemporary urban space.

B. Murgante et al. (Eds.): ICCSA 2013, Part III, LNCS 7973, pp. 524–537, 2013.

2   Spatial Representation and Knowledge

Maps were already used as a tool to assist investigations since the seventeenth century, in Italy, when the later called sanitary maps began to be used to relate disease and its environment [26]. Harvey [8] and Söderström [22] draw attention to the fact that the invention of perspective in the Renaissance and innovation in representations techniques allowed the capture of the entire space, through the recognition of specific geographic features represented in space until then impossible with cartographic techniques from the Middle Ages.

According to Parsons [15], traditional techniques for mapping are limited to expressing geography since these do not represent dynamic features, and employ fixed scales and only two dimensions to represent the three-dimensional reality. The first limitation can be observed in representation of geographic phenomena that do not have a constant shape such as estuarine areas, providing only a frozen in time image of reality. Moreover, the storage of geographic information in paper requires for a complete representation of the area, a great number of maps of different scales and with varying degrees of generalization as in a cartographic Atlas.

Before the emergence of Information and Communication Technologies – ICT – maps lend themselves more to the content presentation than to visualization as a dynamic, interactive and user-controlled process [15], since the manual techniques do not favor the interaction between researcher and data graphing tool. For Wood [26], the new era of visualization in scientific computing and Geographic Information Systems (GIS) has opened new possibilities for innovation in the development and use of cartography as an exploration tool.

Parsons [15] shows that, with visualization tools and multimedia interfaces based on virtual worlds, you can represent information about the space in a qualitative way as well as quantitative information traditionally represented by cartography. Detailed quantitative information express spatial relationships among people and among objects in an absolute and often numeric manner, while qualitative information provides a "sense of place", to be site-specific or regional. In the first case are topographical maps, which transmit information such as length, height, size, population density, etc. Qualitative information, in turn, could express for example, an architectural style of a building, climate, sounds, events, urban characteristics.

Qualitative and quantitative data about geographic information are related to spatial knowledge in different ways. Shum [21] refers to two types of spatial knowledge: "route knowledge" and "map knowledge". The first would depend on the context: it has an individual centric perspective and is capable of driving actions only within a limited range of situations because it relies on tracks in the path [21]. This type of knowledge does not allow the individual to refer to the whole environment since it does not show the spatial relationship of a broader point of view.

The "map knowledge", on the contrary, would operate at the global spatial relations level, it would provide a frame of reference centered in the world and would operate independently of the context [21]. This kind of knowledge usually comes from sources such as maps, and allows selection of alternate routes, generation of new routes, giving more flexibility to the thought, since it is not limited to a specific set of conditions [21]. Peuquet [18] nominates forms of perception of space as direct and indirect acquisition of knowledge, the latter in case of those mediated by various forms of representation (maps, images, documents).

For a better understanding of geographic space it is necessary a descriptive information of the site to facilitate the location in space associated with information about the whole space and relations between objects and space phenomena. Contextualized spatial reference may aid to maintain spatial references socially relevant at both levels of representation, location and context, so that locality can be imbued with meaning. A more conceptual nature than descriptive kind of representation would bring a new meaning to spaces and increases the cognitive map, which would consist by numerous representations of objects in geographical space [15], as well as their relationships, derived from individual experience.

Faced with characteristics of "route knowledge" and "map knowledge" presented by Shum [21], and direct and indirect knowledge acquisition as defined by Peuquet[18], one may expect that knowledge through maps have the power to make known the whole area, unlike qualitative understanding provided by a specific route knowledge.

Sampaio [19] brings two questions to be addressed by urban space representation: one concerns to the apprehension of the urban form that is not limited to the perception of the physical space, and the other about the scale of representation. Regarding the apprehension of space, his work [19] identifies three levels:

a. Level 1 would be the direct apprehension, level at which vision is the privileged sense;

b. Level 2 would be a level at which the direct apprehension is complemented by analysis tools and documents as photographs, maps, thematic mapping, physical models, and representations of space-form in the virtual sense. We complement Sampaio statement citing Geographic Information Systems, and hypermedia computer applications that enable integration of information from images, thematic mapping, three-dimensional virtual models, and remote sensing images, videos, sounds, comments;

c. Level 3 would be related to the most abstract representation field using mathematical simulation models to represent aspects of urban structure or basic quantitative relations among elements. Sampaio [19] mentions the work of Chadwick, McLougin and Echenique published around the 70s, which are complemented by the works developed on cellular models to simulate use and occupation of urban land through dynamic models [1][23].

The first apprehension level is formed by a direct experience with space: such level of spatial cognition is perceived by urban dwellers and shapes their "image of the city" [13]. At the second level with the adoption of GIS technologies and multimedia is possible to construct various representations for different purposes, from three-dimensional models to the representation of socioeconomic population using thematic mapping techniques and computer graphics. In the last level, representations tend to simplify the space represented in a very high level of abstraction such as cellular models cited above.

We often operate systems and models with little or no awareness of the concept of space upon which they are based. When we use GIS – Geographic Information Systems – we produce answers to questions we launched and soon we began to think of objects and phenomena in the real world as polygons, points, lines; sometimes we look at space through a grid whose cells carry attributes from physical and social world.

We are in each case respectively idealizing the space as an empty structure in which we placed discrete objects with well-defined borders or as a mosaic whose intrinsic qualities may be distributed as cells of a grid. Although we consider environment, events, phenomena and objects that populate space and that we are also interested in representing, we almost never are aware of the underlying approaches in our way of representing space and time [3].

Typically, GIS representations are developed from preconceived assumptions regarding homogeneity, uniformity and universality of the properties of its main components, including the space and spatial relationships, time and mathematical model that describes spatial phenomenons [16]. These rigid assumptions are inadequate for modeling dynamic processes, which are anchored in cartographic notions of absolute space mapping. The treatment of space as a space of places – a structural and static vision – or while space of flows – a functional and dynamic vision – reveals different viewpoints: a vision of absolute space, structured and measurable, and a vision of relative space, where objects relate to each other independently of typical spatial relationships [16].

The absolute and relative views of space-time are thus complemented by views of the space-time continuum in apparent opposition to the discrete space-time, whose focus is on objects. Perhaps we can say that such views are close. By adopting an absolute view of space-time we can treat it as an ongoing reality? And, instead, focus on the space-time continuum, we can adopt a discrete vision? The absolute vision seems to ignore human perception as something to be interpreted, in contrast to the relative view. Furthermore, the vision of the continuous space looks familiar to the absolute vision derived from Euclidian geometry, leads to treat space as a space of places. Since the topology is relative and discrete [18], it is the proper representation of space as a space of flows. The path took by technology to represent the space permitted the convergence of both visions: route knowledge X map knowledge and absolute view X relative view of space.

3   Spatial Representation and Technologies

The representation techniques followed the technological development. The first CAD tools (Computer Aided Design and Drawing) mimic the two-dimensional representations used since the Industrial Revolution. Technological development has allowed to build and generate three-dimensional representations of the objects derived by geometric models that can be seen in different views.

In parallel to the development of CAD systems, the emergence and development of GIS brought a step forward in representations of the world. The GIS tools have allowed that non-graphical attributes were linked to geometric representation. Geography is then represented by a database that contains the geometry of objects and its many possible attributes. GIS tools also introduced the idea of representing the environment through grids or matrices, representing not only objects in space, such as CAD based systems did, but the space itself, consistent with the possibility of remote sensors capture every part of the earth's surface [1].

Currently, these models can go further. A Building Information Model, or BIM can be defined as a three-dimensional digital model of building semantically rich and shared. BIM models can also be integrated into a Geographic Information System as argued in [10].

Urban models are idealized representations of the city that aim to describe the urban reality, predict or plan the future city. One of the oldest ways of classifying urban models was created by Ackoff (cited in [11]) and distinguishes three types of models: iconic, symbolic or analogical and was adopted by authors from different areas.

The iconic model is a simplified representation based on similarity to the object he wishes to represent, as architectural models that represent a building. The analog models represent the object in a different system, which reflects the represented object with similar properties. The symbolic systems represent the object using symbols as elements of a formalism that allows rationalization and calculation, as a system of equations that represent the population growth.

City models can be limited to the objects that represent them and their topographical relationships. They can also represent social interactions or economic and symbolic models that can be descriptive, predictive or prescriptive. Models of this type can be dynamic models or events simulation models.

In any case, the representation of the physical reality of architecture and cities – geometric representation, or topographical, as it is called by some authors – is now entirely possible. One result was the emergence of several "virtual cities" created using various techniques and even being published on the web through services like Google Earth, Virtual Earth, World Wind.

Digital representations based on the geometric description of the physical aspects of cities have a huge importance to a large number of activities in Planning, Engineering, Architecture and Urbanism, but fail to capture important aspects of everyday life and the emergence of new types of public space.

The traditional representations of social and demographic data are based on the aggregation of individuals or households in areas defined and uniformly represented (census tracts, for example) or physical address of individuals as in the existing commercial databases. While this form of data and information treatment maintain its relevance, contemporary digital social networks are beginning to play a structural role in society. This brings new challenges for the construction of spatial representations of society and its relations with implications in the areas related with geodemographic data [20] between these, urban planning. After all, for urban planning is necessary to manage and understand the "immobile" – structure and physical infrastructure – and the "flows" – mobility, transportation, and now also add the digital content streams [17].

The kinds of representation of urban information that today are still based on simple mapping techniques can now submit data online in real time: using mash-up of Google Maps, for example, and can be seen not only on the PC screen, how technology allowed some years ago, but in the actual physical place represented digitally on a portable device (Augmented Reality). The representation can be dynamic, showing the evolution of the processes, for instance, the mobility of people, traffic situation, public transport information or the flow in social networks, and can be accessed during the journey and even urban daily updated and modified by citizen thus acts as a consumer and producer of geographic information.

The advent of digital navigation permits to distinguish between a mimetic and a navigational use of maps [14]. In the latter case, beyond the similarity of virtual image, what it is new it is the knowledge of the connection between messages and signals that precede and succeed them. The metaphor it is the network itself. In the digital representation as in the real city the map consists of tracks that connect several units together to meet the needs of those traveling through the city.

The predictions that pointed to a sense of irrelevance of geographic location, with the geographic space replaced by cyberspace, and the dichotomy between the virtual/digital x real/physical were overcome by overlap or convergence between physical and digital environments. Nowadays the emergence of what some authors call "urban computing" or "everyware" [7][12] is the ability to process data and information embedded in urban structure. In this case, unlike the concept of cyberspace – a virtual communication space – the geographical location plays a central role since the interactions are based on "where" you are, named by Gordon [5] as "network locality".

The advent of so-called web 2.0 enabled platforms developed with technologies increasingly ubiquitous, easy to use and economical to be used as auxiliary tools for planning, allowing the user to view and interact with contexts, with data, and incorporates the temporal dimension in the analysis of space. These resources can be used in several ways, such as scenarios evaluation, discussion of planning alternatives, sharing ideas and to establish social interactions, among others.

4   City and Digital Spaces

According to Di Felice [4] the digital multiplication of spaces allowed plural and hybrid experiences of spatiality that surpass the architectural dimensions of urban space and opens the possibility to experience an habitat nomad and plural, partly electronic and partly architectural-communicative-immaterial.

While the city was a written and designed space, the metropolis while digital spatiality, is a dynamic and multiform space which assists the individual, not as an actor-agent, but as dynamic experimenter [4]. However, the metropolis continued to be thought as architectural space, where the landscape remains external to the observer, which appropriates visually of the territory.

The introduction of digital interactive technologies and contemporary forms of network communication changed not only social relations but overcome the concept of territory [4]. The subject and the landscape begin to communicate and interact creatively through digital mediation. As a result we would have an extended territory, composed of spaces and information, ever-changing topographies that allows creative interactions and novel ways of living, possible only through technical interaction [4].

Now we deal with a form of post-urban dwelling in which the landscape is no longer given or not interpreted but achieved through a technological extension-communication, layered, porous, immediately manipulable, which can be instantaneously stretched to move from an interface, a software or system information to another.

There are different definitions of virtual landscapes, i.e. informational territories, according to Di Felice [4], each expressing a theoretical conception of the virtual. The first describes the virtual as a dwelling and as a non-fictional place. According to this view, the virtual creates another dimension, evoking an imaginary and alienated dwelling. This concept finds its origin in Platonic contrast between the real world and the fake world, between the real and the copy. The virtuality and virtual cities, in this view, would be simulacra spaces, nonexistent territories, called "mirror worlds" by Hudson-Smith [9]. A false life and a landscape, a second city, contrary to the real world and therefore lower. An example can be found in cities in Second Life. Amsterdam Digital, on the other hand, delivers urban digitized spaces or information spaces "open to all" without queues or traffic or pollution. The virtual space is not opposed to real world while virtual is real [4]. Rather than create a fictitious geography, this form of scanning appears to broaden and extend the virtual urban space.

Other kind of representation is based on GPS, wireless networks and so-called LBS (Location Based Services). It provides a continuous interaction between information/interfaces and territory. According to Di Felice in [4], the combination between space and information would create another way of living. The simultaneity of information that guide the vehicle or passerby makes the individual as occupant of a multiple space, real and virtual at the same time. This new type of interaction require overcome the dualism between real and virtual, and the very concept of simulacrum. Breaking the relation original/copy, the images do not maintain a relationship of identity with the original [4].

The key conditions for both, networking of space and spatiality of network involve aspects as: the everyday superimposition (and overlapping) of real and virtual space, development of a mobile sense of place, the rise of the network as socio-spatial model, and the growing use of mapping and tracking technologies [24]. Ubiquitous computing and geospatial web are not emerging technologies anymore, but real-time applications that entwine networks and the local, creating possibilities and new questions. This kind of application includes, not only, but mainly, Yahoo Maps, Google Maps, Google Earth, OpenStreet Maps and presents the huge evolution of GIS, making all available to end user. This feature permits real-time information, interactions, interventions and collaborations from users, which influence the way people read and use urban space, but also, it reshapes the way people interact with each other in urban places.

Gordon and Silva [6] used the example of the famous map of John Snow to state that the idea that maps can be produced not only to bring information about the space but to spatialize information became the basis of the activity of spatial analysis. The difference between the two approaches – "information about space" or "information in space" – is that the latter uses geography to give meaning to data and the first uses data to make sense of geography. This use of geography and its representation as a way to assign meaning to the data is behind many contemporary web applications, mobile or not. They address important issues about the spatial references's context and aggregate new social meanings at the local.

5   Digital Urban Spaces: Examples

Several cases and examples of digital urban networks and applications can be observed producing new interactions between citizens and within the city. Such initiatives may provoke a kind of urban *crowdsourcing*, where perceptions, usages, urban paths, causes, events, traditions and problems are shared among people in many-to-many manner and may help to build collective events, meetings and solutions. Or they simply modify the way people commute and read the city.

Example 1: Two friends coming back home at a very cold night try to access buses timetable in Paris to verify how long would they got exposed to french winter. Using a GPS into his mobile, one of the guys visits an Android application of local transport department to check the next bus and the closest station. Using a map combined with the GPS, the application shows the exact location of the guys and blinks to them informing a bus will pass in 1 (one) minute in a bus stop about 500 meters distant. The further bus would only arrive after fifteen minutes awaiting in a 0° C temperature. They realized that the best decision under such a cold night would be run until the bus stop and take the bus that could already be seen coming trough the darkness.

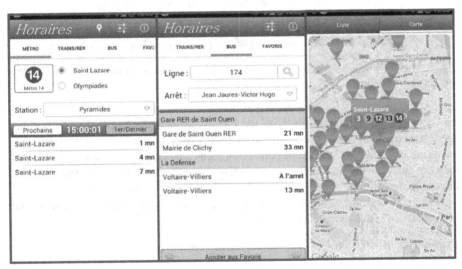

Fig. 1. Three screens of a mobile application that informs about buses, localization and timetable

Example 2: Some days before, the same friends meet others in a pub. After a couple of hours talking, three of the friends decide to get back to the 13th éme using *Velib*, the public bike-sharing system of Paris (Figure 2). They needed to find one of hundreds of stations with three bicycles available. Once more, using the GPS in mobile and the *Velib* system of localization, they could verify into an online map the stations next to them and how many bicycles were available in each of the stations. This application guided them to the correct place according to what they needed at that moment and localization.

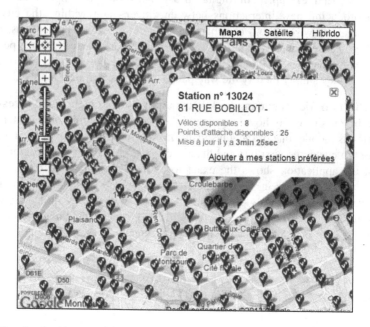

Fig. 2. Map for for bicycle sharing system in Paris with information about bikes availability and places for parking

Another kind of relationship with the city is developed when these applications aggregate free information coming from their user's network describing different locations and usages. The level of perception gains a new dimension and may changes the way citizens uses and interacts with urban space, or even step (?) in the city with physical interventions and mediations. Several sites and Internet applications permit such searches and interactions (with the place itself and with other participants of the networks) as PortoAlegre.cc and BiktIT.org.

Example 3: BikeIT Project is a collaborative project designed to stimulate a good relationship between bicycle users and establishments of their cities and to promote the acceptance of bicycles. The site's user may spread several types of establishments (culture, business, leisure, services) or restaurants that are friendly or hospitable to bicycle users in urban space. The project permits virtual postings with digital georeferenced maps where the user can indicate the *bikefriendly* places. The project certificates the indications trough votes from other users to confirm the published information, granting a seal of *BikeFriendly* to the establishment. This way, this project intends to promote and legitimate the bicycle as a transportation, modifying the urban public and private space usage. This initiative may encourage other establishments to offer an infrastructure to receive bicycles or simply accept them, incrementing the change in the everyday use of the city. A relevant aspect in this project resides in the way used to publish such places with desired characteristics in favor of bicycle users, trough interactions that happen from a users network to indicate them but also to homologate the indications of the *bikefriendly* places using a peer-review method.

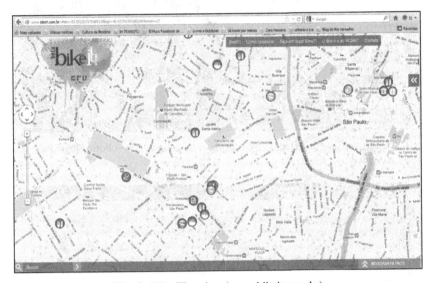

Fig. 3. Bike IT project (www.bikeit.com.br)

Example 4: PortoAlegre.CC – Creative Commons Project. This project was launched in 2011 and presents a digital cartographic application open to all the subscribed in the site allowing each user to create a *cause* with a georeference. The concept of cause in this project encompasses: reports about positive and negative events such as thefts, problems in the streets as holes and leaks, public instruments out of operation, notification and/or organization of sportive and leisure meetings or even for recovery of non-assisted areas by public administration. With all these interactions, it was possible, in a participative and collaborative way, developing a georeferenced mural of problems found into the city of Porto Alegre. This digital platform can guide the public administration to map and fix the reported problems, or plan changes as solutions to the causes. On the other side, citizens without real life relations may create virtual connections with interactions and exchanges of information, allowing these to organize and know about collective events and activities that improve the use and constitution of urban spaces. This project is partially sponsored by the city hall of Porto Alegre in a partnership with a local university and other organizations which, together, execute in parallel several workshops for promoting digital tools utilization.

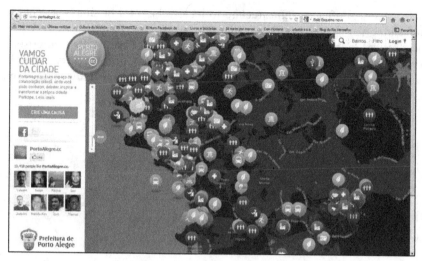

Fig. 4. Cartographic interface for user interactions in PortoAlegre.cc project (www.portoalegre.cc)

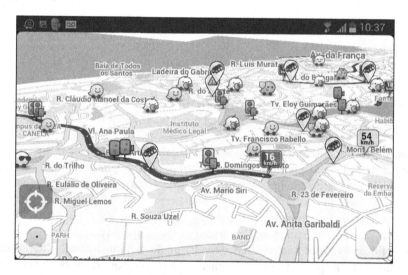

Fig. 5. Interface for Waze application (www.waze.com)

Example 5: Waze.com. Waze is a free GPS navigation software developed by Waze Mobile for GPS enabled smartphones. Waze is available for download and use anywhere in the world. In addition to some features already present in others GPS navigation software such as turn-by-turn voice navigation or specific location alerts, Waze sends anonymous information such the user speed and location to improve the service as a whole. As the web site states "traffic is more than just red lines on the map" (www.waze.com). The application could be characterized as a social navigation software and a crowdsourcing community. The interface use a dynamic map that could be saw in 3D or 2D and some gaming convention to display information. According to Wired magazine Waze is among the big examples of how social media can produce immediate and positive change in our daily lives. "Maps change really slowly. Some providers update every quarter, others every year," CEO Noam Bardin told Wired. "But the trend with mobile is moving closer and closer to real time. And the only way to enact these real-time changes is by engaging people" [25].

6 Final Considerations

The various stages of technologies development are reflected in different manners on urban space and on the composition of cities. The advent of telephone (and fax) reformatted and reconfigured the city, allowing, for example, office and factory to separate itself. A television event is simultaneously in two places: at the source of transmission and all local TV signal receivers [24].

Di Felice highlights that for the first time in history, there is not barrier between sender and receiver, which it was a prerequisite for the development of major contemporary communication practices [4]. Varnelis & Friedberg [24] see cities as the new access interfaces, namely the town itself has the status of great communication device. Technology changes the way someone navigates (browse) through the city, either walking, either by bike, by car, creating a further potential to exploit it. The urban grid of streets, roads and highways was already a network and cars using other forms of communication as radio (such as mobile receiver) and, most recently, GPS, opened new potentials for city reading and use.

The examples in the previous section of this paper are veridical and demonstrate how to access the city trough new interfaces based in geographical information. All of them may be classified in the second level of the scale presented in section 2, where apprehension and usage of urban space is complemented by hypermedia computing applications. Geolocated information about services are made available through portable devices enabling decisions and daily actions based on the current location of the citizen. The city becomes a huge and contemporary interface where those without advanced mobile devices are excluded. The *flaneures* gain a new window to get around the city and experiment it (sometimes, a credit card is very welcome to obtain access to some services). The third and forth examples shows social interactions related to urban space mediated by cartographic representations with the information being produced and shared by network members.

In the case of the navigational use, representation is dynamically modified as the current traveler position changes and it can incorporate information from other travelers beyond the information produced by the user as in the last example. We can say that this kind of representation deliver to the user at the same time "route knowledge" and "map knowledge" presented in section 2.

While interactions between network elements are strengthened expressing themselves to organize urban everyday life, making it more sociable, moreover, it becomes mandatory mobile devices to access interfaces distributed in the urban structure. If we can say that digital spaces are no longer virtual, a kind of new form of exclusion may be underway: access to some facilities of the contemporary city it is only possible to those who have the technological means provided with the characteristics of connectivity and ubiquity that has changed the way people communicate over the last decade. Current social practices present in digital media culture are more self-reflective and participatory as stated in Benkler [2]. For him, the growing capabilities of individuals are at the center of the forces that drive what he calls the networked information economy. The author suggests that we become more able to fulfill some of the functions and creating emotional context traditionally associated with the community and with its network of social ties.

The issues related to representation and spatial knowledge could be summarized in a cycle that leads to the representation of knowledge with the use of technologies that, in turn, enrich the context and local knowledges. Technological convergence and ubiquity of connections would promote particular sociocultural practices that give new meaning to information, with the increasing production of "information in space" mentioned in section 4.

Social interactions of everyday life are in fact permeated by technology. Examples of current uses of spatial representation presented indicate a possible change in the perception of context, through social sharing and a more dynamic information integration. This may enable the convergence of different points of view, by exchanging experience, but also to reveal the various alternative ways to enjoy the territory.

Gordon suggests that network technologies and corresponding practices significantly changes the nature of local situations, not only socially, because of the way we share information geographically located, as phenomenologically, the way we experience what's next. Ongoing research at laboratory LCAD goes in that direction: the use of mobile digital technologies and its influence at the urban environment.

References

1. Batty, M.: Model cities (2007), http://eprints.ucl.ac.uk/3378/
2. Benkler, Y.: The Wealth of Networks: How Social Production Transforms Markets and Freedom. Yale University Press, New Haven (2006)
3. Câmara, G., Monteiro, A.M., Medeiros, J.S.: Representações computacionais do espaço: fundamentos epistemológicos da ciência da geoinformação (2000),
   http://inpe.academia.edu/GilbertoCamara/Papers/504464/
   Representacoes_computacionais_do_espaco_fundamentos_epistemo
   logicos_da_ciencia_da_geoinformacao
4. Di Felice, M.: Paisagens pós-urbanas: o fim da experiência urbana e as formas comunicativas do habitar. Annablume, São Paulo (2009)
5. Gordon, E.: Towards a theory of network locality. First Monday 13(10) (October 6, 2008), http://www.firstmonday.org/htbin/cgiwrap/bin/ojs/index.php/ fm/article/view/2157/2035 (access September 9, 2012)
6. Gordon, E., de Souza e Silva, A.: Net locality: why location matters in a networked world. Wiley-Blackwell, Chichester (2011)
7. Greenfield, A.: Everyware the dawning age of ubiquitous computing. New Riders, Pearson Education, Berkeley CA (2006)
8. Harvey, D.: Condição pós-moderna: uma pesquisa sobre as origens da mudança cultural. Ed. Loyola, São Paulo (1993)

9. Hudson-Smith, A., et al.: The Neogeography of Virtual Cities: Digital Mirrors into a Recursive World. In: Foth, M. (ed.) Handbook of Research on Urban Informatics: the Practice and Promise of the Real-time City. Information Science Reference / IGI Global, London (2009)

10. Isikdag, I., et al.: An opportunity analysis on the future role of BIMs in urban data management. In: Urban and Regional Data Management – UDMS Annual 2011, pp. 25–36. Taylor & Francis, London (2011)

11. Jackson, M.: Some Notes on Models and Modelling. In: Borgida, A.T., Chaudhri, V.K., Giorgini, P., Yu, E.S. (eds.) Conceptual Modeling: Foundations and Applications. LNCS, vol. 5600, pp. 68–81. Springer, Heidelberg (2009),
http://mcs.open.ac.uk/mj665/ModelNotes.pdf

12. Kitchin, R., Dodge, M.: Code space software and everyday life. MIT Press, Cambridge (2011)

13. Lynch, K.: A Imagem da Cidade. Martins Fontes, São Paulo (1997)

14. November, V., Camacho-Hübner, E., Latour, B.: Entering a risky territory: space in the age of digital navigation. Environment and Planning D: Society and Space 28(4), 581–599 (2010)

15. Parsons, E.: GIS visualization tool for qualitative spatial information. In: Hearnshaw, H.M., Unwin, D.J. (eds.) Visualization in Geographical Information Systems, pp. 201–210. John Wiley e Sons, Chichester (1994)

16. Pedrosa, B.: Ambiente computacional para modelagem espacial dinâmica. Universidade de São Paulo, Tese de Doutorado (2003)

17. Pereira, G.C., Rocha, M.C.F.: Spatial representations and urban planning. In: Planning Support Tools: Policy Analysis, Implementation and Evaluation, Roma, pp. 611–623 (2012)

18. Peuquet, D.J.: Representations of space and time. Guilford Press, New York (2002)

19. Sampaio, A.H.L.: Formas Urbanas: Cidade Real & Cidade Ideal – contribuição ao estudo urbanístico de Salvador. Quarteto, Salvador (1999)

20. Singleton, A., Longley, P.: Geodemographics, Visualization, and Social Networks in Applied Geography. Applied Geography 29(3), 289–298 (2009)

21. Shum, S.: Real and Virtual Spaces: Mapping from Spatial Cognition to Hypertext. Hypermedia 2, 133–158 (1990)

22. Söderström, O.: Paper Cities: Visual Thinking in Urban Planning. Ecumene London 3(3), 249–281 (1996)

23. Sullivan, D., Torrens, P.M.: Cellular Models of Urban Systems. In: Bandini, S., Worsch, T. (eds.) Theoretical and Practical Issues in Cellular Automata. Springer, London (2001)

24. Varnelis, K., Friedberg, A.: Place: the networking of public space. In: Networke Publics. MIT Press, Cambridge (2008)

25. Wired.com. Waze Proves the Power of Social Media With Real-Time Map Updates | Gadget Lab | (2013), http://www.wired.com/gadgetlab/2013/02/waze-real-time-map-updates/ (acess february 28, 2013)

26. Wood, M.: Visualization in Historical Context. In: MacEachren, A.M., Taylor, D.R.F. (eds.) Visualization in Modern Cartography. 1a. ed., vol. 2, pp. 13–25. Pergamon (Modern cartography), Grã-Bretanha (1994)

Web 3D Service Implementation

Nuno Oliveira[1] and Jorge Gustavo Rocha[2]

[1] PT Inovação SA, Aveiro, Portugal
nuno-miguel-oliveira@ptinovacao.pt
[2] Universidade do Minho, Braga, Portugal
jgr@di.uminho.pt

Abstract. In this paper we describe an open source implementing of the Web 3D Service (W3DS) based on the OGC's draft proposal. The implementation was developed on top of the open source java-based map server GeoServer, as a community module. With an open source implementation available, test beds can be promoted to better know the strengths and limitations of the current proposal. Without practical interoperability assessments demonstrated, W3DS barely become a 3D standard in urban management applications.

1 Introduction

Massive real world 3D data, from landscape models to detailed indoor textured models, are becoming available. To take advantage of this data, scientific contributions are necessary to be incorporated at all levels of the existing Geographical Information Systems (GIS) software stack.

Technology is enhanced every day, but new paradigms supported by technology enhancements only appear from time to time. For example, the client side AJAX support, enabling asynchronous web applications, was a key feature in the explosion of web mapping applications on the web. Software libraries, like OpenLayers or equivalent, taking advantage of AJAX support with additional supporting logic, raised the base level from which all new web mapping applications were built.

A recent technology development initiated by the Mozilla Foundation might be the key to massive visualization of 3D data: WebGL. Its version 1.0 was released in March 2011. By mixing JavaScript code and shaders (GPU code), sophisticated image processing effects and dynamic physics can be used to provide amazing graphics and sounds directly in the browser. While HTML5 is still nothing more than the next version of HTML, with no real technological breakthrough, the WebGL is the 3D key changer. WebGL enables direct access to the GPU within the browser.

To take advantage of this support on the client side, we need to consider upstream 3D map services in the GIS software stack. These 3D services will mediate data exchange between the large 3D GIS databases and different client types. Two service oriented approaches to 3D portrayal are being considered by the Open Geospatial Consortium (OGC). The Web View Service (WVS) supports

B. Murgante et al. (Eds.): ICCSA 2013, Part III, LNCS 7973, pp. 538–549, 2013.
© Springer-Verlag Berlin Heidelberg 2013

an image based approach. The WVS client requests 3D rendered images from the WVS server. In the other approach, followed by the W3DS, the 3D geometries and textures are rendered on the client side. The W3DS server hosts and manages all requests, but does not render any images. These two approaches are suitable for different types of clients. Thin clients would prefer already prepared 3D imagery, while clients with more computational resources can provide more flexible and powerful visualizations.

In this paper we focus on the development and discussion of the W3DS proposed standard [8]. We contributed with an open source implementation that can be tested in several scenarios, to provide valuable feedback about its interoperability.

We start by reviewing 3D usage in the GIS realm. Afterwards we describe the W3DS specification, introducing its operations. The implementation is described. The W3DS architecture is shown, and the major modules are briefly described. Next, we describe a very simple scenario that can be reproduced to get W3DS up and running. Finally, some conclusions are presented followed by a brief outlook.

2   State of the Art

In 1993 [4] presents a 3D GIS that use CAD models to represent the 3D entities and the DTM. The models had three different kinds of approximation, the first two are used to index and accelerate the rendering process and the last one is a detailed representation. In 1997 [16] describes one of the first 3D Web GIS. The HTML pages were produced dynamically and the 3D data was directly retrieved from the database using SQL. The 3D scenes produced are encoded using the Virtual Reality Markup Language (VRML) which can be interpreted by the browser VRML plug-in. The reference [5] gathers some interesting publications about 3D GIS that can be seen as the state of the art of the 1990s.

Most of the recent works focus on city administration, which have become one of the top use cases for 3D GIS [7, 15, 11, 10]. OSM-3D is one example of 3D GIS projects. Its main objective is to provide a 3D view of OpenStreetMap data integrated with the elevation data of the Shuttle Radar Topography Mission (SRTM). Its implementation is made on top of OGC standards, including W3DS for 3D visualization. The reference [6] makes an overview of the current state of OSM-3D in Germany and provides a good discussion about the generation of 3D building models.

2.1   Web 3D Evolution

VRML was the first web based 3D format, released in 1995 and ISO certified in 1997. The main goal of VRML was to give a way to represent 3D virtual worlds that can be integrated on web pages. A VRML scene is composed of geometric primitives like points, segments and polygons. The scene may also include multimedia content like hyperlinks, images, sounds and videos. The aspect can be customized using lights effects and defining materials properties. VRML scenes can be explored in desktop software or in web browsers, using some compatible plug-in. The reference [13] makes a good overview of the format.

In 2001, the Web3D Consortium, which has become the main supporter of VRML, releases the Extensible 3D (X3D) format, an XML encoding version of VRML [3]. The XML based encoding of X3D makes it more suitable for native integration in HTML pages. X3D also adds new features like the support of shaders, better event handling, new geometric primitives and others short cuts for 3D rendering. X3D brings up the concept of working groups, their job is to extend X3D to custom support of certain areas, like medicine, GIS and Computer Aided Design (CAD). The GIS working group have provided X3D with the capability to natively support the needs of GIS applications. The main features are the full support of georeferenced coordinates and custom events for geographical scenes.

Even if at this time, VRML and X3D stay the most used web based 3D formats, their use has decreased significantly compared to ten years ago. When VRML was released everyone has tried to make use of it, quickly we have seen the appearance of 3D web content everywhere. Some companies have invested large quantities of money to shift their websites to 3D. The same thing happened with GIS applications. Companies and governments have even start buying 3D georeferenced data. But the technology was not already available. Computers with the capability of rendering complex 3D scenes at acceptable frames rates were not common and those that existed were too expensive. With the poor quality offered by 3D web and once the new sensation of a third dimension had worn off, people started looking again at a 2D web.

Around 2009 WebGL appears and the doors to a 3D web are definitively opened [12]. WebGL specification is based on OpenGL ES 2. Even if it is only a draft, it has already been implemented by the majors web browsers and plug-ins have been provided for those that don't natively support WebGL [9]. WebGL gives us the possibility to use 3D hardware acceleration from the JavaScript of web pages, like OpenGL does for desktop 3D applications. Its integration with HTML5 gives the possibility to directly embed complex interactive 3D scenes on web pages.

Recently, a new web based 3D format is being adopted: XML3D. This is the only major web 3D format that is not supported by the Web3D Consortium, however it is a candidate to become a WC3 standard. Unlike the others formats, the main goal of XML3D is to be an extension to HTML5 specification [14]. The authors claim that even if using X3D or VRML we can integrate 3D content in a web page, the separation of the two concepts is well defined. On the other side, XML3D definition is based on other successful standards of W3C like HTML, DOM and CSS. All the interactions with the 3D scenes are made using the web standard path, i.e. using DOM events and JavaScript. XML3D is independent of the 3D rendering API used, in [14] the authors use a modified version of Chromium Browser that uses OpenGL, however the top rendering technology for XML3D is still WebGL.

3   Web 3D Service

The W3DS is a portrayal service proposal for three-dimensional spatial data. The first proposal was presented back in 2005 by Kolbe and Quadt. Since then, some improvements were integrated. In 2009, version 0.4 was accepted as a public discussion paper by the OGC. Afterwards, a version 0.4.1 was rewritten. This is the last version available, and it dates from 2010.

The W3DS service delivers scenes, which are composed of display elements representing real world features. It does not provide the raw spatial data with attributes, like the Web Feature Service (WFS) service does. It only provides a view over the data, according to, for example, the level of detail.

It does not provide rendered images, like Web Map Service (WMS) does. It filters the data to be delivered according to several parameters, like a bounding box, but the result will be a graph of nodes with properties attached to each node, like shapes, materials and geometric transformations.

This graph of display elements must be handled by the client. W3DS clients must implement the necessary logic to take advantage of the W3DS operations. Typically, clients will continuously request scenes from the service, tying to minimize the data delivered to the client, while providing the best user experience.

3.1   W3DS Operations

Like other OGC services, information about the service, the supported operations, available layers and their properties, can be retrieved using the GetCapabilities operation. Two additional operations that return information about features and their attributes are provided: GetFeatureInfo and GetLayerInfo.

Two operations are provided to return 3D data: GetScene and GetTile. These two operations differ essentially in how the features are selected. GetScene allows the definition of an arbitrary rectangular box to spatially filter the features to compose the scene returned to the client. GetTile returns a scene on-the-fly formed by features within a specific delimited cell, within a well-defined grid.

All five proposed operations, tagged as mandatory or optional are listed in Tab. 1.

Interactive scenes with terrain and relevant 3D geographic features will result from multiple GetScene and GetTile requests to the service. These might be mostly called operations. For these two operations, we show how to call them with the most common parameters.

Table 1. W3DS operations

Operation	Use
GetCapabilities	mandatory
GetScene	mandatory
GetFeatureInfo	optional
GetLayerInfo	optional
GetTile	optional

Table 2. GetScene parameters

Operation	Definition	Use
crs	CRS of the returned scene	Mandatory
boundingBox	Bounding rectangle surrounding selected dataset, in available CRS	Mandatory
minHeight	Vertical lower limit for boundingBox selection criteria	Optional
maxHeight	Vertical upper limit for boundingBox selection criteria	Optional
spatialSelection	Indicates method of selecting objects with boundingBox	Optional
format	Format encoding of the scene	Optional
layers	List of layers to retrieve the data from	Optional
styles	List of server styles to be applied to the layers	Optional
lods	List of LODs requested for the layer	Optional
lodSelection	Indicates method for selecting LODs	Optional
time	Date and time	Optional
offset	Offset vector which shall be applied to the scene	Optional
exceptions	Format of exceptions	Optional
background	Identifier of the background to be used	Optional
light	Add light source	Optional
viewpoints	Add viewpoints to choose from	Optional

GetScene. The GetScene operation composed a 3D scene from the available data, according to several parameters. This is the most typical operation of a W3DS service. Table 2 describes GetScene parameters (version, service and request are omitted for simplicity). This extensive list of parameters gives the necessary flexibility to request the desired scene.

In practice, not all parameters are necessary. Clients must be aware of the adequated parameters, prior to any GetScene request. Information about the layer properties and its LOD settings are stated on the GetCapabilities.

GetTile. The GetTile parameters are listed in Tab. 3. These can be submitted either encoded as KVP in a GET request, or as a XML formatted document, in a POST request.

Table 3. GetTile parameters

Operation	Definition	Use
crs	Coordinate Reference System of the returned tile	Mandatory
layer	Identifier of the layer	Mandatory
format	Tile encoding format	Mandatory
tileLevel	Level of requested tile	Mandatory
tileRow	Row index of requested tile	Mandatory
tileCol	Column index of requested tile	Mandatory
style	Identifiers of server styles to be applied	Optional
exceptions	Format of exceptions	Optional

Common GetTile requests, called with key and values pairs, have the following syntax:

```
http://localhost:9090/geoserver/w3ds?
    version=0.4&service=w3ds&request=GetTile&
    CRS=EPSG:27492&
    FORMAT=model/x3d+xml&
    LAYER=guimaraes&
    TILELEVEL=1&TILEROW=5&TILECOL=7
```

Tile Metadata. Clients must know basic service and layer metadata, prior to any request. Tiles can be requested only from layers tagged in the GetCapabilities document with the key: `<w3ds:Tiled>true</w3ds:Tiled>`. They must also have a `<TileSet>` definition associated, as illustrated in the following `<TileSet>` definition.

```
<w3ds:TileSet>
    <ows:Identifier>guimaraes</ows:Identifier>
    <w3ds:CRS>EPSG:27492</w3ds:CRS>
    <w3ds:TileSizes>4000 2000 1000 500 250</w3ds:TileSizes>
    <w3ds:LowerCorner>-17096.156 193503.057</w3ds:LowerCorner>
</w3ds:TileSet>
```

Basically, the `<TileSet>` provides the lower left corner of the grid, and all available tile sizes. Tile sizes are defined as an ordered list decreasing by tile size. The number of levels available is the length of the list. Levels are numbered from 0, starting at the largest tile size. Obviously, tiles are considered of equal size in both axes, accordingly to the draft specification version 0.4.0. The version 0.4.1 introduced the possibility of supporting non-rectangular tiles.

4   Implementation

The implementation of the W3DS service should follow the specification, providing the operations already presented.

There are several OGC compliant open source implementations of web map services, like GeoServer [1], MapServer or Deegree. W3DS ideally should be implemented as a component on top of each of them to become widely available. It should be implemented as an additional component, taking advantage of existing code of these servers to manage the request pool, parsing the requests, handling formats, etc.

We started by developing the W3DS component on top of GeoServer. It has detailed technical documentation, has clear policies regarding new contributions and uses recent development tools. It is written in Java and uses sophisticated technologies making it easier in terms of flexibility, extensibility and maintainability. These three design goals are particularly important in projects like Geoserver, since it implements several different web services and it is used by a large community, there are permanent change requests, requests for new features and new formats, bugs to be fixed, etc. It also has to comply with different versions of the same web service, for example, WMS (1.1.1 and 1.3), WFS (1.0 and 1.1), WCS (1.0 and 1.1).

4.1   Architecture

The GeoServer results from a large open source software stack, taking advantage of other modules, like GeoTools, Maven, Spring Framework, Apache Wicket, and others. The GeoServer itself is made of three types of modules: core modules, extensions and community modules. The core modules provide entry points that are used by extensions and community modules to add extra capabilities to GeoServer. The provided entry points are made available using the proprieties of Java language itself and using some major Java frameworks like Spring and Apache Wicket for example.

The different types of services in GeoServer include WFS, WMS, and WCS, commonly referred to as OWS services. The virtual service OWS can be instantiated to implement any OWS-like service. The W3DS is an instantiation of the OWS service.

Figure 1 provides an overview of the W3DS architecture, in the context of the GeoServer implementation. When a request is sent to GeoServer the responsibility of handling it is delegated to the dispatcher. From the request, the dispatcher identifies the service and the chosen operation, which can be SERVICE=w3ds&REQUEST=GetScene.

The dispatcher then calls the key-value pairs (KVP) parser of the selected operation (ie GetSceneKVPReader). The remaining parameters of the request are decoded and an object that contains all the necessary information to execute the selected operation is created (i.e. GetSceneRequest).

The execution phase begins, gathering all the necessary data to build the response. The response is encoded according to the requested format.

4.2   W3DS Components

The W3DS implementation is composed of five major components: types, styles, responses, service and web.

Types. The types component provides the necessary objects to represent the W3DS service domain model.

Styles. Behind the scenes GeoServer uses GeoTools styling support to handle Styled Layer Descriptors (SLD), but GeoServer doesn't provide any entry point to extend styling. The styles component contains our implementation of 3D SLD support, which extends the native SLD with 3D specific issues.

Responses. GeoServer natively give the possibility of registering a response format producer for each combination of service, request and format. The responses component contains the implementation of our supported response formats.

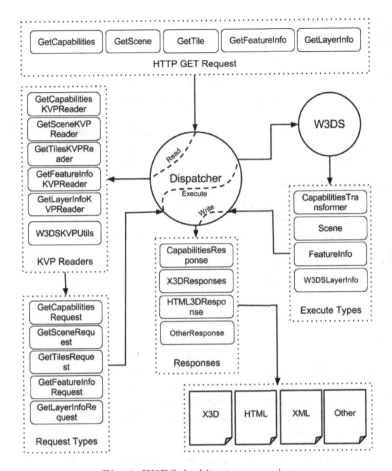

Fig. 1. W3DS Architecture overview

Service. New services need to implement some interfaces and register themselves in the application context to be recognized by GeoServer. The service
component contains that implementation.

Web. The web interface of GeoServer is built on top of Apache Wicket. The web
component extends the GeoServer web interface to include W3DS specific
controls, taking advantage of the existing extension points.

4.3   W3DS Output Formats

The W3DS produces 3D scenes to be consumed by 3D clients. Three formats
were considered: X3D, X3DOM and KML. Currently, only X3D and KML are
produced, by the corresponding packages x3d and kml. The X3D format produces an X3D document or alternatively an HTML 5 document with the X3D
embedded on HTML5. To select the desired output format, clients must use the
following format parameter:

- `application/vnd.google-earth.kml`
- `model/x3d+xml`
- `application/x3dom`

The x3d producer was entirely developed from scratch, in the absence of a proper X3D Java library. The kml producer is built on top of the Java API for KML (JAK) library.

The integration of X3D in the HTML5 is provided by the open source X3DOM framework developed by the Fraunhofer Institute [2]. The HTML5 container will return a minimal HTML document with a header that includes:

```
<linkrel="stylesheet"
 href="http://www.x3dom.org/x3dom/release/x3dom.css"/>
<script type="text/javascript"
 src="http://www.x3dom.org/x3dom/release/x3dom.js">
```

The HTML5 encoding comes in very handy for previewing data directly in the GeoServer administration interface, as we will show.

5   W3DS in Action

To get W3DS up and running, two steps are required.

First, the GeoServer needs to be compiled with the W3DS module. Since it is a community module, it is not included in the GeoServer binary by default. To compile GeoServer with any community module we need to add it to the build explicitly via profiles (-P). Using Maven, we can download the entire project sources and build it with:

```
git clone git://github.com/geoserver/geoserver.git geoserver
cd geoserver/src
mvn clean install -DskipTests=true -P w3ds
```

The previous Maven command generates a `geoserver.war` file.

It can be deployed using Jetty, Tomcat or another application server. By default, the service will be listened on port 8080. The administrative interface can be reached at `http://localhost:8080/geoserver/`.

After building and starting GeoServer, a new layer and the corresponding 3D style must be added to the configuration to take advantage of the W3DS module. This can be done through the GeoServer web administration interface.

To show an example in action, we describe a very simple scenario. A PostGIS table is required to describe the features and its location. Let's create a simple table `interest_points`, and add some points to it:

```
CREATE TABLE public.interest_points
(
   fid integer NOT NULL,
   wkb_geometry geometry,
```

```
description character varying,
CONSTRAINT interest_points_pk PRIMARY KEY (fid)
);
insert into interest_points (wkb_geometry, description)
values (st_geometryfromtext('POINT(-16229 199085 140)'),
  'some street furniture');
```

Besides data, we also need a, SLD style. In the following style we define that

```
<?xml version="1.0" encoding="ISO-8859-1"?>
<StyledLayerDescriptor>
  <NamedLayer>
    <Name>Point Cone</Name>
    <UserStyle>
      <FeatureTypeStyle>
        <Rule>
          <PointSymbolizer>
            <Graphic model="true">
      <href>http://localhost:8080/models/cube.x3d</href>
            </Graphic>
          </PointSymbolizer>
        </Rule>
      </FeatureTypeStyle>
    </UserStyle>
  </NamedLayer>
</StyledLayerDescriptor>
```

The x3d model referenced in the SLD document, cube.x3d, can be as simples as the model showed in the following listing.

```
<?xml version="1.0" encoding="utf-8"?>
<X3D xmlns:xsd="http://www.w3.org/2001/XMLSchema-instance"
  version="3.0" profile="Immersive"
  xsd:noNamespaceSchemaLocation=
    "http://www.web3d.org/specifications/x3d-3.0.xsd">
  <Scene>
    <Shape>
      <Appearance>
        <Material diffuseColor="0.8 0.8 0.8"/>
      </Appearance>
      <Box size='16 16 16'/>
    </Shape>
  </Scene>
</X3D>
```

In this scenario, we have a PostGIS table able to capture 3D points; an X3D model and an SLD style that maps the X3D model to each point. After this data preparation step, the W3DS is able to produce 3D scenes. A simple request would be:

```
http://localhost:8080/geoserver/w3ds?
    version=0.4&service=w3ds&request=GetScene&
    crs=EPSG:27492&
    format=model/x3d+html&
    layers=interest_points&
    boundingbox=-16423,199085,-16229,199336&
    styles=point_cube
```

Such a request would produce an HTML document with the scene in X3D that can be directly shown in the browser, as illustrated in Fig. 2. It is possible to see the 3D in the browser because the HTML document also includes a reference to the X3DOM library.

Managing W3DS layers is as simple as managing any other GeoServer layer. It is possible to manage and preview the W3DS layers from the administrative interface.

W3DS also produces KML documents, which is handy for producing scenes that can be explored in Google Earth.

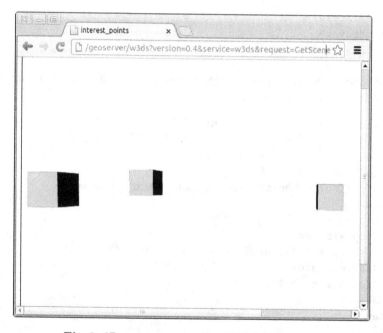

Fig. 2. 3D scene preview directly on the browser

6 Concluding Remarks

While there is much interest in 3D in the urban planning and modelling areas, we still lack a consensus on the approach and a standard, either de jure or de facto, to provide 3D scenes.

In this paper we present one open source implementation of the W3DS draft proposal, as a GeoServer community module. The architecture and the developed modules were discussed. We also showed how a simple layer can be configured and served by W3DS.

The service itself does not provide any urban planning or management oriented facilities. These specific functionalities must be included on the client side, for each use case.

To change the workflow to 3D, both data and tools must address the challenge. While we just addressed the server side tools, moving to 3D has a major impact on the data side. It might be quite challenging, but it is an opportunity to further validate and improve consistency among different datasets.

The availability of services like W3DS also contributes to the development or improvement of existing 3D clients. OpenLayers is a sophisticated client library for displaying maps on the web. Since OpenLayers is being rewritten to support WebGL, it may use a W3DS service as a native data source. That would improve the development of web based urban management applications.

References

[1] GeoServer, open source Java-based map server, http://geoserver.org
[2] Behr, J., Eschler, P., Jung, Y., Zöllner, M.: X3DOM - A DOM-based HTML5 X3D Integration Model (2009)
[3] Brutzman, D., Daly, L.: X3D: extensible 3D graphics for Web authors (2007)
[4] Cambray, B.: Three-dimensional (3D) modeling in a geographical database (1993)
[5] Carosio, A.: La troiséme dimension dans les systémes d'information geographique et la mensuration officielle (1999)
[6] Goetz, M., Zipf, A.: OpenStreetMap in 3D - Detailed Insights on the Current Situation in Germany (2012)
[7] Held, G., Abdul-Rahman, A., Zlatanova, S.: Web 3D GIS for urban environments (2001)
[8] Kolbe, T., Schilling, A.: Draft for Candidate OpenGIS Web 3D Service Interface Standard (2010)
[9] Marrin, C.: Webgl specification (2013)
[10] Moser, J., Albrecht, F., Kosar, B.: Beyond visualization - 3D GIS analyses for virtual city models (2010)
[11] Murata, M.: 3D-GIS Application for urban planning based on 3D city model (2005)
[12] Ortiz, S.: Is 3D Finally Ready for the Web? (2010)
[13] Pesce, M.: VRML: Browsing and Building Cyberspace (1995)
[14] Sons, K., Klein, F., Rubinstein, D., Byelozyorov, S., Slusallek, P.: XML3D: interactive 3D graphics for the web (2010)
[15] Zeile, P., Schildwächter, R., Poesch, T., Wettels, P.: Production of virtual 3D city models from geodata and visualization with 3D game engines. A Case Study from the UNESCO World Heritage City of Bamberg Problem Status - Starting Point (2004)
[16] Zlatanova, S.: VRML for 3D GIS (1997)

The e-Participation in Tranquillity Areas Identification as a Key Factor for Sustainable Landscape Planning

Giuseppe Modica, Paolo Zoccali, and Salvatore Di Fazio

Dipartimento di Agraria, Università degli Studi *Mediterranea* di Reggio Calabria,
Loc. Feo di Vito, 89122 Reggio Calabria, Italy
{giuseppe.modica,paolo.zoccali,salvatore.difazio}@unirc.it

Abstract. Landscapes are multifunctional and adaptive systems in which ecological and human processes interact continuously. Moreover, they are characterised by various functions and values provided to communities in terms of goods and services. The importance and meanings of cultural and perceptual dimensions in landscape assessment are strongly recognised in the European Landscape Convention (ELC) and normally analysed considering the way people appreciate the landscape values. The involvement of the public is one of the important pillars of landscape policy. In this paper, a research case-study based in the natural park of Serre (Italy) and concerning the e-participation in tranquillity areas identification and mapping is reported. The first results of an e-survey carried out in 2012 concerning people's perception of landscape tranquillity are here presented as they emerge from the analysis of 422 questionnaires. The open questions on the use of this issue on sustainable landscape planning are also discussed.

Keywords: Tranquillity areas mapping, e-Participation, Sustainable landscape planning, Questionnaire surveys, Serre Regional Park (Italy), Ecosystem services.

1 Introduction and Objectives of the Work

The on-going transformations of urban- rural systems are one of the most important types of Land Use/Land Cover (LULC) changes, since they are widespread all over the world and often determine negative effects on ecosystem functionality. Study of the evolution trends of landscape, particularly those related to urban/rural relations is crucial for a sustainable landscape planning [1]. Moreover, the expansion of urban areas has typically taken place on former agricultural lands [2–7].

Agricultural intensification in the most suitable agricultural areas and agricultural abandonment in marginal areas are among the most significant landscape transformations recently occurred in Europe [8]. LULC changes can dramatically affect the structure and functioning of ecosystems [9, 10] and, as a consequence, they can alter landscapes' capacity to provide the flow of ecosystem services [11, 12].

Landscape analysis and characterisation are key and basic elements for the interpretation, management, protection and valorisation of a given territory. Landscapes

B. Murgante et al. (Eds.): ICCSA 2013, Part III, LNCS 7973, pp. 550–565, 2013.
© Springer-Verlag Berlin Heidelberg 2013

are complex and adaptive systems [13] characterised by various functions and values provided to communities in terms of goods and services [14, 15]. In turn, landscapes are multifunctional systems in which ecological and human processes interact continuously with each other.

Understanding the diversity of perception and values related to the landscape is fundamental [16] and descends from a thorough consideration of the dualistic relationship between people and the places they live [17].

Usually, ecosystem services are evaluated by ecological and economic approaches [15, 18, 19] not taking into account the values assigned to ecosystem services by people living in a landscape [20]. Since people are the ecosystem service users [16], the economic approaches based on the study of benefit transfers [21, 22] should consider that these benefits are context-dependent and place-related and therefore their evaluation deals with the complex and dynamic relationships between humans and their environment [20].

If compared to the term "ecosystem service", the term "landscape services" appears preferable when referring to sustainable landscape development, [23], because it better captures the set of relationships characterising the spatial patterns and is more interdisciplinary [20].

The European Landscape Convention (ELC) [24] describes the Landscape as 'an area, as perceived by people, whose character is the result of the action and interaction of natural and human factors'. In this definition it is well recognised a clear holistic approach that encompasses both the natural (climate, geomorphology, fauna, etc.) and the cultural (land use, settlements, etc.) landscape factors, as well as perceptual qualities of the landscape, such as tranquillity, which are either sensory and cultural in nature. Participation and sensibilisation of the public are aspects strongly emphasized in the ELC which clearly considers them as important pillars of landscape policy [25]

In the present work, a research case-study conducted in the territory of the natural park of Serre (Italy) and concerning the e-participation in tranquillity areas identification and mapping is reported. In more details, the first results coming from a questionnaire survey carried out in 2012 and concerning people's perception of landscape tranquillity are here presented as they emerge from the analysis of the answers given by 422 respondents. The open questions on the use of this issue on sustainable landscape planning are also discussed.

Referring to the importance and meanings of cultural and perceptual dimensions in landscape assessment it can be noticed that they are normally analysed considering that the way people appreciate the landscape values is strongly linked to land-cover pattern composition and configuration [26–33]. In the present work, they are also considered other aspects which can be detected by analysing landscape preferences gathered from different user groups, also with the use of media favouring the e-participation of people in the planning process.

1.1   Participation and e-Participation in the Sustainable Landscape Planning

A large number of planning experiences reported in scientific literature document several different problems and obstacles that are encountered when trying to put into practice an effective participatory approach [25], and because of which the statement "transdisciplinarity involves participation" often seems to sound as merely axiomatic. When we refer to sustainable landscape planning we have to consider that landscape should be seen as a "living" entity which undergoes continuous change depending on the evolution of the relationship between the local population and its own territory over time. Following the spirit of the ELC, public action should be coupled with patient work to facilitate a direct and wide involvement of the local population to identify, evaluate, protect, plan and manage the landscape.

People's participation in planning should extend scientific expertise by adding local experiences, taking into account public opinions and knowledge as well as recognizing the importance of social judgement. Referring to Stirling's typology of participation [34], in the ELC participation is encouraged from a substantive and even instrumental point of view [25] that means its role in planning should be to increase information or to justify a policy decision to be made.

Citizens increasingly perceive that their actions could be partially or totally accepted by local authorities [35]. Nowadays, the convergence of the Web 2.0 with the arising new geospatial technologies allows for a broader participation of local communities in the planning process. Indeed, the Web 2.0 itself is a participatory platform where people can freely contribute in the creation of new content, and therefore it is an important tool in collaborative decision making. In the planning process, e-participation can be seen as the integration of the new internet-based technologies in the traditional approaches which this way can be implemented so as to reach the higher levels of the Arnstein ladder [36]. Moreover, in wiki-planning it can be recognised a two-way approach in which community not only can express its opinion but can also substantially influence choices.

2   Materials and methods

2.1   The Study-Area

The district of Serra San Bruno is one of the most interesting areas of Calabria because of the presence of many heritage resources of great natural, historic and architectural interest. Nevertheless, owing to its century-old woods, characterised by the prevalence of beech (*Fagus sylvatica* L.) and silver fir (*Abies alba* Miller subsp. *apennina* Brullo, Scelsi and Spampinato) population, it shows a remarkable intrinsic suitability for forestry. In the Serre's district, a great significance is also recognised to the industrial archaeology heritage which is mostly related to the utilization of water and wood and dates between the 18th and the 19th century, i.e. the period when Calabria was under the Bourbon domination.

The high environmental value of this area motivated the institution of two National Natural Reserves in 1977 (now SCIs - Sites of Community Interest - of the Natura 2000 network, the centrepiece of EU nature and biodiversity policy) and of the Serre Regional Park in 2004 (Fig. 1).

In recent years, the development of a tourism based on the valorisation of the rich cultural and natural resources, has been of some help in sustaining the local economy. Tourists search for landscapes with high environmental quality, which should be improved through sensitive landscape planning based on accurate monitoring of the dynamics characterising land use [37]. This is also needed for the maintenance of the environmental balance of these areas, which are particularly fragile due to relevant hydrogeological risk (the last catastrophic flood occurred in 1935, causing death and destruction).

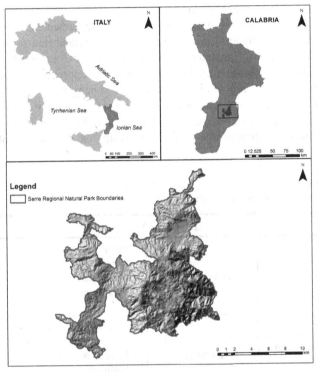

Fig. 1. Geographical location of the study-area: the Serre regional natural park

In the district under examination the local municipalities in many occasions have tried to foster the development of forms of tourism valorising the forestry and natural heritage jointly with the many outstanding cultural sites which are scattered in the peri-urban areas. Some of these sites are historic monastic settlements, such as the Chartreuse of Serra San Bruno (founded by St. Bruno of Cologne in the early 12th century) and the byzantine Monastery of St. John "Therestis". Woods, mountains, nature, religious and pilgrimage sites are entities usually associated by the public with silence and other profound feelings often expressed by- or related to- the idea of "tranquillity". As it is in other areas, the changes that the landscape under study is presently undergoing in many ways seem to threaten tranquillity; tranquillity, in its turn, needs to be defined, identified, monitored, protected and implemented not only to assure the well-being of the local population, but also as an important landscape character and value on which the above mentioned forms of tourism, ever more desirable, should be based.

2.2   Background Concept of Tranquillity Areas

The term 'tranquillity' is widely used in several research fields. Broadly speaking, it can be considered as a state of mind that promotes well-being. Referring to its meaning in landscape planning experience, it is to be considered not just as a holistic sensory experience but as a significant asset of landscape, appearing as an objective attribute in a range of strategies and plans [38].

The first attempt of mapping tranquillity areas was implemented by Simon Rendel who produced a set of tranquil area maps covering England and published by the Campaign to Protect Rural England (CPRE) and the former Countryside Commission. The method followed was criticised due to its structural approach based on the proximity to different sources of potential disturbance. The new approach defined in 2004, added participatory appraisal techniques basing the mapping of tranquillity areas more directly on people's views about the countryside [38–40]; in detail it focused on assessing and understanding people's values, meanings and experiences of tranquillity [38].

Table 1. Categories of ecosystem services and examples of related services based on Millennium Ecosystem Assessment approach (MA, 2005)

Ecosystem Services		
Supporting Soil formation Primary production Nutrient cycling Water cycling ………..	**Provisioning**	Food Genetic resources Bio-chemicals Fresh water ………..
	Regulating	Climate regulation Air and water regulation Erosion regulation Pollination ………..
	Cultural	Cultural diversity Spiritual and religious values Recreation and ecotourism Educational and Aesthetic values ………..

In turn, tranquillity can be considered as an ecosystem service. According to the Millennium Ecosystem Assessment approach [14] ecosystem services can be broadly defined as the benefits and goods provided by different ecosystems which contribute to human wellbeing (Table 1). It is important to notice that the term 'services' is used to encompass the tangible and intangible benefits that humans obtain from ecosystems.

2.3   The Questionnaire Survey

With the aim to favour e-participation on tranquillity areas identification and mapping a questionnaire form was implemented in cooperation with local stakeholders, in order to take into account their expertise, and then tested. To this end, before starting the survey, 25 questionnaires were distributed to a representative sample of respondents in the study area in order to evaluate whether the language and the structure of each question was adequate and easily comprehensible. On the basis of the results of the preliminary test, 10 of the total 21 questions were revised.

The survey was conducted from August to December 2012. Questionnaires were disseminated in Italian and English languages using both the internet (Fig. 2) and the paper form. The internet form presents several advantages: no paper consumption, no postal or face to face interviews are required and a high degree of automation can be achieved for processing the replies [41] linking directly the results with a software package, for example via a simple *.csv file. The paper form, in its turn, can be filled more easily by certain categories of respondents, such as elderly people or tourists. With reference on data quality and validity, as demonstrated in previous studies [30, 41], there are no significant differences in the two type of media.

The questionnaire is structured in three sections and contains a total of 21 questions. In the first section, questions (from 1 to 6) are organised in order to characterise respondents' profile in terms of gender, age, education level, job, place of residence. Respondents not resident in the Park of Serre are also subdivided into subcategories according to the frequency of their visits in the territory of the park.

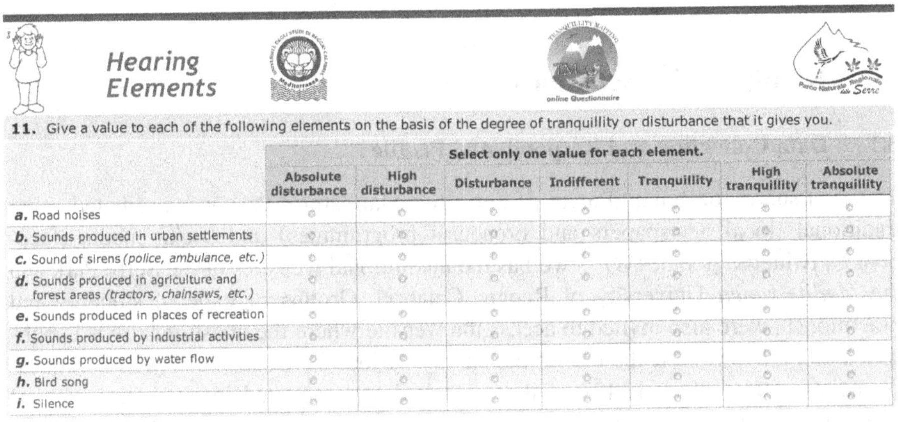

Hearing Elements

11. Give a value to each of the following elements on the basis of the degree of tranquillity or disturbance that it gives you.

	Select only one value for each element.						
	Absolute disturbance	High disturbance	Disturbance	Indifferent	Tranquillity	High tranquillity	Absolute tranquillity
a. Road noises	○	○	○	○	○	○	○
b. Sounds produced in urban settlements	○	○	○	○	○	○	○
c. Sound of sirens (police, ambulance, etc.)	○	○	○	○	○	○	○
d. Sounds produced in agriculture and forest areas (tractors, chainsaws, etc.)	○	○	○	○	○	○	○
e. Sounds produced in places of recreation	○	○	○	○	○	○	○
f. Sounds produced by industrial activities	○	○	○	○	○	○	○
g. Sounds produced by water flow	○	○	○	○	○	○	○
h. Bird song	○	○	○	○	○	○	○
i. Silence	○	○	○	○	○	○	○

Fig. 2. Screenshot referring to the hearing elements section of the online questionnaire on tranquillity areas mapping, implemented using the BOS (Bristol Online Survey) service

The second section of the questionnaire deals with the main concepts and values of the tranquillity areas (questions from 7 to 14); it is split into three subsections containing questions relating to visual, hearing (Fig. 2) and smell elements, either from anthropogenic or natural origin, mostly mappable (37 elements in total). In each of these subsections, respondents are asked to write freely in a blank box other additional elements not contained in the pre-defined lists of elements to select from.

The third specific section is focused on qualitative and open questions (from 15 to 22). Respondents are asked what do they look for, or escape from, when looking for tranquillity. They are also invited to indicate the characteristics that an area should have to be defined "tranquil" or "not tranquil", and the activities they generally undertake in an area they consider as tranquil. Respondents are finally asked to express their idea of tranquillity and, using the place names or alternatively the grid reference, identify a maximum of five areas in which they experienced either tranquillity or disturbance.

In order to collect quantitative data to address the specific purposes of the research and to perform the statistical analysis, preferences and perceptions of respondents were rated using a seven-point Likert scale (Fig. 3), from 1 (absolute disturbance) to 7 (absolute tranquillity). A neutral option (score 4, indifferent) was also considered in the rating scale.

As in previous studies [42], questionnaires (online and paper forms) were filled by respondents without any assistance/presence of the interviewers. The maximum time required for completing the questionnaire was about 15 minutes.

1	2	3	4	5	6	7
Absolute disturbance	High disturbance	Disturbance	Indifferent	Tranquillity	High tranquillity	Absolute tranquillity

Fig. 3. Seven-point Likert scale adopted rating for respondents' preferences and perceptions on tranquillity areas

3   Results and Discussions

3.1   Data Collection and Respondents' Profile

This first survey on tranquil areas of Serre regional natural Park was promoted using traditional (local newspapers and broadcast programmes) and social media (facebook®, twitter®, google+®) as well as the institutional websites of the Serre Park and the *Mediterranea* University of Reggio Calabria. Groups of experts, students and practitioners were also invited to access the website where the questionnaire was published and answer it. At the end of the survey period (31 December 2012) 515 questionnaires were collected; 422 of them resulted as completed in all their parts and therefore were usable for our research purposes, while the remaining 93 had not been completed, mostly due to respondents' misuse of the online form, and were rejected. As it emerged from the e-mails received by respondents, some of them were not able to complete the survey in a single session and did not save correctly the answers already written before leaving the website.

In order to take into account and evaluate the potential divergences in opinion deriving from different grades of knowledge about the Serre Park landscape, respondents were subdivided into two main groups: "residents" and "non-residents" and these last were attributed to two further sub-groups, depending on whether they had already visited the territories of the park or not ("visitors" and "non-visitors").

Respondents' profile according to gender, age group and occupational status, is shown in Table 2. A total of 422 respondents (173 females, 249 males), with an average age of 31.6 ± 13.1 years (10-75 years), completed the questionnaire; 161 were resident in the park territory (65 female, 96 males), while 261 were not (108 females, 153 males). Considering the non-residents, 118 of them had already visited the Serre Park (34 females, 84 males), while 143 had never visited it (74 females, 69 males).

Table 2. Table showing the profile of respondents

| | | TOTAL | | Residents | | Non-residents | | | |
						Visitors		Non-visitors	
Gender		N	%	n	%	n	%	n	%
	Female	173	41	65	40.4	34	28.8	74	51.7
	Male	249	59	96	59.6	84	71.2	69	48.3
Age groups		N	%	n	%	n	%	n	%
	under 18	57	13.5	53	32.9	1	0.8	3	2.1
	18-25	76	18	27	16.8	19	16.1	30	21
	26-30	102	24.2	37	23	30	25.4	35	24.5
	31-40	89	21.1	12	7.5	37	31.4	40	28
	41-50	54	12.8	17	10.6	20	16.9	17	11.9
	51-60	34	8.1	11	6.8	8	6.8	15	10.5
	over 60	10	2.4	4	2.5	3	2.5	3	2.1
Occupational status		N	%	N	%	n	%	n	%
	Student	159	37.7	78	48.4	36	30.5	45	31.5
	Teacher	40	9.5	5	3.1	15	12.7	20	14
	Self-employed	66	15.6	21	13	25	21.2	20	14
	Clerk/Manager/Executive	69	16.4	22	13.6	24	20.3	23	16.1
	Business person	15	3.6	7	4.3	3	2.5	5	3.5
	Unemployed	19	11.6	21	13.6	7	5.9	21	14.7
	Pensioner	10	2.4	4	2.5	2	1.7	4	2.8
	Other	14	3.3	3	1.9	6	5.1	5	3.5

3.2   Data Survey Analysis

The quantitative data were analysed using the SPSS® statistical software, including descriptive statistics and correlation analysis. In more details, analyses measuring the differences in perception of tranquillity between residents and non-residents were carried out on the basis on non-parametric tests for independent sampling.

Different tests were considered for application and among them they were chosen and finally applied the Wilcoxon's signed-rank sum test and the Mann–Whitney U-test (Two-sample test of medians) [43]. The Wilcoxon test is similar to the Mann-Whitney U-test and both are very efficient for a small as well as for a large number of samples [44]. The Mann-Whitney U-test, in particular, is a nonparametric statistical procedure for comparing two independent, or not related, samples. The parametric equivalent to these tests is the *t*-test for independent samples.

In order to show significant differences in the distribution of respondents' answers also box-and-whisker-plots were assessed.

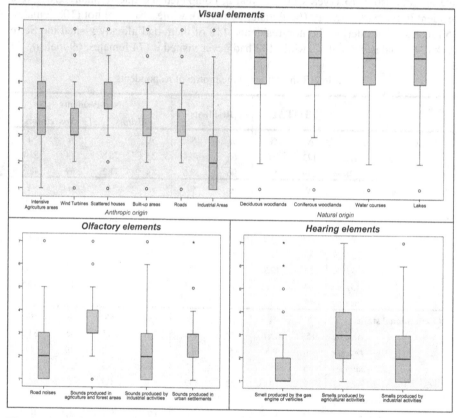

Fig. 4. Box-and-whisker-plots of the median test for the most significant elements analysed for tranquillity areas mapping. The bold line in the middle of the boxes represents the median, the edge of the boxes represent the 25% and 75% quantiles and the whiskers represent the 5% and 95% quantiles. Circles and stars represent outliers beyond the 5% and 95% quantiles.

Considering all the 37 elements about which respondents were asked to express their perception by using the 7-point Likert scale, in the answers given only for 12 of them there isn't significant difference in perception between residents and non-residents. It is important to notice that of all these 12 elements, some are related with aspects of the natural environment for which respondents expressed a general perception of tranquillity (snow, bird songs, glades/meadows, landscapes with closed view), while others relate to traditional activities (sounds and smells of agricultural and forestry activities) or to the green economy (wind turbines and solar panels).

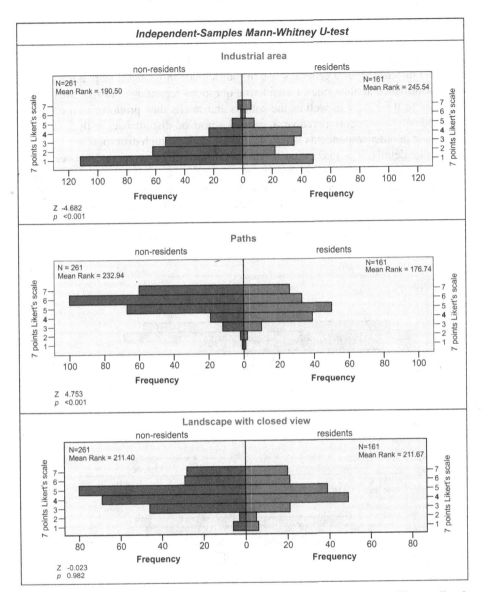

Fig. 5. Results of the Mann–Whitney U-test applied to questions on tranquillity or disturb associated to the following elements: industrial areas, paths, landscape with closed view. Significance of differences between residents and no residents.

If the data are analysed according to the 7 age classes defined, differences in respondents' answers can be found in a few elements. As an example, in order to show the most significant differences in tranquillity perception between the age classes, here we report results for the two extreme age intervals considered: respondents under 18 and over 60 years. Indeed, among the 37 elements analysed only in 8 cases the two samples are completely different with the rejection of the null hypothesis in the

Mann-Whitney U-test. It appears interesting to highlight that these differences mainly concern visual elements with an anthropic origin. In more details, the view of pylons/radio-television and telephone masts, billboard/advertising signs, solar panels and wind turbines gives a general sense of indifference in youngest respondents while gives disturbance in older ones. Considering questions regarding industrial areas, with reference to their view as well as the sounds and smells they produce, as it could be expected all respondents perceive a general sense of disturbance, with a stronger magnitude in older respondents which mostly expressed a high disturbance.

In more details, it is interesting to analyse difference in perception between residents and non-residents, making particular reference to the following elements, each of them contained in a specific survey question: Industrial areas; Paths (foot and bicycle paths); Landscapes with closed view; Intensive agricultural areas (Figg. 5-6).

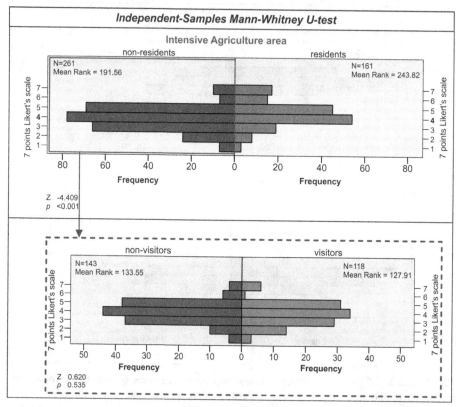

Fig. 6. Results of the Mann–Whitney U-test applied to question on tranquillity or disturb associated to intensive agriculture areas. Significance of differences between visitors and non-visitors of park's territory.

As shown in Figg. 5-7 for the above mentioned elements, compared by Mann–Whitney U-test with the null hypothesis that the two distribution were equal revealed significant differences for Industrial areas and Paths ($p<0.001$) justifying the rejection of the null hypothesis. Thus, for these elements there is a significant difference between residents and non-residents perception of tranquillity/disturbance.

There were significant differences between residents and non-residents in responses given. For residents the economic activities falling in the territory of the Park seem to be perceived as factors having a relatively low influence on tranquillity.

Referring to the question on 'landscapes with closed view' the two samples had the same results ($p=0.982$, $Z=-0.023$), so the null hypothesis is accepted which means that there are no differences in tranquillity perception between the two groups of respondents.

Also the analysis of perception concerning 'Intensive agricultural areas' revealed a significant difference between residents and non-residents ($p<0.001$) justifying the rejection of the null hypothesis, while there is no difference between visitors and non-visitors ($p=0.535$, $Z=-0.620$), justifying the acceptance of the null hypothesis. Box-and-whisker-plots of the median test reveal how, indeed, the distribution of the answers among non-residents (visitors and non-visitors) is the same (Fig. 7).

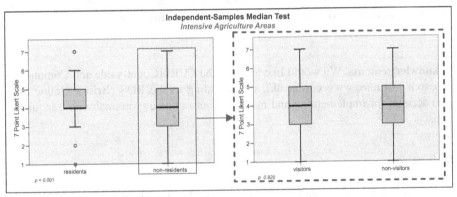

Fig. 7. Box-and-whisker-plots of the median test for intensive agriculture areas

3.3   Discussions and Open Questions

The research presented is part of a wider study aiming at landscape characterisation and interpretation in different Mediterranean areas [1, 8, 45, 46]. It started from the comprehension of the past and on-going landscape transformations in order to better understand the current dynamics affecting the rural landscape and the rural/urban fringe areas. In a subsequent phase, here reported, the issues relating to the perceptual and cultural aspects characterising the landscape were studied in order to put into practice the principles and statements of the ELC for implementing the sustainable landscape-planning process. Within this context and considering its relationship with the perceptual and physical aspects, the theme of tranquillity areas mapping appears as one of the most interesting.

From a methodological point of view, tranquillity areas mapping embraces the assessment of immaterial/intangible landscape elements/values that are strictly linked with the physical/tangible elements recognisable in a territory. These last, not only help the recognition and mapping of the immaterial/intangible elements and values, but when recognised as associated with positive perception on the part of people, their valorisation and protection should become a specific issue in the framework of sustainable landscape planning tools. In studying local community's landscape perception, as it has been shown in the case tranquillity mapping, it is important to take into account the values people attach to different landscape elements depending on a varied range of personal landscape experiences and personal conditions. E-participation can greatly help people's involvement in the planning process since its early stages, such as those concerning landscape identification and evaluation.

Tranquil areas can play an important role in the natural protected areas planning tools. As highlighted in previous studies focusing on urban/rural fringe landscapes [47, 48], nowadays it is widely recognised that valued tranquil areas should be further protected in order to avoid an advancement in the erosion of these wealthy spaces. This appears an important arising question in the current planning strategies not only in the urban space but also in those many rural and natural protected areas, such as those of the Mediterranean basin, where the present challenge is to ensure a durable balance between development and the protection of valuable natural and cultural resources.

Acknowledgements. We would like to thank the CCRI (Countryside and Community Research Institute, www.ccri.ac.uk), for providing us the BOS (Bristol Online Survey) account for implementing and managing our online questionnaire on tranquillity areas.

References

1. Di Fazio, S., Modica, G., Zoccali, P.: Evolution Trends of Land Use/Land Cover in a Mediterranean Forest Landscape in Italy. In: Murgante, B., Gervasi, O., Iglesias, A., Taniar, D., Apduhan, B.O. (eds.) ICCSA 2011, Part I. LNCS, vol. 6782, pp. 284–299. Springer, Heidelberg (2011)
2. Perella, G., Galli, A., Marcheggiani, E.: The Potential of Ecomuseums in Strategies for Local Sustainable Development in Rural Areas. Landscape Research 35, 431–447 (2010)
3. Perchinunno, P., Rotondo, F., Torre, C.: The Evidence of Links between Landscape and Economy in a Rural Park. International Journal of Agricultural and Environmental Information Systems (IJAEIS) 3, 1–14 (2012)
4. Neri, M., Menconi, M.E., Vizzari, M., Mennella, V.G.G.: Propuesta de una nueva metodología para la ubicación de infraestructuras viarias ambientalmente sostenibles. Aplicación en el Tramo Viario de la Pedemontana Fabriano-Muccia. Informes de la Construcción 62, 101–112 (2010)
5. Cerreta, M., De Toro, P.: Urbanization suitability maps: a dynamic spatial decision support system for sustainable land use. Earth System Dynamics 3, 157–171 (2012)
6. Cerreta, M., De Toro, P.: Assessing urban transformations: a SDSS for the master plan of Castel Capuano, Naples. In: Murgante, B., Gervasi, O., Misra, S., Nedjah, N., Rocha, A.M.A.C., Taniar, D., Apduhan, B.O. (eds.) ICCSA 2012, Part II. LNCS, vol. 7334, pp. 168–180. Springer, Heidelberg (2012)

7. Vizzari, M.: Spatio-temporal analysis using urban-rural gradient modelling and landscape metrics. In: Murgante, B., Gervasi, O., Iglesias, A., Taniar, D., Apduhan, B.O. (eds.) ICCSA 2011, Part I. LNCS, vol. 6782, pp. 103–118. Springer, Heidelberg (2011)
8. Modica, G., Vizzari, M., Pollino, M., Fichera, C.R., Zoccali, P., Di Fazio, S.: Spatio-temporal analysis of the urban–rural gradient structure: an application in a Mediterranean mountainous landscape (Serra San Bruno, Italy). Earth System Dynamics 3, 263–279 (2012)
9. Defries, R.S., Foley, J.A., Asner, G.P.: Land-use choices: balancing human needs and eco-system function. Frontiers in Ecology and the Environment 2, 249–257 (2004)
10. Vitousek, P.M.: Human Domination of Earth's Ecosystems. Science 277, 494–499 (1997)
11. Costanza, R., Fisher, B., Mulder, K., Liu, S., Christopher, T.: Biodiversity and ecosystem services: A multi-scale empirical study of the relationship between species richness and net primary production. Ecological Economics 61, 478–491 (2007)
12. De Groot, R.: Function-analysis and valuation as a tool to assess land use conflicts in planning for sustainable, multi-functional landscapes. Landscape and Urban Planning 75, 175–186 (2006)
13. Levin, S.A.: Ecosystems and the Biosphere as Complex Adaptive Systems. Ecosystems 1, 431–436 (1998)
14. MA: Ecosystems and Human Well-being: Biodiversity Synthesis. World Resources Institute, Washington, DC (2005)
15. Daily, G.C.: Nature's services: societal dependence on natural ecosystems. Island Press, Washington (1997)
16. Aretano, R., Petrosillo, I., Zaccarelli, N., Semeraro, T., Zurlini, G.: People perception of landscape change effects on ecosystem services in small Mediterranean islands: A combination of subjective and objective assessments. Landscape and Urban Planning 112, 63–73 (2013)
17. Tress, B., Tress, G.: Capitalising on multiplicity: a transdisciplinary systems approach to landscape research. Landscape and Urban Planning 57, 143–157 (2001)
18. Costanza, R., Arge, R., De Groot, R., Farber, S., Grasso, M., Hannon, B., Limburg, K., Naeem, S., Neill, R.V.O., Paruelo, J., Raskin, R.G., Suttonkk, P.: The value of the world's ecosystem services and natural capital 387, 253–260 (1997)
19. De Groot, R.S., Wilson, M.A., Boumans, R.M.: A typology for the classification, description and valuation of ecosystem functions, goods and services. Ecological Economics 41, 393–408 (2002)
20. Fagerholm, N., Käyhkö, N., Ndumbaro, F., Khamis, M.: Community stakeholders' knowledge in landscape assessments – Mapping indicators for landscape services. Ecological Indicators 18, 421–433 (2012)
21. Wilson, M.A., Hoehn, J.P.: Valuing environmental goods and services using benefit transfer: The state-of-the art and science. Ecological Economics 60, 335–342 (2006)
22. Plummer, M.L.: Assessing benefit transfer for the valuation of ecosystem services. Frontiers in Ecology and the Environment 7, 38–45 (2009)
23. Termorshuizen, J.W., Opdam, P.: Landscape services as a bridge between landscape ecology and sustainable development. Landscape Ecology 24, 1037–1052 (2009)
24. Council of Europe: European Landscape Convention (2000)
25. Sevenant, M., Antrop, M.: Transdisciplinary landscape planning: Does the public have aspirations? Experiences from a case study in Ghent (Flanders, Belgium). Land Use Policy 27, 373–386 (2010)

26. Gulinck, H., Múgica, M., De Lucio, J.V., Atauri, J.A.: A framework for comparative land-scape analysis and evaluation based on land cover data, with an application in the Madrid region (Spain). Landscape and Urban Planning 55, 257–270 (2001)

27. Lewis, J.L.: Perceptions of landscape change in a rural British Columbia community. Landscape and Urban Planning 85, 49–59 (2008)

28. Petrosillo, I., Zurlini, G., Corlianò, M.E., Zaccarelli, N., Dadamo, M.: Tourist perception of recreational environment and management in a marine protected area. Landscape and Urban Planning 79, 29–37 (2007)

29. Pinto-Correia, T., Carvalho-Ribeiro, S.: The Index of Function Suitability (IFS): A new tool for assessing the capacity of landscapes to provide amenity functions. Land Use Policy 29, 23–34 (2012)

30. Rogge, E., Nevens, F., Gulinck, H.: Perception of rural landscapes in Flanders: Looking beyond aesthetics. Landscape and Urban Planning 82, 159–174 (2007)

31. Willemen, L., Verburg, P.H., Hein, L., Van Mensvoort, M.E.F.: Spatial characterization of landscape functions. Landscape and Urban Planning 88, 34–43 (2008)

32. Willemen, L., Hein, L., Van Mensvoort, M.E.F., Verburg, P.H.: Space for people, plants, and livestock? Quantifying interactions among multiple landscape functions in a Dutch rural region. Ecological Indicators 10, 62–73 (2010)

33. Vizzari, M.: Spatial modelling of potential landscape quality. Applied Geography 31, 108–118 (2011)

34. Stirling, A.: Analysis, participation and power: justification and closure in participatory multi-criteria analysis. Land Use Policy 23, 95–107 (2006)

35. Murgante, B., Tilio, L., Lanza, V., Scorza, F.: Using participative GIS and e-tools for in-volving citizens of Marmo Platano-Melandro area in European programming activities. Journal of Balkan and Near Eastern Studies 13, 97–115 (2011)

36. Murgante, B.: Wiki-Planning: The Experience of Basento Park in Potenza (Italy). In: Bor-ruso, G., Bertazzon, S., Favretto, A., Murgante, B., Torre, C.M. (eds.) Geographic Infor-mation Analysis for Sustainable Development and Economic Planning: New Technolo-gies, pp. 345–359. IGI Global (2013)

37. Selicato, M., Torre, C.M., La Trofa, G.: Prospect of integrate monitoring: A multidimen-sional approach. In: Murgante, B., Gervasi, O., Misra, S., Nedjah, N., Rocha, A.M.A.C., Taniar, D., Apduhan, B.O. (eds.) ICCSA 2012, Part II. LNCS, vol. 7334, pp. 144–156. Springer, Heidelberg (2012)

38. Jackson, S., Fuller, D., Dunsford, H., Mowbray, R., Hext, S., MacFarlane, R., Haggett, C.: Tranquillity Mapping: developing a robust methodology for planning support (2008)

39. MacFarlane, R., Haggett, C., Fuller, D., Dunsford, H., Carlisle, B.: Tranquillity Mapping: developing a robust methodology for planning support (2004)

40. Countryside Agency: Understanding tranquillity. The role of Participatory Appraisal con-sultation in defining and assessing a valuable resource (2005)

41. Wherrett, J.: Issues in using the Internet as a medium for landscape preference research. Landscape and Urban Planning 45, 209–217 (1999)

42. Dràbkovà, A.: Tourists in Protected Landscape Areas in the Czech Republic – a Sociologi-cal Survey. Procedia Environmental Sciences 14, 279–287 (2012)

43. Mann, H.B., Whitney, D.R.: On a Test of Whether one of Two Random Variables is Sto-chastically Larger than the Other. The Annals of Mathematical Statistics 18, 50–60 (1947)

44. Corder, G., Foreman, D.: Nonparametric statistics for non-statisticians: a step-by-step ap-proach. John Wiley & Sons, New York (2009)

45. Fichera, C.R., Modica, G., Pollino, M.: GIS and Remote Sensing to Study Urban-Rural Transformation During a Fifty-Year Period. In: Murgante, B., Gervasi, O., Iglesias, A., Taniar, D., Apduhan, B.O. (eds.) ICCSA 2011, Part I. LNCS, vol. 6782, pp. 237–252. Springer, Heidelberg (2011)
46. Fichera, C.R., Modica, G., Pollino, M.: Land Cover classification and change-detection analysis using multi-temporal remote sensed imagery and landscape metrics. European Journal of Remote Sensing 45, 1–18 (2012)
47. Herzog, T.R., Maguire, P., Nebel, M.B.: Assessing the restorative components of environments. Journal of Environmental Psychology 23, 159–170 (2003)
48. Watts, G., Pheasant, R., Horoshenkov, K.: Tranquil spaces in a metropolitan area. In: Proceedings of 20th International Congress on Acoustics, ICA 2010, pp. 1–6. Australian Acoustical Society, NSW Division, Sydney (2010)

Free Web Mapping Tools to Characterise Landscape Dynamics and to Favour e-Participation

Maurizio Pollino[1] and Giuseppe Modica[2]

[1]ENEA National Agency for New Technologies,
Energy and Sustainable Economic Development,
UTMEA-TER "Earth Observations and Analyses" Laboratory,
C.R. Casaccia, Via Anguillarese, 301 – 00123 Rome, Italy
maurizio.pollino@enea.it
[2]Dipartimento di Agraria, Università degli Studi *Mediterranea* di Reggio Calabria,
Loc. Feo di Vito, 89122 Reggio Calabria, Italy
giuseppe.modica@unirc.it

Abstract. GIS methodologies in combination with Remote Sensing data and techniques are fundamental to analyse and characterise Land Use/Land Cover (LULC) and their evolutionary dynamics. The case study here described, has been conducted in two study-areas: the Serre Regional natural Park and the so-called Conca of Avellino (Southern Italy). This study is part of a wider research allowed to understand how the landscape changes dynamics are linked and have been influenced by several causes (demography, economy, transportation network, people preferences, policies, etc.). A multi-temporal set of images (aerial photos and Landsat scenes) has been processed LULC. Then, through a GIS approach, change detection and spatiotemporal analysis has been integrated to characterise LULC dynamics, focusing on urban growth/sprawl phenomenon and loss of rural/natural lands. This paper focuses the attention on the WebGIS application based on free online tools which has been implemented with the aim to publish and to share with local communities all geospatial data produced. This platform will be further implemented in order to favour e-participation in the planning tools.

Keywords: Free Web mapping tools, Spatial analysis, Remote Sensing, Land Cover/Land Use changes (LULCc), WebGIS.

1   Introduction and Objectives of the Work

Identify and map Land Use/Land Cover (LULC) and detect their spatial and temporal changes (Change Detection) are some of the most important topics in Geographic Information Science. Nowadays, Remote Sensing (RS) techniques represent the main source of a wide range of environmental information about landscape and its changes, which is essential for an effective sustainable land planning and management. Moreover, RS has proved to be much better than traditional procedures for LULC mapping and monitoring, in terms of cost effectiveness and timeliness in the availability of information over larger areas [1]. On the whole, the recourse to aerial photos, satellite

B. Murgante et al. (Eds.): ICCSA 2013, Part III, LNCS 7973, pp. 566–581, 2013.

images (e.g., several Landsat data are free available for download), digital maps and - in general - geographic information has become fundamental to contribute to the above mentioned monitoring tasks.

Urban development is a complex process conditioned by factors which vary temporally and is characterised by different spatial patterns of urban growth [2, 3]. The term "sprawl" is used to describe the consciousness of an unsuitable growth, as a disordered expansion of urban areas [4]. Sprawl is the consequence of many individual decisions and among the possible causes of this phenomenon we can find population growth, economic factors, closeness to resources and facilities [5]. Is well known that the development of the urban areas is capable to transform landscapes formed by rural into urban life styles and to make functional changes [6], from a morphological and structural point of view [7, 8]. Such expansion typically takes place on former agricultural or natural areas [9, 11], so that changes in landscape cause significant repercussions for quality of life and natural habitat ecosystems, especially through their impacts on soil, water quality and climatic systems [12]. Collecting detailed information about these dynamics is important not only to analyse the landscape transformation trends, but also to efficiently support planning and monitoring tasks.

Particularly in the Mediterranean area, one of the most significantly altered hotspots on Earth [13], the urbanisation process is still continuing with significant effects in landscape fragmentation and in loss of ecosystem services. The most recent and significant LULC changes are connected not only to urban sprawl, but also to other phenomena such as: agricultural intensification in the most suitable areas and agricultural abandonment in marginal areas; construction of plants for energy production from Renewable Energy Sources (RES) [14]; more frequent and more intense summer forest fires; and rapid expansion of tourism activities and infrastructures, mainly along the coasts [15, 18]. Historically, urban development and agriculture have acted as opposing activities in competition for the use of land [16]. In Italy, in particular, an increase in urban pressure on the countryside can still be observed, owing to land renting and displacement of certain typical urban functions outside the cities [6]. This trend has resulted in the progressive urbanisation of rural areas; in a general increase in the number of inhabitants in the periurban belts; and in a decrease in resident population in towns [19]. Due to these continuous anthropic pressures and dynamics, further research is needed to monitor and analyse, in quantitative and qualitative terms, on-going LULC changes [16, 17, 20, 21] and trends of urban-rural gradients. Data on urban-rural transformations are a valuable source of information in understanding the methods and the effects of urban landscape changes and trends. As a consequence, they can influence the formulation of guidelines and policies to address future growth as well as to put into practice urban green-space protection.

The main goal of the research activities presented in this paper was to implement in a free WebGIS platform a multi-temporal dataset analysed in order to identify the LULC changing pattern occurred during a fifty-year period. So that allowed, on one hand, understanding the changes within the zones of interest, focusing on urban growth/sprawl phenomenon and loss of rural/natural land. On the other hand, it was possible to implement a free accessible platform favouring the e-participation of local communities in the landscape planning process. In particular, the analyses were conducted in two different areas, representative of the Mediterranean landscape: the district of Serra San Bruno (Calabria region) and the Province of Avellino (Campania region), both of them in the southern Italy. As highlighted before, a common task of the two case studies is related to the publishing and sharing phases of the Geospatial data produced by means of a specific WebGIS application.

2   Materials and Methods

2.1   The Two Study-Areas

The first study area (from here indicated as: SA1) is located in the district of Serra San Bruno, which is one of the most interesting areas of Calabria because of the presence of many heritage resources of great natural, historic and architectural interest. The high environmental value of this area motivated the institution of two National Natural Reserves (now SCIs - Sites of Community Interest - of the Natura 2000 network, the centrepiece of EU nature and biodiversity policy) and of the Serre Natural Regional Park (17,700 ha) in 2004 (Fig. 1). Over time, like the many mountain areas of the Mediterranean basin, Serra San Bruno district has developed an economy that strongly depends on local natural resources [22].

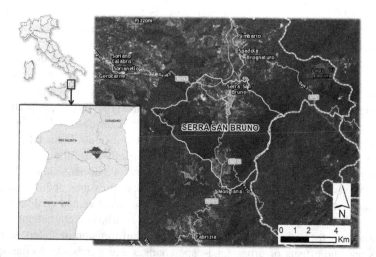

Fig. 1. Geographic location of the study area SA1. In yellow are highlighted the boundaries of the territory of Serra San Bruno municipality.

After the World War II, it suffered progressive depopulation and many traditional economic activities, such as charcoal production, became weak and went through periods of crisis. Recently, the development of forms of tourism, based on the valorisation of the rich cultural and natural resources, has been of some help in sustaining the local economy. Tourists search for high landscape quality, which should be improved through sensitive landscape planning based on accurate monitoring of the dynamics characterising land use [23].

The second study area (SA2) is in the Province of Avellino (Campania region), which is characterised by many small towns and villages scattered across its territory. Its capital city Avellino is situated in a plain called "Conca di Avellino" and surrounded by mountains (Massiccio del Partenio and Monti Picentini). Due to the Highway A16 and to other major roads, Avellino also represents an important hub on the road from Salerno to Benevento and from Naples. Avellino has suffered from seismic activity throughout its history and was struck hard by the disastrous Irpinia earthquake of 23 November 1980. Towns in the Avellino province were hardest hit.

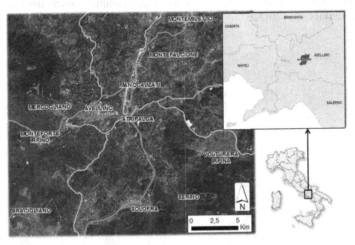

Fig. 2. Geographic location of the study area SA2, "Conca di Avellino"

During the last thirty years, the Italian Government has spent around 30 billion Euros on reconstruction. In order to regulate the reconstruction activities, several specific acts, decrees, zoning laws and ordinances have been issued: the first was the Law n. 219/1981, which entrusted the Campania Region with the coordination of the activities and the damaged Municipalities with urban planning. Since 2006 the urban planning issues of Avellino and of its neighbouring areas have been regulated by two instruments: P.I.C.A. (Italian acronym that stands for Integrated Project for Avellino City) and P.U.C. (Urban Plan for Avellino Municipality). The analysis has concerned the area of the Conca di Avellino (Figure 1), owing to its particular location as a plain between two natural protected areas: the Regional Park of Partenio (14,870 ha) and the Regional Park of Picentini Mountains (62,200 ha).

2.2   Data Processing

For the aims of this research, observation covered a period of intense and radical transformation of the Italian economic, cultural and territorial assets, ranging from the decade immediately after the end of World War II to present years. To characterise the dynamics of changes during the considered fifty year period, has been produced the LULC maps [24, 25] for each of the two areas of interest.

For SA1, LULC changes were analysed using aerial photographs from the Italian Military Geography Institute (IGMI), for years 1955 and 1983, and digital georeferenced orthophotos from the National Geoportal (NG) of the Italian Ministry of the Environment, for years 1994 and 2006. All the aerial photographs were georeferenced by using 20÷30 Ground Control Points (GCPs) and orthorectified by using a Digital Elevation Model (DEM) with 10 m x 10 m of geometrical resolution. Four datasets with a spatial resolution of 1.37 m and an RMSE (Root Mean Square Error) less than 6.5 m were produced. A set of 1998 orthophotos, a 1:10,000 numerical topographic map and the DEM were used as reference material in implementing the Geodatabase. For further details see [26].

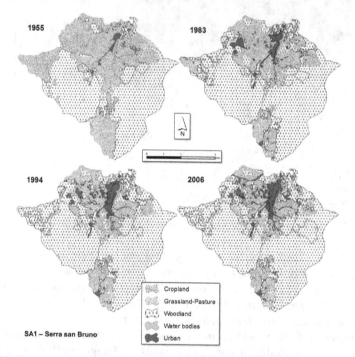

Fig. 3. LULC changes and dynamics within the SA1 for the different time intervals examined

Regarding SA2, to study and analyse LULC changes, the multi-temporal set of remote-sensed used was formed by: aerial photos (surveys carried by IGMI for years 1954, 1974 and 1990), Landsat images[1] (MSS 1975, TM 1993 and ETM+ 2004) and digital aerial orthophotos (1994 and 2006, from the NG). For further details see [27]. Using as thematic reference the NG orthophotos, the earliest information about LULC has been extracted from black-white monoscopic aerial frames taken in 1950s. Then, to obtain information about LULC for the next time interval into SA2, the Landsat

---

[1] Source: Global Land Cover Facility, GLCF, http://glcf.umiacs.umd.edu

images have been processed and classified [28]. The multi-temporal dataset has been geometrically registered, to decrease distortions effects and to reduce pixel errors that could be interpreted as LULC changes. Then, Landsat images have been atmospherically corrected [29]. To mitigate the seasonal effects, which often lead to errors in change detection, have been used only imagery acquired during the summer period, avoiding the uncertainness of inter-annual variability. The used approach was the supervised one (Maximum Likelihood Classification algorithm, MLC).

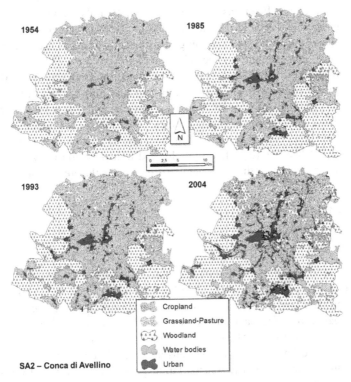

Fig. 4. LULC changes and dynamics within the SA2 for the different time intervals examined

In both cases, using a GIS approach, the results have been synthesised into maps of LULC changes. Spatial analysis tools have been exploited to integrate the data produced with other geographic information, to characterize the landscape dynamics through significant indices and to understand the changes therein. Thanks to the GIS capabilities, it has been possible to manage and to easily evaluate the amount and the directions of changes [30]. Such approach has allowed to determine and describe dynamics in LULC, taking into account the spatial information of each class (area, perimeter, etc.) and the information about amount, location, and typology of change.

2.3   Spatial Analysis of Landscape Transformations

To understand the dynamics of the urban sprawl/growth phenomenon [31] connected to LULC changes in the land surrounding the urban main settlement, different spatial analysis [32] have been performed, using various geographic layers: administrative boundaries, hydrography, main transportation network, Digital Elevation Model (DTM), Census data from the ISTAT (Italian National Institute of Statics, www.istat.it/en). Those layers, then, have been combined with the LULC maps, by means of spatial overlay functions performed using GIS tools and procedures [21].

Analysing and integrating LULC maps with other spatial data, it has been possible to produce specific GIS layers describing transformation and, consequently, to map specific features. For example, by the selection of the typology of LULC representative of urban features, it has been obtained the diachronic expansion of urban areas during the period examined. Then, by comparing this information with the Census data (achieved from ISTAT), the relationships between population growth trends and urban expansion have been evaluated. Further, following the same approach, it has been investigated the relationships between the urban expansion and other geographical and territorial features (e.g., hydrography and transportation network), also considering the interaction with these elements getting across the area of interest.

3   Results and Discussions

3.1   Landscape Transformations

The results coming from the methodologies above described, indicate that the urbanization has considerably modified the LULC in both of the study areas (Fig. 5 and Fig. 6), with significant land conversions. In particular, during the five decades analysed, the urban areas have significantly increased, mostly at the expense of the cultivated areas, which have most suffered the effects of the expansion of the built-up areas.

In general, the driving factors are various and include the effects of natural environment, demography, economy, transportation network, preference (by people) for proximity, neighbourhoods, and central/local policies [33]. Those factors are able to produce a clustering and localised pattern of urbanization, where new development has tended to infill around existing development, as well as a dispersed trend, wherein urban land uses increasingly spread out across the metropolitan area.

In the case of SA2 (Fig. 5), one of those factors is related to the proximity of Avellino to other urban centres (Monteforte Irpino, Mercogliano, Atripalda, Manocalzati and Montefredane). Those towns, situated in the central part of the Conca di Avellino (placed along the SW-NE direction), are currently in a territorial continuity with the main settlement of Avellino. Consequently, the presence and accessibility of transportation routes has forced patterns of urban growth, triggering the so called "linear branch" development [5]. In fact, population in outskirts of Avellino has grown faster than downtown, indicating a certain tendency of the outward expansion of urban areas. This sprawl is linked to the course of the state-road S.S. 7bis that, exactly along the SW-NE direction, connects Avellino to other towns, underlining the relation between place of residence and place of work.

The transformation of urbanised areas is also related to population dynamics [31], that drives the built-up area to expand. The physical growth of a city is normally connected to its population growth, as an increase in population size encourages the agglomeration of businesses and new urban development. Consequently, the city is quickly grown at its fringes, transforming the surrounding rural areas into dense industrial and commercial ones, or less dense suburban developments [34].

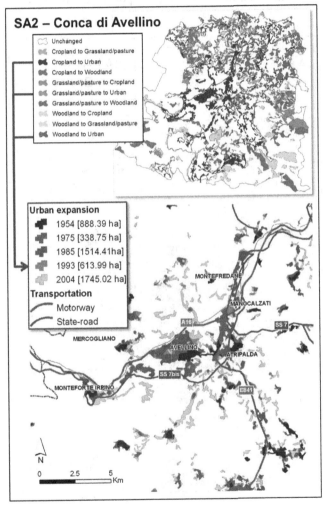

Fig. 5. SA2. Main LULC direction of changes and detail focusing on urban expansion over the fifty years span period

To this end, it was also interesting to compare changes in LULC whit the demographic data. In the case of SA1, the comparison of built-up expansion with ISTAT demographic data, during the last twenty-year period (1983÷2011), showed a growth rate of the urbanised area higher than the population growth rate, as expected. Just to mention a few data, built-up areas increased by 26% between 1983 and 1994 and by 9% between 1994 and 2011. On the other hand, the population increased only by 6% between 1981 and 1991 and by 2% between 1991 and 2011. In the case of SA2, while the total population within the study area increased by 14% between 1971 and 2001, urbanised areas increased by 75% between 1975 and 2004 (85% between 1954 and 2004). These examples of comparison clearly point out that growth rate of urbanised areas is always higher than the growth rate of population, during all the time span considered.

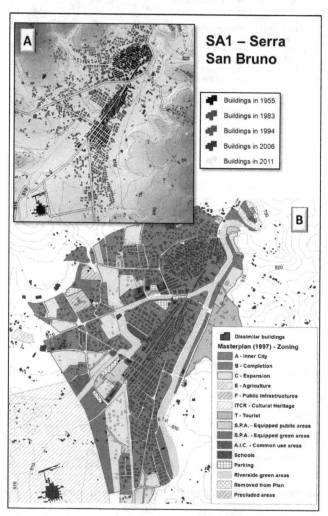

Fig. 6. SA1. (A) Detail of urban expansion detailed as building's growth over the fifty years span period. (B) Detail of General Master Plan (GMP) of the municipality of the Serra San Bruno highlighting (in red) dissimilar buildings according to plan zoning provisions.

Focusing, for instance, to the area surrounding Avellino, between 1981 and 2001 (that is immediately after the 1980 earthquake) it is possible to observe for the Capital city a decreasing trend in population, due to the transfer of many people from the main urban settlement to the neighbouring towns: the consequence is an "extended" urban area, with around 90,000 inhabitants. Another important push to the urban expansion has come from the indications of Master Plans. The analyses carried out were aimed to use LULC maps produced to focus on built-up areas, in order to evaluate their importance and role in shaping the character of the urban areas. In this framework, GIS tools were exploited to support spatial analysis and give suggestions useful to be shared with planners and policy makers. In fact, one of the major benefits of this approach is that highly visual and interactive modelling and analysis can be performed.

For SA1, a comparison of the number of buildings and of the demographic trend in the last decades, which decreased from 8,517 inhabitants in 1961 to 6,873 in 2011, shows that housing construction in the study area has not met residential needs but has rather answered the purpose of investing in the property market in a period of high monetary inflation [26]. A significant number of houses are unoccupied and many of them are let during summer for the so-called "tourism of return" and, to a lesser extent, for exogenous tourism [22]. In more detail, the highest rate of housing construction was recorded between 1983 and 1994, when, owing to the temporary lack of effective planning regulations, many buildings were constructed in unsuitable areas. Between 1994 and 2006, a significant downturn in the construction of new buildings was recorded, which did not affect agricultural areas, as it had happened in the past, but rather open spaces in discontinuous urban areas. The results coming from GIS processing confirm that only very few buildings were built in locations which met GMP zoning definition and standards (Fig. 6).

3.2   Mapping of Spatial Patterns

It is universal recognised the capacity of maps to offer an overview of and insight into spatial patterns and relations [35]. The maps produced in the framework of this study are able to show the structure of the landscape, display the changes in LULC and forecast future development. For this reason, these maps represent a fundamental tool to support, monitor and assess the effectiveness of planning policies [36]. Querying, consulting and sharing the maps - and the Geospatial DataBase (GeoDB) ad hoc developed - implies the adoption of specific and suitable GIS and WebGIS architectures, developed using free/open source (FOSS) packages. Therefore, the WebGIS application on purpose implemented aims at web publication of LULC maps and related geographical layers and represents a useful tool for spatial data sharing.

Various and widespread are the solutions commonly adopted to design and implement similar architectures, generally defined as Spatial Data Infrastructures (SDI). In particular, the implementation of applications in FOSS environment is actually supported by the large availability of software suites and packages, which make possible - among other things - to share and publish the resulting maps by means of suitable web interfaces. To implement the present WebGIS has been adopted the software environment offered by the OpenGeo Suite (http://opengeo.org/products/suite/), that brings together the OpenGeo Architecture into a single, easy-to-install integrated software package. This solution allows to get the geospatial information on the web and is an Open Geospatial Consortium (OGC, http://www.opengeospatial.org/) standards-compliant web mapping platform built on powerful, open source geospatial components. The Suite encompasses a set of specific functions for the automatic generation of thematic maps and allows the developer to play with and change the parameters of many functions. The core module of the WebGIS is GeoServer - an open source server written in Java – that operates as a node within the SDI implemented [37] and allows to share and edit geospatial data. Designed for interoperability (reads a variety of data formats), GeoServer has been exploited to manage the layers stored into the GeoDB and to publish them using open standards. GeoServer also supports the connection to existing web-based maps such as OpenLayers (http://openlayers.org/) and Google Maps (http://maps.google.it/), which have been used as base layers of the WebGIS developed. In addition, GeoServer has been also configured to provide the Geospatial data using Web Feature Service (WFS) standard, Web Map Service (WMS) and Web Coverage Service (WCS) specifications. In this way, users are allowed to access data of study area using any Desktop GIS software enabled to manage WFS, WMS and/or WCS connections. These protocols allow the users to perform their own spatial analysis by means their personal GIS software, used as client. Then, by means of GeoExplorer application (embedded in the OpenGeo Suite) it has been assembled a browser-based mapping module with the typical functionalities of a common desktop GIS. The WebGIS interface developed allows browsing, zooming, and querying the map layers of the study area and makes available a set of basic GIS functions. User options allow to execute many different visually explorative analysis operations, including the connection to additional WMSs. Examples of the implemented WebGIS are depicted in Figure 7.

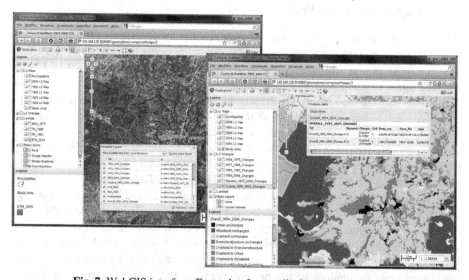

Fig. 7. WebGIS interface. Example of maps display and consultation

3.3   WebGIS Platforms Favouring the Participatory Approach

People's participation in planning should extend scientific expertise by adding local experiences, which reflect opinion and knowledge as well as the importance of social judgement. Referring to Stirling's typology of participation [38], in the ELC participation is encouraged from a substantive and even instrumental point of view [39] that means its role in planning should be to increase information or to justify a policy decision to be made. Citizens are increasingly demanding greater participation in influencing public policy decisions [40] also because they perceive that their actions could be partially or totally accepted by local authorities [41].

Indeed, a large number of planning experiences reported in scientific literature document several different problems and obstacles that are encountered when trying to put into practice an effective participatory approach [39], and because of which the statement "transdisciplinarity involves participation" often seems to sound as merely axiomatic [42]. Citizen participation in the planning process needs new tools in order to put into practise a bottom-up participation process. Nowadays, a variety of participatory tools and community based procedures integrating local knowledge and preferences with the inputs of the experts is [43]. The convergence of the Web 2.0 and the arising geospatial technologies allow a broader participation of local communities in the planning process. Indeed, the Web 2.0 is itself a participatory platform in which people contributing in the creation of new content, thus it represents an essential tool in collaborative decision making. In the planning process, e-participation can be seen as the integration of the new internet-based technologies in implementing the traditional approaches reaching the higher levels of the Arnstein ladder [44].

Thanks to WebGIS tools, information sharing is able to improve the effectiveness of analyses carried out, in order to support the decision and policy making, as well as of the planning debate. Moreover, in wiki-planning it can be recognised a two-way approach in which community not only can express its opinion but can also substantially influence choices. WebGIS coupled with Web 2.0 technologies allow to overcome most of the limitations and critical aspects of traditional methods of public participation inherent to their synchronous and place-based nature [40, 43]. WebGISs offer additional solutions accessible also to non-experts such as specific online tools (e.g. web-based discussion fora, online questionnaires, etc.). In turn, WebGISs provide an alternative to the traditional place-based planning (e.g. public meetings/hearings) without time or location restriction [45, 46].

4   Conclusions

The methodologies developed in the framework of this study have concerned the Land Cover classification and changes in two different study areas of Mediterranean landscape in southern Italy. Characterising LULC dynamics by means spatial analysis and GIS approach, the research focuses on urban areas change patterns in relation to morphology, transportation network, population growth, etc. The results confirm the capability of multi-temporal RS data to provide accurate and cost-effective tools to understand changes, through detailed spatiotemporal analysis. Maps and geospatial data produced have been stored and structured in a suitable free WebGIS application, that allows to access and share the results through the Web. The main advantage of this methodology is not only to easily manage and update GIS databases, but also to provide a very useful tool to improve land planning and monitoring tasks and to support environmental policy and decision making favouring the e-participation of citizen since the earlier stages of the planning process. Further, this approach is applicable to studies at various locations, and represents a valid contribution to cope with matters related to the sustainable rural and urban development.

This research also offered an example of how GIS techniques and methodologies can be used to support land-use planning and decision making in urban growth management [47, 48], by providing helpful tools to spatially analyse and study landscape patterns, spatial variations and their most significant correlations.

Finally, mapping at times the patterns of urbanisation and using ad hoc Geospatial tools is fundamental to forecast future development and in order to monitor and to assess the effectiveness of planning policies. In particular, maps provided in a WebGIS environment, thanks to their interactive and dynamic appearance, will support the user in solving geospatial analysis problems. Moreover, using a WebGIS platforms allow citizens with different values in different locations at the same or different time, to collaborate using shared data, alternatives, and negotiations towards agreements [45]. A fortiori, in case of free web mapping tools which do not require additional costs for end users. The future developments of this research will also take into account the currently increased availability of mobile devices, proper software and abundant data also produced in the framework of a specific Volunteered Geographical Information (VGI) [49] a version of crowdsourcing in which members of the general public can create georeferenced data synthesised into geodatabases. The use of VGI has proven very successful in the past few years and presents several advantages considering that the information created is free and can provide vast amounts of types of data that have never figured before in mapping practice [50]. However, VGI carries no assurance of quality, normally highly variable and undocumented, and its coverage is incomplete [50]. Thus, the quality of data provided by volunteers should be checked and assured before their use in the planning process. In the next steps of our research, the role of such novelties in geospatial information will be take into account to integrate the maps traditionally produced with layers of knowledge arisen by local stakeholders and people living the places. Indeed, cconsidering that neogeography has generally opened up the collection and use of information to a larger section powerful part of the society [51], it is necessary to empower participation of marginalised groups since the first phases of the participation process, giving them the opportunity to contribute in the design and development of technological platforms. In other words, it is necessary to increase participation of local communities and bring new voices that are unheard or ignored [51].

In this sense, a benchmark for the next developments of the present research activities could be represented by the experience carried out within the Geo-Wiki Project [52].This is a global network of volunteers, aiming to help the quality improvement of land cover maps (at a global scale). In the framework of a specific WebGIS application (mainly based on Google Earth data and APIs), volunteers are asked to review hotspot maps of global LULC discrepancy and fix, exploiting both Google Earth views and their local knowledge, if the maps are correct or incorrect. Their input is stored in a database, along with uploaded additional information (e.g., photos), so that it is possible to use them in the future to update/improve exiting maps or to create new ones. An approach, conceived as a thoughtful combination of the above mentioned experiences and methodologies, will further favour and improve the proactive participation of the local communities in the planning process.

References

1. Dewan, A.M., Yamaguchi, Y.: Using remote sensing and GIS to detect and monitor land use and land cover change in Dhaka Metropolitan of Bangladesh during 1960-2005. Environmental Monitoring and Assessment 150, 237–249 (2009)
2. He, C., Okada, N., Zhang, Q., Shi, P., Li, J.: Modelling dynamic urban expansion processes incorporating a potential model with cellular automata. Landscape and Urban Planning 86, 79–91 (2008)
3. Kong, F., Yin, H., Nakagoshi, N., James, P.: Simulating urban growth processes incorporating a potential model with spatial metrics. Ecological Indicators 20, 82–91 (2012)
4. Sudhira, H.S.S., Ramachandra, T.V.V., Jagadish, K.S.S.: Urban sprawl: metrics, dynamics and modelling using GIS. International Journal of Applied Earth Observation and Geoinformation 5, 29–39 (2004)
5. Hoffhine Wilson, E., Hurd, J.D., Civco, D.L., Prisloe, M.P., Arnold, C.: Development of a geospatial model to quantify, describe and map urban growth. Remote Sensing of Environment 86, 275–285 (2003)
6. Murgante, B., Danese, M.: Urban Versus Rural: The decrease of agricultural areas and the development of urban zones analyzed with saptial statistics. International Journal of Agricultural and Environmental Information Systems 2, 16–28 (2011)
7. Antrop, M.: Changing patterns in the urbanized countryside of Western Europe. Landscape Ecology 15, 257–270 (2000)
8. Antrop, M.: Landscape change and the urbanization process in Europe. Landscape and Urban Planning 67, 9–26 (2004)
9. Hidding, M., Needham, B., Wisserhof, J.: Discourses of town and country. Landscape and Urban Planning 48, 121–130 (2000)
10. Perella, G., Galli, A., Marcheggiani, E.: The Potential of Ecomuseums in Strategies for Local Sustainable Development in Rural Areas. Landscape Research 35, 431–447 (2010)
11. Perchinunno, P., Rotondo, F., Torre, C.: The Evidence of Links between Landscape and Economy in a Rural Park. International Journal of Agricultural and Environmental Information Systems (IJAEIS) 3, 1–14 (2012)
12. Antrop, M.: Why landscapes of the past are important for the future. Landscape and Urban Planning 70, 21–34 (2005)
13. Myers, N., Mittermeier, R., Mittermeier, C.: Biodiversity hotspots for conservation priorities. Nature 403, 853–858 (2000)
14. Caiaffa, E., Marucci, A., Pollino, M.: Study of sustainability of renewable energy sources through GIS analysis techniques. In: Murgante, B., Gervasi, O., Misra, S., Nedjah, N., Rocha, A.M.A.C., Taniar, D., Apduhan, B.O. (eds.) ICCSA 2012, Part II. LNCS, vol. 7334, pp. 532–547. Springer, Heidelberg (2012)
15. Antrop, M.: Landscape change and the urbanization process in Europe. Landscape and Urban Planning 67, 9–26 (2004)
16. EEA (European Environment Agency): Urban sprawl in Europe: The ignored challenge. EEA Report n. 10/2006. Publications Office of the European Union, Luxembourg (2006)

17. EEA (European Environment Agency): Landscape Fragmentation in Europe. EEA Report n. 2/2011, joint EEA-FOEN Report. Publications Office of the European Union, Luxembourg (2011)

18. Neri, M., Menconi, M.E., Vizzari, M., Mennella, V.G.G.: Propuesta de una nueva metodología para la ubicación de infraestructuras viarias ambientalmente sostenibles. Aplicación en el tramo viario de la pedemontana Fabriano-Muccia. Informes de la Construcción 62, 101–112 (2010)

19. Alberti, M., Tsetsi, V., Solera, G.: La città sostenibile: analisi, scenari e proposte per un'ecologia urbana in Europa, Franco Angeli, Milano (1994)

20. Girard, L.F., De Toro, P.: Integrated spatial assessment: a multicriteria approach to sustainable development of cultural and environmental heritage in San Marco dei Cavoti, Italy. Central European Journal of Operations Research 15, 281–299 (2007)

21. Fichera, C.R., Modica, G., Pollino, M.: GIS and Remote Sensing to Study Urban-Rural Transformation During a Fifty-Year Period. In: Murgante, B., Gervasi, O., Iglesias, A., Taniar, D., Apduhan, B.O. (eds.) ICCSA 2011, Part I. LNCS, vol. 6782, pp. 237–252. Springer, Heidelberg (2011)

22. Di Fazio, S., Laudari, L., Modica, G.: Heritage interpretation and landscape character in the forestry district of Serra San Bruno (Calabria, Italy). In: XVII World Congress of the International Commission of Agricultural Engineering (CIGR) on Sustainable Biosystems through Engineering, Québec City, Canada (2010)

23. Selicato, M., Torre, C.M., La Trofa, G.: Prospect of integrate monitoring: A multidimensional approach. In: Murgante, B., Gervasi, O., Misra, S., Nedjah, N., Rocha, A.M.A.C., Taniar, D., Apduhan, B.O. (eds.) ICCSA 2012, Part II. LNCS, vol. 7334, pp. 144–156. Springer, Heidelberg (2012)

24. Lucas, R., Rowlands, A., Brown, A., Keyworth, S., Bunting, P.: Rule-based classification of multi-temporal satellite imagery for habitat and agricultural land cover mapping. ISPRS Journal of Photogrammetry and Remote Sensing 62, 165–185 (2007)

25. Yuan, F., Sawaya, K.E., Loeffelholz, B.C., Bauer, M.E.: Land cover classification and change analysis of the Twin Cities (Minnesota) Metropolitan Area by multitemporal Landsat remote sensing. Remote Sensing of Environment 98, 317–328 (2005)

26. Di Fazio, S., Modica, G., Zoccali, P.: Evolution Trends of Land Use/Land Cover in a Mediterranean Forest Landscape in Italy. In: Murgante, B., Gervasi, O., Iglesias, A., Taniar, D., Apduhan, B.O. (eds.) ICCSA 2011, Part I. LNCS, vol. 6782, pp. 284–299. Springer, Heidelberg (2011)

27. Fichera, C.R., Modica, G., Pollino, M.: Land Cover classification and change-detection analysis using multi-temporal remote sensed imagery and landscape metrics. European Journal of Remote Sensing 45, 1–18 (2012)

28. Lillesand, T., Kiefer, R.W., Chipman, J.: Remote sensing and image interpretation. Wiley and sons, New York (2003)

29. Song, C., Woodcock, C., Seto, K., Lenney, M., Macomber, S.: Classification and Change Detection Using Landsat TM Data When and How to Correct Atmospheric Effects? Remote Sensing of Environment 75, 230–244 (2001)

30. Petit, C.C., Lambin, E.F.: Integration of multi-source remote sensing data for land cover change detection. International Journal of Geographical Information Science 15, 785–803 (2001)

31. Bhatta, B.: Analysis of urban growth pattern using remote sensing and GIS: a case study of Kolkata, India. International Journal of Remote Sensing 30, 4733–4746 (2009)

32. Fotheringham, S., Rogerson, P.: Spatial analysis and GIS. Taylor & Francis ltd., London (1994)

33. Mayer, C.J., Somerville, C.T.: Land use regulation and new construction. Regional Science and Urban Economics 30, 639–662 (2000)
34. Huang, B., Zhang, L., Wu, B.: Spatiotemporal analysis of rural–urban land conversion. International Journal of Geographical Information Science 23, 379–398 (2009)
35. Kraak, M.-J.: The role of the map in a Web-GIS environment. Journal of Geographical Systems 6 (2004)
36. Hardin, P., Jackson, M., Otterstrom, S.: Mapping, measuring, and modeling urban growth. In: Jensen, R.R., Gatrell, J.D., McLean, D. (eds.) Geo-Spatial Technologies in Urban Environments, pp. 141–176. Springer, Heidelberg (2007)
37. Steiniger, S., Hunter, A.: Free and open source GIS software for building a spatial data infrastructure. In: Bocher, E., Neteler, M. (eds.) Geospatial Free and Open Source Software in the 21st Century, pp. 247–261. Springer, Heidelberg (2012)
38. Stirling, A.: Analysis, participation and power: justification and closure in participatory multi-criteria analysis. Land Use Policy 23, 95–107 (2006)
39. Sevenant, M., Antrop, M.: Transdisciplinary landscape planning: Does the public have aspirations? Experiences from a case study in Ghent (Flanders, Belgium). Land Use Policy 27, 373–386 (2010)
40. Boroushaki, S., Malczewski, J.: ParticipatoryGIS: A Web-based Collaborative GIS and Multicriteria Decision Analysis. URISA Journal 22, 23–32 (2010)
41. Murgante, B., Tilio, L., Lanza, V., Scorza, F.: Using participative GIS and e-tools for involving citizens of Marmo Platano-Melandro area in European programming activities. Journal of Balkan and Near Eastern Studies 13, 97–115 (2011)
42. Modica, G., Zoccali, P., Di Fazio, S.: The e-participation in tranquillity areas identification as a key factor for sustainable landscape planning. In: Murgante, B., et al. (eds.) ICCSA 2013, Part III. LNCS, vol. 7973, pp. 550–565. Springer, Heidelberg (2013)
43. Jankowski, P.: Towards Participatory Geographic Information Systems for community-based environmental decision making. Journal of Environmental Management 90, 1966–1971 (2009)
44. Murgante, B.: Wiki-Planning: The Experience of Basento Park In Potenza (Italy). In: Borruso, G., Bertazzon, S., Favretto, A., Murgante, B., Torre, C.M. (eds.) Geographic Information Analysis for Sustainable Development and Economic Planning: New Technologies, pp. 345–359. IGI Global (2013)
45. Dragićević, S., Balram, S.: A Web GIS collaborative framework to structure and manage distributed planning processes. Journal of Geographical Systems 6, 133–153 (2004)
46. Boroushaki, S., Malczewski, J.: Implementing an extension of the analytical hierarchy process using ordered weighted averaging operators with fuzzy quantifiers in ArcGIS. Computers & Geosciences 34, 399–410 (2008)
47. Cerreta, M., De Toro, P.: Integrated spatial assessment for a creative decision-making process: a combined methodological approach to strategic environmental assessment. International Journal of Sustainable Development 13, 17–30 (2010)
48. Cerreta, M., Panaro, S., Cannatella, D.: Multidimensional spatial decision-making process: Local shared values in action. In: Murgante, B., Gervasi, O., Misra, S., Nedjah, N., Rocha, A.M.A.C., Taniar, D., Apduhan, B.O. (eds.) ICCSA 2012, Part II. LNCS, vol. 7334, pp. 54–70. Springer, Heidelberg (2012)
49. Goodchild, M.F.: Citizens as sensors: the world of volunteered geography. GeoJournal 69, 211–221 (2007)
50. Goodchild, M.F., Li, L.: Assuring the quality of volunteered geographic information. Spatial Statistics 1, 110–120 (2012)
51. Haklay, M.(M.): Neogeography and the delusion of democratisation. Environment and Planning A 45, 55–69 (2013)

Improving EU Cohesion Policy: The Spatial Distribution Analysis of Regional Development Investments Funded by EU Structural Funds 2007/2013 in Italy

Francesco Scorza

Laboratory of Urban and Regional Systems Engineering, University of Basilicata,
10, Viale dell'Ateneo Lucano, 85100, Potenza, Italy
francescoscorza@gmail.com

Abstract. The incoming European Programming Period 2014-2020 addresses to the New Cohesion Policy the role of effective promotion of regional development. The principle of concentration on few strategic issues appears to be reached a mature approach through local Managing Authorities. This concentration affects several dimension of cohesion policy: economic development, social inclusion, labour market, place based instances, infrastructures, private investments. The spatial dimension of EU Regional Policies, since the first European programming cycle 1989s-1993, has not been analysed in deep due to the lack of data concerning programs implementation at regional level. During last years new tools have become available under the umbrella phenomena of open-data, VGI etc. In particular we refer to the project 'opencoesione' by Italian Ministry for Territorial Cohesion providing open data on Operative Programs funded by EU Structural funds. The service, with the general objective of improving Citizen Engagement on investments policies, offers a data set with specific information concerning each project (operative program, beneficiary, budget, funds etc.). The paper, after a general position of the actual issues of improving effectiveness of public investments in regional developing programs, describes the spatial data analysis process for the evaluation of spatial concentration of investments in the 2007-2013 programming period. The results are interesting for the opportunity to evaluate 'ex-post' aggregation of public and private investments. Conclusion regards possible application in developing incoming New Cohesion Policies at national and regional level.

Keywords: Regional Development Programmes, New Cohesion Policy, Spatial concentration of investments, Open-Cohesion.

1   Introduction

The Reformed Cohesion Policy, developed in the context of Europe 2020 agenda, opens to an integrated place-based approach for territorial and social cohesion. Smart growth, sustainable growth and inclusive growth for EU 2020 represent overall goals to be achieved under the comprehensive approach defined by Barca [1] as 'place based approach'. The paper gives a short review of New Cohesion Policy issues

B. Murgante et al. (Eds.): ICCSA 2013, Part III, LNCS 7973, pp. 582–593, 2013.

assumed to be relevant for defining procedures in spatial evaluation of the impact of EU regional programs investments.

A relevant aspects for improving effectiveness of cohesion policy investments regards the principle of concentration. There are different interpretations of such instance connected to the multidimensional structure of EU intervention strategy at regional and local scale. The spatial concentration is considered as a relevant condition in order to measure effects and impacts at local scale. In the third chapter the principle of concentration is described under multidimensional positions, comparing thesis and innovative approaches.

The contribution of open data to the impact assessment of EU Operative Programs appears to be mature in concept but still week in accuracy of available data bases. We used for the research data from the project 'opencoesione' by Italian Ministry for Territorial Cohesion. The Italian Ministry engaged with this unstoppable process of collecting and sharing data for improving citizens commitment on public policies. It developed a web service distributing data on investments policies developed by National and Regional Operative Programs 2007/2013 matching together data from regional and national administrations. The results are analysed in the paragraph number four of the paper with the application of spatial analysis techniques for the evaluation of spatial effects.

Conclusions regards possible application and perspectives for improving and supporting regional development planning considering the exploitation of open data sources and spatial analysis in order to define specific territorial targets coherent with place based approach of EU new cohesion agenda.

2   New Cohesion Policy: Expectations and Reforms

EU cohesion policies include different areas of intervention and generally are carried out in order to promote the principle of redistributing opportunities among European regions and territories. It is the largest area of expenditure for European Union and it is possible to affirm that policy analysis tends to overlook the evaluation stage of such complex strategies while a proper assessment practice [2].

The EU Cohesion Policy is actually interpreted as the main tool in order to achieve the Europe 2020 target addressing a wide range of EU economic, environmental and social objectives. It represents a driven tool toward a new concept of Europe with smart, sustainable and inclusive growth. It currently offers both examples of significant economic and environmental "win-wins" and of "tradeoffs" that fail to offer net added value.

The reform of cohesion approach can be highlighted in two main concept areas including a wide spread of arguments and objectives:

- Investment choices: "where to spend more, where to spend less"
- Investment better - via improved Cohesion Policy governance and tools

The two key EU reference strategies for the next decades are defined by 'Europe 2020' & the Territorial Agenda (TA) 2020.

Europe 2020' is aimed at providing more jobs and better lives 'by stimulating smart, sustainable and inclusive growth' over the coming decade. It involves EU Member States to integrate efforts related to socio-economic development through greater coordination of national and European policies. This strategy was approved by the European Council in June 2010 after three months of elaboration and consultation.

The TA 2020 also puts forward an ambitious strategy, though applying specifically here to EU territorial development. Although this document is also designed for a very wide audience, it has received a lower level of public recognition than Europe 2020' strategy. This probably stems from its elaboration process, which was essentially intergovernmental in nature, i.e. a collaboration between the national authorities responsible for spatial planning and territorial development in the EU. The TA 2020 has not been formally adopted by any EU body. TA 2020 was adopted in May 2011 at the informal ministerial meeting.

Europe 2020 and the TA 2020 thus originate from different political processes, and have a different political status, but the aim is to reinforce each other integrating territorial development and inclusion.

The Europe 2020 strategy is mainly focussed on economic development, in particular the recovery from the 2008 financial crisis and the strengthening of the development opportunities in the EU. Europe 2020 has replaced the Lisbon strategy. It puts forward three mutually reinforcing priorities:

1. Smart growth: developing an economy based on knowledge and innovation.
2. Sustainable growth: promoting a more resource efficient, greener and more competitive economy.
3. Inclusive growth: fostering a high-employment economy delivering social and territorial cohesion.

The TA 2020 is the action-oriented policy framework of the ministers responsible for spatial planning and territorial development in support of territorial cohesion in Europe. It aims to provide strategic orientations for territorial development, fostering integration of the territorial dimension within different policies across all governance levels while overseeing implementation of the Europe 2020 strategy in accordance with the principles of territorial cohesion.

Six main territorial priorities for the development of the EU have been set out in the TA 2020:

1. Promoting polycentric and balanced territorial development as an important precondition of territorial cohesion and a strong factor in territorial competitiveness.
2. Encouraging integrated development in cities, rural and specific regions to foster synergies and better exploit local territorial assets.
3. Territorial integration in cross-border and transnational functional regions as a key factor in global competition facilitating better utilisation of development potentials and the protection of the natural environment

4. Ensuring global competitiveness of the regions based on strong local economies as a key factor in global competition preventing the drain of human capital and reducing vulnerability to external development shocks

5. Improving territorial connectivity for individuals, communities and enterprises as an important precondition of territorial cohesion (e.g. services of general interest); a strong factor for territorial competitiveness and an essential condition for sustainable development

6. Managing and connecting ecological, landscape and cultural values of regions, including joint risk management as an essential condition for long term sustainable development

In the figure below IEEP summarized the relationships between economic and environmental outcomes from policy interventions and investments in a win-loss diagram.

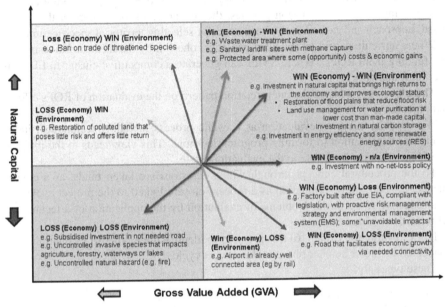

Fig. 1. Relationships between economic and environmental outcomes from policy interventions and investments [3]

This complex policy framework is based on the key objective of "achieving greater economic and social cohesion in the European regions". Anyway some critical consequences could be derived from the point of view of convergence process for lagged regions: the imbalance between regional objectives and financial resources; the existence of serious difficulties for complying with earmarking; and the unknown effects of other policies on regional convergence [4].

3   Strategic 'Place' Concentration

The principle of concentration is widely stressed within New Cohesion Policy framework.

The concentration of EU efforts is contra-posed to the indiscriminate distribution of funding ('raining models'). This interpretation could intend that investment promoted by Regional Operative Programs (ROPs) should be focused on specific and circumscribed instances generating effective local development processes. We are in the case of redistributing investment effectiveness and positive outcomes on local communities, instead of realizing a well balanced €/citizen rate within a region.

This synthetic and not exhaustive remark allows us to highlight two relevant aspects of concentration principle: the concentration on objectives and the territorial concentration of investments. If the first appears to be not so far from traditional behavior of managing EU cohesion policies, the second level looks more at the 'place based' approach and expresses the importance of selecting territorial specification.

Where agriculture and farmers can generate cohesion by the means of investments and funding from ROPs? Where SMEs can generate a competitive cluster in EU production framework?

These questions change the interpretation pattern on the evaluation of ROPs effects and objectives.

We consider program evaluation as a cyclic process including techniques, judgments and contribution to forming program decisions. This view leads to the integration of program/project cycle to the evaluation cycle.

To the procedural concept through which decisions are taken binds, as a consequence, the concept of evaluation as a process closely linked to the project cycle [5]. It is reductive to solve the problems of evaluation by the application of a techniques set related to limited issues.

In the "evaluation cycle" [6] it is possible to identify three types of evaluation each connected to one or another phase of the project cycle.

These types are:

- the evaluation of the state of the art;
- the benchmarking evaluation of strategies, plans or alternative projects;
- the achieved results evaluation.

The evaluation cycle is entitled to respond to the questions that arise at each stage of the project cycle and in particular:

- the assessment of the state of the art is closely related to the formulation of the "raison d'etre" and purpose of the project;
- the assessment of the pre-feasibility of the project highlights the need to put an assumption on the occurrence of conditions not dependent on the project so that the expected benefits can be achieved;
- the feasibility assessments leads to the definition of costs and expected benefits from the project, it identifies the sources of greatest concern dependent on various components of the work and it limits the field of choices;

- the benchmarking evaluation, between different intervention alternatives, leads to the choice to intervene and what changes have to be made to the project;
- the evaluation of the state of the art of the project defines whether to continue the project and what corrective should be provided;
- the final evaluation (and the ex-post evaluation) determines whether the assumptions made ex-ante were reliable and permits do decide if the experience can be repeated.

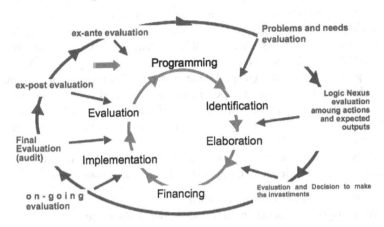

Fig. 2. "Project cycle and evaluation cycle" [7]

Connected to the interpretation depending on evaluation approach to local development processes, appeared the thesis of "renationalization" of cohesion policies [8]. This approach reinforced the role of national administration in driving the implementation of ROPs at regional ad local level. While the negotiation of Member States at the top level of Cohesion Policy hierarchy was previously considered the primary role of National Authorities, now the importance is mainly focussed on the implementation phase. This idea fits more with the 'place based' approach in terms of local specific needs interpretation. We intend that managing authorities and public administrations generate a progressive perception of territorial capacity and/or territorial needs under the program structure of ROPs. This process of reinforcing territorial knowledge includes tools and procedure of analysis in order to define the spatial specification of ROPs.

Data availability is at the base of such process and data sources comes from different organizations entitled of ROPs management.

In following section we describe the use of an open data service provided by Italian Ministry for Territorial Cohesion. The project 'opencoesione'[1] collected and distributed data on Operative Programs implementation in Italy for the programming period 2007/2013.

---

[1] http://www.opencoesione.gov.it/

4   Open Data for Effective Spatial Evaluation of Cohesion Policies

Today data availability is not a problem in terms of lack of data sources but mainly in terms of data management, certification and standard exchange protocols. Many people and organizations collect a wide range of different data in order to perform their own tasks.

The Open data, and in particular on open government data, are an immense resource still largely untapped. The Government role is particularly important in this sense, not only for the quantity and the centrality of the data collected, but also because most of the government data are public by law, and therefore should be made open and available for anyone to use.

According to the Open Knowledge Foundation Italia [9] there are many circumstances in which we can expect that the open data have significant value. There are also several categories of individuals and organizations that can benefit from the availability of open data, including public administration. At the same time it is not possible to predict how and where value is created.

We can identify a large number of areas where public open data contribute to create value for user knowledge building and participation, among them:

- Transparency and democratic control
- Participation and increasing influence in the public discussion
- Improvement or creation of products and services for private sector
- Innovation and R&D
- Improving the efficiency of public services
- Improving the effectiveness of public services
- Measuring the impact of public policies

We are interested in the last point of application as the extraction of new knowledge by combining different sources of data and the identification of regularities that emerge from the analysis of large masses of data represent the core of the application we propose for the evaluation of ROPs impact at local scale.

In Italy there are many initiatives opening of information assets undertaken by public central and local administrations. The portal dati.gov.it, (available since 2011) is a milestone in the process of opening a new era for innovation and transparency in the public administration.

Other most famous experienced in this route are: American Data.gov was launched by the Obama administration as a result of the Directive on Open Government in December 2009; Data.gov.uk strongly backed and sponsored by Tim Berners-Lee "the inventor of the World Wide Web"; Australia Data.gov.au; Canada Data.gc.ca; Norway Data.norge.no; France Data.gouv.fr; European portal beta opendata.europa.eu.

Actually we can affirm that the practice of open data and data stores government has extended, but a lot of work and efforts should still be pay in order to get affective services for data integration.

The project Open Cohesion provide an open data service concerning cohesion policies effects with a orientation toward planning processes. The initiative was strongly supported by the Minister for Territorial Cohesion, which warns the urgency of a more active participation of citizens in decision-making relating to planning decisions and in the process of social vigilance on the use of collective resources.

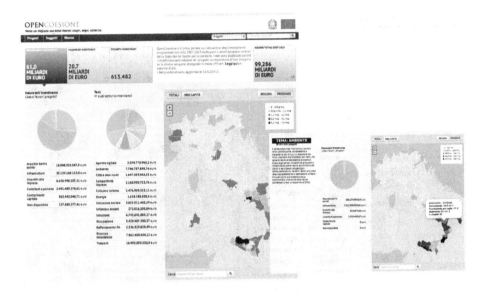

Fig. 3. Opencoesione screen shots

The publication of the data in an accessible format and reusable on their corporate websites shows the willingness of the government to move in a systematic way towards a structure of transparency that encourages the active participation of citizens and the re-use of data. The service pursues the objective of improving Citizen Engagement on investments policies, and offers a data set with specific information concerning project funded by the current programming period 2007-2013 matching implementation data from regional and national administrations entitled of Ops management.

We develop our analysis exploiting this data set. We worked with a data set of project funded by the whole of regional and national operative programs in Italy updated at the 31-12-2012.

The data set includes more than 600.000 records with information on the classification of projects per thematic and programmatic objectives, the amount of resources classified by EU relevant categories (admitted investment, certified expenses etc.), the beneficiary, etc.

The data set includes also a relevant information concerning the address of each project. This aspect allowed us to overcome the standard representation od aggregated data per administrative units in order to obtain a more expressive analysis based on the spatial punctual distribution of the interventions.

In the following figures are represented the traditional representation of aggregated data provided by www.opencoesione.gov.it and our elaboration through a geo-coding of the project address developed using google fusion table (we consider the Basilicata Region as target area).

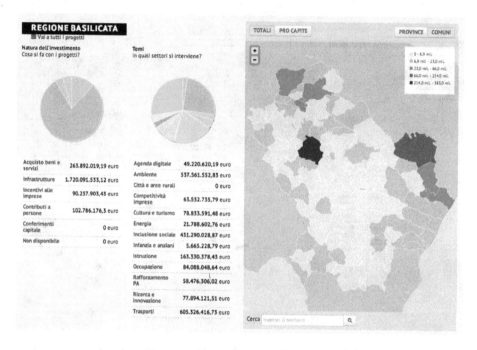

Fig. 4. ROPs investments in Basilicata from Opencoesione

Apart from the visual representation it is important to underline how this data could improve the analysis of territorial impacts of cohesion policy. In fact the analysis of spatial distribution of investments can give information on the territorial specialization adopting standard spatial geostatistics applied on a point pattern dataset.

In this direction, previous researches remarked the value that a proper spatial dataset give to the citizen involvement and for supporting the public programming and managing activities. We refer to the experience developed by PIT Marmo Platano Melandro (Basilicata – IT) during the EU programming period 2000-2006 with a web gis service for the spatialization of development policies [10].

While in that experience the main effort was in territorial data production we have to affirm that today it is possible to develop accurate spatial analysis concerning the distribution of EU funded investments with public open data.

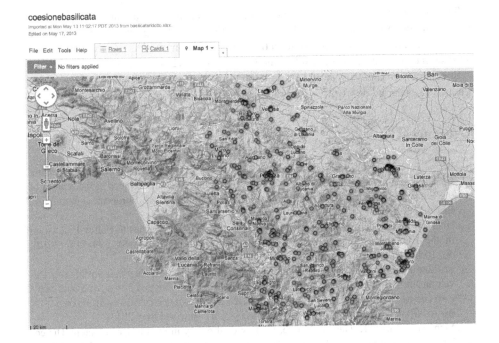

Fig. 5. Point pattern data set from geocoding open data from Opencoesione project

5   Conclusions

Place based approach will bring to innovations in EU cohesion management. Where outcomes indicators measure the implementation of cohesion operative program [11] other efforts should be addressed to the identification of local specialization. It could generate not a fix picture of a context, places and communities evolve continuously especially as a reaction to the huge changes brought by economic crisis.

Main issues connected with the instances of the New Cohesion Policies are:

1. The need of a clear identification of the combined place-specific characteristics in each region;
2. a clear identification of the appropriate territorial context in order to implement effectively "smart specializations".

Open data phenomena represent an useful process that already driven the research from data production to exploitation of the informative value of several data sources available for everybody. But data and data analysis technique cannot bring to useful information. Regional science has the task to produce effective 'places' interpretation in order to support public decision in incoming generation of EU ROPs. We are in the case in which it is relevant to use numerous data sources and indicators assuming a variable rate of approximation in the accuracy of the datasets.

The information management and exchange implies problem in interoperability between sources, procedures and technologies. In the field of Regional development the ontological approach provided alternative interpretation models of the interaction between the context, the program and the beneficiaries [12] [13] [14].

Specialization analysis should be developed through an integrated set of technique oriented to generate descriptive geographies of the EU region at a variable scale.

References

1. Barca, F.: An agenda for a reformed cohesion policy: a place-based approach to meeting European union challenges and expectations. Independent report prepared at the request of the European Commissioner for Regional Policy, Danuta Hübner, European Commission, Brussels (2009)
2. Hoerner, J., Stephenson, P.: Theoretical Perspectives on Approaches to Policy Evaluation in the EU: the Case of Cohesion Policy. Public Administration 90, 699–715 (2012), doi:10.1111/j.1467-9299.2011.02013.x
3. Institute for European Environmental Policy (IEEP) 'Cohesion Policy and Sustainable Development, Executive Summary (2011), http://ec.europa.eu/regional_policy/sources/docgener/studies/pdf/sustainable_development/sd_executive_summary.pdf (retrived May 2013)
4. Mancha-Navarro, T., Garrido-Yserte, R.: Regional policy in the European Union: The cohesion-competitiveness dilemma. Regional Sci. Policy & Practice 1, 47–66 (2008), doi:10.1111/j.1757-7802.2008.00005.x
5. Las Casas, G.B.: Processo di piano ed esigenze informative. In: Clemente, F. (ed.) Pianificazione del Territorio e Sistema Informativo. F.Angeli, Milano (1984)
6. Lombardo, S. (ed.): "La valutazione del processo di piano. Contributi alla teoria e al metodo" Collana Scienze regionali. Franco Angeli – Milano (1995)
7. Las Casas, G., Scorza, F.: Un approccio "context-based" e "valutazione integrata" per il futuro della programmazione operativa regionale in Europa. In: Bramanti, A., Salone, C. (eds.) Lo Sviluppo Territoriale Nell'economia Della Conoscenza: Teorie, Attori Strategie. Collana Scienze Regionali, vol. 41. FrancoAngeli, Milano (2009)
8. Bachtler, J., Mendez, C.: Who Governs EU Cohesion Policy? Deconstructing the Reforms of the Structural Funds. JCMS: Journal of Common Market Studies 45, 535–564 (2007), doi:10.1111/j.1468-5965.2007.00724.x
9. Open Knowledge Foundation Italia, Open data handbook, http://opendatahandbook.org/ (retrived May 2013)
10. Murgante, B., Tilio, L., Lanza, V., Scorza, F.: "Using participative GIS and e-tools for involving citizens of Marmo Platano – Melandro area in European programming activities" special issue on "E-Participation in Southern Europe and the Balkans". Journal of Balkans and Near Eastern Studies 13(1), 97–115 (2011) ISSN:1944-8953, doi:10.1080/19448953.2011.550809
11. Barca, F., McCann, P.: Methodological note: outcome indicators and targets-towards a performance oriented EU cohesion policy and examples of such indicators are contained in the two complementary notes on outcome indicators for EU2020 entitled meeting climate change and energy objectives and improving the conditions for innovation, research and development (2011), http://ec.europa.eu/regional_policy/sources/docgener/evaluation/performance_en.htm (accessed October 1, 2011)

12. Scorza, F., Las Casas, G.B., Murgante, B.: That's ReDO: Ontologies and Regional Development Planning. In: Murgante, B., Gervasi, O., Misra, S., Nedjah, N., Rocha, A.M.A.C., Taniar, D., Apduhan, B.O. (eds.) ICCSA 2012, Part II. LNCS, vol. 7334, pp. 640–652. Springer, Heidelberg (2012)
13. Las Casas, G.B., Scorza, F.: Redo: applicazioni ontologiche per la valutazione nella programmazione regionale. Italian Journal of Regional Science - Scienze Regionali 10(2), 133–140 (2011), doi:10.3280/SCRE2011-002007
14. Scorza, F., Casas, G.L., Murgante, B.: Overcoming Interoperability Weaknesses in e-Government Processes: Organizing and Sharing Knowledge in Regional Development Programs Using Ontologies. In: Lytras, M.D., Ordonez de Pablos, P., Ziderman, A., Roulstone, A., Maurer, H., Imber, J.B. (eds.) WSKS 2010. CCIS, vol. 112, pp. 243–253. Springer, Heidelberg (2010)

A Web-Based Participatory Management and Evaluation Support System for Urban Maintenance

Ivan Blecic, Dario Canu, Arnaldo Cecchini, and Giuseppe Andrea Trunfio

DADU, Department of Architecture, Design and Urban Planning
University of Sassari, Alghero, Italy
{ivan,cecchini,trunfio}@uniss.it, dario.canu@gmail.com

Abstract. The participatory support system for urban maintenance we present here: (1) lets citizens directly report neighbourhood issues requiring attention from urban maintenance services, (2) evaluates the priority of reported issues, (3) allows the allocation and management of resources and workforce on solving issues and (4) permits public tracking of their status. Unlike many existing platforms for collecting citizens reports, our system incorporates an explicit, transparent and publicly accessible evaluation model to prioritise issues and assign resources for their solution. This, we argue, is a crucial element to assure that the principles of transparency, publicity, accountability and equity are observed by the municipal government. After presenting the system's standard workflow, the evaluation model and an example application, we discuss its possible more general implications for citizen participation.

Keywords: management and evaluation support system, participation, citizen reporting, priority sorting, urban maintenance.

1   Introduction

All is good on *Axiom*. People consume, robots work. Few things break down or get dirty, and when they do, nobody needs to report it, decide what to do and where to start from. Robots are everywhere, cleaning, repairing, maintaining everything, effortless and snappy.

On *Axiom*, the starliner from Pixar's film WALL-E, this paper is useless. Here on Earth, city affairs are a little different: things break down and get dirty all the time, there are no all-present and all-seeing robots patrolling, fixing and cleaning things, it isn't an effortless and snappy job, and those who are there to do it cannot see everything, have limited resources and time, need to decide where to start first, possibly explaining why to citizens. A day will come when WALL-Es will be around, but in the mean time we should put to service existing technology to assist us in these urban chores.

We here present one such web-based support system for urban maintenance. The easiest way to start is to list what the system does and lets users do. It:

B. Murgante et al. (Eds.): ICCSA 2013, Part III, LNCS 7973, pp. 594–605, 2013.

1. makes citizens city's "eyes and ears", by allowing them to report via Web the neighbourhood issues (damaged roads, signs, buildings; abandoned waste; untidy places; acts of vandalism, etc.) requiring action from urban maintenance services;
2. lets back-office operators from maintenance services check and validate (if necessary by sending out inspectors) the information reported by citizens;
3. evaluates and assigns a rating to issues, establishing their level of priority;
4. lets services' operators assign tasks and resources (e.g. maintenance and cleaning workforce, tools, machinery, etc.) to solve issues, and to flag their status and progress;
5. evaluates past distribution of resources among different neighbourhoods, which should hint operators to also follow inter-neighbourhoods distributional concerns for future assignment of resources on issues.
6. allows citizens to transparently see all this information and monitor the progress on reported issues.

We designed and integrated these features following a set of more general principles we believe municipal governments should observe. These principles are:

- *Openness and inclusiveness.* Citizens should be given a clear and uniform access to report issues and propose solutions.
- *Transparency.* All the proposals, alternatives, constraints and any other information relevant for decision-making should be known to citizens, easily accessible, clearly presented and made understandable.
- *Publicity principle.* In the general Rawlsian conception [1], the publicity principle bans government from selecting a policy that it would not be able or willing to defend publicly to its own citizens. In our case, the reasons for a decision to prioritise one rather than another reported issue should be explicit and as much as neutral and nonarbitrary as possible.
- *Accountability.* The decision-makers should openly acknowledge and assume the responsibility when they exercise their discretionary power of choice and decision.
- *Equity.* Distributional considerations among neighbourhoods should count. Given the constraints of limited resources and time, these should be distributed among citizens living in different neighbourhoods according to some principle of equal treatment.

Our attempt through the support system was to take these principles seriously and to try implement them.

In the following section we briefly discuss some recent trends and experiences which were useful sources of inspiration for us, but we also point at their shortfalls. In Section 3 we describe the system and the standard workflow from citizen's report to issue resolution, and present examples of interfaces and system's outputs. In Section 4 we specify the rating evaluation model used to prioritise the issues. Conclusions are in Section 5.

2   The Missing Link

The wave of Web 2.0 has produced countless systems and platforms for collecting citizens reports and suggestions on issues of urban maintenance. The most mature and widely used solutions often share a common set of features allowing citizens to describe, classify and sometimes place issues on a city map, to comment, vote and track them. The system administrators can then usually flag the issues to signal their status (received, in progress, solved) and thus permit their public monitoring. Some systems also allow dialogue and exchange of comments between the administrators and the citizens.

They are available on a variety of platforms and are using different hosting and application providing models. There are nation-wide services the municipalities can opt into, like multi-platform multi-device *City sourced* [2] in the USA, the web-based *decorourbano.org* [3] in Italy and *Cidade Democrática* [4] in Brazil. Some are stand-alone applications directly hosted by the municipalities, like Boston's multi-device *Citizens Connect* [5] and City of Venice's web-based IRIS system [6]. Few interesting experiments are starting to emerge around popular social networks, like the Brazilian *Urbanias* [7] developed for Facebook.

Among all the things they have in common, these platforms also share a common shortfall. While they have by and large successfully settled how the citizens should report, comment, vote and track issues, the missing link is the lack of an explicit, transparent and publicly accessible evaluation model to prioritise issues and assign resources for their solution. To speak in terms of the general principles we presented above, the systems mentioned may well grant greater openness and inclusiveness, possibly a somewhat better transparency, but the publicity principle, accountability and equity may only be assured if the criteria and the constrains for choosing which issues to fix when are publicly known (possibly after a public debate).

The system we present here is an attempt to show how this missing link – the evaluation model for prioritising issues – may be provided. What is relevant in our case is both that there *is* an explicit evaluation model, and that it is made *publicly* known.

3   The Standard Workflow of the System

The standard workflow around an issue is made of the following five steps: (1) citizen's report of the issue, (2) data validation by a back-office operator, (3) issue evaluation and rating of priority, (4) allocation of resources and workforce, (5) issue tracking. In Fig. 1 below we show a sample of front-end interfaces and in Fig. 2. a detail of the back-office issues management control panel.

Citizen's Report. Through an online form (cfr. top-right in Fig. 1), citizens can report the location and the type of the issue, provide a description and upload photos. The types of issues currently contemplated in the online form are: *waste* (uncollected or damaged waste containers, littering, unauthorised dumps, abandoned vehicles), *infrastructures* (water and sewage pipes damage or leaks), *transportation* (unauthorised parking, damaged, incorrect or missing signs and traffic lights), *maintenance* (fallen branches and trees, damaged flowerbeds, damaged roads and footpaths), *acts of vandalism* (graffiti, unauthorised billposting), *environment* (pollution, request for disinfestations, bad smells, stray animals).

In the online form we further ask citizens to answer several multiple-choice questions, reported in Table 1., which are essential for the subsequent evaluation and rating of the issue.

Table 1. Multiple-choice questions used for the evaluation and priority rating of issues

Questions	Possible answers (values used in the evaluation model in square brackets)
1. Is there a serious hazard for human health and security?	Yes, I'm certain [4] – Probably yes [3] – Probably not [2] – Certainly not [1] –I don't know
2. Are there waste and materials hazardous for the environment?	Yes, I'm certain [4] – Probably yes [3] – Probably not [2] – Certainly not [1] –I don't know
3. Does the issue obstruct natural flows and functions (e.g. water streams)?	Yes, I'm certain [4] – Probably yes [3] – Probably not [2] – Certainly not [1] –I don't know
4. Is there a risk the issue to cause traffic incidents?	Yes, it happened / was about to happen [4] – It is possible [3] – Probably not [2] – Certainly not [1] – I don't know
5. Does the issue obstruct the circulation of vehicles?	The final destination is completely inaccessible [4] – It is necessary to take alternative route to reach a destination [3] – The circulation is not obstructed but only slowed down [2] – It doesn't obstruct the circulation in any way [1] – I don't know
6. Does the issue obstruct the pedestrian routes and footpaths?	The path is completely obstructed [4] – The path must be avoided [3] – It is possible to transit but the circulation is slowed down [2] – No, it doesn't significantly obstruct the transit [1] – I don't know
7. How many people daily visit the place on average?	A lot (more than 500) [4]– Quite many (from 200 to 500) [3] – Not many (from 50 to 200) [2] – A little (less than 50) [1] – I don't know
8. How visible is the issue?	It can be immediately seen and it's very extended [4] – It can be seen if looked at [3] – It's hard to see [2] – It's barely visible [1] – I don't know

There are, of course, two standard problems with this approach. One is related to the inevitable uncertainty of interpretation and fuzziness when expressing evaluative judgements, so different citizens may give different meanings to questions and scales of answers, classifying and describing differently the same issue. The other problem is the possibility, even a strong likelihood, of strategic behaviour: knowing that different answers will induce different responses and actions by the maintenance services provides incentive to citizens to overemphasise the gravity and urgency when reporting issues.

These are hard problems hard to eradicate entirely. On the long-run it requires social learning and development of trust in the institutions and among citizens. In the mean time, we have devised two practical countermeasures. First, we tried to construct the questions and the possible answers (scale) in as natural and comprehensible a language as possible. We're not sure about how successful we were, and no doubt there is space for improvements, but it is a good general principle to follow. Second, the information provided by citizens are not directly feed into the evaluation model: the system operators serve as arbiters who validate, interpret, uniform and re-codify the information submitted to the system by citizens.

Visual tutorial

Detail of the online form for issue reporting

Map of reported issue

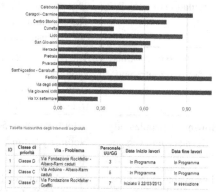

Past distribution of workforce among neighbourhoods (above) and issues' priority classes and statuses (below)

Fig. 1. Samples of the public web front-end

Data Validation by Back-Office Operators. As just said, all reported issues have to be assessed by operators before being processed by the evaluation model. Operators can check if the issue has already been reported, ask for further clarifications and discuss the report with citizens, and if necessary, send out inspectors for direct observation on the field. All this leads to a validated record of the report, which is then made publicly available and rated by the evaluation model. (cfr. below-left and below-right in Fig. 1).

ID	Via + Intervento	Quartiere	UU/GG	UU/GG in esecuzione	UU/GG classi A e B	CLASSE DI APPARTENENZA	Domanda UU/GG Offerta UU/GG	In esecuzione	Eseguito
1	Via Fondazione Rockfeller - Albero-Rami caduti	Via Giovanni XO	**A** 3	0		**B** Classe D	**C** 1,13	**D**	No
2	Via Arduino - Albero-Rami caduti	Centro Storico	5	0	u	Classe C	0,76	No	No
3	Via Fondazione Rockfeller - Graffiti	Via Giovanni XXIII	7	7	0	Classe D	1,13	Si	No
4	Via Roma - Veicolo abbandonato	Centro Storico	1	0	0	Classe C	0,76	No	Si
5	Via Antonio de Curtis - Veicolo abbandonato	Via XX Settembre	2	2	0	Classe C	0,28	Si	No
6	Via Fabra Pompeu - Discarica abusiva	Lido	6	0	6	Classe B	0,87	No	No
7	Via XX Settembre - Tombino (Perdita)	Via XX Settembre	10	10	10	Classe B	0,28	Si	No
8	Via Pietro Nenni - Albero-Rami caduti	Calabona	3	3	0	Classe C	0,63	Si	Si
9	Via Fiume - Tombino (Perdita)	Carmine	10	10	10	Classe B	0,94	Si	No
10	Via Mozart - Strada sporca	Cunetta	5	5	5	Classe B	0,39	Si	No
11	Via Pola - Disinfestazione	Fertilia	4	4	0	Classe C	1,16	Si	No
12	Via Malta - Segnaletica (Errata)	Lido	2	0	0	Classe C	0,87	No	validation l'intervento
13	Via Lo Frasso - Parcheggi abusivi frequenti	Mercede	1	1	1	Classe B	0,59	Si	
14	Via Lo Frasso - Parcheggi abusivi frequenti	Mercede	1	1	1	Classe B	0,59	Si	Si
15	Via Napoli - Graffiti	Pietraia	4	4	0	Classe C	0,58	Si	No
16	Via Cataiogna - Strada dissestata	Pivarada	20	20	20	Classe B	0,41	Si	No
17	Via Don Luigi Sturzo - Impianto semaforico non funzionante	Sant'Agostino	7	0	0	Classe C	0,34	No	No
18	Via Enrico Costa - Affissioni abusive	Via degli Orti	5	0	0	Classe C	0,45	No	No
19	Via Caprera Atti vandalici	Lido	40	40	40	Classe B	0,87	Si	No
20	Via Macciotta - Discarica abusiva	Via XX Settembre	25	25	25	Classe A	0,28	Si	No
21	Via Sari - Segnaletica (Mancante)	Carmine	25	25	25	Classe B	0,76	Si	No
22	Via Ospedale - Edificio fatiscente	Centro Storico	30	30	30	Classe B	0,64	Si	Si
23	Via fratelli Cervi - Tombino (Perdita)	San Giovanni	5	5	5	Classe B	0,34	Si	No
24	Via Carrabufas - Discarica abusiva	Sant'Agostino	10	10	10	Classe A	0,63	Si	No
25	Via Corso - Inquinamento	Calabona	22	22	22	Classe A	0,63	Si	No
26	Via Pisa - Marciapiede danneggiato	Pietraia	20	20	20	Classe B	0,58	Si	No
27	Via Pola - Discarica abusiva	Fertilia	10	10	10	Classe A	1,16	Si	No

Fig. 2. A detail of the back-office control panel: the workforce (in man-day) assigned to resolving issues in column A, priority class in column B, workforce demand/supply for issues' neighbourhood in column C, status flags in columns D.

Issue Evaluation and Rating of Priority. Based on the information provided by citizens and validated by operators, the evaluation model assigns a priority rating to each issue, following the evaluation procedure described below in Section 4. Again, once attributed, the priority class of each issue is made publicly visible (cfr. below-right in Fig. 1).

Allocation of Resources and Workforce. Here the system functions as a simple resource and workforce management platform. Our design decision was not to automatically allocate resources to issues based on their priority rating. Operators need to do that manually, by assigning resources and workforce from an available pool to reported issues, specifying the start date and the allocated workdays (cfr. Fig. 2). The idea behind this design decision was that operators should be given the possibility to take into account concerns, information and motivations not explicitly codified in the issue data record. Among these concerns the operator should keep in view the past per capita allocation of resources and workforce among different neighbourhoods and should try to keep them in a reasonable balance. In any case, the operator is called to give an explicit written explanation of her decision, a thing also made visible to citizens on the public front-end.

Issue Tracking. Citizens can track the status of issues on the public front-end as they are being processed by the operators and evaluated by the system. Once the maintenance works have been completed, the operators flag the issue as resolved and the system registers the end date. This then allows the system to update the total workforce allocated to that neighbourhood (cfr. below-right in Fig. 1).

4   The Evaluation Model

As we said, the purpose of the evaluation model is not to automatically provide a complete ordering of issues. It is rather a guidance and a hinting tool. That is why we held it more appropriate to have an evaluation model for the *classification* (rating) of issues in a series of priority classes. Among the methods for multiple criteria evaluation of ratings [10], the so called ELECTRE TRI model [8,9] is a prominent approach, and a natural candidate for our evaluation. This modelling has several desirable properties: it allows a complete sorting of issues in priority classes and the aggregation of criteria is fairly flexible, permitting to account for their importance (weights), coalitions (majority rule and threshold) and possible veto powers.

We define four classes of priority from lowest to highest, C^1, C^2, C^3 and C^4, with the following respective meanings: "little relevant issues", "notable issues", "pressing issues" and "urgent issues". Our problem is then to assign each issue evaluated on eight criteria h_1, h_2, ..., h_8 to one and only one of the four classes of priority. The eight criteria correspond to the questions in Table 1., each using an ordinal scale of four possible values (plus "I don't know" to which by convention we assign value 2).

Each priority class C^k is defined by a limiting profile π^k on eight criteria; in our model we define $\pi^k = (k, k, k, k, k, k, k, k)$. To assign an issue a to a class we then apply the following two rules [10]:

1. if an issue a has the same or higher priority than π^k, it should at least belong to priority class C^k;
2. if π^{k+1} has the same or higher priority than a, the issue a should at most belong to priority class C^k.

Formally:

$$a \in C^k \Leftrightarrow a P \pi^k \wedge \pi^{k+1} P a \qquad (1)$$

where P is the binary outranking relation meaning "has the same or higher priority than".

We define the binary outranking relation P using a crisp relation based on a concordance-discordance principle, that is, an issue a outranks a limiting profile π^k if there is a "significant" coalition of criteria for which "a has the same or higher priority than π^k" (concordance principle) and there are no "significant opposition" against this proposition (discordance principle). In other words:

$$a P \pi^k \Leftrightarrow C(a, \pi^k) \wedge \neg D(a, \pi^k) \tag{2}$$

where:
- $C(a, \pi^k)$ means there is a majority of criteria supporting the proposition that a outranks ("has the priority at least as") π^k;
- $D(a, \pi^k)$ means there is a strong opposition, that is to say a veto, to the proposition that a outranks ("has the priority at least as") π^k.

Following Roy [11], we use the following definitions of $C(x, y)$ and $D(x, y)$:

$$C(x, y) \Leftrightarrow \frac{\sum\limits_{i \in H(x, y)} w_i}{\sum\limits_{j=1}^{n} w_j} \geq \gamma \tag{3}$$

$$D(x, y) \Leftrightarrow \exists h_i : h_i(y) - h_i(x) > v_i \tag{4}$$

where:
- h_i, $i = 1, \ldots, n$ are the criteria (the higher the value the higher the priority, cfr. Table 1.)
- w_i are the importance coefficients (weights) associated to each criterion;
- $h_i(x)$ is the evaluation of x on the criterion h_i;
- $H(x, y)$ is the set of criteria for which x has the same or higher evaluation than y, that is, for which $h_i(x) \geq h_i(y)$;
- γ is the majority threshold;
- v_i is the veto threshold on criterion h_i.

There is another advantage of the ELECTRE TRI method. it is also reasonably easy to communicate and be intuitively understood by citizens. For things are simpler than they seem. We will try to show this through an example.

4.1   An Example Prototype

We have developed a working prototype application for the city of Alghero (Italy) mashing up free Google tools: Docs (for online forms and spreadsheets), Maps and Blogger. Ideally, the model parameters – weights of criteria, majority and veto thresholds – should be defined by decision-makers and subject to public debate. For our prototype, we have interviewed the Town Councillor of Alghero responsible for the Environment and Waste Management, who has helped us determine the following model parameters:
- weights of criteria w_i (following the order in Table 1): 0, 0.2, 0.1, 0.15, 0.15, 0.1, 0.15, 0.15 (note that the first criterion has zero weight but, see below, a decisive veto power);

- majority thresholds $\gamma = 0.6$;
- veto power by the first two criteria (human health/security, and environmental hazard), with veto thresholds $v_1 = 0$ (meaning that issues should at least be assigned to the priority class of the health/security criterion) and $v_2 = 1$ (which assures that issues are classified at least one class below the value of the environmental hazard criterion).

Let us illustrate the classification procedure using a simple example. Suppose there are three reported issues with the evaluations on the eight criteria given in Table 2.

Table 2. Example evaluations of four issues on eight criteria; criteria weights in parenthesis

	h_1 (0)	h_2 (0.2)	h_3 (0.1)	h_4 (0.15)	h_5 (0.15)	h_6 (0.1)	h_7 (0.15)	h_8 (0.15)
a_1	4	2	3	1	2	3	4	3
a_2	1	2	1	3	4	3	2	4
A_3	1	4	3	1	1	2	1	3

These issues show three distinct situations that may arise.

The issue a_1 is classified in the highest-priority class C^4 ("urgent issues") because, no matter the evaluations on other criteria, the issue's belonging to any other class would be discordant due to the veto power of the criterion h_1.

The issue a_2 belongs to the class C^2 ("notable issues"). In fact, the sum of weights of the criteria for which a_2 belongs to the class C^1 or higher is of course 1, of those for which a_2 belongs to C^2 or higher is 0.9, of those for which a_2 belongs to C^3 or higher is 0.55, and of those for which a_2 belongs to C^4 or higher is only 0.3. Therefore, according to the rule (1) and given the majority rule with the threshold of $\gamma = 0.6$, the issue a_2 belongs to the class C^2. No veto power is violated with this attribution, given that a_2 is evaluated 1 (lowest priority) on the first and 2 (second-lowest priority) on the second criterion.

The issue a_3 is classified in the second-highest-priority class C^3 ("pressing issues"), Here, the sum of weights of coalitions for the four classes are respectively 1, 0.55, 0.45 and 0.2. So, according just to the majority rule, a_3 would belong to the lowest-priority class C^1. However, the veto power of the second criterion h_2 with the threshold $v_2 = 1$ imposes the issues be classified at least one class below the value of that criterion. Since $h_2(a_3) = 4$, therefore $a_3 \in C^3$.

5   Conclusions

We presented a concrete proposal of a system for citizens' reporting, evaluation and management of issues for urban maintenance. It is important to place this tool within the complicated and interesting debate on public participation [12,13], even more so if we think about the so-called e-participation [14,15,16].

A starting point is the much cited quote by Sherry Arnstein [17]: «[Participation] is the redistribution of power that enables the have-not citizens, presently excluded from the political and economic processes, to be deliberately included in the future. It is the strategy by which the have-nots join in determining how information is shared, goals and policies are set, tax resources are allocated, programs are operated, and benefits like contracts and patronage are parceled out. In short, it is the means by which they can induce significant social reform which enables them to share in the benefits of the affluent society.»

In that paper Arnstein proposed the by now renowned *ladder* of citizen participation, so many times debated and revisited [18, 19, 20, 21, 22].

Following Cecchini [23], we will also use a ladder (Fig. 3), slightly revisited from the Arnstein's original.

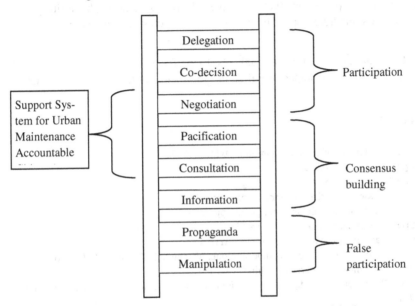

Fig. 3. Arnstein's ladder of citizen participation revisited

In the ladder in Fig. 3, we have placed our system in an area between consensus building and participation. In fact, a fundamental feature of the system, we hold, is its adherence to the principles of accountability and publicity, which makes of citizens' reporting of issues not only consultation and pacification, but also – indirectly and directly – a negotiation. It would not be difficult to think of possible developments to also have, in specific situations, forms of co-decision.

A crucial question, also in this case, is how to develop a communication strategy for effective involvement of all citizens. Our system makes an attempt in that direction, even if questions remain of what to do about those who do not participate, how to involve the Arnstein's "have-nots" in the democratic process, and which strategies to devise to reach them.

There are three groups of people in general who don't participate. Those how do not show interest, do not feel like participating, do not have the necessary capacities nor tools (among whom we find Arnstein's "have-nots" citizens). Then, there are those who hold that the "system" doesn't deserve people's involvement and that the only right way to fight it is to "stay out" of it. Third, there are those who don't participate because they have no interest to make decision-making mechanisms more transparent and account-able, quite the opposite, their true interests would not be safeguarded in democratic processes.

Precisely for it doesn't require adhesions and commitments to predefined and pre-charted processes, the mechanism we proposed in this paper may be successful in in-volving some from the first and the second group. There remains the problem of how to support and kindle the participation of those in the first group who don't have the necessary tools and capabilities (yet may have a deep knowledge of the territory).

But this isn't something impossible to come about with something about.

Acknowledgements. We whish to thank Ms. Elena Riva, the Town Councillor of the City of Alghero, who has helped us define the parameters of the evaluation model.

References

1. Rawls, J.: A Theory of Justice. Claredon, Oxford (1971)
2. http://www.citysourced.com
3. http://www.decorourbano.org
4. http://www.cidadedemocratica.org.br/
5. http://www.cityofboston.gov/apps/
6. http://iris.comune.venezia.it/
7. http://apps.facebook.com/urbanias
8. Yu, W.: Electre Tri: Aspects méthodologiques et manuel d'utilisation. In: Document du Lamsade 74, Université Paris-Dauphine (1992)
9. Roy, B., Bouyssou, D.: Aide multicritère à la décision: méthodes et cas, Economica, Paris (1993)
10. Bouyssou, D., Marchant, T., Pirlot, M., Tsoukiàs, A., Vincke, P.: Evaluation and Decision Models: Stepping stones for the analyst. Springer, Berlin (2006)
11. Roy, B.: Classement et choix en présence de points de vue multiples (la méthode ELECTRE). La Revue d'Informatique et de Recherche Opérationelle (RIRO) (8), 57–75 (1968)
12. Irvin, R.A., Stansbury, J.: Citizen Participation in Decision Making: Is It Worth the Effort? Public Administration Review 64(1), 55–65 (2004)
13. Burby, R.J.: Making plans that matter: Citizen involvement and government action. Jour-nal of the American Planning Association 69(1), 33–49 (2003)
14. Coleman, S.: e-Democracy: The history and future of an idea. In: Mansell, R., Avgerou, C., Quah, D., Silverstone, R. (eds.) The Oxford Handbook of Information and Communi-cation Technologies, pp. 362–382. Oxford University Press, Oxford (2007)
15. Charalabidis, Y., Koussouris, S., Kipenis, L.: The 2009 Report on the Objectives, Structure and Status of eParticipation Initiative Projects in the European Union. MOMENTUM Whitepaper (June 2009)

16. Castells, M.: Networks of Outrage and Hope: Social Movements in the Internet Age. Polity Press, Cambridge (2012)
17. Arnstein, S.R.: A Ladder of Citizen Participation. Journal of the American Planning Association 35(4), 216–224 (1969)
18. Connor, D.M.: A New Ladder of Citizen Participation. National Civic Review 77(3), 249–257 (2007)
19. Wiedemann, P.M., Femers, S.: Public Participation in waste management decision making: analysis and management of conflicts. J. of Hazardous Materials 33(3), 355–368 (1993)
20. Dorcey, A., Doney, L., Rueggeberg, H.: Public Involvement in government decision making: choosing the right model. B.C. Round Table on Environment and Economy (1994)
21. Pretty, J.N.: Participatory Learning For Sustainable Agriculture. World Development 23(8), 1247–1263 (1995)
22. Rocha, E.M.: A Ladder of Empowerment. Journal of Planning Education and Research 17(1), 31–41 (1997)
23. Cecchini, A.: Il paesaggio come bene comune (e un esempio di gioco per imparare i beni comuni), Quaderni dell'Istituto Alcide Cervi n.6 "Il paesaggio agrario italiano protostorico e antico" (2010)

Web 3.0 and Knowledge Management: Opportunities for Spatial Planning and Decision Making

Beniamino Murgante[1] and Vito Garramone[2]

[1]University of Basilicata, 10,Viale dell'Ateneo Lucano, 85100 Potenza, Italy
[2] Veneto Region, Calle Priuli 99, Cannaregio, 30123, Venice, Italy
beniamino.murgante@unibas.it, garramonevito@gmail.com

Abstract. The overabundance of information produced by new technologies, if on one side can be considered as a knowledge enrichment in planning process, on the other side it has not improved neither reality understanding nor possibilities of intervention. Old forms of citizens participation to planning process, generally based on assemblies, have been replaced by continuous discussions on social networks, blogs, etc..

The attempt to take into account the huge data flow produced everyday, it is not an easy task for planners. An ontologies based approach can represent an important support to such activities.

"Comelicopedia" an European project between Italy and Austria, probably is one of the first experiences in applying ontologies to spatial planning process.

All potentialities in planning and decision making fields will be analyzed and tools, such as "comelicopedia", can become usual in supporting a regulatory dialogue between decision makers and citizens.

Keywords: Spatial planning, Decision making, Knowledge management, Ontology, Citizens participation.

1 New Technologies and Information Overload: Problems and Opportunities

The problem of overabundance of information sources, not always seen as a cultural enrichment, has always been discussed, even in ancient times, when spread of knowledge was totally restricted to specific social categories. Seneca [1] in ancient Rome, affirmed that the abundance of books is a distraction. Too many books are dispersive: since you cannot read all the volumes you may have, you should just own the right number that you can read. It is better to prefer authors with recognized value and if occasionally you can think to switch to other authors, then return to the first ones. Reading a huge number of different authors and all kinds of books is a sign of inconstancy and volubility: if you would like to have durable advantages it is important to insist only on certain writers.

The term "Information overload" introduced in social science by Myron Gross [2] is related to information overabundance which generates barriers in an effective

B. Murgante et al. (Eds.): ICCSA 2013, Part III, LNCS 7973, pp. 606–621, 2013.

understanding of reality, limiting the information process and producing a cognitive inability. Toffler [3] introduces also the term over-choice when diversity advantages are dissolved by decision-making process complexity.

In recent times, there has been a transition from traditional web pages to a new emerging Internet model, the web 2.0, based on extensive content generation by users and on collaboration. In addition, there is also an impressive increase of geographical information production, due to the growth of spatial data infrastructures, such as Google, which transformed geographical information from a specialist interest to a mass phenomenon and the great GPS diffusion in mobile devices, so that every person owns at least one GPS [4] [5].

Today we are in the information society, based on information opportunities, where the object is not simply knowledge but the possibility of accessing it producing and diffusing new knowledge. This means that different informal or tacit knowledge can emerge, finding an expression channel, but also incomplete, erroneous and tendentious information which can be diffused.

We live in a world with an overabundance of information, which produces a lack of everything that information consumes, mainly the attention of people who receive the information itself. Therefore, wealth of information creates a poverty of attention and a need to efficiently allocate that attention within the overabundance of information [6].

There are not solutions, but dilemmas. Among the various information dilemmas, two metaphors can be suitable in describing overabundance of information. The puzzle metaphor: knowledge is fragmented and everyone should build it time to time and as needed. The filter metaphor: if mesh is too narrow, we know what we are looking for but we lose time to discard; if mesh is too large, everything passes and we find big misunderstandings, failures and information should be filtered again.

Collective intelligence occurs wherever there is a huge human interaction and new technologies can easily encourage synergies even among geographically distant people. Someone who lives in a remote part of the world can interact with other people with complementary knowledge, living very distant, continuously communicating with each other, exchanging their experiences, cooperating, etc. [7]. Collective intelligence can be defined as an information mixed with different points of view producing synergies and developing complementary aspects.

2   New Technologies in Citizens Participation to Planning Processes

These forms of collective intelligence can be particularly useful in citizens participation to planning processes, enriching the construction of cognitive frameworks, improving knowledge management and supporting decision making.

In first experiences, the support of participatory planning processes served mostly the purpose of consensus building, within the combined action of technicians and politicians. A triangulation of knowledge (technical, political and local) has been built in a circular relationship, constantly looking for spaces of mediation, negotiation and sharing as well as the co-construction of a chain choices-responsibilities actions [8].

The groupware concept [9] represents one of the first attempts to adopt technologies in participatory planning processes, building actors network, increasing production and quality of information, improving interoperability and data access. This concept allows to identify a family of techniques to support multi-actor cooperation, leading to conflicts resolution and agreements between parties in building future projects. Groupware success consists of an equation which considers all of the following elements: technology, culture, economics and politics [10].

The term "participatory GIS" has been coined to highlight a trend towards democratization of techniques. Subsequently, this definition evolved considering other aspects, such as critical evaluation of uses of GIS in society [11], to foster a grassroots involvement in policy decision-making [12] and an analytic-deliberative approach to policy decision-making for situations with high decision equity [13].

Very important was the introduction of argumentation mapping tools [14], also.

The advent of web 2.0 increased a transition from a one-way approach, where citizens are only informed, to a two-way approach where citizens can express their opinion in a wiki-way [15]. Blogs and social networks are collectors of citizens instances, leading to virtual deliberative arenas [16] and to wiki-planning [17].

The same diffusion of deliberative acting in several cases challenged the role of representative democracy, creating more space for participatory democracy and its tools. Also, occasional participatory events are information and education moments, producing different nature of knowledge in participants, who become wide diffusion elements of social change. This change is added to the traditional innovation due to local associations activism and political actions.

A lot of new terms have been coined to define various aspects of collaborative actions. Volunteered Geographic Information [18] identifies a mass collaboration to create, manage and disseminate spatial data; Crowdsourcing means the possibility of obtaining suggestions, services, ideas, support in decisions by actions of online communities [19]; Neogeography [20] describes a bottom-up approach to geography integrating maps with geo-tagged photos, videos, blogs, Wikipedia, etc..

In Iceland, for instance, after banks collapse, the government allows citizens to discuss and propose their ideas through crowdsourcing. UK Prime Minister David Cameron adopts a sort of dashboard, which synthesizes main data concerning the Country, polls and Twitter feed. New York Major Michael Bloomberg defined Twitter as a source of everyday referendum. If social networks collaborative approach can increase a "planning through debate" [21], it is also important to consider the "rational ignorance" [22]. Citizens often trivialize the concepts or manifest inertia in understanding technical issues. The main barrier in knowledge increase using technologies is due to semantic discrepancies. In planning participation process this barrier is more evident in the transition from the technical sphere to more shared levels, such as political and common (social) knowledge.

It is very common that a well defined concept in the technical sphere does not match with a concept with the same name in political and social sphere. Often all actors involved in a planning process do not adopt the same language in describing the same concept [23] [24]. The most adopted approach to overcome semantic barriers is represented by ontologies.

3   Ontologies and Spatial Planning

The term Ontology originates from philosophical disciplines, reaching a large spread in the field of artificial intelligence. A definition, shared by philosophers and computer scientists, considers ontology as "the theory of objects and their relationships" [25]. In this section the attention will be focused on the evolution of informative bases of the ontological nature for planning purposes. Great interest will be paid to test technical tools for knowledge management and decisions support related to urban and regional planning aspects.

After the first phase, during which ontologies were relegated in the field of philosophy and artificial intelligence, the attempt was to put into practice what previously theorized, trying a first modeling and a geographical declination.

Mark and Smith [26] define ontologies related to the places different from all other ontologies related to objects of everyday life. Furthermore, they criticize the typical approach in developing an ontology, too close to expert issues, also highlighting that folks categorizations are important because they are transparent. Their ontology is the first case of an implementation shared to non-experts.

Hopkins et al. [27] mainly focus his research on the semantic of processes. His aim is to establish and define terms and areas of action to build, on these, a relational system, as much as possible discursive and narrative. The ontology has been developed to represent urban development processes, and elaborated for land use regulations. Plans are intentional actions directed towards change, and explicitly recognize relationships-agenda, design, policy and strategy among actions. This model is built on processes, strategies and time, in a few words on complexity, perhaps so much to make it an exhaustive and sophisticated model, but difficult to implement.

Laurini et al. [28] adopt a completely different approach, considering relations between elements and their transformation for interoperability improvement and a ready to use tool in Geographic Information Systems. In urban domain, two main objectives have been pursued: interoperability of urban information systems and clarification of main urban concepts. Ontologies, with the aim of ensuring systems interoperability, allow concepts clarification and deepening, often considered known with a certain superficiality [29] [30].

In seeking semantic interoperability, it is necessary to reach an agreement among all actors, choosing a definition which is a sort of compromise. This is the case of post consensus ontologies adopted in most cases. Once reached the agreements, the definition can be translated into Ontology Web Language (OWL). The problem is how to reach this consensus. In the case of well-standardized domains, as in the field of mathematics, chemistry, physics, etc., an agreement is fairly simple [31].

In a context with abundant definitions, with a very articulate vocabulary, achieving consensus is not an easy task. The domain analyzing territorial phenomena falls in the latter case, especially if one examines the context in different nations. Therefore, in this domain achieving consensus before ontology construction is not a trivial thing at all. For this reason Preconsensus Ontologies have been preferred, previously collecting all existing definitions, and then seeking consensus among all actors [32].

Another interesting ontology application in planning domain is a Land Use Planning Ontology [33] where Land Based Classification Standards (LBCS) developed by the

American Planning Association have been transferred in OWL2 producing a land use ontology adopted as a basic structure for the City Information Model. Other interesting experiences have been developed in Strategic Environmental Assessment procedure applied to city Masterplans in Sardinia Region [34], in Regional Development Programs [35] and in disaster and emergency management [36].

4   The Case Study: Comelicopedia

Today we have a lot of examples of ontology applications. The domain investigated in the case study is quite complex, the use of ontology in supporting decision-making process. In particular, Comelicopedia experience was developed within the European Project Susplan (http://www.susplan.info/) "Sustainable planning in mountain areas" (Interreg IV Italy - Austria 2007-13).

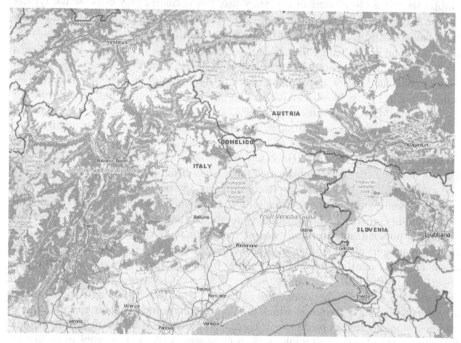

Fig. 1. Study area location (Source OpenStreetMap)

The term Comelicopedia comes from the term Comelico, pilot area of Susplan project. Comelico, also known as Comelico Valley, is an Italian mountainous region in Belluno Province close to the Austrian border. Comelico, from the administrative point of view, is also a consortium of communes in mountain areas with a population of 8908 inhabitants and it covers an area of 280 km², including five municipalities with Ladino language (Comelico Superiore, Danta di Cadore, San Nicolò di Comelico, San Pietro di Cadore, Santo Stefano di Cadore) and one municipality with German language (Sappada).

Comelicopedia experience is mainly based on three needs: developing an analysis of sustainable development concept evolution; attempting to reinforce, compare, and go beyond the assessment of policies, programs, plans and projects of territorial development (paying particular attention to Strategic Environmental Assessment); organizing a more effective and efficient knowledge base of political agendas to support decision-making and to build new common cross-border spatial development strategies.

These three main needs have been satisfied through two transversal and interrelated approaches: new information technologies (wikis, semantic web, ontologies, folksonomies, etc.) and participation.

In addition, a common spatial information system, among all partners, has been implemented, according to INSPIRE Directive and adopting Open Geospatial Consortium (OGC) standard ISO 1915.

This experience has been applied in a mountain area with great ecological and environmental values, characterized by marginalization and depopulation, as well as cross-border (inter-regional and international fundamental Interreg IV program requirements) territorial and ecological uniformity.

The peculiarity of the Italy-Austria cooperation project is the characteristic of three regions: Veneto ordinary statute, Friuli Venezia Giulia extraordinary statute and Carinzia foreign region. This area well represents all the major issues of mountain and its related planning problems [37]:

a) high degree of approximation in defining a conceptual framework that usually inspires political actions [38] [39];

b) elevated limits, both in expertise and in knowledge and in instruments that influence public decision-makers and other actors involved in the processes [40] [41];

c) poor ability demonstrated so far by both experts and citizens in delivering suggestions or information available (objectives, ideas, projects) to the decision level, where they can be conveniently collected and taken into account [42] [43];

d) perception of dissatisfaction, distrust and separation from the plans, programs and public policies.

For these reasons, Comelicopedia group worked to define methods and tools for "knowledge management", which was both opportunity and instrument both for knowledge dissemination and building learning community [44] [45], necessary conditions to achieve the objectives of sustainable local development.

Comelicopedia working group developed the ontological scheme [46] considering other experiences [27].

In parallel to the simplification and the operative translation of this ontological scheme, knowledge domain has been defined, more particularly types and forms of knowledge to be introduced in the system and reference sources.

Starting from the principle that all sources are equal in dimensions and importance three types-forms of sources have been considered.

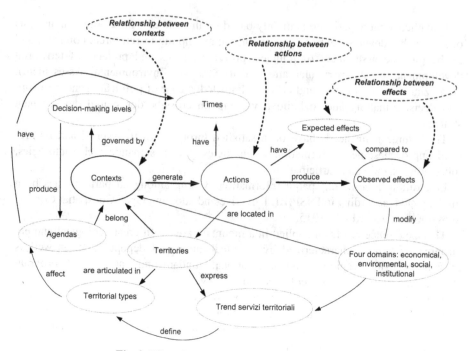

Fig. 2. "Comelicopedia" Ontological scheme [46]

First of all, formal written official sources have been considered, main political agendas defined by means of programming and planning documents, related to the study area, at different territorial level (regional, provincial, consortium of communes in mountain areas), accompanied by documents and national and international strategies concerning sustainable development. In order to build the ontological model, sources, producing top down knowledge, have been more useful. In order to balance knowledge from a non-institutional perspective, collecting local knowledge, many interviews have been realized adopting a semi-structured interview protocol [47]. This step is important to stimulate a debate and building a future local voice in order to produce a bottom-up knowledge. These two types of knowledge have been integrated with technical written sources, academic reports, specialist and informative, and practical source of knowledge based on experiences, examples of recognized and certified best practices tested in pilot areas (Fig. 2).

The objective of the introduction of these two new sources was to balance the information, taking into account knowledge and experiences of scientific communities and good practices to promote social learning and stimulating participation.

Participation could also provide assessment and checking of official sources, analyzing the correspondence between real needs and various analyses of the context and strategies put in place.

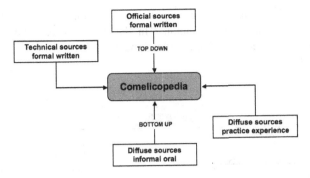

Fig. 3. Types and forms of knowledge adopted in Comelicopedia experience

The built ontology (Figure 3) is based on a knowledge tree structure, by means of Category tree and Semantic Network.

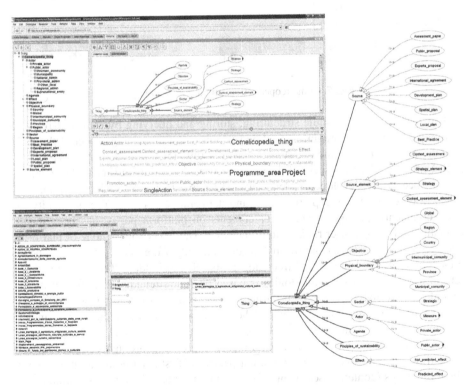

Fig. 4. Ontology structure realized with Protégé software

For descriptive purposes, each information source was divided into two trees (Fig. 4): "context assessment" tree, where all forms of analysis and representation of reality were collected (description, SWOT analysis, trend analysis, etc.), and "strategy" tree based on logical-operative chains, which contains aims and objectives and the various ways and actions (Measures, Projects, etc.) to achieve them. The whole volume of Comelicopedia information was contained in the technological platform semantic wiki [48]. It was a collection of textual documents and hyperlinks, easily viewable and updateable by users.

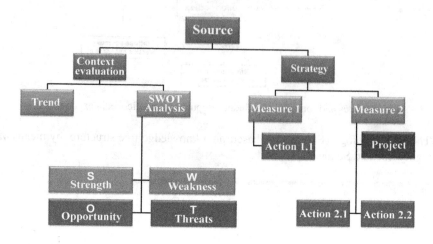

Fig. 5. Comelicopedia tree-structured knowledge

This technology can handle structuring, storing and querying by means of related ontology and via search engines based on semantics. An interesting aspect of the tool was the possibility to combine social networks [49] [50] with Semantic Web [51] [52]. Unfortunately, this aspect of the project, because of limits of time and resources, has not been fully developed. Social networks allow to involve a large number of users, in order to lead to a participatory and deliberative opening. The possibility to write comments, judgments and other contents on platform pages, has been made available, following a user registration procedure. The never solved node of platform management authority and platform free use, related to the needs of the European project have affected simple and free cooperation of stakeholders. The second aspect, related to Semantic Web technologies is based on four strengths. Information is accessible and available on the Web through simple and informative pages, like Wikipedia (hence the name Comelicopedia). The system allows a knowledge management, not simply based on an alphanumeric approach, but it works using meanings of the language, semantics. The system allows to capture, destructure and compare knowledge from multiple sources on two main aspects: the first is based on what we know about the issues (context assessment), the second one is based on what we want to do (strategy). But we had a common core ontology, which was the constant element which allowed us to compare variables that we wanted to consider. In other words, the platform was an opportunity to open both database and public policies, in order to create new cognitive frameworks for future local actions.

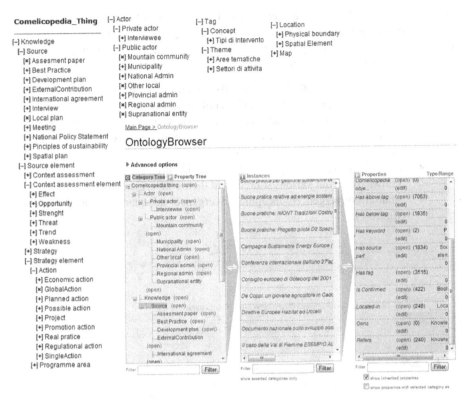

Fig. 6. Comelicopedia Taxonomy and its implementation in ontology browser

Data entry was accompanied by a system of information labelling, annotations aimed at returning contents through two classes: keywords and semantic tags. The first class, keyword, was defined a priori by the authors and arranged into two distinct hierarchical levels in the ontology.

First level keys identify thematic macro areas (environment, community, territory, economic sectors and innovation spheres), second level keys are specifications of first level keys. Both levels were extracted from Gemet Thesaurus, a technical multilingual thesaurus widely adopted in the European Union. Semantic tags could be freely inserted, by both authors and users when keywords were not enough to define page contents. The defined tags were managed by the system producing folksonomies, taxonomies with a bottom up categorization of information, integrating the original ontology [53] [54]. Finally, Comelicopedia also considered a basic level of geographic information on the places mentioned by sources.

The small encyclopedia-database was composed by: (i) ten documents of territorial policies produced by the competent authorities at various levels; (ii) interviews (or individual contributions) to experts and recognized leaders of community; (iii) a catalogue of good practices from which decision makers can take inspiration for the action; (iv) research reports prepared by the working group; (v) comments expressed by individual or associated citizens. Comelicopedia beta version consisted of approximately 800 pages; 76 keywords were defined by experts, while some users simulations had already introduced about 100 semantic tags [37]. The system is able to provide both classic type of query, related to the contents, and geographical queries. The objective of geographical queries is to satisfy requests concerning:

- Spatial distribution of contents: distribution in space of instances belonging to ontological categories, object of selection.
- Convergence between actions and context evaluations: coherence between an action or a goal located at a given point and context evaluation, into a single source.
- Spatial convergence points of view: identifying, for given locations, knowledge from various sources and their convergence.
- Spatial convergence/conceptual divergence: adequacy of various sources (figure 7).

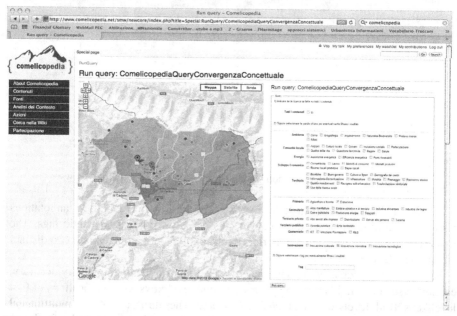

Fig. 7. An example of spatial query through spatial convergence/conceptual divergence (http://www.comelicopedia.net/smw/newcore/index.php?title=Special:RunQuery/Comelicopedi aQueryConvergenzaConcettuale)

5   Results, Limits, Potentialities and Possible Future Developments

Despite Comelicopedia was a temporal limited experience and needed further research, it was an opportunity to build and manage knowledge related to decision making and to define local development policies.

Comelicopedia also allowed to evaluate some agendas and check the level of internal coherence to various instruments. It has to be noticed, for example, that the level of correspondence between context assessment and programmed actions decreases, reducing the scale of the explored source (from global to local). Tools at local scale seem inadequate to local challenges of issues identified and established at a global scale.

Comelicopedia analyzed the congruence between instruments. How much and how policies pursued at the global level dialogue with planned interventions at the local level. There are many mismatches and gaps between policies and actions, since more distant are problems identified in official documents and assessments expressed by specialists. In addition local plans completely disregard the recommendations contained in global politics.

Other times, documents at the local scale are collections of actions that have a consistency with a formal and/or ex post built strategy, following sectoral clustering of detailed actions that take place in the area, without following aims and theories of sustainability. It is possible to notice that sectoral logic, prevailing in plans, tried to integrate economy and environment over time.

The innovation found in plans is strictly related to technologies and not associated to strategies. Innovation in lifestyle and in local regulatory systems is almost completely absent. Finally, the ability of Comelicopedia to analyze semantic aspects allowed to investigate also contents assumed coherent with the definition of sustainable development. In agendas, a vision of sustainable development of the mountain is missing and the few references do not highlight an implicit pattern.

Also, use and frequency of terms related to sustainable development highlight just a theoretical approach, with weak operative feedbacks. The theme of development is mostly seen as a correction or compensation within existing institutional frameworks, rather than re-thinking institutional actions in a perspective of a new governance phase [55].

Tools such as Comelicopedia can have great potentialities in planning and decision-making, producing huge advantages achievable by a continuous use in normative dialogue between administrators and administered.

The wiki-semantic applied to the project needs to be more tested by all possible users (public employees, experts, citizens, associations, etc.): first tests show that this frontier technology can really aid to achieve a more effective stakeholders interaction in a process of policies modeling [37].

Comelicopedia can be a more effective communication tool in order to improve the dialogue between parties, if it is used not only by analysts and external experts, but also by local technicians and by those who produce data, information and knowledge.

It would be preferable to use the platform in regular daily activities, and to carry out data entry by information producers, providing directly to a continuous update. Plans and programming documents directly inserted in Comelicopedia by technical and administrative staffs of various local authorities coupled with citizens information and comments could allow, with accessible costs, the implementation of cognitive frameworks and to create permanent consultative and deliberative arenas.

There would be a better relationship between administrators and administrated producing more transparent political choices and actions and building a broader consensus on choices and a wider and informed participation.

One of the main problems occurred in project was the complexity of data entry activities, which involved few people with huge tasks. It is important to simplify this process, adopting "the principle of source proximity": the source is introduced by the producer or someone close to him, who adopts the same use of terms and semantic basin. Only in the case of historical or oral sources, specific users will be involved in data entry.

The text has to be open to future implementations by selected users, in order to avoid that a low quality of information is added. In other words, "collective intelligence" and "social control" are needed to build and implement knowledge base processes.

For a more project completeness transfer of competences to local authorities technicians for system management and creation of a core community able to stimulate discussions and implementing, updating and revising the basic information are also necessary [56]. Not great attention has been paid to GIS interface without considering a widespread use of the tool in the project.

Another error in project development, due to the limited economic resources and time, was the lack of connection and integration of platform with social networks and forums of our pilot area. Also, simply providing information about weather, local events and folklore, could attract potential users who can take part to Comelicopedia participation process. Finally, we should highlight two basic issues on which launching a debate at the international level. The first issue deals with the "linguistic dilemma". Can we have comprehensive thesauri able to work effectively at local and global level? Can these thesauri be an opportunity for knowledge management despite the semantic difference related to various languages and information sources? The second issue is related to time factor [55].

Considering that an ontology is a "specification of explicit and shared conceptualization of a domain", the question is: how to make the ontology dynamic and able to self-update? Ontology always needs a management group or it is simply enough self-organization of users. How can we change over time knowledge structures?

Acknowledgements. This study has been supported by the European Project Susplan "Sustainable planning in mountain areas" (Interreg IV Italy - Austria 2007-13).

Authors are grateful to Franco Alberti, Head of Planning and Landscape Bureau of Veneto Region and to project participants Igor Jogan, Claudio Chiapparini, Andrea Mancuso, Alessio Gugliotta, Francesca Borga, Valentina De Marchi, Mauro Nordio, Viviana Ferrario, Mauro De Conz and all other Italian and Austrian partners involved in Susplan project.

References

1. Senecae L. Annaei: Ad Lucilium Epistularum Moralium Quae Supersunt. Kessinger Pub. Co., Montana (1898), Hense, O. (ed.)
2. Myron Gross, B.: The managing of organizations: The administrative struggle. Free Press of Glencoe, New York (1964)
3. Toffler, A.: Future Shock. Random House Publishing Group, New York (1970)

4. Murgante, B., Borruso, G., Lapucci, A.: Geocomputation and Urban Planning. In: Murgante, B., Borruso, G., Lapucci, A. (eds.) Geocomputation and Urban Planning. SCI, vol. 176, pp. 1–18. Springer, Heidelberg (2009)

5. Murgante, B., Tilio, L., Scorza, F., Lanza, V.: Crowd-Cloud Tourism, New Approaches to Territorial Marketing. In: Murgante, B., Gervasi, O., Iglesias, A., Taniar, D., Apduhan, B.O. (eds.) ICCSA 2011, Part II. LNCS, vol. 6783, pp. 265–276. Springer, Heidelberg (2011)

6. Simon, H.A.: Designing Organizations for an Information-Rich World. In: Martin Greenberger, Computers, Communication, and the Public Interest. The Johns Hopkins Press, Baltimore (1971)

7. Levy, P.: L'intelligence collettive. Pour une anthropologie du cyberspace, Paris, la Découverte (1994)

8. Garramone, V.: Una svolta per la pianificazione nella chiarificazione morale? Riflessioni sulla deontologia e su altre questioni rilevanti, Urbanistica, INU Edizioni, Roma, 150 (in press, 2013)

9. Laurini, R.: Groupware for urban planning: an introduction. Computers, Environment and Urban Systems 22(4), 317–333 (1998)

10. Coleman, D., Khanna, R. (eds.): Groupware: Technology and Applications. Prentice-Hall, Englewood Cliffs (1995)

11. Rinner, C., Keßler, C., Andrulis, S.: The use of Web 2.0 concepts to support deliberation in spatial decision-making. Computers, Environment and Urban Systems 32, 386–395 (2008)

12. Sieber, R.: Public participation geographic information systems: A literature review and framework. Annals of the Association of American Geographers 96(3), 491–507 (2006)

13. Stern, P.C., Fineberg, H.V.: Understanding risk: Informing decisions in a democratic society. National Academy Press (1996)

14. Rinner, C.: Argumentation maps – GIS-based discussion support for online planning. Environment and Planning B: Planning and Design 28(6), 847–863 (2001)

15. Evans-Cowley, J., Conroy, M.M.: The growth of e-government in municipal planning. Journal of Urban Technology 13(1), 81–107 (2006)

16. Garramone, V.: Uno strumento dove i cittadini sono la città. Metodo e ruolo degli attori nei Town Meeting elettronici. In: Garramone, V., Aicardi, M. (eds.) Democrazia Partecipata ed Electronic Town Meeting, Incontri ravvicinati del terzo tipo, pp. 137–160. Franco Angeli, Milano (2011)

17. Murgante, B.: Wiki-Planning: The Experience of Basento Park in Potenza (Italy). In: Borruso, G., Bertazzon, S., Favretto, A., Murgante, B., Torre, C. (eds.) Geographic Information Analysis for Sustainable Development and Economic Planning: New Technologies, pp. 345–359. Information Science Reference IGI Global, Hershey (2012), doi:10.4018/978-1-4666-1924-1.ch023

18. Goodchild, M.F.: Citizens as Voluntary Sensors: Spatial Data Infrastructure in the World of Web 2.0. International Journal of Spatial Data Infrastructures Research 2, 24–32 (2007)

19. Howe, J.: Crowdsourcing: Why the Power of the Crowd is Driving the Future of Business. Crown Publishing Group, New York (2008)

20. Turner, A.: Introduction to Neogeography. O'Reilly Media, Sebastopol (2006)

21. Healey, P.: Planning through debate: the communicative turn in planning theory. The Town Planning Review 63, 142–162 (1992)

22. Krek, A.: Rational ignorance of the citizens in public participatory planning. In: In Proceedings of the CORP 2005 & Geomultimedia Conference, Vienna (April 2005)

23. Murgante, B., Scorza, F.: Ontology and spatial planning. In: Murgante, B., Gervasi, O., Iglesias, A., Taniar, D., Apduhan, B.O. (eds.) ICCSA 2011, Part II. LNCS, vol. 6783, pp. 255–264. Springer, Heidelberg (2011)

24. Murgante, B.: Interoperabilità semantica e pianificazione territoriale. Italian Journal of Regional Science 10(3), 135–144 (2011), doi:10.3280/SCRE2011-003008

25. Gruber, T.R.: The Role of Common Ontology in Achieving Sharable, Reusable Knowledge Bases. In: Allen, J.A., Fikes, R., Sandewall, E. (eds.) Principles of Knowledge Representation and Reasoning: Proceedings of the Second International Conference, Cambridge, MA, pp. 601–602 (1991)

26. Mark, D.M., Smith, B.: Ontology and geographic kinds. In: International Symposium on Spatial Data Handling, Vancouver, Canada, July 12-15, pp. 308–320 (1998)

27. Kaza, N., Finn, D., Hopkins, L.: Updating plans: A historiography of decisions over time. Electronic Journal of Information Technology in Construction 15, 159–168 (2010)

28. Arara, A., Laurini, R.: Formal contextual ontologies for intelligent information systems. In: Proceedings (WEC 2005) 3rd World Enformatika Conference, vol. 5, pp. 303–306 (2005)

29. Laurini, R., Murgante, B.: Interoperabilità semantica e geometrica nelle basi di dati geografiche nella pianificazione urbana. In: Murgante, B. (a cura di) L'informazione Geografica a Supporto Della Pianificazione Territoriale, pp. 229–244. FrancoAngeli, Milano (2008)

30. Murgante, B.: L'informatica, i Sistemi Informativi Geografici e la Pianificazione del Territorio. In: Murgante, B. (a cura di) L'informazione Geografica a Supporto Della Pianificazione Territoriale, pp. 7–37. FrancoAngeli, Milano, (2008)

31. Gómez-Pérez, A., Fernández-López, M., Corcho, O.: Ontological Engineering. Springer, Berlin (2004)

32. Laurini, R.: Pre-consensus Ontologies and Urban Databases. In: Teller, J., Lee, J., Roussey, C. (eds.) Ontologies for Urban Development. SCI, vol. 61, pp. 27–36. Springer, Heidelberg (2007)

33. Montenegro, N., Gomes, J.C., Urbano, P., Duarte, J.P.: A Land Use Planning Ontology: LBCS. Future Internet 4(1), 65–82 (2012), doi:10.3390/fi4010065

34. Lai, S., Zoppi, C.: An Ontology of the Strategic Environmental Assessment of City Masterplans. Future Internet 3(4), 362–378 (2011), doi:10.3390/fi3040362

35. Scorza, F., Las Casas, G.B., Murgante, B.: That's ReDO: Ontologies and Regional Development Planning. In: Murgante, B., Gervasi, O., Misra, S., Nedjah, N., Rocha, A.M.A.C., Taniar, D., Apduhan, B.O. (eds.) ICCSA 2012, Part II. LNCS, vol. 7334, pp. 640–652. Springer, Heidelberg (2012)

36. Murgante, B., Scardaccione, G., Las Casas, G.B.: Building ontologies for disaster management: seismic risk domain. In: Krek, A., Rumor, M., Zlatanova, S., Fendel, E.M. (eds.) Urban and Regional Data Management, pp. 259–269. CRC Press, Taylor & Francis, London (2009) ISBN: 978-0-415-055642-2, doi:10.1201/9780203869352.ch23

37. Alberti, F., Garramone, V., Jogan, I.: Comelicopedia: una wiki semantica per lo sviluppo sostenibile delle aree montane. In: Varotto, M., Castiglioni, B. (eds.) Di chi sono le Alpi? Appartenenze Politiche, Economiche e Culturali nel Mondo Alpino Contemporaneo, pp. 85–100. Padova University Press, Padova (2012)

38. Bina, O.: A critical review of the dominant lines of argumentation on the need for strategic environmental assessment. Environmental Assessment Review 27, 586–606 (2007)

39. Boothroyd, P.: Policy assessment. In: Vanclay, F., Bronstein, D. (eds.) Environmental and Social Impact Assessment, pp. 83–126. John Wiley, Chichester (1995)

40. Connelly, S., Richardson, T.: Value-driven SEA: time for an environmental justice perspective? Environ. Impact Assess Review 25, 391–409 (2005)
41. Owens, S., Rayner, T., Bina, O.: New agendas for appraisal: reflections on theory, practice and research. Environment and Planning A 36, 1943–1959 (2004)
42. Garramone, V., Aicardi, M. (eds.): Democrazia partecipata ed Electronic Town Meeting, Incontri ravvicinati del terzo tipo. Franco Angeli, Milano (2011)
43. Petts, J.: Barriers to deliberative participation in EIA: learning from waste policies, plans and projects. Journal of Environmental Assessment Policy and Management 5, 269–293 (2003)
44. Hopkins, L.D.: Urban development: the logic of making plans. Island Press, Washington DC (2001)
45. Innes, J.E., Booher, D.E.: Planning with complexity: an introduction to collaborative rationality for public policy. Routledge, NY (2010)
46. Jogan, I.: Preliminary report to the Project SUSPLAN- Sustainable planning in mountain areas, WP 2 "Methods and models for sustainable planning and for strategic environmental assessment", Interreg IV Italia-Austria (2010)
47. De Marchi, V.: Un contributo socio-antropologico alla Comelicopedia. In: Alberti, F., Garramone, V., Jogan, I. (eds.) Ripensare la Montagna nel Web 3.0. Soluzioni di e-government e Knowledge Management Per Gli Interventi Locali People-centred in Ambito Montano, pp. 67–76. Franco Angeli-Regione del Veneto, Milano (2012)
48. Kamel Boulos, M.N.: Semantic Wikis: A Comprehensible Introduction with Examples from the Health Sciences. Journal of Emerging Technologies in Web Intelligence 1(1), 94–96 (2009)
49. Weiss, A.: The Power of Collective Intelligence. netWorker 9(3), 17–23 (2005)
50. Andrus, D.C.: The Wiki and the Blog: Toward a Complex Adaptive Intelligence Community. The Social Science Research Network (SSRN) (2005)
51. Antoniou, G., van Harmelen, F.: A Semantic Web Primer. MIT Press, Massachusetts (2004)
52. Della Valle, E., Celino, I., Cerizza, D.: Semantic Web. Modellare e condividere per innovare. Pearson Paravia Bruno Mondadori (2008)
53. Peters, I.: Folksonomies indexing and retrieval in web 2.0. Walter de Gruyter, Berlin (2009)
54. Smith, G.: Tagging: people-powered metadata for the social web. New Riders, Brekley (2008)
55. Gugliotta, A., Jogan, I.: Comelicopedia: una wiki semantica per lo sviluppo sostenibile delle aree montane. In: Alberti, F., Garramone, V., Jogan, I. (eds.) Ripensare la Montagna nel Web 3.0. Soluzioni di e-Government e Knowledge Management Per gli Interventi Locali People-centred in Ambito Montano, pp. 47–66. Franco Angeli-Regione del Veneto, Milano (2012)
56. De Michelis, G., Garramone, V.: La gente, le storie ed i sistemi informativi. Intervista a Giorgio De Michelis. In: Alberti, F., Garramone, V., Jogan, I. (eds.) Ripensare la Montagna nel Web 3.0. Soluzioni di e-Government e Knowledge Management Per gli Interventi Locali People-centred in Ambito Montano, pp. 115–124. Franco Angeli-Regione del Veneto, Milano (2012)

Enhancing the Spatial Dimensions of Open Data: Geocoding Open PA Information Using Geo Platform Fusion to Support Planning Process

Francesco Izzi[1], Giuseppe La Scaleia[2], Dimitri Dello Buono[1],
Francesco Scorza[2], and Giuseppe Las Casas[2]

[1] CNR IMAA – geoSDI - Direzione Tecnologie e Sviluppo
C.da S. Loja, 85050, Tito Scalo (PZ), Italy
`francesco.izzi@geosdi.org, giuseppe.lascaleia@geosdi.org`
[2] Laboratory of Urban and Regional Systems Engineering, University of Basilicata,
10, Viale dell'Ateneo Lucano, 85100, Potenza, Italy
`francescoscorza@gmail.com`

Abstract. The complexity of planning process exponentially increased during last decades matching together a wide range of instances deriving from the evolution of national and local regulations and laws, an heterogeneous methodological framework, the contribution of technologies and especially the affirmation of web and social communities as relevant dimensions for citizen' participation. The general increase of data availability strongly forced planning process and today the planner has mainly the task to select, to organize and to share data in order to support decisions at different scales. The technological wide spread, open data, 2.0 approach and social-network interactions generates data continuously. We can affirm that data are everywhere, but how to get good information? It is the case of several open data services by P.A.s distributing numbers of file not fully exploitable by final users. The paper investigates some relevant examples from the Italian case in order to demonstrate the benefits of data territorialisation and the opportunity to use some specific tools developed within an open source framework: Geo Platform by GeoSDI. In particular we refer to the geocoding process translating a physical property address such as for a house, business or landmark into spatial coordinates. Geocoding intelligence implies the overcoming of semantics barriers in data code and the 'ex-ante' definition of the specific purpose of spatial application in order to accept variables accuracy levels in the final output. Conclusions regard potential application and methodological recommendation for data coding optimization.

Keywords : Geocoding, Spatial Data Infrastructures, open-data, webgis.

B. Murgante et al. (Eds.): ICCSA 2013, Part III, LNCS 7973, pp. 622–629, 2013.
© Springer-Verlag Berlin Heidelberg 2013

1   Introduction

The "sharing" principle in planning processes revealed various issues in technological and methodological application.

While participation processes started very positive experience based on mass collaboration generated Wikinomics [1], which, following the advent of Web 2.0, have become Socialnomics [2], where Citizens are voluntary sensors [3], the technological dimension of planning opened to a new generation of tools promoting and supporting the use of geographical information through web and internet.

A wide series of e-tools has been developed, considering different kinds of communication possible through internet, affirming that the great internet potential allows citizens to "visualize issues and concepts, participate in dialogue, and gain knowledge by interacting" [4].

The ability to explore geographic area on a map and to see the resources in that place, monitoring the associated information, to control parameters, etc., represents a tremendous opportunity for the management of activities in all sectors: from Government to Tourism, from the monitoring networks of services to the study of trade dynamics, the creation of sites for land management, fleets control, humanitarian aid, etc ....

GeoSDI research group has developed and implemented high-level operational solutions for the entire chain of geospatial information management and use, dealing with all the components that make possible the process, providing customers with complete solutions and adhering to specific needs.

In this paper, after describing the relevance of open data fro territorial interpretation, we describe the overall infrastructure of GeoSDI technologies and in particular GeoSDI Fusion, the wigget oriented to geocoding operation. Conclusions regard perspectives in application of open data and GeoSDI technologies for planning pourpose.

2   From Weak Data Availability to Open-Data and Volunteering Information

If during the past decades the main problem in GIS implementation was the lack of spatial data availability, nowadays the wide diffusion of electronic devices containing geo-referenced information generates a great production of spatial data.

Some authors refers to this phenomena as "GIS wikification" [5], where mass collaboration plays a key role in all main components of spatial information (hardware, software, data, and people). Mass collaboration in many cases represents a threat for a lot of professions and new terms have been coined, such as citizen journalism, citizen science, citizen geography, etc. [6]. The term "neogeography" [7] is often adopted to describe people activities when using and creating their own maps, geo-tagging pictures, movies, websites, etc. [8]. It could be defined as a new approach to geography without geographer [9]. Considering that this activity is mainly developed by keens, it is possible to reach good levels of accuracy in the same way as Wikipedia has reached quality levels comparable to Encyclopaedia Britannica [10].

Volunteered geographic information activities (e.g. Wikimapia, OpenStreetMap), public initiatives (e.g. Spatial Data Infrastructures, Geo-portals) and private projects (e.g. Google Earth, Microsoft Virtual Earth, etc.) produced an overabundance of spatial data. While technologies (e.g. GPS, remote sensing, etc.) can be useful in producing new spatial data, volunteered activities are the only way to update and describe such data.

If on one side spatial data have been produced in various ways, remote sensing, sensor networks and other electronic devices generate a great flow of data concerning diverse aspects of human activities or monitoring of environmental phenomena.

Kitsuregawa et al. [11] called "Information-Explosion Era" the amount of information produced by human activities and automated systems; the manipulation of this information is called ubiquitous computing and represents a sort of bridge between computers into the real world, accounting for the social dimension of human environments [12]. If this technological evolution produced a new Paradigm of Urban Development, called u-City [13].

The Open data, and in particular on open government data, represents a new trend to be exploited in building information for planning and territorial issues. The Government role is particularly important in this sense, not only for the quantity and the centrality of the data collected, but also because most of the government data are public by law, and therefore should be made open and available for anyone to use. According to the Open Knowledge Foundation Italia [14] there are many circumstances in which we can expect that the open data have significant value.

We can identify a huge number of areas where public open data contribute to create value for user knowledge building and participation, among them:

- Transparency and democratic control
- Participation and increasing influence in the public discussion
- Improvement or creation of products and services for private sector
- Innovation and R&D
- Improving the efficiency of public services
- Improving the effectiveness of public services
- Measuring the impact of public policies
- Risk and disaster management

Actually it is possible to affirm that the practice of open data and data stores government has extended, but a lot of work and efforts should still be pay in order to get affective services for data integration.

A relevant example in Italy is the project 'Open Cohesion' providing open data service concerning cohesion policies effects.

3   Managing Open-Data: GeoSDI Solution

GeoSDI is a research group of the Institute of Methodologies for Environmental Analysis of the Italian National Research Council (CNR IMAA) who studies, develops and distributes geospatial web based software systems, using an open source approach.

Since 2007 GeoSDI collaborated with maior Public Administrations in Italy, developing GIS solution for risk management, turism, planning etc..

GeoSDI has designed and launched the open source project **Geo-Platform Framework**, the first pure java opensource framework to develop Rich Web GIS Application.

Geo-Platform allows to extend webgis applications adding widgets, software plugins that perform specific functions: in this way every geo-portal is different from the others and it realizes an exact reflection of the functional needs of the end user.

In the following figure we describe the architecture of GeoSDI technology.

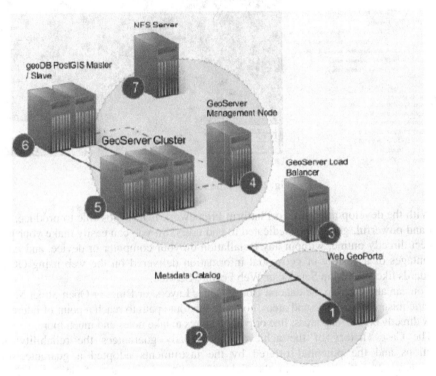

Fig. 1. Basic Architecture Schema for HP & HA

The Basic Architecture Schema for HP & HA includes: 1. Web GeoPortal Server (geo-platform fusion stack); 2. Metadata Catalog; 3. GeoServer Load Balancer; 4. GeoServer Management Node; 5. Geoserver Cluster; 6. PostGIS GeoDB (Master/Slave); 7. NFS Server.

In this solution GeoServer is used as Open Geospatial Consortium compliant server: GeoSDI has realized complex architecture using clustered servers, so as to achieve high performance and high reliability.

Geo-Platform includes the main web technologies, such as Google Web Tollkit, Openlayers, Hibernate, and adds to the versatility of javascript on the web, the power, security and control that Java technology can give.

Fig. 2. Geoportal screen-shot

With the development of Geo-Platform Framework is now possible to produce, so fast and powerful, geoportals dedicated to end users. So you can easily make your gis project directly online, without any installation on your computer or device, and take advantages of a variety of geospatial information delivered on the web using OGC standards like Web Map Service or Web Feature Service.

You can also overlap the data on Google Maps Layers, or Bing, or Open Street Map, execute misure of distance and areas, compute the route path to reach a point of interest, draw directly on the map areas, line or point features to take notes and much more.

The Case History of the achievements already guarantees the reliability of solutions and the potential offered by the instruments adopted a guarantee of efficiency and effectiveness in the field. The use of a series of open and free components makes the cost-effective solutions, as a series of costs normally allocated for the purchase of licenses and royalties of various kinds they fall away.

GeoSDI approach looks to open data and concentrates effort in order to support open data management. Recently GeoSDI development is concentrated to simplify the use of the sistem in order to support voluntering data production through web.

Geo-Platform offers skills on a wide basis, to be expanded according to specific requirements, taking advantage of the development of tailor-made widgets. Its components are two big modules, which make it scalable and powerful:

- **Geo-Platform Services** allow the business logic to be separated from the user's interface. As far as concerns the cluster, Geo-platform can be scaled according to several needs. An API client allows the connection with the interface.

- **Geo-Platform GUI** is the structure taking care of whatever concerns the interface and the stack of services. GUI section contains several under-modules offering functions on different levels.

4   Geo Platform Fusion

Geo Platform Fusion is a Geo Platform feature allowing and supporting the localization of any king of documents.

We thought that all of common internet users are creating and managing standard files formats. Word, excel, powerpoint are currently the most used extension in the world and on the web. Just thinking the you can geo-locate your proper excel file would be something really fantastic for the end user but at the same time it contributes to generate geo-located information according to open and volunteering approach.

The functionality works on a given input file, it parses the file checks to see if any columns contain addresses and starts a geo-location algorithm for single line.

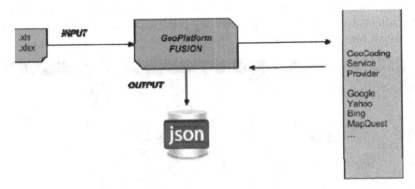

Fig. 3. Geo Fusion schema

The diagram simplify the Geo Platform Fusion architecture. The output returned by geo-platform fusion algorithm is stored in a NO SQL database. In our experiments we chose to use MongoDB.

The service of fusion can work in combination with other widgets that allow instant viewing of the map.

The diagram simplify the Geo Platform Fusion architecture. The output returned by geo-platform fusion algorithm.

Geo Platform Fusion algorithm is very simple. Uses a round robin algorithm on geo-location services configured. At each geo-location is also calculated geo-code rank that can be used to verify the goodness of the geo location made.

The data that is stored in json, will also contain additional information such as: the service provider used for geo-coding, the geocode rank etc. In some cases, localization is not performed correctly, in this case the document is marked as json NOT geolocalized.

PSEUDO CODE of GeoPlatform Fusion Algorithm

```
→ INIT (); // Check geo-platform fusion service is UP

→ Boolean CHECK_INPUT_FILE(File file); // Check if input file
contains addresses information

If (checkInputFile) {
    for (all row) {
        Json address = roundRobingGeoCodeRow(row);
        putJsonIntoMongoDB(address);
    }
}
```

Fig. 4. Pseudo Code of Geo Platform Fusion Algorithm

The applications of such system allow to obtain geo-loacted point pattern data set from traditional data sources with increasing opportunity in data management and data analysis for territorial issues.

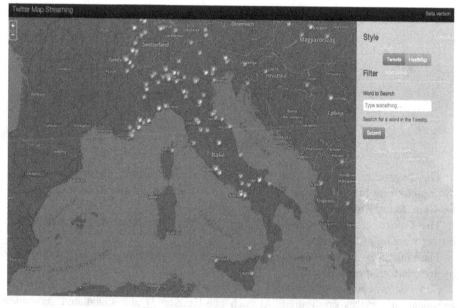

Fig. 5. Twitter mapping by GeoSDI

5   Conclusions

Open data and open software are the actual scenario for territorial analysis. While the web users are growing exponentially the new relevant trend is based on production of geo spatial information. This allow to keep information for territorial purposes in a wide range of data sources. An interesting example comes from social network and in the following figure we show an application fo GeoSDI mapping technologies to the stream of social network (in particular to the twitter geolocated stream).

These should be considered as new opportunities for planners and geographer in order to achieve a deeper territorial interpretation sometimes suggested by internet users.

GeoSDI research provided effective tools for geospatial data management developed in an open community and adopting open source approach. This appears no more as a perspective but it represents a concrete support for researchers, technicians, and public organizations in order to include the use of geo-spatial information in practice.

References

1. Tapscott, D., Williams, A.D.: Wikinomics: How Mass Collaboration Changes Everything. Penguin Group, New York (2006)
2. Qualman, E.: Socialnomics: How social media transforms the way we live and do business. John Wiley & Sons, Hoboken (2009)
3. Goodchild, M.F.: Citizens as sensors: the world of volunteered geography. GeoJournal 69, 211–221 (2007)
4. Macpherson, L.: Joystick Not Included: New Media Technologies are Ideal Tools for Gaining Stakeholder Interest, Acceptance. Water Environment and Technology 11(9), 51–53 (1999)
5. Sui, D.S.: The wikification of GIS and its consequences: Or Angelina Jolie's new tattoo and the future of GIS. Computers, Environment and Urban Systems 32(1), 1–5 (2008)
6. Goodchild, M.F.: Citizens as Voluntary Sensors: Spatial Data Infrastructure in the World of Web 2.0. International Journal of Spatial Data Infrastructures Research 2, 24–32 (2007)
7. Turner, A.: Introduction to Neogeography. O'Reilly Media, Sebastopol (2006)
8. Hudson-Smith, A., Milton, R., Dearden, J., Batty, M.: The NeoGeography of Virtual cities: Digital mirrors into a recursive world. In: Foth, M. (ed.) Handbook of Research on Urban Informatics: The Practice and Promise of the Real-Time City. Information Science Reference. IGI Global, Hershey (2009)
9. Goodchild, M.F.: NeoGeography and the nature of geographic expertise. Journal of Location Based Services 3, 82–96 (2009)
10. Giles, J.: Internet encyclopedias go head to head. Nature 438, 900–901 (2005)
11. Kitsuregawa, M., Matsuoka, S., Matsuyama, T., Sudoh, O., Adachi, J.: Cyber Infrastructure for the Information-Explosion Era. Journal of Japanese Society for Artificial Intelligence 22(2), 209–214 (2007)
12. Greenfeld, A., Shepard, M.: Urban Computing and Its Discontents. The Architectural League of New York (2007)
13. Hwang, J.S.: u-City: The Next Paradigm of Urban Development. In: Foth, M. (ed.) Handbook of Research on Urban Informatics: The Practice and Promise of the Real-Time City. Information Science Reference. IGI Global, Hershey (2009)
14. Open Knowledge Foundation Italia, Open data handbook (May 2013),
http://opendatahandbook.org/ (retrived)

Cities and Smartness: A Critical Analysis of Opportunities and Risks

Beniamino Murgante[1] and Giuseppe Borruso[2]

[1] University of Basilicata, 10,Viale dell'Ateneo Lucano, 85100 Potenza, Italy
[2] University of Trieste, P. le Europa 1, 34127 Trieste, Italy
beniamino.murgante@unibas.it, giuseppe.borruso@econ.units.it

Abstract. The term "Smart City" is to-date widely used, but little clarity appears in the definition behind it. Several approaches led to a growing emphasis on the combined use of geographic information and communication technology to build cognitive frameworks in city planning and management. The present paper tackles an effort to define 'smart cities' and to identify both elements of smartness, and critical aspects related to the current interpretation of the term. In particular, the risk of considering the technological layer of Smart City as an innovative element has been observed, highlighting, on the contrary, the need to consider Smart Cities in terms of a major urban planning effort to coordinate and harmonize different urban players, sustained by ICT instruments.

Keywords: Smart city, Smart communities, Neogeography, Open data, Citizens as sensors, Governance.

1 Smart City: A Non-unique Definition

Although the term "Smart City" is to-date widely used, little clarity appears in the definition behind it and particularly on its actual meaning.

The idea behind a Smart City is that in the current digital age, not only physical infrastructures and endowment of a city characterize an urban area and its functions, but something less 'hard' and not so easy to identify, as quality of knowledge communication and 'social infrastructure', or social and intellectual capitals. In such an (urban) environment, mood and attitude, the concept of Smart City arises, as a device or, better, as a framework where 'traditional' urban production factors are coupled with the social, cultural capital, by means of a massive use of ICTs.

The stress – quite agreed to-date – tends to be on 6 main axes of 'smartness' including economy, mobility, environment, people, living and governance (Table 1). Such axes include the concepts behind neoclassical theories of urban growth, sustainable development, ICT and citizens' participation in urban governance.

In these terms, a smart city is something more than 'just' a digital or an intelligent city, where the attention is mainly drawn on the ICT components, as enabling connection and exchange of data and information within an urban environment. Given the '6 axes' and the attention to growth, sustainability, ICT and citizens' governance and

B. Murgante et al. (Eds.): ICCSA 2013, Part III, LNCS 7973, pp. 630–642, 2013.

participation, a smart city appears more like a new 'urban utopia', although not too difficult to be realized, and basically as the evolution of the sustainable city, in terms of combining economic, social and environmental aspects to elements of social and cultural capital, as well as to the power of ICT technologies and applications. Going back to the beginning of this paragraph, if it is true that a city's physical infrastructure, as well as its endowments, are the result of a process of interaction between humans and (urban) environment, it is also true that physical infrastructures (buildings, roads, utilities) are built by humans to ease urban growth and development, while their presence and essence give also a direction for future development and evolutions or represent a constraint. So there is a mutual exchange of influences and causal relations. As De Biase states, reminding Winston Churchill's words: "We shape our buildings; thereafter they shape us". Smart cities are not so different in this sense. Of course, buildings and infrastructures are still being built in cities, but to-date, such buildings and infrastructures are also those not immediately visible and 'fix' in space and time. ICTs infrastructures, as well as devices based on them, shape structures and functions of cities, being XXI century equivalent of medieval cathedrals, ordered Renaissance' squares and XIX century railway stations.

Other definitions containing an attribute coupled with the term 'city' in the (also recent) past provided quite a concrete and almost precise orientation and meaning. Without going back to utopian cities population, various periods of human times or ideal cities dating back to the Renaissance, also the recent concept of sustainable city hold a quite strong and well-defined set of attributes describing its characters.

This is not true, at present, for the locution 'smart city'. The different definitions meanings provided in different areas of the World have in common the implication of technology, and, particularly, the wide use of ICT infrastructures and devices. However, such elements can represent either the most important and relevant part or just a component of an overall meaning.

2   Virtual Cities, Computable City and Ubiquitus City

The concept of Smart city can derive from several approaches, sometimes slogans, which lead to a growing emphasis on the combined use of geographic information and communication technology to build cognitive frameworks in city planning and management.

Since the late '90s, with the growing diffusion of the internet, the experience of Virtual Cities beginning [1] has focused on construction and representation of urban scenarios. The use of Virtual Reality Modeling Language (VRML) allowed the creation of virtual environments and three-dimensional models of cities usability on the internet. This experience is not only restricted to simulation fields, but, using the large internet diffusion, it has been used to create online participatory experiences, allowing part of the population to take part in urban policies creation. In other cases, citizens were allowed to contribute to a neighbourhood renewal project choice [2] [3] [4] [5] simply by means of electronic vote.

Batty [6] considered the huge possibilities deriving from a massive convergence of computer and communications through various forms of media.

Initially computers were used as a deeper support in city planning and programming. In subsequent years, interest has been moved on how computers and information technologies are changing cities. The result is the concept of Computable City [6], focused on the simultaneous analysis of both aspects. This concept examined both the ways in which computers were changing methods for city understanding and changes in city structure and dynamics. Later on, other types of computing with strong impact on the city have been adopted, such as ubiquitous computing, pervasive computing, physical computing, tangible media, each as facet of an interaction coherent paradigm, which Greenfield [7] (2006) defines "everyware". At the end of 1990s, Openshaw [8] [9] coined the term Geocomputation, considering two main issues: intensity of the process and increase of knowledge and intelligence. This expression has been interpreted according to several meanings. Ehlen et al. [10] analyzed four aspects of Geocomputation: from a high performance computing point of view, as a set of spatial analysis methods, as the essential aspects of Geocomputation and as their relationship with GIS [39] [40]. In some cases there is a transition from a vision based on a computing power to a distributed environment where computers, seen in their traditional sense, disappear.

Consequently, the concept of computable city assumed increasing importance with the growth of electronic devices in our physical environment [11].

The transition towards a not only virtual environment, i.e. an environment with a deep human and social interaction through computers, characterizes urban computing [12]. These theories take into account the social dimension of human environments, placing computers at the background. Shepard and Greenfield's [12] theories on urban computing coupled with ubiquitous computing research developed at the Xerox Palo Alto Research Centre [13] promoted the first experiences of ubiquitous cities [14], mainly concentrated in Asia. The objective of an ubiquitous city (U-city) is to create an integrated environment, where citizens can get any type of services, in all places, at any time and with all kinds of ICT devices [15]. These applications are based on infrastructures with the aim to support local needs by improving daily life of local communities.

The possibility of using real time acquired data, allowing continuous monitoring of main urban phenomena, can substantially improve the effectiveness of spatial planning and urban management. There is a transition from a traditional approach, based on the sequence real city, computer, virtual representation, to the sequence, computer, real city, ubiquitous city.

The traditional sequence considered many people working on one or on a few computers, while in U-city sequence only one person handles much computers and electronic devices [16].

3   Open Government and Gov. 2.0

A large amount of information produced by human activities and automated systems Information-Explosion Era [17] is available, not only in Asia, where experiences of U-city are mostly concentrated.

In the last five years, acceleration occurred, supported by the diffusion of GPS devices and 3G connections in mobile phones, which has led to a large production of geo-localized or social networks based applications. This has led to a huge activity of Crowdsourcing [18], where suggestions services, ideas and any decision support can be achieved by online communities' actions. Population directly provides certain services that government is not interested to develop and private sector does not consider convenient to realize.

There are more and more initiatives (OpenStreetMap, WikiMapia, Google Map Maker, Geo-Wiki) of Volunteered Geographic Information [19], based on mass collaboration to create, manage and disseminate geographic data where citizens are voluntary sensors [20]. The huge production of data on the web has led to "Neo-geography" [21], defined as a new approach to geography without geographers [22] which describes the bottom-up production of maps with geo-tagged photos, videos, blogs, Wikipedia, etc. [23].

Another important tendency in progress in recent years is open government. Such an approach is based on a more participative method of government and it starts from the assumption that ideas of citizens have always to be collected, not only before elections. Consequently, public involvement, getting ideas and suggestions, is a daily activity, aiming to have a wider inspiration in managing and to collect feedback in already started actions. Obama's administration has given a great impetus to this approach, implementing such a policy and enlarging the possibility to capture public imagination by means of social networks, blogs and all possible solutions to directly interact with citizens.

This new approach is often called Gov. 2.0. Open government without a 2.0 approach is still based on a direct action. "Providers" are a sort of Right to Information, where the administration tries to inform people, but interacting just with main stakeholders. Gov. 2.0 is a more open approach, which "enables" citizens to have an important role in defining policies. Social media and all 2.0 platforms are a key element in generating a direct contact with citizens. Extensions of 2.0 philosophy changed completely the relationship between citizens and administration [24].

It is a type of governance where aspects related to participatory decision-making are central and the transition from Government to Governance is combined with visioning techniques.

Since early '90s a transition occurred from an approach where local authorities directly provide to problem solutions (Government), to another approach, where local authorities tend to accompany the process (Governance). In the latter one, administrations enable and facilitate the search of different solutions, in collaboration and agreement with other public and private stakeholders [25] [26]. In the same years visioning methods were adopted in order to develop bottom-up contributions, fundamental in planning process. This technique emphasizes plan communication aspects, highlighting the importance of social imagination as a contribution to the definition of a scenario of desirable actions in planning process [27].

In a lot of cases traditional participatory approaches, based on public meetings, proved to be unsuccessful, due to restricted number of participants who did not represent a significant sample. Electronic participation goes beyond space and time dimensions, allowing all citizens, who may be working during the meeting time, or live in a distant place, or are embarrassed of public speaking, to express their opinions and producing a significant contribution in improving ideas.

Ten years ago, Kingston [28] adapted Arnstein [29] ladder to electronic era, defining E-participation Ladder, adopting several levels from a simple web site to online decision-making. Haklay [30], considering citizens cooperation, distinguishes four levels of citizens science, where crowdsourcing is the lowest level and the highest level is a sort of collaborative science, where citizens can have the responsibility to define problems and to find possible solutions. Today we are living in wikification era, with many successful initiatives based on mass collaboration [31] [32], which may also lead to a wiki approach to decisions and planning [33] [34].

4   City Sensing and Smart City

City sensing is based on electronic and human sensors or on the combination of both [35], on voluntary or unconscious actions [36], and it is a key component in Smart City.

It is central to correctly define the relationship between city sensing and smart city, because these are new concepts without a precise and unambiguous definition.

Considering also that the application domain is the city, whose elements are rooted in our daily lives, there is a risk, in analogy with what happened with the concept of sustainability, that after many years we have collected a lot of words and few results. The correct relationship between city and sensing Smart city must be based on equal dignity of all aspects. It could happen to forget the city, focusing the attention only on technology. The main risk would be represented by a fall of electronic devices on the city, which does not have a direct relationship with its main problems.

In analogy with the beginnings of geographic information systems, when the market was mainly determined by supply more than by demand, the risk is to invest significant resources in purchasing hardware and software without having a clear idea of administration needs and their possible use in city management.

The European experience differs from U-city in giving less importance to computational aspects and in paying more attention to the potential of technologies for the improvement of city quality. Great attention has been paid to digital citizenship that leads to new forms of social organization related to information technology.

A shared definition identifies smart cities in a synthesis of physical and social infrastructures [37], where the first one can represent a catalyst for knowledge communication, increasing social and intellectual capital. A superficial approach combined with a rush to be included under "smart umbrella", can lead to ignore these aspects, mainly focusing on improving devices and technological systems which quickly get old. A city can be considered smart if it can quickly integrate and synthesize data produced by each type of sensor, to improve efficiency, equity, sustainability and quality of life [38]. It is important to consider the big impact of technologies on new forms of policy and planning. In analyzing smart cities, Batty et al. [38] identify seven points on which the attention should be focused, analyzing key problems of cities, using information and communication technologies:

1. a new understanding of urban problems;
2. effective and feasible ways to coordinate urban technologies;
3. models and methods to use urban data across spatial and temporal scales;
4. developing new technologies for communication and dissemination;
5. new forms of urban governance and organisation;
6. defining critical problems about cities, transport, and energy;
7. risk, uncertainty and hazard in the smart city.

It is important to give priority to the construction of cognitive frameworks and to a wider knowledge in supporting decisions in urban planning, compared to approaches based on procedural efficacy. Today, especially in Europe, compliance with procedures is mainly considered the production of a bureaucratic truth, in most cases very far from reality, when analysing urban phenomena. Recently, a lot of reports have been published in order to define variables to classify smartness level of municipalities in a hypothetical path to smarter cities. Table 1 is an attempt to synthesize the main variables adopted in reports which analyze smart cities.

Table 1. Synthesis of the main variables adopted in reports analyzing smart cities

Dimension	Variables
Smart Economy	Employment rate; presence of innovative enterprises, presence and quality of universities and research institutes; infrastructures (roads, railways, airports, electronic infrastructures, etc.).
Smart Environment	Air quality, percentage of separate collection of municipal waste (also electrical and electronic equipment waste), presence of green spaces in the city, efficiency and quality of water supply (water leakage and water treatment).
Smart Governance	Not only related to e-government, percentage of ecological cars, use of recycled paper, energy saving, adoption of ecological policies for city planning and development, ability to network with other municipalities.
Smart Living	Investments in culture and welfare providing several services, from childcare facilities to community libraries, from counselling structures for old people to cinemas, number of people below poverty level, hospital emigration rate, immigrants social integration, criminality rate.
Smart Mobility	Extensive and efficient public transportation network, park and ride, great diffusion of ecological cars, limited traffic areas, cycle paths, bike and car sharing.
Smart People	Education and early school leaving level, number of women working and holds positions within the administration, presence of foreign students, political participation, involvement in voluntary associations, newspapers diffusion and level of participation to cultural events.

In most cases they are traditional indicators, concerning the city based on old variables, with the addition of the "smart" attribute. If we delete this last term in the above table we achieve typical socio-economic or environmental sustainability indicators.

Indicators concerning smartness level of our cities should consider the following aspects:

1. adoption of OpenData and OCG Standard;
2. free wifi;

3. projects implementation of augmented reality for tourism;
4. crowdfunding initiatives;
5. decisions taken by crowdsourcing;
6. implementation of INSPIRE Directive;
7. quantity of public services achievable through App.

5   Smart City: The Pillars

Identifying what makes a city smart is related to the different dimensions, which are connected to concepts quite consolidated in references dealing with urban topics. In the *smart* meaning, the technological component is particularly related to ICT features and infrastructures. These play an important role, in particular as facilitators of processes of innovation, sharing and active participation by citizens/users, as well as of the development of elements typical of knowledge economics. Following some of the most interesting interpretations [42], smart cities are cities in which a 'technological layer' is overlaid onto the existing urban structure and fabric, allowing its citizens and users to connect to the net, interact among them and with other different players – public administration, suppliers of goods and services, etc., actually optimizing a city and its spaces. Since world population is growing and such growth is expected to be particularly concentrated in cities, technology can play an important role in limiting soil consumption and enhancing quality of life.

However, one of the risks today is that decision makers, politicians, citizens, enterprises focus just on the fashion of the technological side of "smartness", with little attention to insert it into a process of urban planning and project.

In a *smart city* the technological infrastructure related to ICT is central, in the same way as in the past the realization of new buildings, roads, railways, telephone and energy distribution lines and networks was. Such infrastructures both supported population needs and influenced how such population interacted with the urban space. Infrastructures of a smart city should play a similar role, therefore needing a focused planning, as their use must not be limited to the short terms but it should persist and, actually, persists, having in mind that to-date settings will influence how citizens will interact with the city in present and future times. In a smart city, the network metaphor is overlaid onto the urban metaphor; in such sense acting as a new, different infrastructure capable of channelling relations and interactions and to be influenced and shaped by such interactions, similarly to a public transport network developing in an embryonic city to connect and serve places and then evolving and giving birth to 'new' places.

The city should therefore set as an "enabling platform for the activities that citizens are able to develop, linking those inherited from the past to those that can be realized in the future, so it is not focused on just applications but on the possibility that citizens realize them" [41].

A smart city should therefore be passed on different pillars, elements to be organized and linked together. These can be summarized [41] in three main elements (Figure 1):

1. connections - as networks and technological infrastructures;
2. data – open and public or public interest data to allow the development of innovative solutions and the interaction between users/citizens and the city;
3. sensors - these including citizens [19] [20] [22] able to actively participate in a bottom up way to city activities.

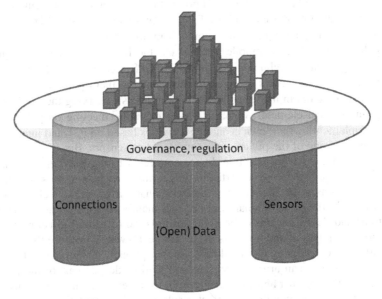

Fig. 1. The Pillars sustaining the Smart City and its Governance (graphical elaboration, after concepts in De Biase, 2012).

These pillars must be coupled with a governance capable of linking them together, giving a direction and a vision to the city. Such governance should regulate the smart city in a neutral way, without entering into the details for applications and contents.

A Smart city therefore appears as an urban project, as a big infrastructure and as a metaphor of the net in an urban context. In a sentence, a smart city becomes an environment where a definite set of elements, as the ones above reported – sensors, data and connections – harmonized by a limited set of basic rules, gives public bodies, citizens, enterprises the possibility of developing applications and solutions able to improve life of the city itself, leaving actually the initiative of doing that to people, groups, firms, etc., allowing also to create new markets and solutions also where the public sector is not able to move.

6   Are Cities Smart?

Finally, are cities smart? Twenty years ago we would have asked: are cities sustainable? In that period, that was the paradigm of the moment – actually it still is – as cities are the places where main human actions take place and therefore the places where to set policies aimed at a sustainable future in terms of adequate and respectful exploitation of resources from an economic, environmental and social point of view.

How does 'smart' differ from sustainable? And why is it different? What elements were added? Smart cities – and communities! - aim at sustainable development. Actually the six dimensions of smart cities share the basic dimensions of sustainability in development: environmental, economic and social. Of course a difference is in the presence of a 'techy layer' as Ratti [42] pointed out – see above –particularly characterized by the revolution occurred in ICT, that allows an unprecedented opportunity of interactions among places, individuals, organizations. This is the real revolution, coupled with the spreading of mobile devices and the increasing precision in location allowed by geospatial technologies (embedded GPS receivers, etc.). Therefore, the role of citizens or city users changed in time, making them potential and powerful influencers and actors in the urban arena, both in terms of serving their communities, highlighting critical elements, or participating to public meeting on policy choices, but also implementing their own economic activities based on ICT and interaction.

Citizens – as one of the pillars – are considered as sensors. But what sensors? Are sensors only citizens with a mobile device connected to the Internet? A Smart City holds a strong social dimension, particularly in terms of inclusion of its citizens and in enabling solutions to be implemented to tackle that. However, a 'techy' orientation and particularly the view of smartness just and mainly focused on developing smart apps, tools and devices seems to be going towards a direction of affecting just a part of the urban population and users, as those 'Hi-Tech aware', or those that to-date are constantly connected using mobile devices – smartphones, tablet pc, etc. In doing so, digital divide issues can arise. At present and worldwide just part of the population has access to the Internet and to IT devices. In these terms a 'smartness' just limited to a 'rainfall of apps' would only affect a subset of the population, thus worsening social disparities rather than reducing them. Talking about citizens as sensors, we could say that this is not completely new, just faster, simpler and wider. Citizens have been participating to urban issues since the emerging of various media. Letters to newspapers, local municipalities, phone calls, have always been ways of pointing out faults in urban fabrics rather than bad services. Of course at present that can be done by means of a geo-tagged photo shared among social media and networks and therefore more easily reaching a vast amount of users and bodies.

So a Smart City, as an enabling platform, should allow both the development and hosting of 'rainfall of apps' but also including other less-techy users – phone callers, etc. – and in that lays the difficulty: that of building a real network and making things work. What is the point in having cutting edge mobile applications that, say, allow you communicating to your municipality about a sewage leak close to a primary school, if behind that the public body did not set any infrastructure, procedure and habit to tackle such an issue? So smartness should act as a cultural product other than

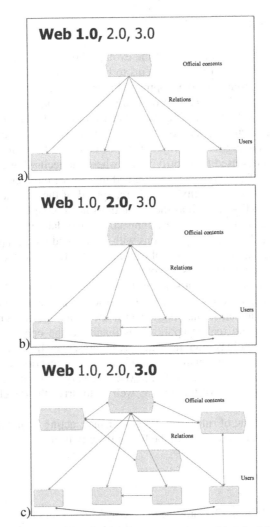

Fig. 2. The evolution of the Web. a) Web 1.0, b) Web 2.0, c) Web 3.0

a technological feature. Also, attention should be put in the interaction between public bodies and public utility bodies, in that allowing the interaction not just in the Web 2.0 approach, but going to the 'Web 3.0' one, in which institutions share their data and contents not just with users but between them, thus generating misunderstandings and mismatching (Figure 2).

The problem of governance and setting common rules becomes the real question in the smartness of smart cities. Thinking about the 'smart infrastructure', a code of rules should be agreed, in a similar way of the highway code that allows us driving on a road network and avoiding – in most of the cases – problems respecting minimal restrictions.

7   Conclusions

Smart city as a paradigm is the result of the evolution of thinking and reasoning over the city and its issues. In particular, it seems to be a combination of concepts related to sustainability and sustainable development, in terms of its urban application. Also, the idea of 'locally acting' originally proposed for urban sustainability, presents some of the suggestions that few years later have been introduced in participation of citizens and the web 2.0. Furthermore, Smart city derives from the evolution of technology and thinking in the digital era. Digital City, Computable City and Virtual City are just a few of the names used to identify a city where the technological component is strongly present and affects how citizens use and interact with the city.

The revolutions of sustainability, digital era, spread of the Internet, of mobile devices and data availability, as well as the revolutions in Geographical Information, led to a widespread availability of devices, connections and data and the opportunity to link them together and develop applications with high added value capable to enhance quality of urban life. An attention to applications and to 'techy' aspects related to city therefore arose, opening new issues and opportunities.

The debate is still on-going, but some reflections lead to think to Smart Cities as a revolution intervening in terms of a new infrastructure and platform, made of both virtual and physical elements, enabling citizens, users and all different urban players to carry on activities and realize applications thanks to the opportunity allowed by improvements in technology and its widespread presence. In such terms, we talk about an infrastructure conceptually not different from transport ones, developed in the past years and centuries, that both allowed to enlarge the city extension and to connect places once not part of the city, as well as to drive the development of new urban areas.

Vital is also the setting of rules and of a governance, acting as an highway code for city users, with little interference with the life of the city itself.

References

1. Smith, A.: Virtual Cities - Towards the Metaverse, Virtual Cities Resource Centre (1998), http://www.casa.ucl.ac.uk/planning/virtualcities.html
2. Levy, R.M.: Visualisation of Urban Alternatives. Environment and Planning B: Planning and Design 22, 343–358 (1995)
3. Batty, M., Doyle, S.: Virtual regeneration. (CASA Working Papers n. 06). Centre for Advanced Spatial Analysis (UCL): London, UK (1998) ISSN: 1467-1298
4. Hudson-Smith, A., Dodge, M., Doyle, S.: Visual Communication in Urban Planning & Urban Design. GIS and Urban Design. (CASA Working Papers n. 02). Centre for Advanced Spatial Analysis (UCL): London, UK (1998) ISSN: 1467-1298
5. Batty, M, Dodge, M, Jiang,B., Hudson-Smith, A.: GIS and Urban Design. (CASA Working Papers n. 03). Centre for Advanced Spatial Analysis (UCL): London, UK (1998) ISSN: 1467-1298
6. Batty, M.: The computable city. In: Fourth International Conference on Computers in Urban Planning and Urban Management, Melbourne, Australia, July 11-14 (1995)
7. Greenfield, A.: Everyware: The dawning age of ubiquitous computing. New Riders, Berkeley (2006)

8. Openshaw, S.: Building automated Geographical Analysis and Explanation Machines. In: Longley, P.A., Brooks, S.M., McDonnell, R., Macmillan, B. (eds.) Geocomputation, a Primer. John Wiley and Sons, Chichester (1998)

9. Openshaw, S.: GeoComputation. In: Openshaw, S., Abrahart, R.J. (eds.) GeoComputation (2000)

10. Ehlen, J., Caldwell, D.R., Harding, S.: GeoComputation: what is it? Comput. Environ. and Urban. Syst. 26, 257–265 (2002)

11. Hudson-Smith, A., Milton, R., Dearden, J., Batty, M.: Virtual Cities: Digital Mirrors into a Recursive World. (CASA Working Papers n. 125). Centre for Advanced Spatial Analysis (UCL): London, UK (2007) ISSN: 1467-1298

12. Shepard, M., Greenfield, A.: Urban Computing and its Discontents. The Architectural League of New York, New York (2007)

13. Weiser, M.: Hot Topics: Ubiquitous Computing. IEEE Computer 26(10), 71–72 (1993)

14. Jang, M., Suh, S.-T.: U-City: New Trends of Urban Planning in Korea Based on Pervasive and Ubiquitous Geotechnology and Geoinformation. In: Taniar, D., Gervasi, O., Murgante, B., Pardede, E., Apduhan, B.O. (eds.) ICCSA 2010, Part I. LNCS, vol. 6016, pp. 262–270. Springer, Heidelberg (2010)

15. Lee, S.H., Han, J.H., Leem, Y.T., Yigitcanlar, T.: Towards ubiquitous city: concept, planning, and experiences in the Republic of Korea. In: Yigitcanlar, T., Velibeyoglu, K., Baum, S. (eds.) Knowledge-Based Urban Development: Planning and Applications in the Information Era, pp. 148–170. Information Science Reference, Hershey (2008), doi:10.4018/978-1-59904-720-1.ch009

16. Lee, B.G., Kim, Y.J., Kim, T.H., Yean, H.Y.: Building Information Strategy Planning for Telematics Services. In: KMIS International Conference, Jeju Island, Korea, November 24-26 (2005)

17. Kitsuregawa, M., Matsuoka, S., Matsuyama, T., Sudoh, O., Adachi, J.: Cyber infrastructure for the information-explosion era. Journal of Japanese Society for Artificial Intelligence 22(2), 209–214 (2007)

18. Howe, J.: Crowdsourcing: Why the Power of the Crowd Is Driving the Future of Business. Crown Publishing Group, New York (2008)

19. Goodchild, M.F.: Citizens as Voluntary Sensors: Spatial Data Infrastructure in the World of Web 2.0. International Journal of Spatial Data Infrastructures Research 2, 24–32 (2007)

20. Goodchild, M.F.: Citizens as sensors: the world of volunteered geography. GeoJournal 69(4), 211–221 (2007), doi:10.1007/s10708-007-9111-y

21. Turner, A.: Introduction to neogeography. O'Reilly Media, Sebastopol (2006)

22. Goodchild, M.F.: NeoGeography and the nature of geographic expertise. Journal of Location Based Services 3, 82–96 (2009)

23. Hudson-Smith, A., Milton, R., Dearden, J., Batty, M.: The neogeography of virtual cities: digital mirrors into a recursive world. In: Foth, M. (ed.) Handbook of Research on Urban Informatics: The Practice and Promise of the Real-Time City. Information Science Reference, IGI Global, Hershey (2009)

24. Murgante, B., Tilio, L., Lanza, V., Scorza, F.: Using participative GIS and e-tools for involving citizens of Marmo Platano – Melandro area in European programming activities. Journal of Balkans and Near Eastern Studies 13(1), 97–115 (2011) ISSN:1944-8953, doi:10.1080/19448953.2011.550809

25. Balducci, A.: Pianificazione strategica e politiche di sviluppo locale. Una relazione necessaria? Archivio di Studi Urbani e Regionali 64 (1999)

26. Gibelli, M.C.: Tre famiglie di piani strategici: verso un modello "reticolare" e "visionario". In: Gibelli, M.C., Curti, F. (eds.) Pianificazione Strategica e Gestione Dello Sviluppo Urban, Alinea, Firenze (1996)

27. Gibelli, M.C.: Riflessioni sulla pianificazione strategica. In: Rosini, R. (ed.) L'urbanistica Delle Aree Metropolitane, Alinea, Firenze (1992)

28. Kingston, R.: The role of e- government and public participation in the planning process. In: Proceedings of XVI AESOP Congress, Volos, Greece (2002)

29. Arnstein, S.R.: A ladder of citizen participation. Journal of the American Planning Association 35(4), 216–224 (1969)

30. Haklay, M.: Citizen Science as Participatory Science (November 27, 2011), http://povesham.wordpress.com/2011/11/27/citizen-science-as-participatory-science/ (retrieved)

31. Tapscott, D., Williams, A.D.: Wikinomics: How Mass Collaboration Changes Everything. Penguin Group, New York (2006)

32. Qualman, E.: Socialnomics: How Social Media Transforms the Way we Live and do Business. John Wiley, Hoboken (2009)

33. Noveck, B.S.: Wiki Government: How Technology Can Make Government Better, Democracy Stronger and Citizens More Powerful. Brookings Institution Press, Harrisonburg USA (2009)

34. Murgante, B.: Wiki-Planning: The Experience of Basento Park In Potenza (Italy). In: Borruso, G., Bertazzon, S., Favretto, A., Murgante, B., Torre, C. (eds.) Geographic Information Analysis for Sustainable Development and Economic Planning: New Technologies, pp. 345–359. Information Science Reference IGI Global, Hershey (2012), doi:10.4018/978-1-4666-1924-1.ch023

35. Bergner, B.S., Exner, J.P., Memmel, M., Raslan, R., Dina Taha, D., Talal, M., Zeile, P.: Human Sensory Assessment Methods in Urban Planning – a Case Study in Alexandria. In: Proceedings REAL CORP 2013, Rome (Italy), May, 20-23, pp. 407–417 (2013)

36. Manfredini, F., Pucci, P., Tagliolato, P.: Mobile Phone Network Data: New Sources for Urban Studies? In: Borruso, G., Bertazzon, S., Favretto, A., Murgante, B., Torre, C. (eds.) Geographic Information Analysis for Sustainable Development and Economic Planning: New Technologies, pp. 115–128. Information Science Reference, Hershey (2013), doi:10.4018/978-1-4666-1924-1.ch008

37. Caragliu, A., Del Bo, C., Nijkamp, P.: Smart cities in Europe. Research Memoranda Series 0048 (VU University Amsterdam, Faculty of Economics, Business Administration and Econometrics). CRC Press, Boca Raton (2009)

38. Batty, M., Axhausen, K.W., Giannotti, F., Pozdnoukhov, A., Bazzani, A., Wachowicz, M., Ouzounis, G., Portugali, Y.: Smart cities of the future. The European Physical Journal Special Topics 214(1), 481–518 (2012)

39. Murgante, B., Borruso, G., Lapucci, A.: Geocomputation and Urban Planning. In: Murgante, B., Borruso, G., Lapucci, A. (eds.) Geocomputation and Urban Planning. SCI, vol. 176, pp. 1–17. Springer, Heidelberg (2009)

40. Murgante, B., Borruso, G., Lapucci, A.: Sustainable Development: concepts and methods for its application in urban and environmental planning. In: Murgante, B., Borruso, G., Lapucci, A. (eds.) Geocomputation, Sustainability and Environmental Planning. SCI, vol. 348, pp. 1–15. Springer, Heidelberg (2011)

41. De Biase, L.: L'intelligenza delle Smart Cities (2012), http://blog.debiase.com/2012/04/intelligenza-delle-smart-city/

42. Roche, S., Nabian, N., Kloeckl, K., Ratti, C.: Are 'Smart Cities' Smart Enough? In: Global Geospatial Conference 2012. Global Spatial Data Infrastructure Association (2012), http://www.gsdi.org/gsdiconf/gsdi13/papers/182.pdf

Author Index

Abreu, Mário I-304
Acharyya, Rashmisnata II-73
Agrawal, Dharma P. V-143
Aguilar, José Alfonso III-59
Ahnert, Tobias V-91
Akdag, Herman I-204
Akter, Taslima III-408
Alam, Md. Shafiul V-48
Albertí, Margarita I-1
Aleb, Nassima II-487, II-574
Allegrini, Elena II-160, II-231, II-288
Almeida, José João II-443
Álvarez, Mabel III-43
Alves, Gabriela II-559
Amaolo, Marcelo III-43
Anderson, Roger W. II-46, II-60
Ando, Takahiro III-114
Andrighetto, Alberto I-84
Ang, Kenneth Li-Minn I-464
Anjos, Eudisley III-199
Aquilanti, Vincenzo II-32, II-46, II-60
Areal, Janaina II-559
Arezzo di Trifiletti, Michelangelo II-160
Aromando, Angelo II-652
Arroyo Ohori, Ken I-526
Asche, Hartmut II-635, IV-221
Assaf, Rida II-129
Attardi, Raffaele IV-541
Auephanwiriyakul, Sansanee III-246
Azevedo, Luiz III-230
Azzato, Antonello IV-304

Bae, Kang-Sik V-127
Bae, NamJin III-287
Bae, Sueng Jae I-120, I-131
Baiocchi, Valerio IV-136, IV-150
Balena, Pasquale IV-528, IV-587, IV-600
Balucani, Nadia I-47
Barazzetti, Luigi I-608, IV-328
Barbier, Guillaume I-253
Bartocci, Alessio I-69
Bärwolff, Günter V-17, V-91
Basappa, Manjanna II-73

Bastianini, Riccardo I-96
Beccali, Marco II-344
Bedini, Roberto II-299
Behera, Dhiren Kumar III-258
Belanzoni, Paola I-57
Belviso, Claudia II-652
Bencivenni, Marco I-84
Beneventano, Domenico I-194, V-462
Benner, Joachim III-466
Ben Yahia, Nour I-683
Bergamaschi, Sonia I-194, V-462
Berger, Ágoston II-529
Bernal, Roberto III-59
Berres, Stefan V-17
Bhatia, Shveta Kundra II-498
Bhattacharya, Indira IV-108
Bhowmik, Avit Kumar IV-120
Bhuruth, Muddun V-77
Bimonte, Sandro IV-253
Biondi, Paolo II-220, II-288
Birkin, Mark IV-179
Bitencourt, Ana Carla Peixoto II-1, II-46, II-60
Blecic, Ivan III-594, IV-284
Boada-Oliveras, Immaculada IV-17
Bocci, Enrico II-256, II-271
Bollini, Letizia III-481
Bona, Luis Carlos Erpen V-281
Bonifazi, Alessandro IV-528
Borfecchia, Flavio III-422
Borg, Erik II-635
Borriello, Filomena IV-515
Borruso, Giuseppe III-630, IV-375, IV-389
Boubaker, Karemt II-220, II-288
Braga, Ana Cristina I-573, I-585
Bravo, Maricela I-452, V-636
Brettschneider, Matthias III-128
Brumana, Raffaella IV-328
Brummer, Stephan II-99
Bruns, Loren V-364
Brus, Jan IV-166
Buarque, Eduardo II-391

Cabral, Pedro   IV-120
Caccavelli, Dominique   IV-358
Caiaffa, Emanuela   III-422
Calderini, Danilo   II-32
Campobasso, Francesco   IV-444
Candori, Pietro   I-69
Canu, Dario   III-594
Cappuccini, Andrea   II-231
Caprini, Luca   I-708
Carlini, Maurizio   II-160, II-176, II-242
Carone, Paola   IV-515
Carswell, James D.   IV-61
Carvalho, Nuno   II-443
Casagli, Nicola   II-693
Casavecchia, Piergiorgio   I-47
Castellucci, Sonia   II-160, II-242
Castor, Fernando   III-199
Cavalcante, Francesco   II-652
Cavalini, Luciana Tricai   V-475
Cecchini, Arnaldo   III-594, IV-284
Cecchini, Massimo   II-192, II-231
Ceppi, Claudia   IV-600
Cerofolini, Gianfranco   I-57
Cerreta, Maria   IV-572
Cesini, Daniele   I-84
Chaim, R.   II-559
Chang, Fu-Min   V-270
Chatterjee, Kakali   III-187
Chaturvedi, K.K.   II-408
Chen, ChongCheng   II-623
Chen, Minjie   V-91
Chintapalli, Sahithi   V-143
Cho, YongYun   III-287
Choi, Bum-Gon   I-120
Choi, Young-Hyun   I-382, III-175
Choo, Hyunseung   I-157, I-347
Christara, Christina C.   V-107
Chung, Min Young   I-120, I-131
Chung, Tai-Myoung   I-372, I-382,
    III-175
Ciavi, Giovanni   I-96
Ciulla, Giuseppina   II-344
Cividino, Sirio R.S.   II-192
Cocchi, Silvia   II-242
Colantoni, Andrea   II-220, II-231, II-288
Coletti, Cecilia   II-32
Conforti, Massimo   IV-473
Congiu, Tanja   IV-284
Correa, Luiz Alberto R.   V-295
Corrias, Alessandro   IV-77

Coscia, José Luis Ordiales   II-475
Costa, M. Fernanda P.   I-333
Costantini, Alessandro   I-84
Crasso, Marco   II-475
Crawford, Broderick   III-98, V-452
Crawford, Jeff   I-16
Cruz, Carla   I-293
Cuca, Branka   IV-358
Cucchi, Véronique   I-253
Cunha, Jácome   II-459
Cwayi, Qhayisa S.   II-677

Dai, H.K.   I-562
Dang, Anh Tuan   V-437
Dang, Duy-Minh   V-107
Dang, Tran Khanh   V-437
Dang, Van H.   IV-629
Das, Gautam K.   II-73
das Neves, Carlos Rafael Gimenes
    V-531
De Blasi, Liliana   IV-501
De Bonis, Luciano   III-340
De Cecco, Luigi   III-422
Decker, Hendrik   II-543
de Doncker, Elise   II-129
De Falco, Marcello   II-256
de Felice, Annunziata   IV-489
de la Barra, Claudio León   III-98
Delame, Thomas   II-113
de la Rosa-Esteva, Josep-Lluís   III-324
Del Fatto, Vincenzo   I-241
Dell'Antonia, Daniele   II-192
Dell'Era, Alessandro   II-256
Dello Buono, Dimitri   III-622
de Macedo Mourelle, Luiza   I-500, I-511
De Mare, Gianluigi   II-359, III-493,
    III-509, IV-457
Dembogurski, Bruno   I-646
Dembogurski, Renan   I-646
de Mendonça, Rafael Mathias   I-500
de Moraes, João Luís Cardoso   V-475
de Oliveira, Hugo Neves   III-214
De Palma, Rinaldo   III-481
De Paolis, Lucio Tommaso   I-622, I-632
De Rosa, Fortuna   IV-541, IV-572
DeSantis, Derek   I-216
De Santis, Fortunato   II-652
De Silva, Lasanthi N.C.   I-264
Desjardin, Eric   I-204
de Souza, Wagner Dias   III-378

de Souza, Wanderley Lopes   V-475
de Souza da Silva, Rodrigo Luis   I-646
de Souza Filho, José Luiz   V-332
Dias, Joana M.   I-279
Dias, Luis S.   I-304
Di Carlo, Andrea   II-256, II-271
Di Fazio, Salvatore   III-550
Di Giacinto, Simone   II-220
Dinh, Thang Ba   V-307, V-391, V-558
Dinh, Tien Ba   V-307, V-391, V-558
Di Palma, Diego   II-328
Di Palma, Maria   IV-541, IV-572
Dixit, Veer Sain   II-498
Doan, Dung A.   V-321
Doan, Nghia Huu   V-391
Dolz, Daniel   III-43
Dominici, Donatella   IV-136, IV-150
do Prado, Antônio Francisco   V-475
dos Anjos, Eudisley Gomes   III-214
dos Santos, Gerson Rodrigues   III-378
dos Santos, Rafael Duarte Coelho   V-295
Drogoul, Alexis   I-662
Duong, Anh-Duc   V-502
Duong, Trong Hai   V-607
Duran, Christopher   V-364

Eldredge, Zachary   I-16
El-Zawawy, Mohamed A.   III-82, V-516
Eom, Jung-Ho   I-382, III-175
Eom, Young Ik   I-173
Estima, Jacinto   IV-205
Evans, Christian   II-86

Faginas Lago, Noelia   II-17
Falcão, M. Irene   I-293
Falcinelli, Stefano   I-47, I-69
Falvo, Maria Carmen   II-271
Fanizzi, Annarita   IV-444
Farhan, Md. Hasan Murshed   III-394
Farooq, Fatima   I-396
Faudot, Dominique   II-113
Feng, Wenya   V-166, V-574
Ferenc, Rudolf   II-513, II-529
Fernandes, Clovis Torres   II-375, II-391,
    III-230, IV-614, V-531
Fernandes, Edite M.G.P.   I-333
Fernandes, Florbela P.   I-333
Fernandes, João Paulo   II-443, II-459
Ferreira, Afonso Pinhão   I-585
Ferreira, Brigida C.   I-279

Ferreira, João   I-304
Ferreira Pires, Luís   V-475
Fichtelmann, Bernd   II-635
Fischer, Manfred M.   IV-1
Florentino, Pablo Vieira   III-524
Fornauf, Leif   III-452
Franzoni, Valentina   IV-643, IV-657
Fritsi, Daniel   II-513
Fukuda, Akira   III-114
Fukushi, Masaru   V-197
Fülöp, Lajos Jenő   II-529
Fusari, Elisa   V-462
Fúster-Sabater, Amparo   V-33, V-407

Gaido, Luciano   I-84
Galante, Guilherme   V-281
Galli, Andrea   IV-315
Ganser, Andreas   III-160
Gao, Shang   III-33
Garnero, Gabriele   IV-77, IV-193
Garramone, Vito   III-606
Gavrilova, Marina L.   II-140
Geng, Peng   V-547
Gervasi, Osvaldo   I-708
Ghazali, Rozaida   I-427
Ghosh, Soumya K.   IV-108
Ginige, Athula   I-228, I-264
Giorgi, Giacomo   I-57
Giorgio, Emidio   I-84
Gkadolou, Eleni   IV-268
Glazier, Paul   I-396
Goonetillake, Jeevani S.   I-264
Greco, Ilaria   IV-45
Grignard, Arnaud   I-662
Gröger, Gerhard   III-466
Grossi, Gaia   II-32
Gubiani, Rino   II-192
Guerra, Eduardo Martins   II-375, II-391,
    III-230, IV-614, V-295, V-531
Guimarães, Ana Paula Nunes   III-214
Gulinck, Hubert   IV-315
Gupta, Daya   III-187
Gyimothy, Tibor   II-513

Häberlein, Tobias   III-128
Häfele, Karl-Heinz   III-466
Hanke, Timothy   II-86
Haque, Antora Mohsena   III-394
Harland, Kirk   IV-179
Hasan, Md. Mehedi   III-408

Hasan, Osman   I-358
Hassani, Marwan   V-181
Hill, David R.C.   I-253
Hiraishi, Kunihiko   III-17
Hisazumi, Kenji   III-114
Hossain, Md. Rifat   III-394, III-408
Hsu, Bi-Min   V-380
Hu, Szu-Ying   V-270
Hu, William   V-364
Huang, Henry H.M.   II-140
Huang, Ying-Fang   V-380
Huth, Frank   V-17, V-91
Huyen, Phan Thi Thanh   III-17
Hwang, Boram   I-347
Hwang, Sungsoon   II-86

Ialongo, Roberta   IV-136
Iannaccone, Giuliana   IV-358
Imtiaz, Sahar   I-396
Inglese, Pasquale   IV-572
Inoguchi, Yasushi   V-197
Iqbal, Wafa   I-396
Ismail, Mohd Hasmadi   II-611
Itamiya, Yoshihiro   V-348
Izzi, Francesco   III-622

Jackson, Kenneth R.   V-107
Jaiswal, Shruti   III-187
Jamal, Amna   I-396
Janda, Florian   II-99
Jayaputera, Glenn   V-364
Jemni, Mohamed   I-683
Jin, Shichao   V-166, V-574
Jung, Jun-Kwon   I-372
Jung, Sung-Min   I-372

Kah-Phooi, Jasmine Seng   I-464
Kang, Myoung-Ah   IV-253
Kao, Shang-Juh   V-259, V-270
Karathanasis, Charalampos   IV-268
Kechadi, M-Tahar   II-623
Kechid, Samir   II-487, II-574
Khan, Abdullah   I-413, I-438
Khuzaimah, Zailani   II-611
Kim, Dongho   I-697
Kim, Dong-Hyun   I-120, V-127
Kim, Dongsoo S.   I-347
Kim, Jeehong   I-173
Kim, Keonwoo   I-173
Kim, Mihui   I-142, I-347
Kim, Okhee   V-166, V-574

Kim, Tae-Kyung   I-372
Klimova, Anastasia   IV-489
Kogeda, Okuthe P.   II-677
Kong, Weiqiang   III-114
Kwak, KyungHun   III-287

Laganà, Antonio   I-1, I-31, I-84, I-96, II-17
Lago, Noelia Faginas   I-1, I-69
Laha, Dipak   III-258
Lankes, Stefan   V-181
Lanorte, Antonio   II-652
La Porta, Luigi   III-422
Lasaponara, Rosa   II-652, II-663
La Scaleia, Giuseppe   III-622
Las Casas, Giuseppe   III-622
Le, Bac   I-540, V-321
Ledoux, Hugo   I-526
Lee, Jungha   I-131
Lee, Junghoon   I-110, III-368
Lee, Sungyoung   I-396
LeKhac, NhienAn   II-623
Lemus, Cuauhtémoc   III-144
Le Ngo, Anh Cat   I-464
Leonard, Kathryn   I-216
Le Thi, Kim Tuyen   V-437
Le-Trung, Quan   III-271
Leung, Clement H.C.   IV-657
Leung, Yee   IV-93
Li, Yuanxi   IV-657
Lichter, Horst   III-160
Lisboa-Filho, Jugurta   III-378
Littlejohn, Robert G.   II-60
Liu, Jiang B.   V-590
Lo Brano, Valerio   II-344
Loconte, Pierangela   IV-556
Lombardi, Andrea   I-1, I-69, II-17
Longo, Leonardo   II-220
Lopes, Maria do Carmo   I-279
Lorbacher, Matias Ruiz   III-452
Löwner, Marc-O.   III-466
Lucentini, Marco   II-328

Ma, Jianghong   IV-93
Maccherani, Gabriele   I-708
Maier, Georg   II-99
Malleson, Nicolas   IV-179
Malonek, Helmuth R.   I-293
Mancini, Annagrazia   IV-473
Manfredini, Fabio   III-438

Manganelli, Benedetto   II-359, III-509, IV-304

Mangialardi, Giovanna   IV-528

Manigas, Luisa   IV-77

Maniruzzaman, Khandoker M.   IV-294

Mansor, Shattri   II-611

Manzo, Alberto   II-242

Manzolaro, Mattia   I-84

Marcheggiani, Ernesto   IV-315

Marghany, Maged   II-587, II-599

Marinelli, Dimitri   II-46

Marsal-Llacuna, Maria-Lluïsa   III-324, IV-17

Marsh, Jesse   III-294

Martini, Sandro   III-422

Martins, Pedro   II-443

Martirano, Luigi   II-271

Marucci, Alessandro   III-422

Marucci, Alvaro   II-192, II-231, II-299

Masini, Nicola   II-663

Mateos, Cristian   II-475

Mazzarolo, Claynor   II-559

Mazzei, Mauro   IV-419

Md. Said, Abas   I-596

Mendes, Jorge   II-459

Menghini, Giuseppina   II-220

Michelotto, Diego   I-84

Michels, Dominik Ludewig   II-150

Milani, Alfredo   IV-643, IV-657

Milone, Maria Vittoria   IV-136, IV-150

Min, Changwoo   I-173

Min, Jae-Won   III-175

Minh, Thai Tieu   I-485

Minucciani, Valeria   IV-193

Misra, Sanjay   II-427, II-475, III-59, III-70, III-98

Mitarai, Hiroko   V-348

Mitre, Hugo A.   III-144

Mizzelli, Luca   II-160

Modica, Giuseppe   III-550, III-566

Mohmad Hassim, Yana Mazwin   I-427

Molinari, Francesco   III-294

Monarca, Danilo   II-231

Monetti, Alberto   I-84

Monfroy, Eric   III-98, V-452

Montrone, Silvestro   IV-501

Morano, Pierluigi   IV-433, IV-457

Mormile, Martina   IV-136, IV-150

Mosconi, Enrico Maria   II-160

Mota, Gabriel   I-304

Mourão, Maria Filipa   I-573

Mueller, Markus   IV-221

Muhi, Kornél   II-529

Mukhopadhyay, Asish   V-48

Mukunoki, Daichi   V-211

Müller, Hartmut   III-309

Munklang, Yutthana   III-246

Murgante, Beniamino   III-606, III-630, IV-304, IV-473

Nagy, Csaba   II-513

Naso, Vincenzo   II-312

Nawi, Nazri Mohd.   I-413, I-438

Nedjah, Nadia   I-500, I-511

Neema, Meher Nigar   III-351, III-394, III-408, IV-294

Nesticò, Antonio   II-359, III-493, III-509

Ngo, Son Hong   V-154

Nguyen, Anh Vu   II-427

Nguyen, Cuong Duc   V-224, V-607

Nguyen, Dung Tien   V-224

Nguyen, Hai-Trieu   V-307

Nguyen, Hoang Anh   I-697

Nguyen, Hong-Quang   V-232

Nguyen, Minh-Son   III-271

Nguyen, Ngoc-Hien   V-307

Nguyen, Nam Vinh   I-540

Nguyen, Phong Hoang   V-558

Nguyen Quang, Khanh   I-157

Nguyen, Thanh-Lam   V-380

Nguyen, Thuc Dinh   IV-629

Nguyen, Trung Dung   I-157

Nguyen, Tuan Ngoc   V-437

Nguyen, Van Duc   I-157

Nguyet, Tran Thi Nhu   I-485

Nota, Rossella   III-481

Ochimizu, Koichiro   III-17

Ohgai, Akira   IV-294

Oliveira, Antonio   V-332

Oliveira, José A.   I-304

Oliveira, Nuno   III-538

Oliveira, Pedro Nuno   I-550, I-573

Oreni, Daniela   IV-328, IV-344, IV-358

Pacifici, Leonardo   I-31

Painho, Marco   IV-205

Palazzetti, Federico   II-17

Pallottelli, Simonetta   I-96

Palma, Armando Luigi   IV-419

Palo, Andi   II-299

Panaro, Simona   IV-515
Panigrahi, Satish Chandra   V-48
Pantazis, Dimos N.   IV-268
Paolino, Luca   I-241
Parisi, Serena   IV-473
Park, Chang-Sup   V-620
Park, Gyung-Leen   I-110, III-368
Park, JangWoo   III-287
Park, Min-Woo   I-382
Parker, Gregory A.   I-16
Partipilo, Valeria   IV-556
Pascale, Stefania   IV-473
Pascual, Jorge   V-636
Patton, Robert M.   V-491
Pazzola, Myriam   IV-284
Peer, Arshad Ahmud Iqbal   V-77
Perchinunno, Paola   IV-501
Pereira, Gilberto Corso   III-524
Pereira, Guilherme A.B.   I-304
Pergher, Gianfranco   II-192
Pesantes, Mery   III-144
Pessanha, Fábio Gonçalves   I-511
Pham, Cuong   I-673
Pham, Duy   V-391
Pham, Quoc Trung   II-427
Pham, Thi Thanh Hiên   IV-238
Pham Thi, Thanh Thoa   IV-61
Pham, Van Cu   IV-238
Pham, Van-Hau   V-224
Phuong, Tu Minh   I-673
Pietruska, Franz   IV-221
Pirani, Fernando   I-69, II-17
Piscitelli, Claudia   IV-541
Plaue, Matthias   V-91
Poli, Giuliano   IV-572
Pollino, Maurizio   III-422, III-566
Pontrandolfi, Piergiuseppe   IV-304
Popelka, Stanislav   IV-166
Previtali, Mattia   I-608
Prudente, Frederico Vasconcellos   II-1
Pucci, Paola   III-438
Puttini, Ricardo   II-559

Qaisar, Saad   I-396
Qiu, Guoping   I-464
Quintas, Artur   I-304

Raba, Nikita O.   V-248
Ragni, Mirco   II-1, II-46, II-60
Rahman, M.M. Hafizur   V-197

Raja, Debasish Roy   III-351
Recanatesi, Fabio   II-288
Rehman, Mohammad Zubair   I-413, I-438
Renhe, Marcelo Caniato   V-332
Reynoso, Luis   III-43
Ribeiro, Fabíola Goncalves C.   III-70
Roberts, Steven   IV-403
Robertson, Colin   IV-403
Rocha, Ana Maria A.C.   I-318
Rocha, Humberto   I-279
Rocha, Jorge Gustavo   I-550, III-538
Rocha, Maria Célia Furtado   III-524
Rodríguez, José   I-452, V-636
Romagnoli, Manuela   II-288
Romero, Manuel   I-452
Roncoroni, Fabio   I-608, IV-328
Rosi, Marzio   I-47, I-69
Rotondo, Francesco   IV-556
Rottenberg, Flavio   II-312
Roudet, Céline   II-113
Ruhé, Martin   III-452

Saib, Aslam Aly El-Faïdal   V-77
Sannicandro, Valentina   IV-587
Sanwal, Muhammad Usman   I-358
Saraiva, João   II-443, II-459
Saraswathi, Sivamani   III-287
Sattler, Kai-Uwe   V-421
Sbordone, Danilo   II-271
Schiattarella, Marcello   IV-473
Schindler, Andreas   II-99
Schirone, Dario A.   IV-489
Schoeneman, Larry   V-590
Schoier, Gabriella   IV-375
Schwandt, Hartmut   V-17, V-91
Scorza, Francesco   III-582, III-622
Sdao, Francesco   IV-473
Sebillo, Monica   I-241
Seidl, Thomas   V-181
Sesana, Marta Maria   IV-358
Shah, Asadullah   V-197
Shakhov, Vladimir V.   I-184
Shao, Jianwei   V-547
Shin, ChangSun   III-287
Shon, Minhan   I-347
Shu, Jian-Jun   V-65
Shu, Ming-Hung   V-380
Silva, Glauco de Sousa e   III-214
Silva, Jefferson O.   IV-614

Silva Júnior, Luneque Del Rio de Souza e
   I-511
Silveira, Fábio   II-391
Singh, Manoj Kumar   IV-33
Singh, V.B.   II-408
Sinnott, Richard O.   V-364
Skouteris, Dimitrios   I-47
Smokty, Oleg I.   V-1
Soares, Michel S.   III-70
Sobottka, Gerrit Alexander   II-150
Sole, Aurelia   IV-473
Sorrentino, Serena   I-194, V-462
Sosnin, Petr   III-1
Soto, Ricardo   III-98, V-452
Stahl, Chris G.   V-491
Stankova, Elena N.   V-248
Stankute, Silvija   IV-221
Steed, Chad A.   V-491
Stell, Anthony   V-364
Stoter, Jantien   I-526
Stratakis, Panagiotis   IV-268
Suraci, Vincenzo   II-299
Szőke, Gábor   II-529

Tagliolato, Paolo   III-438
Tajani, Francesco   III-493, IV-433,
   IV-457
Takahashi, Daisuke   V-211
Tapete, Deodato   II-693
Tarakji, Ayman   V-181
Tasso, Sergio   I-96
Tavares, Tatiana Aires   III-214
Thamm, Hans-Peter   III-452
Theera-Umpon, Nipon   III-246
Toffanello, Andre   II-559
Tong, Thi Huyen Ai   IV-238
Torre, Carmelo Maria   IV-587, IV-600
Tortora, Genoveffa   I-241
Tran, Dang-Hoan   V-421
Tran, Hoang Viet   I-697, V-154
Tran, Khoi-Nguyen   V-232
Tran, Minh-Triet   V-502
Tran, Ngoc-Trung   V-321
Tran, Ngoc Viet   III-160
Tran, Thanh-Toan   V-127
Trapani, Ferdinando   III-294
Treadwell, Jim N.   V-491
Trujillo, Juan   IV-253
Trunfio, Giuseppe Andrea   III-594,
   IV-284

Truong, Toan-Thinh   V-502
Truong-Hong, Linh   IV-61
Tucci, Andrea O.M.   II-176

Uddin, Mohammed Nazim   V-607
Urzal, Vanda   I-585

Van, Ha Duc Son   V-437
Van Hoai, Tran   I-485
Vaucheret, Claudio   III-43
Vecchiocattivi, Franco   I-69
Vecchione, Luigi   II-256
Venkatachalam, Parvatham   IV-33
Verdicchio, Marco   I-31
Veronesi, Paolo   I-84
Vidal-Filho, Jarbas Nunes   III-378
Vidigal, Armando   II-559
Vieira, Marcelo Bernardes   I-646, V-332
Vijaykumar, Nandamudi L.   V-295
Vilaça, Rita   I-318
Villarini, Mauro   II-256, II-271
Vitiello, Giuliana   I-241
Vo, Dinh-Phong   V-321
Voženílek, Vit   IV-166

Walisadeera, Anusha Indika   I-228
Wang, Hsiu-Lang   V-259, V-270
Wang, Z.   I-562
Weragama, Nishan   V-143
Wikramanayake, Gihan N.   I-228, I-264
Wu, Bo   II-623
Wu, Tianjun   IV-93
Wu, Wei   V-547
Würriehausen, Falk   III-309

Yatsu, Hirokazu   III-114
Yin, Junjun   IV-61
Yoon, Hee-Woong   I-131
Yoshitaka, Atsuo   V-321, V-348
Yutuc, Wilfredo   II-207

Zaldívar, Anibal   III-59
Zaman, Akter Uz   III-408
Zayrit, Karima   I-204
Zedda, Stefania Valentina   IV-77
Zenha-Rela, Mário   III-199
Zoccali, Paolo   III-550
Zuccaro, Letterio   II-299
Zucker, Jean-Daniel   I-662
Zunino, Alejandro   II-475